An Introduction to Cosmology
Third Edition

The third edition of this successful textbook is fully updated and includes important recent developments in cosmology. The book begins with an introduction to cosmology and general relativity and then goes on to cover the mathematical models of standard cosmology. The physical aspects of cosmology, including primordial nucleosynthesis, the astroparticles physics of inflation and the current ideas on structure formation, are discussed. Alternative models of cosmology are reviewed, including the model of quasi-steady-state cosmology, which has recently been proposed as an alternative to big-bang cosmology. The final chapters discuss observational cosmology and describe the debate on theory versus observational methods. The book includes 400 problems and numerous worked examples.

This introductory textbook describes modern cosmology at a level suitable for advanced undergraduates and graduates who are familiar with mathematical methods and basic theoretical physics. It is intended for use on courses in theoretical physics, astrophysics or applied mathematics that include modern cosmology.

JAYANT NARLIKAR graduated from Banaras Hindu University, India, in 1957. He then studied mathematics at Cambridge University, graduating with the highest honours and the Tyson Medal for Astronomy. He continued in Cambridge, obtaining first a Ph.D. and then the Sc.D. degree of Cambridge University.

Professor Narlikar was a founder staff member of the Institute of Theoretical Astronomy in Cambridge in 1966 and later, in 1988, set up and became director of the Inter-University Centre for Astronomy and Astrophysics in Pune, India. He has received world-wide acclaim for his research in gravitation and cosmology.

An Introduction to Cosmology

Third edition

Jayant V. Narlikar

Inter-University Centre for Astronomy and Astrophysics, Pune, India

CAMBRIDGE
UNIVERSITY PRESS

PUBLISHED BY THE PRESS SYNDICATE OF THE UNIVERSITY OF CAMBRIDGE
The Pitt Building, Trumpington Street, Cambridge, United Kingdom

CAMBRIDGE UNIVERSITY PRESS
The Edinburgh Building, Cambridge CB2 2RU, UK
40 West 20th Street, New York, NY 10011-4211, USA
477 Williamstown Road, Port Melbourne, VIC 3207, Australia
Ruiz de Alarcón 13, 28014 Madrid, Spain
Dock House, The Waterfront, Cape Town 8001, South Africa

http://www.cambridge.org

First published 2002
First South Asian edition 2002
Reprinted 2004, 2007, 2010

Printed in India at Gopsons Papers Ltd., NOIDA

Typeface Times 10.25/13.5pt.   *System* LATEX 2_ε [DBD]

A catalogue record of this book is available from the British Library

Library of Congress Cataloguing in Publication data

Narlikar, Jayant Vishnu, 1938–
   An introduction to cosmology / Jayant Vishnu Narlikar. – 3rd ed.
      p.   cm.
   Includes bibliographical references and index.
   ISBN 0 521 79028 X – ISBN 0 521 79376 9 (pbk.)
   1. Cosmology. I. Title.
   QB981.N3 2002   523.1–dc21   2001035268

ISBN 10: 0 521 531411   paperback
ISBN 13 : 978 0 521 53141 2   paperback
Special edition for Sale in South Asia only, not for export elsewhere.

Contents

Contents   vii

Foreword

This is an important book, which I hope will be studied by everybody concerned with physics and astronomy. I can guarantee that the student who works steadfastly through the many splendid examples will end by knowing a very great deal about relativity and cosmology. I can also guarantee that the practised expert will find much that is a surprise and a delight.

Backed by many years of distinguished research, the book is a masterpiece of clarity. From his earliest days as a graduate student, Jayant Narlikar has always been an incisive writer and lecturer. The mystery about a lecture by Narlikar is to understand how he manages to go at such an apparently leisurely pace, to write on a blackboard with extreme precision and without haste and yet at the end of an hour to have covered an immense amount of ground. The solution to the mystery has to be that he wastes less time than most of us on irrelevances, which is just what the reader of this book will find from the first page to the last. Author, publisher and reader are all to be congratulated.

I wrote the above at the time of the first edition of this book. No word needs changing but a few need adding. This is not only an important book. It is the best book and, I believe, by a considerable margin. Pity the student who doesn't work from it.

Sir Fred Hoyle

(Sir Fred Hoyle passed away shortly before the publication of this book.)

Preface to the first edition

The progress of modern cosmology has been guided both by observational and by theoretical advances. The subject really took off in 1917 with a paper by Albert Einstein that attempted the ambitious task of describing the universe by means of a simplified mathematical model. Five years later Alexander Friedmann constructed models of the expanding universe that had their origin in a big bang. These theoretical investigations were followed in 1929 by the pioneering work on nebular redshifts by Edwin Hubble and Milton Humason, which provided the observational foundations of present-day cosmology. In 1948 the steady-state theory of Hermann Bondi, Thomas Gold and Fred Hoyle added a spice of controversy that led to many observational tests, which are essential for the healthy growth of the subject as a branch of science. Then in 1965 Arno Penzias and Robert Wilson discovered the microwave background, which not only revived George Gamow's concept of the hot big bang proposed nearly two decades before but also prompted even more daring speculations about the early history of the universe.

The landmarks mentioned above have led to many popular and technical books on cosmology. In particular, the rapid growth of interest in the areas of general relativity and cosmology during the 1970s was reflected in a number of classic textbooks that came out in the early 1970s. The purpose of the present textbook is to introduce the reader to the state of the subject in the early 1980s. The approach adopted here is, however, different from that found in most other texts on the subject, so it is perhaps desirable to state what the differences are and why they have been introduced.

For example, it is usual to find cosmology appearing at the end of a text on general relativity, introduced more as an appendage than as a subject in its own right. Perhaps this is one reason why cosmology still stands apart from the rest of astronomy, where it really belongs. The astronomer tends to regard cosmology as a playground for general relativists rather than as a logical extension of extragalactic astronomy. To correct this tendency, the relative importance of cosmology and general relativity has been inverted in this text. Chapter 2 introduces general relativity more as a tool for studying cosmology than as a subject in its own right. Thus the relativist may find many topics dealt with at a superficial level or not at all. This chapter covers only those topics that are really necessary for understanding the large-scale geometrical properties of the universe. I have taken this approach in the hope that the relatively elementary treatment of general relativity will not put a newcomer off, as a more exhaustive treatment might well do. The expert relativist may skip this chapter and refer to it only for fixing the notation.

Chapters 3 and 4 introduce the standard models of cosmology as solutions of Einstein's equations. The tools developed in Chapter 2 will be found applicable here; and the reader will find the pace more relaxed than in Chapter 2.

Chapters 5–7 concentrate on the physical aspects of standard cosmology. Gamow's idea of primordial nucleosynthesis, the current state of ignorance on galaxy formation, the properties of the microwave background and the various recent contributions of particle physics to our understanding of the early universe are discussed here.

Perhaps this would have been the appropriate stage to move on to observational cosmology. However, I felt that the reader should also be taken on a short excursion into non-standard cosmology. Contrary to the view propagated (unfortunately) by many experts in cosmology today, the subject is not a closed book; neither is standard cosmology the only answer to the problem of the origin and the evolution of the universe. Part III of this book introduces some alternatives to the standard models.

Although some readers may prefer to see an observational test discussed immediately after the theoretical prediction, I have left observations to the last part of the book. This approach has made an overall assessment of the various models possible. A survey of cosmological observations shows how better techniques and a better appreciation of errors and uncertainties have led to frequent reassessments (a classic example being the value of Hubble's constant, which is still uncertain!). I have therefore not gone into very many observational details, but have emphasized how the observations are made and the likely sources of errors. In any case it would be unwise to go into too many details in an introductory text.

In spite of many remarkable advances, cosmology is still very much an open subject. On the observational side, the launching of the space telescope in the mid-1980s is likely to revolutionize our view of the universe. On the theoretical side, the grand unified theories (GUTs) are still grappling with the problem of the early universe, while quantum cosmology is in a rudimentary state. Cosmologists have yet to appreciate the problems posed by life in the universe. How did life come into existence? Is it confined to the Earth or is it widespread in the universe? A text of the future may well devote a large part of its discussion on cosmology to contributions from biology.

It is assumed that the reader is familiar with standard mathematical methods like differential equations, vector analysis, Fourier series and transforms, the calculus of variations and so on. A knowledge of basic physics including mechanics, elementary thermodynamics, electromagnetic theory, atomic structure and fluid dynamics is also assumed. Similarly, basic knowledge of elementary astronomy will be useful. The text is intended for advanced undergraduates, graduate students and teachers of astronomy and cosmology.

This book contains over 400 exercises, of which over 80% are of a computational nature. Many of them are designed to illustrate or amplify the material described in the text. It is hoped that they serve their intended purpose.

I thank Art Bartlett for encouraging me to write the book. Comments received from Bob Gould, Bob Wagoner, Dimitri Mihalas, Richard Bowers and Geoff Burbidge were of great help during the preparation of the manuscript. Last, but not least, it was Fred Hoyle who introduced me to the fascinating field of cosmology as a graduate student and I am indebted to him for agreeing to write a foreword to this book.

I began writing this book while visiting the Department of Applied Mathematics and Astronomy at the University College, Cardiff, Wales. I am grateful to the head of the department, Chandra Wickramasinghe, for the facilities extended to me at Cardiff. For the prompt typing of the manuscript I am indebted to Ms Suzanne Ball and Mr P. Joseph. It is also a pleasure to acknowledge the help I received from the Drawing Office and Xerox Facility of the Tata Institute of Fundamental Research.

Bombay, India   Jayant Narlikar

Preface to the second edition

I am happy that the revised second edition of *An Introduction to Cosmology* is seeing the light of the day. The motivation and format of this edition continue to be the same as for the earlier edition and hence this preface only supplements the more detailed preface of the first edition given above.

The changes incorporated in this edition broadly reflect the new developments in cosmology that came in the eighties, e.g., inputs from particle physics including the inflationary universe, new attempts at structure formation, recent observations of the large-scale structure and the improved (more sensitive) limits on the intensity fluctuations of the microwave background. The observational sections have been updated although no textbook can really keep pace with the rapid advances in cosmological observations.

A comparison of the two editions will reveal a slight rearrangement of the chapters, including a streamlining of the part devoted to alternative cosmologies. The final chapter is perhaps more critical of standard cosmology than before. This is necessary, in my opinion, in order to correct the prevailing impression that the standard hot-big-bang model describes the universe so well that no significant new or alternative inputs are required.

I thank Simon Mitton for encouraging me to proceed with the job of revising the book for Cambridge University Press. Thanks to speedy typing by Santosh Khadilkar and help with artwork by Arvind Paranjpye, the job could be completed within the time frame set by Simon. I also thank the numerous reviewers of the first edition whose constructive comments helped in preparing the revised manuscript.

Jayant V. Narlikar
Inter-University Centre for
Astronomy and Astrophysics,
Pune, India

Preface to the third edition

It is a measure of the advances that cosmology has made in the last decade that a third
edition of this book was felt to be necessary. The second edition had come out just
as the COBE satellite had revealed for the first time small-scale inhomogeneities in
the cosmic microwave background. Although the COBE findings came as a relief to
the big-bang theorists who were becoming uncomfortable with the apparent absence
of any evidence for a primordial interaction between matter and radiation (that any
scenario for the formation of large-scale structure would inevitably require), they
demanded considerable modifications of the existing paradigms. Since COBE the
number of findings of inhomogeneities on various angular scales has grown and,
at the time of preparing this edition, it has just been possible to take note of the
BOOMERANG and MAXIMA findings.

The third edition also takes note of the observations of type-I supernovae in
arriving at the magnitude–redshift relation for redshifts as high as ∼0.8. It also
relates the story of how the inflation paradigm has gradually been changing ground
and admitting a greater role for the cosmological constant. Thus the chapters on
observational cosmology have been revised substantially and expanded.

On the theoretical side, the basic ideas in standard cosmology have remained
essentially unchanged. The major thrust has been on the problem of understanding
structure formation, which can very well claim to be the stiffest challenge for the
big-bang cosmology. Does dark matter play a crucial role in forming the visible
structures? Is there a significant non-baryonic component in the dark matter? Can the
baryonic component be made consistent with the observed abundances of light nuclei, especially deuterium and lithium? Are theoretical models able to accommodate
the very old stars observed today? At the time of writing there are no definitive 'yes'-

or 'no'-type answers to these questions. The final chapter examines the big-bang scenario critically from various points of view and raises issues of relevance to the future developments in cosmology.

It is because of the relatively open-ended nature of these issues that the book reviews in Chapters 8 and 9 past attempts to understand the universe through alternative ideas and concentrates on one that claims to offer viable competition to the big-bang cosmology today. Even though alternative cosmologies do not feature prominently in any cosmology curriculum today, I strongly urge students (and readers in general) to read these chapters, if only to see how a minority of scientists of repute have felt uncomfortable enough with the standard paradigm to be inspired to think of alternatives.

I thank Simon Mitton for his persuasive push to work on this third edition. It is a pleasure to thank Arvind Paranjpye for help with obtaining images and Prem Kumar for illustrations, most of which are new to this edition. I also acknowledge the assistance of Vyankatesh Samak for help in getting this manuscript into a LaTeX-ready form.

Jayant V. Narlikar
Inter-University Centre for
Astronomy and Astrophysics,
Pune, India

Chapter 1

The large-scale structure of the universe

1.1   Astronomy and cosmology

No branch of science can claim to have a bigger area of interest than cosmology, for cosmology is the study of the universe; and the universe by definition contains *everything*. Although, because of its profound implications, cosmology has traditionally excited the imaginations of poets, philosophers and religious thinkers, our approach to the subject will be through the science of astronomy. Astronomy started as a study of the properties of planets and stars and gradually reached out to include the limits of the Milky Way system, which is our Galaxy. Modern astronomical techniques have taken the subject beyond the Galaxy to distant objects from which light may take billions of years to reach us.

Cosmology is concerned mainly with this extragalactic world. It is a study of the large-scale structure of the universe extending to distances of billions of light years – a study of the overall dynamical and physical behaviour of billions of galaxies spread across vast distances and of the evolution of this enormous system over several billion years.

At first such a study may appear an ambitious task. Are our tools of observation good enough to provide sufficient scientific information about the large-scale structure of the universe? Is our knowledge of the laws of nature sufficiently advanced and mature to interpret this information? By way of an answer to these questions, we quote a remark of Albert Einstein: 'The most incomprehensible thing about the universe is that it is comprehensible'. Although our observing techniques are far from perfect and our knowledge of physical laws leaves considerable room for improvement, we are still in a position to make some sense out of what we observe about the universe, which is why we can begin to study cosmology as a branch of science. That is, we will take the large-scale structure of the universe as a physical system, subject to the known laws of physics, and follow the usual procedure of science, that of observing features of relevance to this study and then trying to interpret them and making predictions that can be tested by further observations. This is what this book is all about.

Nevertheless, our successes in interpreting the universe through physical models should not blind us to the enormity and depth of the basic questions that emerge. The models may, at best, represent a first approximation to reality, which could still be elusive.

We will begin with a brief survey of some of the features of the universe that are pertinent to the subject of cosmology. Before coming to the present understanding, however, it is instructive to present a brief historical review of how this understanding was arrived at. In what follows and, indeed, throughout the book, we will use the following units of length, mass and time that are common in cosmology.

- *Length* The physicist uses the metre (for the SI units) or centimetre (for the c.g.s. units). While we shall use the latter in deriving quantitative expressions from physical formulae, they are too short for cosmic distances. The physicist would have preferred the *light year*, that is the distance travelled by light in one year (9.46×10^{17} cm $\approx 10^{18}$ cm), while the astronomer prefers the *parsec*, a unit that naturally arises when one is determining stellar distances by using the method of parallax. One parsec is defined as the distance at which half diameter of the Earth's orbit subtends an angle of 1 *arcsecond* when it is looked at symmetrically from the normal direction. Larger units constructed from the parsec (pc) are, the *kiloparsec* (1 kpc = 10^3 pc), the *megaparsec* (1 Mpc = 10^6 pc) and the *gigaparsec* (1 Gpc = 10^9 pc).

- *Time* The basic unit of time is of course the 'second', but on occasions cosmology demands much longer time scales. The typical unit is the gigayear (1 Gyr $\equiv 10^9$ years $\approx 3 \times 10^{16}$ s).

- *Mass* The physicist may use the 'kilogramme' (SI units) or gramme (c.g.s. units), but masses of astronomical objects are best expressed in the mass unit of the Sun's mass $M_\odot \approx 2 \times 10^{33}$ g.

1.2   From Herschel to Hubble

In the history of astronomy, the era of William Herschel (1738–1822) stands out as one when astronomy expanded its frontiers beyond the Solar System, to look at stars in the Milky Way. Although Herschel (see Figure 1.1) shot into fame largely through his discovery of the planet Uranus (the first ever new planet to be discovered

in recorded history), it was his study of stars that gave a first understanding of what
the Milky Way is, namely that it is a distribution of stars in a disc, spread all around
the Sun. Such a distribution of stars on a large scale is now known as a galaxy, with
the Milky Way being called the Galaxy. Figure 1.2 shows the map of the Galaxy
prepared by Herschel. Notice that the Sun is shown to be at the centre of the Galaxy.

The Herschelian picture with the Sun at the centre of the Galaxy, remained the
accepted picture until the beginning of the nineteenth century. The distances to
stars in the Galaxy were determined in the early days by the trigonometric method.

Figure 1.1 William
Herschel (1738–1822).
Photograph by courtesy of
the Royal Astronomical
Society, London.

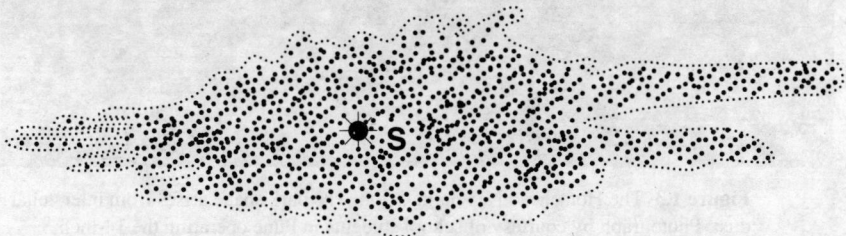

Figure 1.2 A map of the Milky Way prepared by William Herschel *ca.* 1788, shows
the Sun (S) at the centre.

Unfortunately this method loses accuracy beyond ∼50–100 pc. Furthermore, the dark patches in the Milky Way are not due to an absence of stars; rather they arise from absorption of starlight by particles of dust en route. Dark nebular patches like the famous Horsehead Nebula (see Figure 1.3) show regions of concentration of dust. Interstellar dust may exist in several forms, such as graphite, silicates and solid hydrogen. The effect of dust is to reduce the intensity of light from distant stars in the Galaxy through absorption and scattering. In the early days astronomers overestimated stellar distances in the Galaxy because they failed to correct for interstellar extinction. (Without correction, the faintness of a star was assumed to be wholly due to its distance from us.) The early astronomers also mistook dark regions for 'holes' or empty regions in the Galaxy.

A more reliable method that made use of the variable stars called Cepheids became available in 1912. Harlow Shapley used this method to measure the distances of remote stars in our Galaxy and showed that our Galaxy was much larger than it had previously been thought to be. The revised picture of the overall size of the

Figure 1.3 The Horsehead Nebula in Orion. The dark shape arises from interstellar dust. Photograph by courtesy of school students in Pune operating the 14-inch telescope at Mount Wilson Observatory through the internet. Facility provided by the Director, Gilbert A. Clark Telescope in Education Foundation.

Galaxy also meant a change in the perceived location of the Sun and its planets *vis-à-vis* the centre of the Galaxy, but not before several debates between those such as J. C. Kapteyn, who believed that the Sun was more or less at the Galactic Centre (GC), and a small minority represented by Shapley, who placed it considerably far from the GC. We have to allow for the fact that, unlike other galaxies, which we can look at from outside, the shape and size of the Milky Way have to be determined by observations *from within*.

Figure 1.4 shows a schematic representation of the Milky Way as it is perceived today. We see it face-on in Figure 1.4(a) and edge-on in Figure 1.4(b). The face-on picture shows the spiral structure of the Galaxy, whereas the edge-on picture demonstrates that it is a disc with a central bulge. The disc is also referred to as the galactic plane.

The diameter of the disc is estimated to be ~30 kpc and its thickness ~1 kpc. The Sun, together with all its planets, is located ~10 kpc from the centre. The Galaxy rotates about its polar axis, although not as a rigid body. Figure 1.5 shows how the circular orbital velocity varies at different distances from the GC. The Sun, for example, takes ~200 million years to make one complete orbit. Like the Sun, many other stars also go round the Galactic Centre (GC), whereas some stars have highly eccentric orbits that take them out of the galactic plane and also very close to the GC. (Figure 1.6 illustrates the two types of orbits.) Stars of the former type (like our Sun) with nearly circular orbits in the disc (shown by continuous lines) are called population-I stars, whereas stars of the latter type are called population-II stars. Their orbits are shown by dotted lines in Figure 1.6. From the metal contents of the two types of stars and the theory of nucleosynthesis it is possible to argue that population-II stars are older than population-I stars.

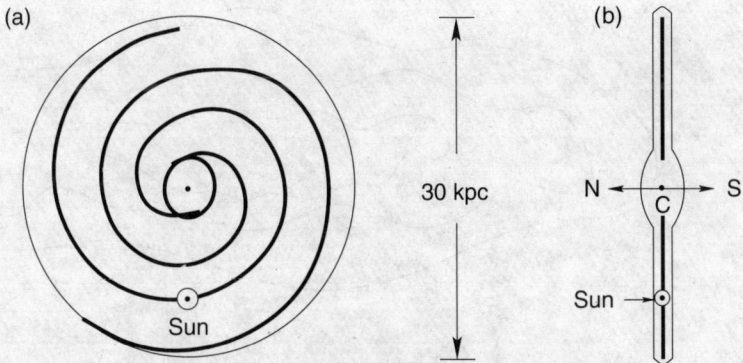

Figure 1.4 A schematic picture of the Milky Way, seen on the left (a) face-on as a circular system with spiral arms and on the right (b) edge-on as a disc with a central bulge. The Sun and its planets are located about two thirds of the way out from the centre. The Galaxy rotates about a central axis, with N and S the galactic North and South poles. C is the centre of the Galaxy.

Astronomers also refer to an even earlier generation of stars called the population-III stars, which were supposedly very massive and are now extinct, having burnt out quickly.

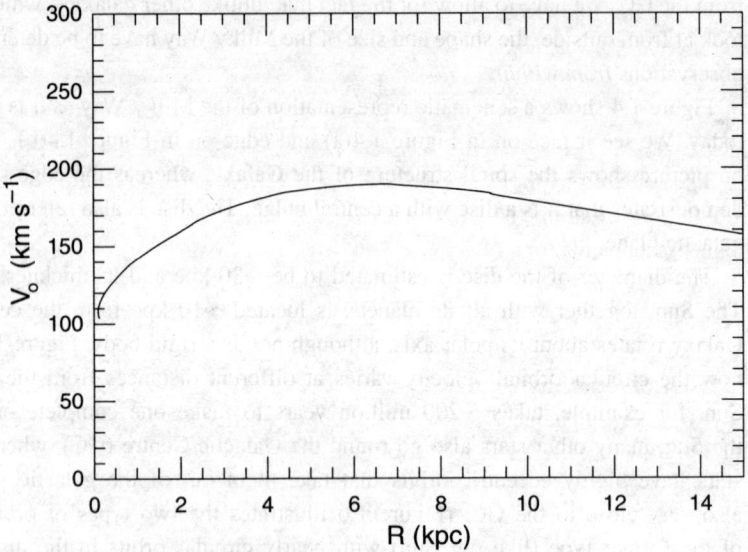

Figure 1.5 The diagram shows a model of the speed of rotation of the Galaxy about its polar axis, around the GC, which is clearly not that of a rigid disc rotating about its axis. Figure adapted from J. N. Bahcall & R. M. Soneira (1980), *Ap. J. Suppl.*, **44**, 73.

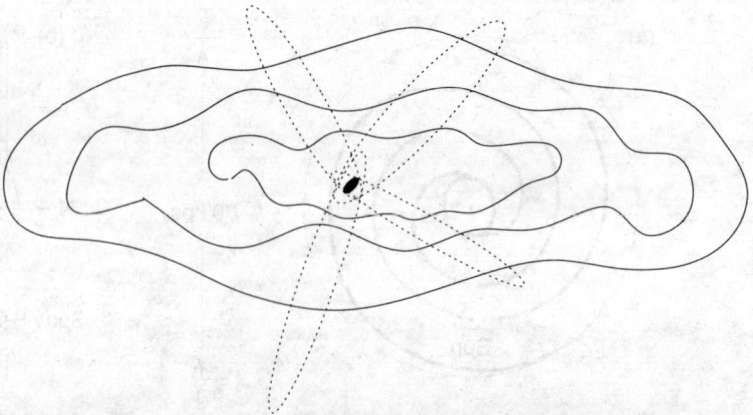

Figure 1.6 Two types of orbits of population-I and population-II stars. The former (like the Sun, as in Figure 1.5) move in the disc in nearly circular orbits, whereas the latter go into the halo and also close to the GC in highly eccentric orbits.
Population-I orbits also wobble up and down the galactic disc.

The mass of our Galaxy is estimated at $\sim 1.4 \times 10^{11} M_\odot$ and it is estimated that there are upwards of 10^{11} stars in the Galaxy. However, stars alone do not make up the whole of the Galaxy. The dark lanes in Figure 1.7 show that obscuring matter is also present.

Absorption lines in the spectra of galactic stars show that absorbing gases are present in the interstellar medium. Gas appears in various forms – atomic and molecular, hot and cold. Emission nebulae around stars are made of gas that absorbs the ultraviolet radiation from stars and radiates it as visible light in spectacular colours. The so-called H II regions are hot regions near stars and contain hydrogen gas that has been ionized by the ultraviolet light of the stars. In contrast the H I regions are cool regions of atomic hydrogen. The 21-cm observations in radio-astronomy were largely responsible for detecting neutral hydrogen in the Galaxy. Moreover, since the 1960s radio and microwave studies have revealed the existence of several complex molecules in the interstellar gas clouds. Figures 1.8 and 1.9, respectively, illustrate the H I and H II regions.

It took a few years longer for astronomers to appreciate the extragalactic nature of the universe. The general belief at the turn of the twentieth century was that all the hazy nebulae were located within the Milky Way. The idea, proposed by a few astronomers, that the nebulae were galaxies of stars similar to the Milky Way was ridiculed or ignored. One reason for doubting that they were distant galaxies was the claim by A. van Maanen that they possessed significant proper motions. Spiral nebulae such as M33, M81, M101, NGC 4051 and NGC 4736 exhibited, according to van Maanen, proper motions as great as $15\,000$ km s^{-1}, if they were indeed as far as a few million light years away. Dynamical theories ruled out such large motions on the grounds of stability of spiral structures. However, other observers could not repeat van Maanen's measurements and gradually the claim lost credibility.

It is interesting to note that Shapley, who had contributed so much towards our understanding of the structure and size of the Milky Way, took the conservative view

Figure 1.7 A composite picture of the Milky Way obtained by joining together pictures taken in different directions. The bright portions are due to stars, whereas the dark lanes indicate where interstellar dust is extinguishing the starlight *en route*. Courtesy of NASA/Ames Research Center.

Figure 1.8 A map of the Milky Way, using atomic hydrogen; that is, highlighting the H-I regions. Courtesy of NASA/Ames Research Center.

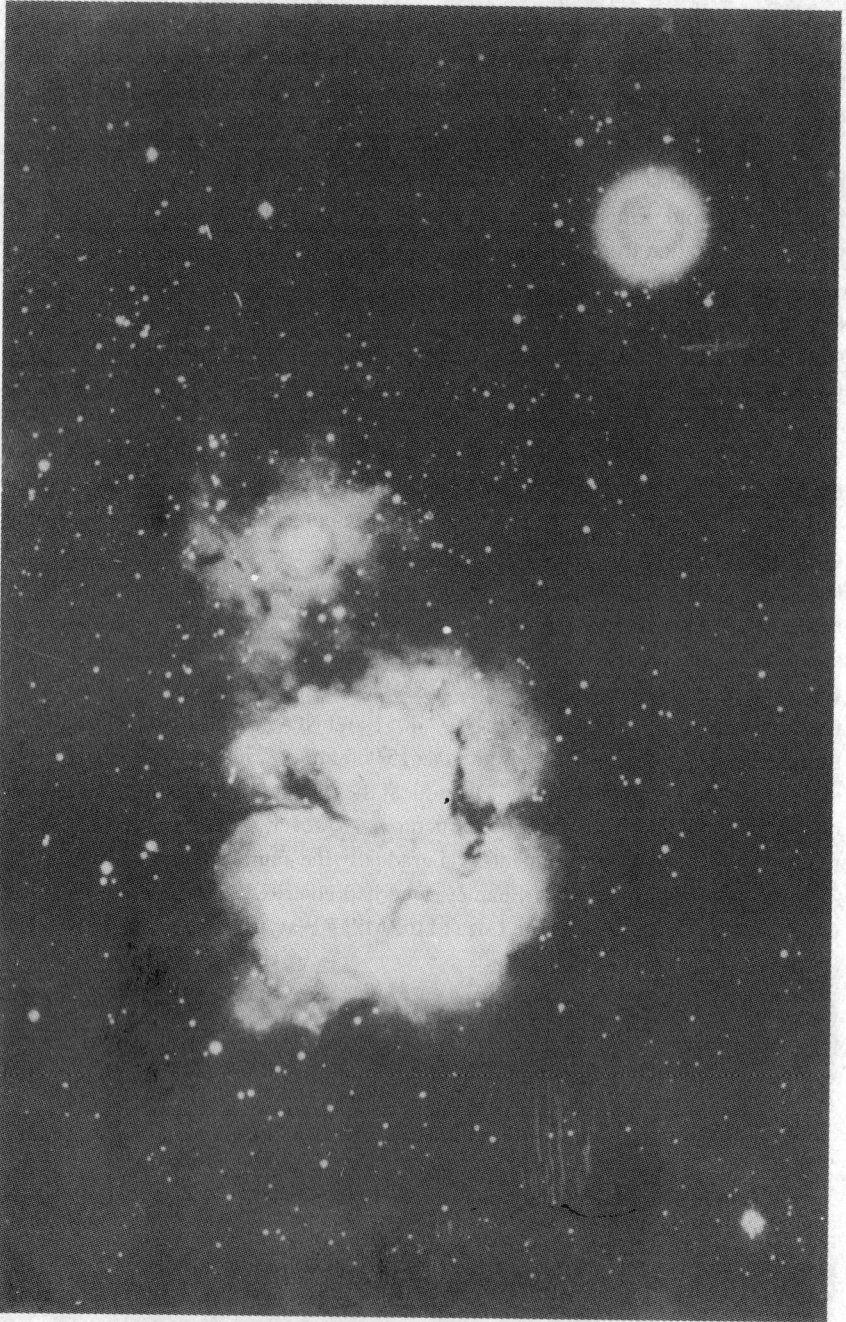

Figure 1.9 A typical H-II region in the Galaxy, the Trifid Nebula. Photograph by Arvind Paranjpye.

(shared by the majority) that all nebulae are part of the Milky Way Galaxy. Towards the end of the second decade of the twentieth century, there was a famous debate on this issue between Shapley and Curtis. As it soon turned out, Shapley's viewpoint was untenable.

In the 1920s, Hubble discovered that certain bright nebulae previously considered part of the Galaxy were actually remote objects lying well beyond it. Hubble's discovery finally laid to rest the belief that the whole of the observable universe was contained in our Milky Way, an island floating in infinite space. The nebulae that Hubble had proved to be extragalactic turned out to be galaxies in their own right. The idea that our Galaxy may be one amongst many populating the universe and that all such galaxies are like islands in vast (otherwise empty) space had in fact been proposed by Immanuel Kant in the eighteenth century and had been known as the 'island-universe hypothesis'. It received confirmation through such observations.

Today the astronomer has a much better perspective on the vastness of the extra-galactic world. The following section describes broad features of various types of galaxy known today. There we shall also see that the galaxies appear to contain dark matter that extends substantially beyond their visible boundaries.

1.3 Types of galaxy

The spiral structure of our Galaxy shown in Figure 1.4(a) was difficult to establish observationally, since we view it from within. It is easier to see this structure in other galaxies, unless we are viewing them edge-on. Our nearest large galaxy, labelled M31 (see §1.7 for the meaning of this label), in the Andromeda constellation, has a similar spiral structure (see Figure 1.10). *Spiral galaxies*, as such galaxies are called, are probably the most numerous amongst the various types of bright galaxy (see Figures 1.11 and 1.12). Like our Galaxy they exhibit rotation, flattening with a central bulge and dark lanes of absorbing matter.

In 1926 Hubble classified the various types of galaxy in the following way. The various classes of spiral galaxies are called Sa, Sb, Sc and so on. The sequence is in decreasing order of the importance of the central nucleus or bulge in relation to the surrounding disc (see Figure 1.13). Along the sequence, the central spheroid has decreasing luminosity and the spiral arms become more loosely wound. Our Galaxy and M31 are of type Sb. Some spirals have bars in the central region. These are called *barred spirals* and are categorized as SBa, SBb and so on. See Figure 1.14 for an illustration of a barred spiral.

While spirals are the most numerous amongst bright galaxies, the most numerous amongst *all* galaxies are those classified as *ellipticals*. These are ellipsoidal in shape (see Figures 1.15 and 1.16), exhibit very little rotation and have very little gas and dust. The various types of ellipticals are placed in the sequence E0, E1, ... , E7. This sequence describes progressively flattened profiles of galaxies, E0 being nearly spherical and E7 of markedly flattened lenticular form. If a and b are the major and minor axes of the ellipse, then the n for the galaxy En is given by

$$n = 10\frac{a-b}{a}.$$ (1.1)

Figure 1.10 The Great Galaxy in Andromeda, a spiral galaxy of type Sb. Photograph of Palomar Observatory/Caltech.

Figure 1.11 A galaxy of type Sa, the well-known 'Sombrero Hat'. Photograph by courtesy of William C. Keel, Department of Physics and Astronomy, University of Alabama, Tuscaloosa, USA.

These types are schematically illustrated in Figure 1.17. Unlike star images, which tend to be pointlike, galaxies have nebulous shapes like those described above. Astronomers can measure the distribution of light across a galaxy with great accuracy using solid-state instruments such as the charge-coupled device (CCD). The distribution of light is conveniently described in terms of *isophotes*, or contours of

Figure 1.12 A galaxy of type Sc, called the 'Whirlpool'. Photograph by courtesy of school students in Pune operating the 14-inch telescope at the Mount Wilson Observatory through the internet. Facility provided by the Director, Gilbert A. Clark Telescope in Education Foundation.

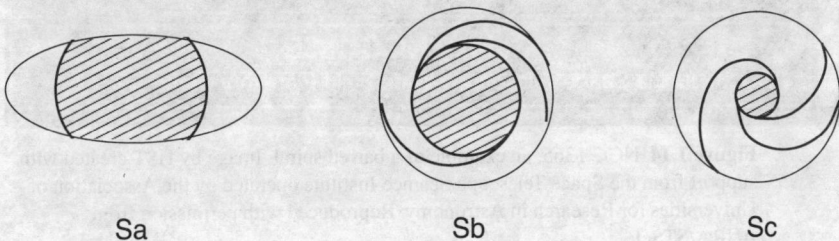

Sa   Sb   Sc

Figure 1.13 The sequence of types of spiral galaxy. The shaded region represents the nucleus.

equal intensity. For many galaxies, especially the ellipticals, increasing sensitivity of measurement shows that the boundary of a galaxy does not come to an abrupt end; rather, there is a gradual diminution of intensity of light outwards from the centre. In this connection astronomers often use the so-called *Holmberg radius*, which corresponds to the isophote at which the surface brightness drops to $26.5m_{pg}$ (photographic magnitude)[1] per square arcsecond, to indicate the optical boundary of a galaxy.

For many decades since the discovery of galaxies it was believed that they extend as far as they are visible. Thus the Holmberg radius was taken as the extent of a typical galaxy. However, in the seventies the orbits of neutral hydrogen clouds circling round a spiral galaxy indicated that the masses enclosed within them far exceeded the visible mass of the galaxy.

Figure 1.18 shows the typical rotation curve of a spiral galaxy. At a distance r from the centre O of the galaxy, a circular Keplerian orbit will have velocity

$$v = \sqrt{\frac{GM(r)}{r}}, \tag{1.2}$$

Figure 1.14 NGC 1365, an example of a barred spiral. Image by HST created with support from the Space Telescope Science Institute operated by the Association of Universities for Research in Astronomy. Reproduced with permission from AURA/STScI.

[1] Magnitude is a measure of the brightness of a celestial object. For quantitative details see §3.6.

where $M(r)$ is the galactic mass out to radius r from the centre. The point A represents the visible extent of the galaxy. If all the mass were visible then $M(r)$ = constant beyond A, and v should have dropped as $r^{-1/2}$ as shown by the dotted

Figure 1.15 An elliptical galaxy of type E0 in Virgo, M87. Its nucleus is believed to contain a highly collapsed mass of the order of $5 \times 10^9 M_\odot$. Photograph courtesy of Palomar Observatory/Caltech.

Figure 1.16 An elliptical galaxy of type E2 in Andromeda, M32. Photograph courtesy of William C. Keel, Department of Physics and Astronomy, University of Alabama, Tuscaloosa, USA.

curve. In reality v is more or less constant as far as point B which may easily be two or three times further away from O than A.

If Newtonian laws of gravity and mechanics hold then we have to conclude that $M(r)$ keeps on increasing beyond A; in other words, there is unseen matter present well beyond the visible radius of the galaxy. This dark matter poses many problems for, as well as introducing new elements into, cosmological theories, which we shall encounter later in this book.

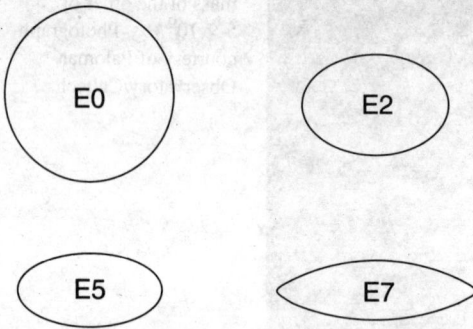

Figure 1.17 The sequence of types of elliptical galaxy shown schematically. Not all types between E0 and E7 are shown.

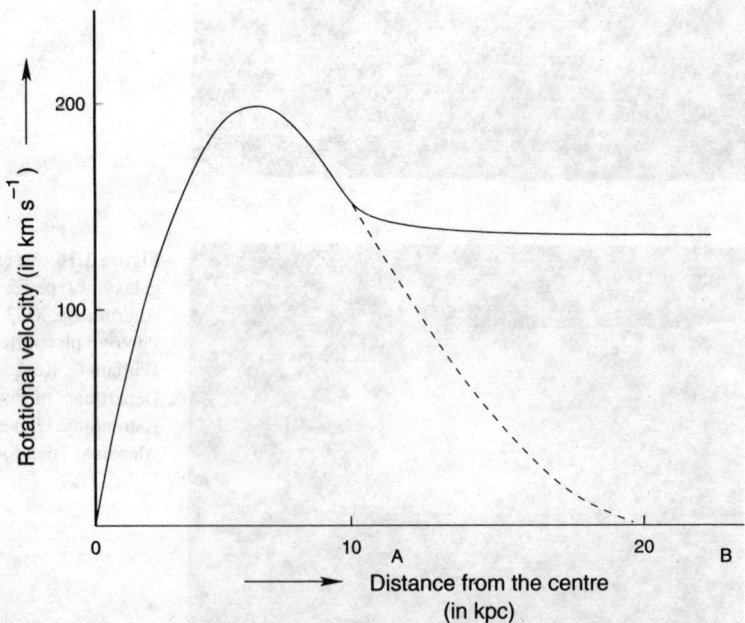

Figure 1.18 The rotation curve of a spiral galaxy is flat right up to point B, well beyond the visible extent up to A. The dotted curve shows how the rotation curve should have sloped down, had the entire mass of the galaxy been confined to its visible boundary.

Another type of galaxy, called S0, is intermediate between the ellipticals and the spirals (see Figure 1.19). Like the ellipticals, the S0 galaxies have little gas and dust, but their isophotes are more like those of the spirals (see Figure 1.20). These galaxies may have formed from collisions of spirals and ellipticals. Collisions between galaxies are not uncommon, especially in rich clusters of galaxies. Stars may go through a collision relatively unscathed, since they are widely spaced, but

Figure 1.19 Galaxy of type S0 in Virgo, M84. Image from AURA/NOAO/NSF (http://www.noao.edu/image_gallery/html/im0101.html).

interstellar gas and dust may be spewed out into the intergalactic space. In such a case the isophotes (which arise from starlight) may remain intact.

Figure 1.21 shows Hubble's 'tuning-fork'-type classification and sequence of galaxies. Although a certain order is seen in this picture, we are still not able to

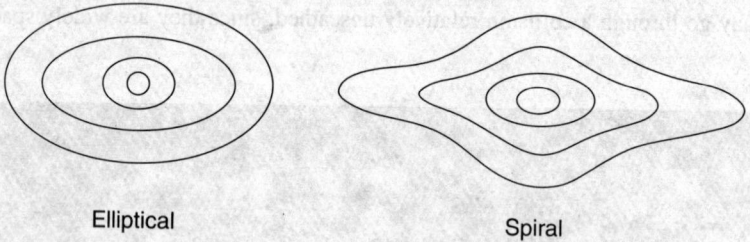

Elliptical   Spiral

Figure 1.20 The isophotes (contours of equal brightness) of an S0 galaxy are more like those of an elliptical (shown on the left) than those of a spiral (shown on the right).

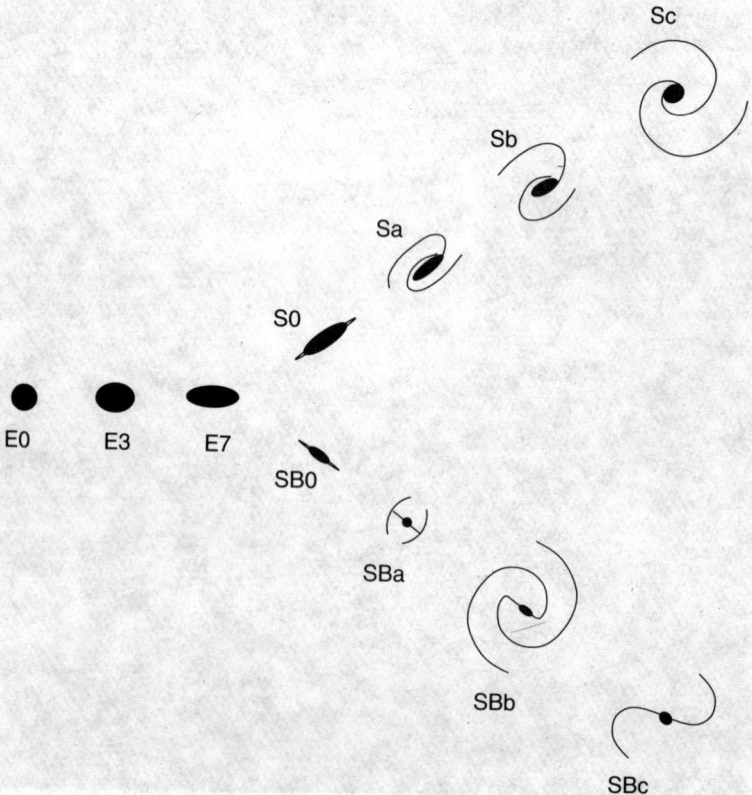

Figure 1.21 The sequence of types of galaxy envisioned by Hubble.

say whether this is some kind of evolutionary sequence, for our understanding of the formation and evolution of galaxies is still very rudimentary.

In addition to the types of galaxies already mentioned, there are others that are broadly classified as 'irregular'. Figure 1.22 shows a galaxy in this class. As the name implies, these galaxies have no set structural pattern. However, some rarer species of galaxies in this group stand out because of certain special features. For example, in 1943 Seyfert investigated a class of galaxies in which the nuclei exhibit many features common to stars, such as broad *emission lines*. (The spectra of galaxies as a rule exhibit *absorption lines* from interstellar gas.) The Seyfert galaxies also have a large amount of infrared emission; in some cases the infrared luminosity may be as much as 100 times the visual luminosity of our Galaxy (see Figure 1.23). There is considerable similarity between Seyferts and another class of astronomical objects, the quasars (described in §1.5).

Apart from these morphological types, galaxies are also classified in terms of their spectral features and luminosities. W. W. Morgan introduced the formal spectral classification, while van den Bergh introduced the so-called luminosity classes. We will not go into the details of these classifications here. It is hoped that, when the

Figure 1.22 NGC 4449, an irregular galaxy. © Sven Kohle and Till Credner, Observatory Hoher List, Bonn, Germany.

subject of the structure and evolution of galaxies is better understood, such classifications can be better appreciated. Meanwhile, these may be regarded as empirical indicators of galactic physics, indicators that may help us understand it better, just as empirical classification of stars helped bring about the eventual understanding of stellar structure and evolution.

1.4   Radio sources

The advent of radio-astronomy led to the discovery of strong sources of radio emission outside the Galaxy. As we shall see in Chapter 11, these radio sources also serve as useful probes of the structure of the universe. The first extragalactic radio source, Cygnus A, was discovered by J. S. Hey, S. J. Parsons and J. W. Phillips in 1946. When the position of the radio source in the sky could be accurately specified, Walter Baade and Rudolf Minkowski, at the Mt Wilson and Palomar Observatories, located

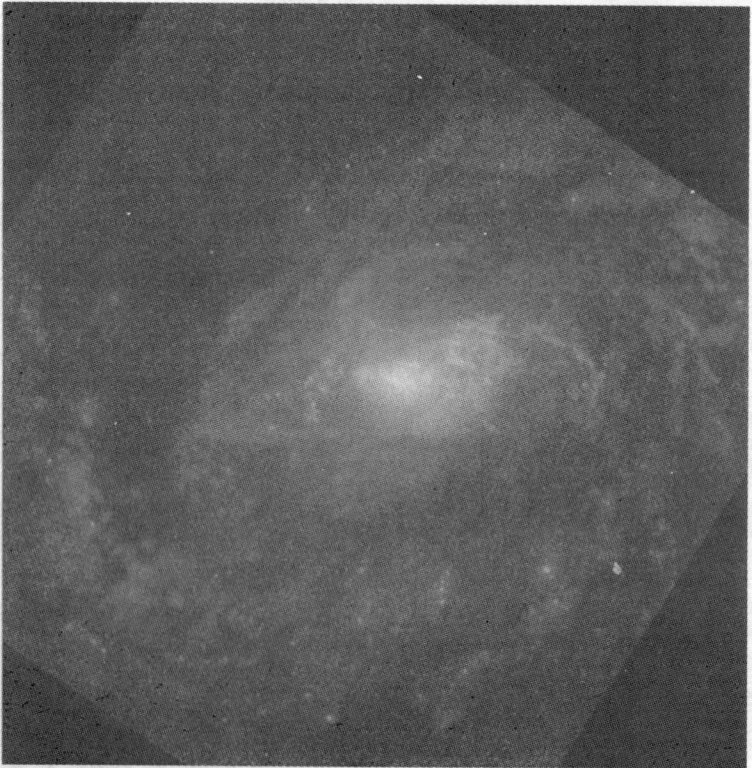

Figure 1.23 The central region of the Seyfert galaxy NGC 1068 imaged by the Hubble Space Telescope. Image by courtesy of Fred C. Bruhweiler and Cherie L. Miskey at IACS/Catholic University of America and the *Astrophysical Journal*.

what looked like a pair of colliding galaxies at the position of the radio source (see Figure 1.24). This process of identifying an object on the photographic plate at (or very close to) the position of the radio source is known as *optical identification* of the radio source. It implies that the optical and radio-astronomers are looking at the same object, but at different wavelengths. The discovery of Cygnus A led

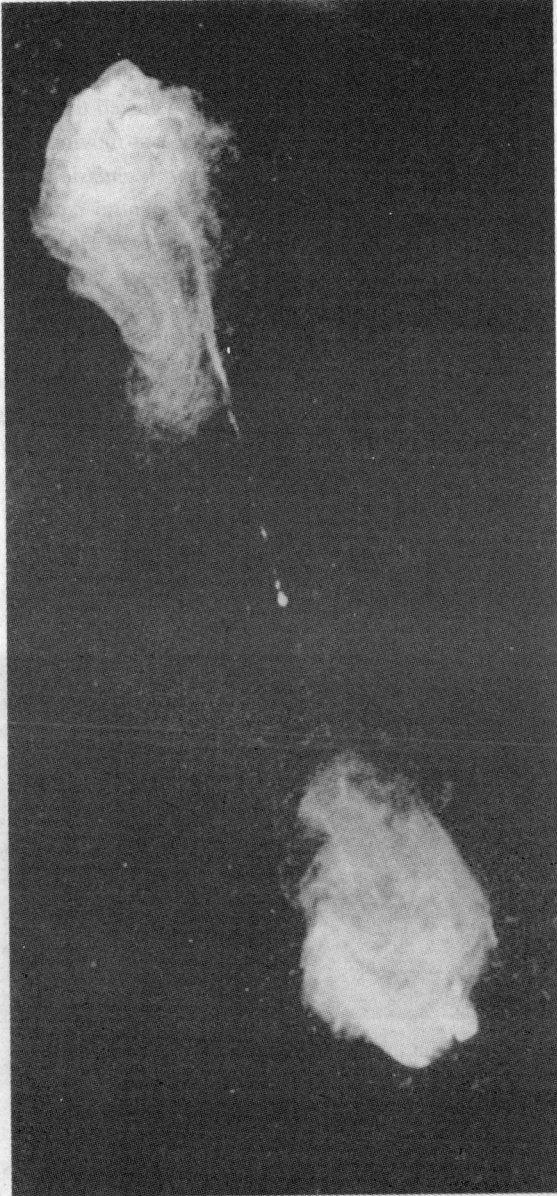

Figure 1.24 The radio source Cygnus A is located around the optical object at the centre of the photograph. Image by courtesy of R. Perley, J. Dreher and J. Cowan.

to the early view advocated by Baade that radio sources arise from collisions of galaxies.

Eventually, however, it turned out that Baade was wrong in considering the optical object in Cygnus A an example of colliding galaxies. In the seventies it became possible to study structures of radio sources in great detail. (Very-long-baseline interferometry can detect structures at the angular scale of less than a milli-arcsecond.) The picture that has emerged not only for Cygnus A but for a majority of extragalactic radio sources is shown in Figure 1.25.

Here we have two radio-emitting blobs on opposite sides of a central component, usually located close to and on opposite sides of a galaxy. It is believed that radio emission from the blobs takes place owing to the acceleration of fast-moving electrons by ambient magnetic fields, a process known as *synchrotron emission*. The particles themselves may have been fired in an explosion in the central region of the object. The source of the explosion is still a mystery. In 1963 Fred Hoyle and William A. Fowler suggested that gravitational energy in a collapsed object may somehow have been converted into kinetic energy of the electrons. In the late 1970s several scenarios involving a supermassive black hole of mass $\sim 10^8 M_\odot$ were proposed. As was first pointed out by Geoffrey Burbidge in 1958, theoretical estimates demand a powerful energy machine to generate energy reservoirs of 10^{58}–10^{62} erg in these radio sources. The potential energy of two colliding galaxies falls far short of this target.

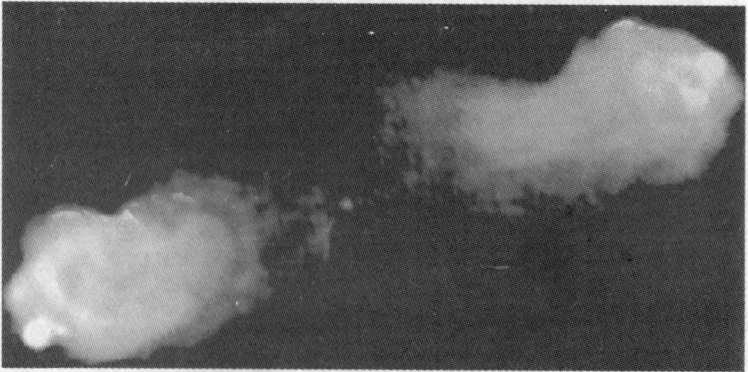

Figure 1.25 This radio map of Cygnus A illustrates the most common type of extragalactic radio source, made up of two radio-emitting blobs located symmetrically on the opposite sides of a central region. The central region is believed to be the source of activity that generates fast particles moving out along jets, which collide in the intergalactic medium, producing radio emission in blobs. Image by courtesy of R. Perley, J. Dreher and J. Cowan.

1.5 Quasars

The term *quasar* was first used as a short form for the full name 'quasi-stellar radio source'. Quasars were first discovered in 1963 as a result of an optical identification programme. The position of the radio source 3C 273 (see §1.7 for the meaning of these catalogue numbers) was accurately determined by lunar occultation.[2] The optical identification of this object (see Figure 1.26) and of another radio source, 3C 48, revealed starlike objects with emission lines; and it was originally assumed that these were radio stars in the Galaxy. When their spectra were carefully examined, however, it became clear that the wavelengths of emission lines were strongly redshifted.

If the wavelength of an emission line in the laboratory is λ_0 and its observed wavelength in the source of the light is $\lambda > \lambda_0$, then the line is said to be redshifted by a fraction z given by

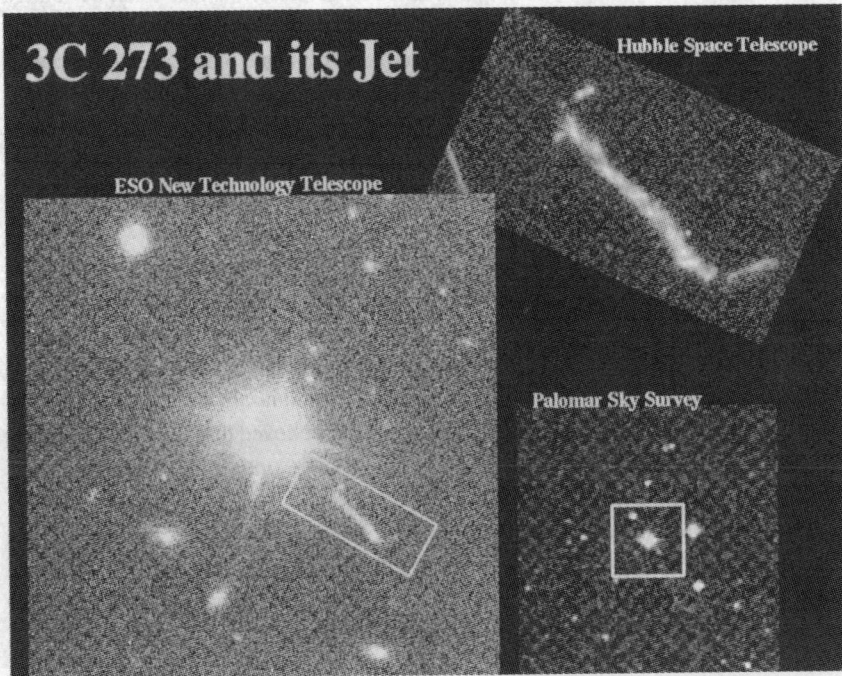

Figure 1.26 The quasar 3C 273, the first to be discovered and identified as an extragalactic object. The jet is highlighted. Image courtesy of Herman-Joseph Roser.

[2] If the Moon happens to cross the line of sight to a source, the source is said to be *occulted*. The drop in the intensity of a radio source as it is blocked by the Moon and the rise when the Moon has moved out of its way give an accurate indication of when it was occulted and, since the Moon's position in the sky is accurately known, it is possible to have an accurate estimate of the direction to the radio source.

$$z = \frac{\lambda - \lambda_0}{\lambda_0}. \qquad (1.3)$$

It is usual to call z the *redshift* of the object. (The name *redshift* is used to indicate a shift to the red end of the visual spectrum.) For 3C 273, $z = 0.158$; for 3C 48, $z = 0.367$.

These were high values of z for stars in the Galaxy, which tend to have values $<10^{-3}$. What were these objects? In 1964 Terrell suggested that they were high-velocity stars ejected from the Galaxy. The commonly accepted interpretation, however, has been that the redshifts arise from the expansion of the universe, a concept we will introduce in §1.8.

If the latter interpretation is correct, it implies that quasars are very distant objects and, since from such large distances they look bright enough to be mistaken for stars, they must be intrinsically very powerful. Many quasars exhibit rapid variation in their light and radio output. This fact places a limit on their physical size; for, if an object exhibits variability on a characteristic time scale T, its size must be limited by $\sim cT$, where c is the speed of light. This limitation, arising from the special relativistic result that no physical disturbance can propagate with a speed $>c$, makes quasars very compact indeed. We saw in §1.2 how big our Galaxy is. A quasar, in comparison, may emit a comparable amount of energy per unit time from a volume whose linear extent may be only a few light hours!

More than 10 000 quasars are now known. Only a few per cent of the total quasar population emit radio waves. Thus the early qualification 'radio source' is not applicable to the bulk of the quasar population and the term 'quasar' today is used also for radio-quiet objects. The purist may, however, prefer the term 'quasi-stellar object' (QSO).

In the early 1980s, the X-ray astronomy satellite *Einstein Observatory* revealed that X-ray emission is also a common feature amongst quasars, much more common than radio emission (see Figure 1.27). It is generally believed that the X-ray emission comes from the innermost region of a quasar while the optical and radio emissions come from progressively outward-lying regions.

In recent times, multi-wavelength studies of QSOs have become common, as has also the linking of the QSO phenomenon with compact and active nuclei of some galaxies like the Seyferts. The hope is that in this way one may have a better handle on the models of energy sources in these systems and their evolution.

1.6   Structures on the largest scale

A galaxy that is not a member of a group of galaxies is called a *field* galaxy. Other galaxies are members of groups or *clusters* that may contain from a few to hundreds of big galaxies. Our Galaxy, for example, is a member of a group of $\sim$28 galaxies, known as the *Local Group*, that are separated by distances of up to $\sim$1 Mpc. The nearest members of the group are the *Large and Small Magellanic Clouds*, which are located ~50 kpc from us. While a group may contain a few to a few tens of galaxies, a *cluster* contains several hundreds of them.

Table 1.1 lists a few of the larger clusters (see the image of one of them in Figure 1.28). The distances quoted in Table 1.1 are not exact because of the uncertainty surrounding the measurements of extragalactic distances. The extragalactic distance scale is related to the magnitude of Hubble's constant (see §1.8). Despite progress in measurement techniques having been made, there continues to be disagreement amongst astronomers regarding the true value of this constant. The ratios of these numbers should, however, give us reliable estimates of the relative distances from Earth of these clusters.

Figure 1.27 An X-ray map of the quasar 3C 273 made by the Chandra X-ray telescope. Image courtesy of NASA/CXC/Smithsonian Astrophysical Observatory and H. Marshall *et al.*

Table 1.1 Some rich clusters of galaxies

Name of the cluster	Distance from Earth[a] (Mpc) $\times h_0^{-1}$
Virgo	10.5
Pisces	41
Perseus	50
Coma	61
Hercules	95
Gemini	215
Hydra II	555

[a] Distances corresponding to $H_0 = 100h_0$ km s^{-1} Mpc^{-1}.

George Abell has catalogued clusters out to distances of the order of that of Hydra II using strict criteria for what constitutes a cluster. In order to pick out a cluster one has to look for an enhancement of the number density of galaxies within a specified region relative to the overall background density. The order of 'richness' of a cluster is accordingly fixed by specifying the size, brightness range and background. In general a rich cluster will have a denser population of galaxies. F. Zwicky had also catalogued clusters of galaxies, but with less strict criteria than those adopted by Abell.

How much matter is contained in a cluster? We will attempt to answer this question in §10.4. For the time being we may say that a large cluster may contain a mass of the order of $\sim 10^{14} M_{\odot}$. Furthermore, if we try to estimate the mean density of matter in the universe by taking account of how much matter we see in clusters of galaxies, we come up with a figure that is a few times 10^{-31} g cm^{-3}. However, as we shall see later in this book, even clusters may have dark matter in substantial amounts. This is because the dynamical activity in a cluster, estimated by calculating the kinetic energy of member galaxies, far exceeds the gravitational potential energy of the galaxies and, if one *assumes* that the clusters have reached dynamical equilibrium, then this discrepancy becomes serious. To resolve it one has to assume that a typical cluster contains a considerable quantity of dark matter with relatively low velocity. If one

Figure 1.28 The Coma cluster of galaxies. Image by Omar Lopez-Cruz and Ian Shelton, NOAO/AURA/NSF.

includes dark matter, then these masses and estimates of density may have to be increased.

The mean density of matter in a galaxy, on the other hand, is $\sim 10^{-24}$ g cm^{-3}. Thus the volume occupied by galaxies is $\leq 10^{-6}$ of the total volume of the universe. This also explains why galaxies are considered as points in constructing the simplest cosmological models.

Apart from optical emission, clusters of galaxies also exhibit radio and X-ray emission. These emissions arise not only from individual sources located in the clusters but also in a diffuse fashion throughout the clusters.

Does a structure larger than clusters exist in the universe? This can be decided by studying the distribution of clusters across the sky and looking for non-randomness (that is, grouping or clumping) on larger and larger scales. Such studies have revealed the existence of larger structures on the scale of $\sim$50 Mpc, compared with cluster scales of $\sim$ 5 Mpc. These larger units are referred to as *superclusters*.

For example, in the late 1950s, G. de Vaucouleurs found that our Local Group is a member of a supercluster centred on the Virgo cluster of galaxies. C. D. Shane and coworkers at the Lick Observatory found similar clumpiness in other regions of the sky. Abell also found clumpiness in an analysis of the plates in the National Geographic–Palomar Sky Survey.

In the seventies and eighties there were considerable improvements in the techniques for observing discrete extragalactic objects. With distances determined by Hubble's law (§1.8) it has become possible to have three-dimensional perspectives of distributions of matter in the universe. These are beginning to indicate a large-scale inhomogeneity in the distribution of matter. For example, the following features are revealed by such redshift surveys.

1. *Superclusters* As seen above, these are on scales of $\sim$50 Mpc or more and contain several thousand galaxies. For example, the Local Supercluster containing the Local Group is shaped like a flattened ellipsoid with a plane of symmetry called the *supergalactic plane*. It passes through the centre of the Virgo cluster and the centre of our own Galaxy.

2. *Voids* These are gaps in the distribution of large superclusters, with sizes that can be as large as 100–200 Mpc. There are apparently no (or very few) galaxies in these regions (see Figure 1.29).

3. *Filaments* The boundaries of voids are filamentary distributions of galaxies in clusters and superclusters. Figure 1.29 indicates this feature clearly.

4. *The Great Attractor and the Great Wall* In the late eighties it became apparent that galaxies in and around the Local Group seem to have a large-scale streaming motion towards the Hydra–Centaurus supercluster in the southern sky. The typical streaming velocity against the so-called *cosmological rest frame*, which we shall discuss in detail in Chapter 3, was

estimated to be around 600 km s^{-1}. This is also the reference frame in which the cosmic microwave background (see §1.9) is isotropic. This motion is believed to have been caused by a 'Great Attractor' (GA); that is, a huge mass of some tens of thousands of galaxies. The volume of the attractor is as large as 10^6 Mpc^3. The GA is not seen but its existence is inferred from its gravitational effect. Such massive structures may be present elsewhere in the universe also.

Mapping of the universe on a large scale reveals the presence of a large but thin sheet of mass. This is known as the 'Great Wall' and it has an area of 60×170 Mpc^2 (using a Hubble constant of 100 km s^{-1} Mpc^{-1}). Unlike the GA, the Great Wall is seen and Figure 1.30 clearly shows its existence. These structural inhomogeneities therefore span distances as large as 10% of the characteristic size of the universe defined in §1.10.

Some studies also suggest a cellular structure for the universe; that is, matter distributed preferentially in a grid-like fashion, with a characteristic cell size of ∼125 Mpc. Some analyses claim that there is a fractal hierarchy in large-scale structure extending beyond the 200-Mpc scale. However, at the time of writing this text, evidence of this kind needs to be checked further.

In the 1920s and 1930s the general belief was that the universe is homogeneous on a sufficiently large scale. The cosmological models which were proposed in those days made this simple assumption. In Chapters 3 and 4 we will outline these models. However, it is now becoming apparent that these models were too simplistic. They face the problem of explaining how structure inhomogeneous on such a large scale came into existence. The problem has become more difficult,

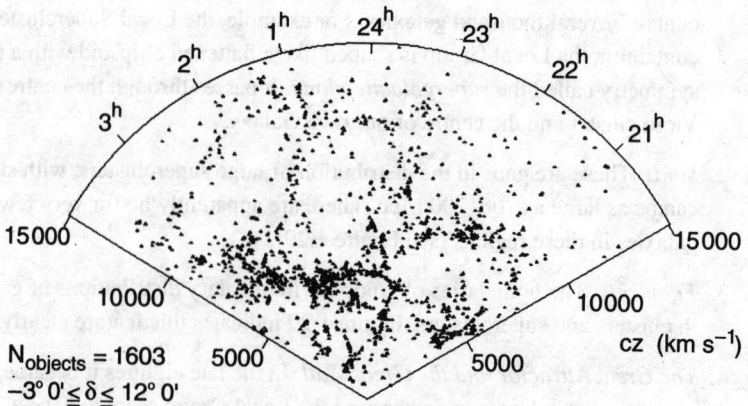

Figure 1.29 Galaxy distributions exhibit giant voids and long filaments, as mapped in this survey by the Center for Astrophysics. Courtesy of M. G. Geller and J. P. Huchra of the Smithsonian Astrophysical Observatory.

since, as we shall see later in this chapter, the radiation background is relatively much smoother.

1.7 Coordinates and catalogues of astronomical objects

Before proceeding further we will describe how the astronomer locates the position of a heavenly body in the sky. In general the astronomer does not know the distance of the body from us; he sees it projected on the sky, on what is known as the *celestial sphere*. Two coordinates, akin to longitude and latitude, are therefore needed in order to specify the position of the body on this sphere.

Figure 1.31 shows two different coordinate systems, both of which are useful to the astronomer in different contexts. The system in Figure 1.31(a) uses *right ascension* (RA, denoted by α) and *declination* (δ), coordinates related to the geometry of the Sun–Earth system. Here the poles are the points N and S on the celestial sphere where the Earth's axis of rotation intersects it. The *celestial equator* is the great circle on the celestial sphere whose plane is perpendicular to NS. The plane in which the Sun appears to go round (as seen from the Earth) intersects the celestial sphere in another great circle called the *ecliptic*. The ecliptic and the celestial equator intersect at two points γ (the 'first point of Aries') and Ω (the 'first point of Libra'), corresponding to the position of the Sun on March 21 and September 22, respectively. Now α and δ are the longitude and latitude of a celestial object measured with respect to the celestial equator and the great circle through N, γ, S and Ω. This circle, known as the *celestial meridian*, plays the role of the

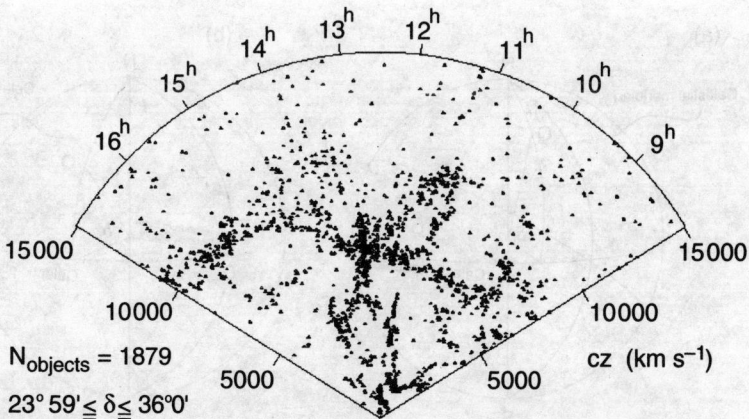

Figure 1.30 The distribution of galaxies in this survey by the Center for Astrophysics shows a long stream of galaxies within the velocity range 7000–10 000 km s^{-1} that is often referred to as the Great Wall. Courtesy of M. G. Geller and J. P. Huchra of the Smithsonian Astrophysical Observatory.

Greenwich meridian on the Earth, with γ the point of zero α. It is customary to measure α in hours and minutes, with the range of 360° corresponding to 24 h. The declination is written in degrees, minutes and seconds, with + for North, − for South.

While the (α, δ) coordinates are convenient for measurements made from the Earth, the cosmologist is often interested in knowing how the object is located *vis-à-vis* the plane of the Galaxy. For such purposes the *galactic coordinates* are useful. These are illustrated in Figure 1.31(b). The *galactic equator* is the great circle where the plane of the Galaxy intersects the celestial sphere. N and S are the North and South galactic poles, while the 'zero' meridian is the one passing through the points N, S and C, where the direction from Earth to the centre of the Galaxy meets the celestial sphere. This meridian is also called the *galactic meridian*. The galactic longitude is denoted by l and latitude by b. In terms of the (α, δ) system, the point C has the coordinates $\alpha \cong 17^h 42^m.4$, $\delta \cong -28°55'$. In galactic coordinates, of course, C has $l = 0$. It is possible to convert from one coordinate system to another using the standard formulae of spherical trigonometry.

Astronomical objects are catalogued in many ways. Table 1.2 lists some of the catalogues referred to in this book and their code letters. This list is by no means exhaustive, but is given as an illustration of how sources are numbered and listed. A more systematic method common in recent compilations is to list the object by its (α, δ) values in the form $\alpha(\pm)\delta$. Thus the object 1143−245 has right ascension $11^h 43^m$ and declination $-24°30'$ ($\equiv -24.5°$).

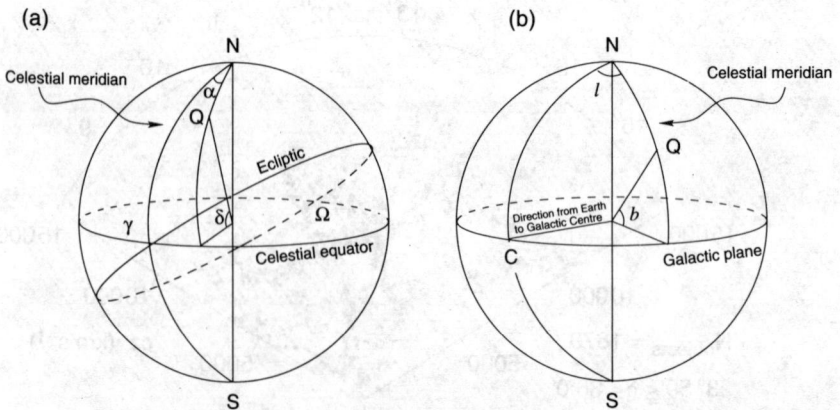

Figure 1.31 This figure demonstrates how to measure (α, δ) and (l, b) for an object Q in the sky using two different coordinate systems: (a) the coordinate system based on the geometry of the Sun–Earth system; (b) the coordinate system based on the geometry of our Galaxy.

1.8   The expansion of the universe

We now come to the observations that launched modern cosmology, leading to the important law enunciated by Edwin Hubble (Figure 1.32). Between 1912 and 1925, V. M. Slipher measured the shifts in the spectra of more than 20 objects that subsequently were identified as galaxies. Slipher was surprised that almost all shifts were towards the red end. (Our near neighbour the Andromeda galaxy shows a blue shift.) If we use the Newtonian Doppler-shift formula, we get the radial velocity of recession of the galaxy as $V = cz$, where z is the redshift defined in relation (1.3). (Since $z \ll 1$ in these observations, the Newtonian Doppler-shift formula can be applied.)

Later Edwin Hubble and Milton Humason extended Slipher's list of observations to more galaxies and to the brightest cluster galaxies. An example of the pattern that emerged when the redshift was plotted against the distance from Earth of a galaxy is shown in Figure 1.33. Also, see Figure 1.34, wherein we have photographs of cluster galaxies against their spectra. The dark lines progressively shift more to the right (the red end) as we move down the list. Also, the galaxies get smaller and fainter, consistently with the distance effect.

Hubble, in fact, plotted the radial velocity of a galaxy against its apparent magnitude. If all galaxies seen are equally bright, then their magnitudes are proportional to the logarithms of their distances from Earth (see Chapter 3 for a definition of

Table 1.2 Some catalogues of use in cosmology

Name	Type of object	Catalogue code
Messier	Nebulae and galaxies	M followed by catalogue number
New General	Nebulae and galaxies	NGC followed by catalogue number in increasing RA
Abell	Clusters	A followed by catalogue number in increasing RA
Cambridge (third, fourth and fifth surveys)	Radio sources	3C, 4C, 5C followed by catalogue number in increasing order of RA
Ohio Source	Radio sources	O followed by a letter (B to Z omitting O) and a number (the letter gives hours of RA, the first digit the declination in $10°$ intervals and the last two digits the decimal part of the RA to two places), thus $1442 + 101$ is OQ-172

'magnitude'). Thus the straight line drawn through the cluster of points corresponds to the linear relation

$$V = cz = H_0 D, \tag{1.4}$$

where D is the distance from Earth of the galaxy and z its redshift. The constant H_0 is now known as *Hubble's constant*.

If, instead of plotting z against the distance D, $\log z$ is plotted against the apparent magnitude m of the galaxy, then another straight-line relation shows up. For, using equation (3.58) of Chapter 3, we get

$$m = 5 \log D + \text{constant}, \tag{1.5}$$

Figure 1.32 Edwin Hubble (1889–1953) standing in front of the Palomar 48-inch Schmidt Telescope. Courtesy of Palomar Observatory/Caltech.

and (1.4) implies

$$m = 5 \log z + \text{constant}. \qquad (1.6)$$

Since the distances from Earth of remote galaxies are determined through their apparent magnitudes (as discussed in Chapter 10), (1.6) is the practical form of Hubble's linear relation (1.4).

The relation (1.4) is called *Hubble's law*. It was published as a linear law by Hubble in 1929 and it caused great excitement. For the *prima facie* interpretation of Hubble's law seemed to be that there was a great explosion in our neighbourhood from which galaxies were thrown out. However, the linearity of Hubble's law shows that we need not consider ourselves in any special position in the universe. If we viewed the population of galaxies from any other galaxy, we would notice the same form of Hubble's law. This fact and the homogeneity and isotropy of the distribution of galaxies suggests a highly regular structure of the universe with no preferred position or direction.

Imagine a piece of dough with self-raising flour being baked in the oven and suppose that we have spread caraway seeds uniformly throughout the dough. As the dough bakes it expands and the seeds move away from each other. The phenomenon of the recession of galaxies might be looked upon in the same light. They are points

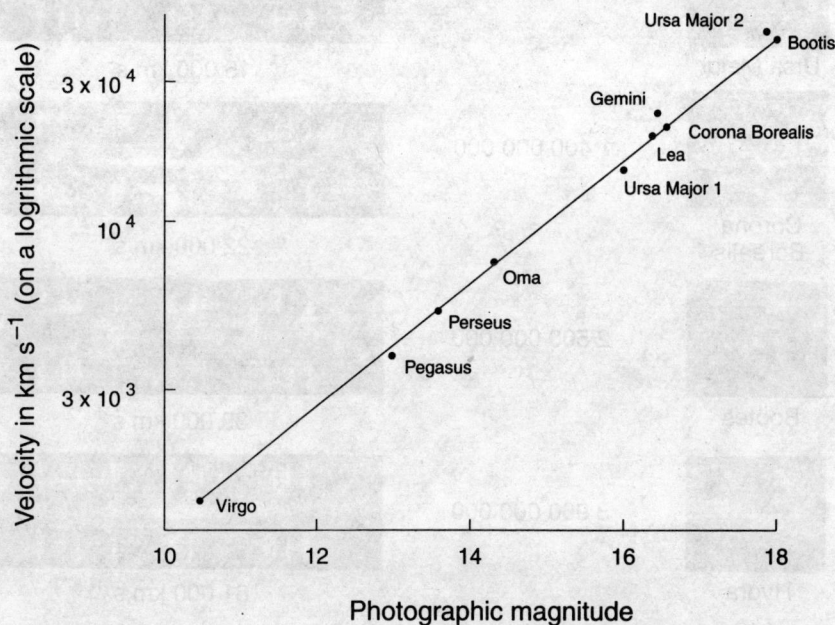

Figure 1.33 Hubble's plot for the fifth-brightest member in clusters of galaxies. The magnitudes are photographic. In Chapter 10 we will see how to convert magnitudes into distances. The velocities are obtained by multiplying the observed redshifts by *c*.

embedded in space that is expanding. This is how the concept of the *expanding universe* was born. Notice that, as in the analogy, the caraway seeds don't expand; here too the galaxies do not expand but their mutual separations increase.

The rate of expansion is characterized by Hubble's constant. Hubble obtained a value for H_0 in the neighbourhood of 530 km s^{-1} Mpc^{-1}. (Note that these units arise because H_0 is velocity divided by distance. The dimensions of H_0^{-1} are simply those of time.) As we will discuss in §10.2, Hubble had grossly underestimated the galactic distances, with the result that his value of H_0 was too high. The value of

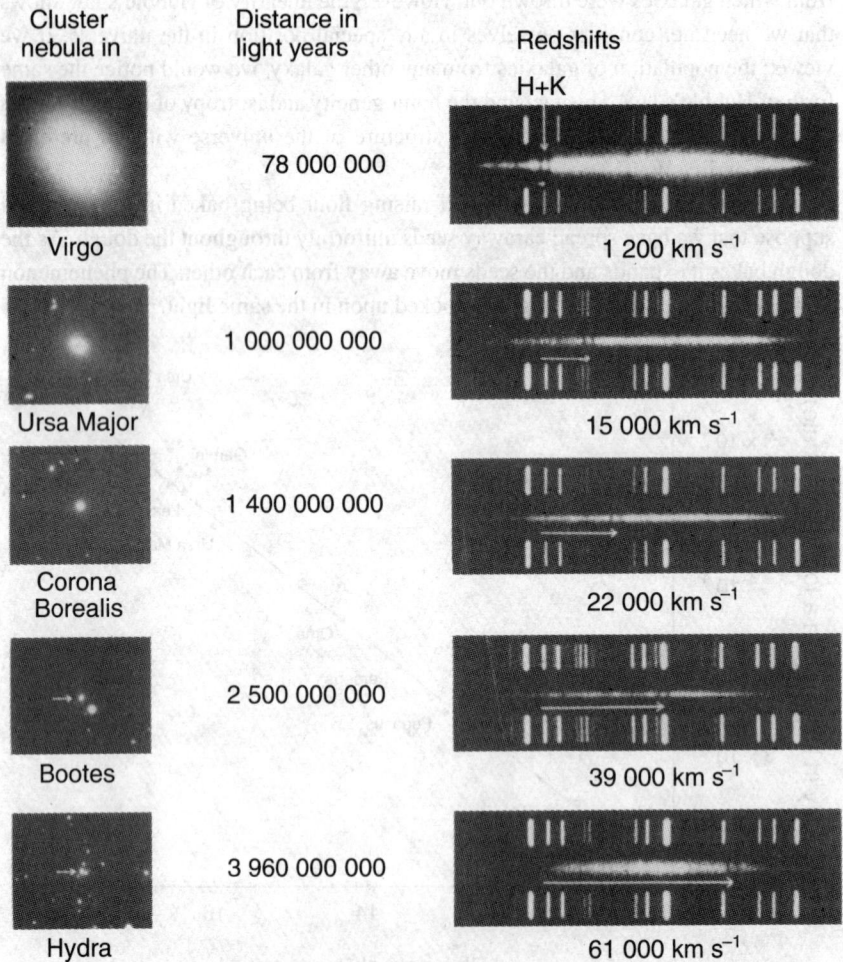

Cluster nebula in	Distance in light years	Redshifts H+K
Virgo	78 000 000	1 200 km s^{-1}
Ursa Major	1 000 000 000	15 000 km s^{-1}
Corona Borealis	1 400 000 000	22 000 km s^{-1}
Bootes	2 500 000 000	39 000 km s^{-1}
Hydra	3 960 000 000	61 000 km s^{-1}

Figure 1.34 The relation between redshift and distance for extragalactic nebulae. Redshifts (z) are expressed as velocities, cz. Arrows indicate shifts of the H and K lines of calcium. Distances are based on a rate of expansion of 50 km s^{-1} Mpc^{-1}. Courtesy of Palomar Observatory/Caltech.

H_0 is now believed to lie in the range 50–80 km s^{-1} Mpc^{-1}. We will write it as $100h_0$ km s^{-1} Mpc^{-1}, where h_0 lies between 0.5 and 0.8, perhaps in the narrower range 0.55–0.75. Notice that, if we assume the validity of Hubble's law, we can estimate the distance from Earth of an extragalactic object from its redshift given by equation (1.4).

1.9   Radiation backgrounds

Apart from matter in its visible form, the universe also contains radiation at various frequencies. In general, measurements of the electromagnetic radiation at a given frequency (or in a given range of frequencies) reveal peaks in specific directions that are associated with relatively nearby discrete sources, many of which can be identified. However, after these peaks are eliminated, there is still a residual background of radiation. This background radiation could also arise from discrete sources that are located much further away and therefore cannot be resolved, or it could arise from processes in the interstellar and intergalactic space. Table 1.3 gives a rough estimate of the energy densities in the various wavelength ranges. Figure 1.35 shows the same information in a pictorial form. It should be remembered that the measurements in X-rays, γ-rays and so on became possible only in the early 1960s with the advent of space astronomy.

Compared with the estimates of the density of matter, the total radiation energy density at all wavelengths is less by at least three orders of magnitude. Recall that the observed density of matter in the forms of galaxies, clusters, etc. is a few times 10^{-31} g cm^{-3}, i.e., an equivalent energy density of a few times 10^{-10} erg cm^{-3}. This observation is often expressed by the statement that the universe is at present 'matter-dominated'.

Table 1.3 or Figure 1.35 also tells us that the most dominant form of radiation background is in the microwaves. The spectrum of the microwave background is very nearly that of blackbody radiation of temperature $\sim$2.7 K. Since blackbody

Table 1.3 The radiation background at various levels

Type of radiation	Wavelength λ, Frequency, ν Energy range, E	Energy density (erg cm^{-3})
Radio	$\nu \leq 4080$ MHz	$\leq 10^{-18}$
Microwaves	λ in 80 cm to 1 mm	$\sim 4 \times 10^{-13}$
Optical	λ in 4000–8000 Å	$\sim 3.5 \times 10^{-15}$
X-rays	E in 1–40 keV	$\sim 10^{-16}$
γ-rays	$E \geq 100$ MeV	$\leq 2 \times 10^{-17}$

radiation wipes out individual details of specific sources, it is not surprising that no discrete source of this radiation is identifiable. Moreover, the extreme homogeneity and isotropy of this radiation on small angular scales rules out the possibility that it could have arisen from galaxy-sized discrete sources. As we shall see in Chapter 5, the most popular interpretation of this radiation is that it is a relic of an early hot epoch when the universe was much denser than it is now. Unlike the distribution of matter, this relic radiation is extremely homogeneous. This contrast further exacerbates the difficulty of relating the origin of discrete structures of matter to tiny fluctuations observed against a smooth radiation background.

1.10 Theoretical models of the universe

We now come to the theoretical revolution initiated by Albert Einstein. In cosmology, theory and observations have proceeded hand in hand. If Hubble's observations are seen to have launched modern observational cosmology, it is Einstein's general theory of relativity that can be credited with laying the foundations of modern

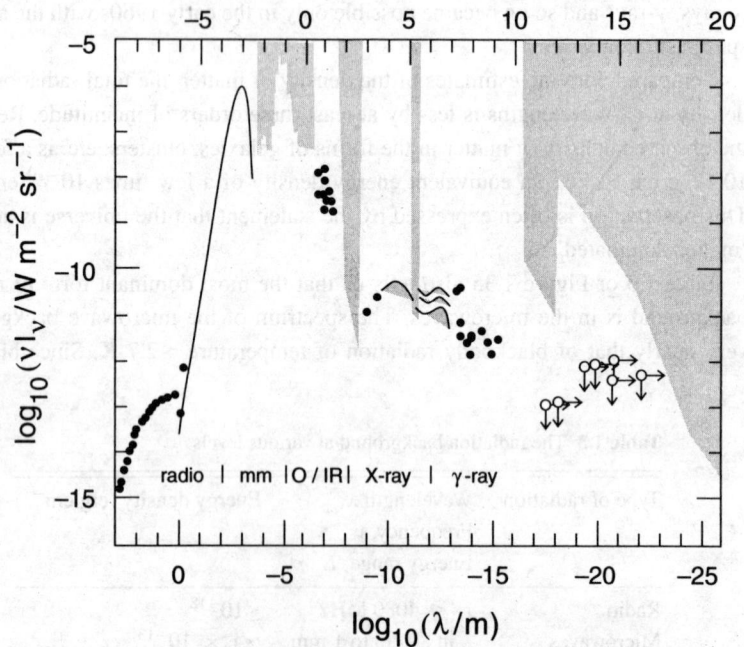

Figure 1.35 This graph shows intensities of the cosmic radiation background at various wavelengths. The plot is on a log–log scale. The shaded region indicates upper limits only. O/IR stands for optical and infrared. Notice the dominance of millimetre wavelength, i.e., microwaves.

theoretical cosmology. We will discuss in Chapter 3 the details of how theoretical developments in cosmology actually began more than a decade before Hubble's exciting observations. We conclude the present chapter by considering the general question of why relativity is taken to be so important for cosmology, when Newtonian gravitation has generally served astronomy and astrophysics well right from the study of the Solar System to the Milky Way.

Table 1.4 shows the orders of magnitude involved in the large-scale structure of the universe. The last entry refers to the characteristic distance scale c/H_0 that emerges from Hubble's constant and the mass contained in the 'observable' volume of radius c/H_0 if the density were that seen for visible matter in our neighbourhood. Similarly, the time scale characteristic of the universe is $H_0^{-1} \approx 10^{10}$ years.

What interaction in physics is likely to be influential over such long distances and for such large masses? Of the four known interactions, only gravity and electromagnetism are of long range. Although the electromagnetic interaction is much stronger than gravity on the scale of atoms, it is ineffective at determining the large-scale structure of the universe, since all indications are that an electric charge balance is preserved in galaxies, clusters and the intergalactic space. Neither is there any evidence for large-scale electric currents that could interact with the magnetic fields in the universe to produce large forces. In contrast, the enormous masses of astronomical objects generate huge gravitational fields. Gravitation, which is usually neglected in micro-physics as being too weak to be of any significance, is therefore the most relevant interaction in cosmology.

Given that we need a theory of gravitation for cosmology, what is wrong with the Newtonian framework? It has worked well in the theory of stellar structure. It is even used for stellar dynamics in the Galaxy. Why not use it in cosmology? Let us try to understand the answer with the help of the entries in Table 1.4. Newtonian gravity is a theory of instantaneous action at a distance. As such it is inconsistent with the special theory of relativity, in particular with the limit (c) placed by that theory on the speed with which any interaction can propagate across space. In those parts of astronomy where the distances across which gravitation is supposed to act

Table 1.4 Spatial dimensions and masses of astronomical systems

Object	Linear size	Mass
Sun	7×10^{10} cm (radius)	2×10^{33} g $\equiv M_\odot$
Galaxy	$\sim$15 kpc	$\sim 10^{11} M_\odot$
Cluster	$\sim$5 Mpc	$\sim 10^{13}$–$10^{14} M_\odot$
Supercluster	$\sim$50 Mpc	$\sim 10^{15} M_\odot$
Universe[a]	$\sim$3000 Mpc	$\sim 10^{21} M_\odot$

[a] For $h_0 = 1$.

are relatively small, the use of Newtonian gravity is permissible. As can be seen in Table 1.4, however, the distances in cosmology are so large that action at a distance with infinite speed becomes an unrealistic and unreliable concept. This is not so with objects of stellar dimensions or even for individual galaxies.

However, special relativity itself is an imperfect theory in the presence of gravity, as Einstein himself had realized. The concepts of the inertial frame and the inertial observer (on whom no force acts), which are so basic to special relativity, are unrealizable in the presence of gravity. Gravity seems to be an ever-present force that cannot be switched off altogether. Since all matter attracts gravitationally, an inertial observer cannot exist at all over extended regions of space and time! Which is why Einstein was motivated towards formulating the general theory of relativity by incorporating the above-mentioned properties of gravitation. General relativity has the merit that it reduces to Newtonian gravitation and also to the special theory of relativity when the gravitational effects are relatively weak.

Nevertheless, it was shown in 1934 by W. H. McCrea and E. A. Milne that, with suitable compromises, Newtonian gravitation and dynamics can describe cosmology in an adequate manner. Newtonian cosmology is simple to understand and the McCrea–Milne approach has its merits in presenting a simplistic picture of cosmology. We shall refer to it in Chapters 3 and 4. Nevertheless, our approach here is intended to prepare the reader for the more advanced ideas in cosmology and so it is preferable for us to resort to a framework that is free of conceptual difficulties and compromises. Since general relativity provides a framework that is free of conceptual difficulties, cosmologists feel at home with the use of this theory. It is therefore appropriate that we begin our discussion of cosmology by first outlining the general theory of relativity. The treatment of this theory given in Chapter 2 is at an introductory level. Even so, readers already familiar with the theory may find this chapter useful, if only to familiarize themselves with the notation which will be used in the rest of this book.

Exercises

1 Take the diameter AB of the Earth's orbit as 3×10^8 km and consider a star S at a distance d, such that SA = SB and the angle ASB = 2 arcseconds. Calculate d. This is the distance unit of one *parsec*. Relate it to one light year.

2 Given that the Sun takes $\sim 2 \times 10^8$ years to make a circular orbit around the Galactic Centre, staying at a distance of 10 kpc, estimate the mass of the Galaxy contained within the solar orbit, assuming that it is spherical and ignoring any effects of mass lying outside the orbit. (You may use Newtonian dynamics and gravitation.)

3 A galaxy has a visible mass of $10^{11} M_\odot$ and a flat rotation curve extending to 25 kpc at the level of 150 km s^{-1}. What is the ratio of its dark-matter mass to its visible mass?

4 Compute the gravitational energy of a pair of colliding galaxies, each of mass $10^{11} M_\odot$, separated by a distance of ~ 10 kpc and compare it with the energy requirements of a

powerful radio source. What conclusion do you arrive at from such a comparison?

5 Using a Hubble constant given by $h_0 = 0.6$, place an approximate upper limit on the angular size of a quasar of redshift $z = 0.2$ if it exhibits a time variability on the scale of 1 h. (Use the argument that no physical effect travels in a material object faster than light.)

6 Use the energy density of the microwave background given in Table 1.3 to estimate its temperature, assuming that it has a blackbody form.

Chapter 2

The general theory of relativity

2.1   Space, time and gravitation

Every major scientific theory has its own distinctive characteristic. The distinctive feature of Newtonian gravitation is the radial inverse-square law. To those uninitiated in the laws of dynamics, the fact that a planet goes *around* the Sun under a force of attraction *towards* the Sun comes as a surprise. The major achievement of Maxwell's electromagnetic theory was the unification of electricity and magnetism and the demonstration that light itself is an electromagnetic wave. The unique place held by the speed of light characterizes Einstein's special theory of relativity, while quantum mechanics can point to the uncertainty principle as the crucial feature that sets it apart from classical mechanics.

To what distinctive feature can general relativity lay its own special claim? A clue to the answer to this question is provided in the title of this section.

Let us compare gravitation with electricity. We know that two unlike electric charges attract each other through the Coulomb inverse-square law, just as any two masses attract each other gravitationally according to the Newtonian inverse-square law. To this extent, electricity and gravitation are similar. However, we can go no further! We also know that two like electric charges repel each other and that this property seems to have no parallel in gravitation. Every bit of matter attracts every other bit and, as yet, we know of no instance of gravitational repulsion.

We can express this difference between electricity and gravitation in another, more practical, way. The existence of repulsion as well as attraction enables us to construct a closed chamber whose interior is completely sealed from any outside electric influence. This is not so with gravitation! We cannot point to any region of space as being totally free of external gravitational influences. Gravitation is permanent: it cannot be switched off at will.

This ever-present nature of gravitation plays the key role in Einstein's general theory of relativity. Einstein argued that, because of its permanence, gravitation must be related to some intrinsic feature of space and time. With a master stroke of genius, he identified this feature as the *geometry* of space and time. He suggested that any effects we ascribe to gravitation actually arise because the *geometry* of space and time is 'unusual'. Let us now try to understand what is meant by the word 'unusual' and how this property of space and time leads to gravitational effects – for therein lies the distinctive characteristic that sets general relativity apart from other physical theories.

The 'usual' geometry of space, the geometry that we learn at school and learn to apply in so many ways, is the geometry whose foundations were laid by the Greek mathematician Euclid *ca.* 300 BC. Euclidean geometry is a logical structure wherein theorems about triangles, parallelograms, circles and so on are proved on the basis of postulates that are taken as self-evident. Thus the results shown in Figure 2.1 follow as theorems in Euclidean geometry, which is based on the original postulates of Euclid. The validity of these results appears to be borne out by measurements of lengths and angles in physical space.

It was only in the last century that mathematicians realized that there is nothing sacrosanct about Euclid's postulates. Provided that they are not mutually contra-dictory, any new set of postulates can lead to a new type of geometry. Indeed, as the work of such mathematicians as Gauss (1777–1855), Bolyai (1802–1860), Lobatchevsky (1793–1856) and Riemann (1826–1866) showed, a host of such new geometries can be constructed. These are collectively called *non-Euclidean geometries*. For instance, the geometry on the surface of a sphere is non-Euclidean. If we define a straight line on the surface of a sphere as the line of shortest distance between two points, it is easy to see that these lines are arcs of great circles. Because

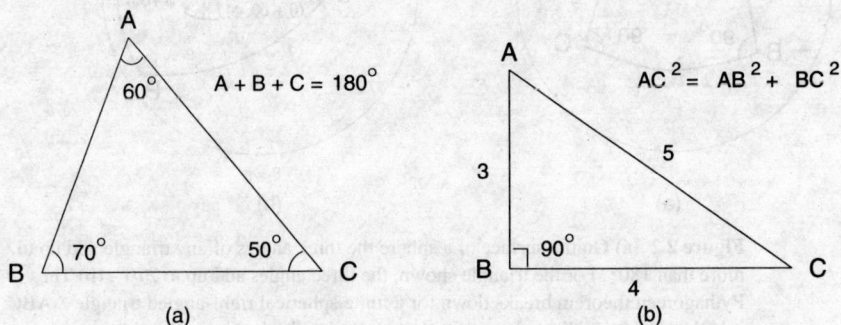

Figure 2.1 (a) The three angles of any triangle ABC add up to 180°. (b) The well-known theorem of Pythagoras for a typical right-angled triangle ABC.

any two great circles intersect, there are no parallel lines in this geometry. Figure 2.2 demonstrates how the theorems of Figure 2.1 break down when they are applied to the non-Euclidean geometry of the surface of the sphere.

The concept of geometry of space can be extended to the geometry of space and time, thanks to the foundations laid by Einstein's special theory of relativity. Let us first recall a familiar result from special relativity in the following form. Let (x, y, z) denote a Cartesian coordinate system and t the time measured by an observer O at rest in an inertial frame, that is by an observer who is acted on by no force. We will return to a discussion of such observers later. Let two neighbouring events in space and time be labelled by the coordinates (x, y, z, t) and $(x+dx, y+dy, z+dz, t+dt)$. The resulting analogue of the Pythagorean theorem shown earlier in Figure 2.1(b) is as follows. The square of the 'distance' between the two events is given by

$$ds^2 = c^2 \, dt^2 - dx^2 - dy^2 - dz^2. \tag{2.1}$$

The distance ds is invariant in the sense that another inertial observer O' using a different coordinate system (x', y', z', t') to measure this distance will find the same answer.

However, when we make a transition from special to general relativity and quantify Einstein's idea that the geometry of space and time is unusual in the presence of gravitation, we abandon the simple form of (2.1) in favour of a more complicated form. This is comparable to the transition from Figure 2.1(b) to Figure 2.2(b). The more complicated form is still quadratic and we may state it formally as follows:

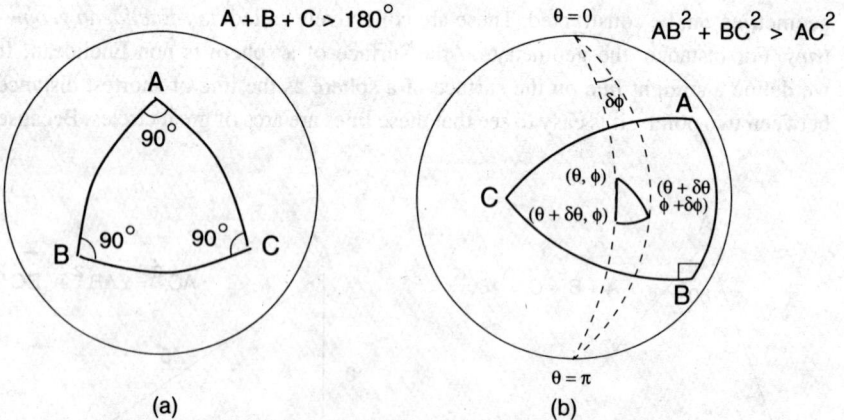

Figure 2.2 (a) On the surface of a sphere the three angles of any triangle add up to more than 180°. For the triangle shown, the three angles add up to 270°. (b) The Pythagorean theorem breaks down for a finite spherical right-angled triangle $\triangle ABC$: it looks more complicated even for the small triangle shown inside with vertices in spherical coordinates (θ, ϕ): $ds^2 = a^2(d\theta^2 + \sin^2\theta \, d\phi^2)$, where a is the radius of the sphere.

$$ds^2 = \sum_{i,k=0}^{3} g_{ik}\, dx^i dx^k. \tag{2.2}$$

Here we have modified the notation as follows. The coordinates are now called x^i, with $i = 1, 2, 3$ representing the three space coordinates and $i = 0$ the time coordinate. The coefficients g_{ik} are functions of x^i with the property that the matrix $\|g_{ik}\|$ has the signature -2. It is convenient to refer to this unified structure of space and time as *spacetime*.[1]

Clearly, the geometry of spacetime in which the basic invariant distance is given by (2.2) instead of by (2.1) is going to be more complicated to describe. Its properties will depend on the functions g_{ik}. Do these complications arise simply because of a choice of coordinates, or do they indicate a spacetime with a geometry genuinely different from that used in special relativity? How can we express the familiar laws of physics such as Maxwell's equations in this spacetime? More specifically, how do these properties tell us about the presence of gravitation? In what way, for example, can we interpret manifestly gravitational phenomena like the motion of planets as effects of geometry? The remainder of this chapter attempts to answer these questions.

2.2   Vectors and tensors

Let us consider again the example of geometry on the surface of a sphere Σ of radius a. If we consider the sphere as embedded in a three-dimensional space with the Cartesian coordinates x, y and z, we may write the equation of the surface of the sphere as

$$x^2 + y^2 + z^2 = a^2. \tag{2.3}$$

For describing the geometry on the surface of the sphere it is, however, more convenient to use coordinates intrinsic to the surface of the sphere. Such coordinates are available and are like the latitude and longitude used to locate a point on Earth. More specifically,

$$x = a\sin\theta\cos\phi, \qquad y = a\sin\theta\sin\phi, \qquad z = a\cos\theta, \tag{2.4}$$

so that, for any (θ, ϕ) with $0 \le \theta \le \pi$ and $0 \le \phi < 2\pi$, we can locate a point (x, y, z) on the surface of the sphere. Spherical trigonometry tells us how to measure and relate the angles, sides and so on of triangles drawn on this surface. The rules of Euclid's geometry do not apply to these measurements.

---

[1] This means that, if the quadratic equation (2.2) is diagonalized, it has one square term with a positive coefficient and three square terms with negative coefficients. The signature equals the number of positive terms minus the number of negative terms.

In our example above, the square of the distance between two neighbouring points (θ, ϕ) and $(\theta + d\theta, \phi + d\phi)$ is given by

$$d\sigma^2 = [dx^2 + dy^2 + dz^2]_\Sigma = a^2(d\theta^2 + \sin^2\theta \, d\phi^2). \tag{2.5}$$

Thus we have here an example of g_{ik} that are not all constants.

However, neither the inconstancy of g_{ik} nor its non-diagonal nature guarantees that we are dealing with a non-Euclidean geometry. For example, in three-dimensional Euclidean space, the transformation

$$x = r\sin\theta\cos\phi, \qquad y = r\sin\theta\sin\phi, \qquad z = r\cos\theta, \tag{2.6}$$

with (θ, ϕ) as defined before and $0 \le r \le \infty$, gives

$$d\sigma^2 = dx^2 + dy^2 + dz^2 = dr^2 + r^2(d\theta^2 + \sin^2\theta \, d\phi^2). \tag{2.7}$$

Again we have g_{ik} as functions of r and θ. However, now we *know* that we are dealing with Euclidean geometry and that the dependence of g_{ik} on r and θ is purely a coordinate effect.

Thus we clearly have to devise a means of extracting essential geometrical information as distinct from pure coordinate effects. In a qualitative way we can see that the essential information must survive even when we change from one coordinate system to another. In order to extract such information, we must devise machinery that tells us *what things remain unchanged under coordinate transformations*. Such machinery is provided by the invariants, the vectors and the tensors, which we shall now study.

Let us first introduce a summation convention. We will frequently encounter sums like

$$\sum_{i=0}^{3} A_i B^i, \qquad \sum_{k=0}^{3} A_{ik} B^k, \qquad \sum_{i,k=0}^{3} P_{ik}\xi^i\xi^k, \dots$$

It is convenient in such cases to drop the summation symbol and write these quantities as

$$A_i B^i, \qquad A_{ik} B^k, \qquad P_{ik}\xi^i\xi^k, \dots$$

the rule being that *whenever an index appears once as a subscript and once as a superscript in the same expression, it is automatically summed over all the values* (from 0 to 3). Thus we can rewrite (2.2) in the more compact form

$$ds^2 = g_{ik}\, dx^i\, dx^k. \tag{2.8}$$

A note of caution is needed here: the summation convention does not apply under any other circumstances. Thus it does not apply to quantities like

$$A_i B_i, \qquad A_{ik} B_i C_i, \dots,$$

wherein repeated indices do not follow the rule by appearing only twice, once up and once down. However, such expressions fortunately do not arise in most relativistic calculations. Indeed, the appearance of such 'monster' expressions is a warning that we have made a mistake in our manipulation of indices.

We will assume that the Latin indices $i, j, k, \ldots$ will run over all four values 0, 1, 2 and 3. On some (rather infrequent) occasions we may want to refer to index values 1, 2 and 3 only, which are usually reserved for space components, and we will use Greek indices $\mu, \nu, \ldots$ to represent these.

It is worth pointing out here that many other textbooks use the convention of denoting the spacetime coordinates by Greek indices λ, μ, ν, etc. and the space coordinates by Latin indices i, j, k, etc. Also many authors prefer to write (2.1) with the opposite sign for the right-hand side. Likewise, in some texts, time is treated as coordinate number 4 instead of 0, as it is here. These differences are of a 'cosmetic' nature and do not affect the 'physics' being described.

2.2.1 Scalars

A *scalar* or an *invariant* does not change under any change of coordinates. Thus, if $\phi(x^i)$ is a function of coordinates, then it is invariant provided it retains its value under a transformation from x^i to new coordinates x'^i:

$$\phi(x^i) = \phi[x^i(x'^k)] = \phi'(x'^k). \tag{2.9}$$

Note that the form of the function may change, but its value does not. Note that the infinitesimal square of distance (2.2) is a scalar quantity.

2.2.2 Contravariant vectors

Suppose that we are given a curve Γ in space and time, which is parametrized by λ (see Figure 2.3). Thus, the points along the curve have coordinates

$$x^i \equiv x^i(\lambda), \tag{2.10}$$

where x^i are given functions of λ. The direction of the tangent to Γ at any point on it is given by a vector with four components

$$A^i \equiv \frac{dx^i}{d\lambda}. \tag{2.11}$$

Notice that the direction of a tangent to the curve is an invariant concept: a change of coordinates should not alter this concept, although its four components in the new coordinates will be different. Suppose that the new coordinates are x'^i and the new components are A'^i. Then

$$A'^i \equiv \frac{dx'^i}{d\lambda}. \tag{2.12}$$

Unless stated otherwise, we will assume that the transformation functions

$$x^i = x^i(x'^k), \qquad x'^k = x'^k(x^i) \tag{2.13}$$

are continuous and possess at least second derivatives. It is then easy to see that A'^i and A^i are related by the linear transformation

$$A'^k = \frac{\partial x'^k}{\partial x^i}\frac{\mathrm{d}x^i}{\mathrm{d}\lambda} = \frac{\partial x'^k}{\partial x^i} A^i. \tag{2.14}$$

We use (2.14) as the transformation law for *any* vector A^i. Quantities in general that transform according to the above linear law are called *contravariant vectors*. The four components of a contravariant vector are specified by a superscript.

For example, consider the curve parametrized by

$$x^0 = \text{constant}, \qquad x^1 = \text{constant}, \qquad x^2 = \lambda, \qquad x^3 = \lambda^2. \tag{2.15}$$

The tangent to this curve is specified by the contravariant vector A^i with components

$$A^0 = 0, \qquad A^1 = 0, \qquad A^2 = 1, \qquad A^3 = 2\lambda. \tag{2.16}$$

2.2.3   Covariant vectors

Consider next a scalar function $\phi(x^k)$. The equation

$$\phi(x^k) = \text{constant} \tag{2.17}$$

describes a hypersurface (that is, a surface of three dimensions) Σ, whose normal has the direction given by the four quantities

$$B_i = \frac{\partial \phi}{\partial x^i} \tag{2.18}$$

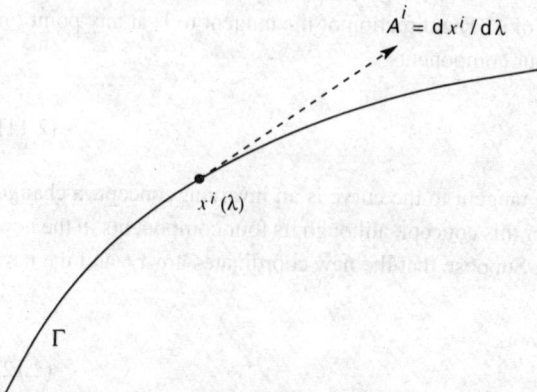

$A^i = \mathrm{d}x^i/\mathrm{d}\lambda$

$x^i(\lambda)$

Γ

Figure 2.3 The tangent vector to the curve Γ transforms as a contravariant vector.

(see Figure 2.4). Again, the concept of a normal to a hypersurface should be independent of the coordinates used. Under the coordinate transformation (2.13), the new components are

$$B_i' = \frac{\partial \phi}{\partial x'^i}.$$

It is easy to see that $B_i' \leftrightarrow B_i$ is a linear transformation:

$$B_k' = \frac{\partial x^i}{\partial x'^k} B_i. \tag{2.19}$$

Again, we generalize (2.18) as a transformation law of any vector B_i. Quantities that transform according to this rule are called *covariant vectors*. For example, the normal to the unit sphere given by

$$\phi \equiv (x^1)^2 + (x^2)^2 + (x^3)^2 = 1$$

has the covariant components

$$B_0 = 0, \qquad B_1 = 2x^1, \qquad B_2 = 2x^2, \qquad B_3 = 2x^3.$$

2.2.4 Tensors

The concept of a vector can be generalized to that of a tensor. Thus a contravariant tensor of rank 2 is characterized by the following transformation law:

$$T'^{ik} = \frac{\partial x'^i}{\partial x^m} \frac{\partial x'^k}{\partial x^n} T^{mn}. \tag{2.20}$$

A covariant tensor of rank 2 is similarly characterized by the transformation law

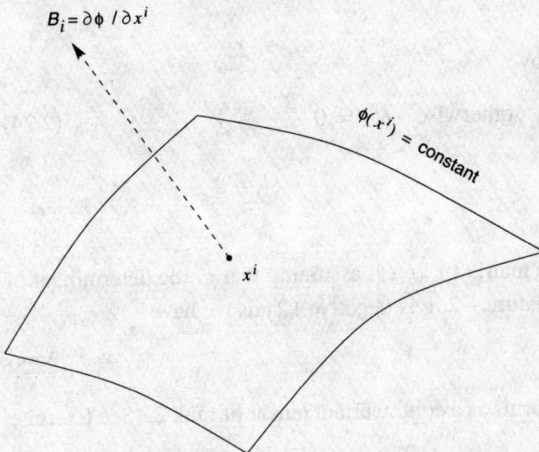

$B_i = \partial \phi / \partial x^i$

$\phi(x^i) = \text{constant}$

x^i

Figure 2.4 The normal to a three-dimensional surface Σ transforms as a covariant vector.

$$T'_{ik} = \frac{\partial x^m}{\partial x'^i} \frac{\partial x^n}{\partial x'^k} T_{mn}.$$

(2.21)

It is also possible to have *mixed* tensors. Thus $T^i{}_k$ is a mixed tensor of rank 2, with one contravariant index and one covariant index. It transforms as

$$T'^i{}_k = \frac{\partial x'^i}{\partial x^m} \frac{\partial x^n}{\partial x'^k} T^m{}_n.$$

(2.22)

Again, these concepts are easily generalized to tensors of rank higher than 2. The rule is to introduce a transformation factor $\partial x'^i / \partial x^m$ for each contravariant index i and a factor $\partial x^n / \partial x'^k$ for each covariant index k.

Example 1

The quantities g_{ik} transform as a covariant tensor. This result follows from the assumption that $\mathrm{d}s^2$ as given by (2.8) is invariant. For,

$$\mathrm{d}s^2 = g_{ik} \, \mathrm{d}x^i \, \mathrm{d}x^k$$

$$= g_{ik} \left(\frac{\partial x^i}{\partial x'^m} \, \mathrm{d}x'^m \right) \left(\frac{\partial x^k}{\partial x'^n} \, \mathrm{d}x'^n \right)$$

$$= \left(g_{ik} \frac{\partial x^i}{\partial x'^m} \frac{\partial x^k}{\partial x'^n} \right) \mathrm{d}x'^m \, \mathrm{d}x'^n$$

$$= g'_{mn} \, \mathrm{d}x'^m \, \mathrm{d}x'^n;$$

that is,

$$g'_{mn} = \frac{\partial x^i}{\partial x'^m} \frac{\partial x^k}{\partial x'^n} g_{ik}.$$

(2.23)

This tensor is called the *metric tensor*. The quadratic expression for $\mathrm{d}s^2$ is called the *line element* of spacetime or the *spacetime metric*.

Example 2

The *Kronecker delta* defined by

$$\delta^i{}_k = 1 \quad \text{if} \quad i = k, \quad \text{otherwise} \quad \delta^i{}_k = 0$$

(2.24)

is a mixed tensor of rank 2.

Example 3

Define $\|g^{ik}\|$ to be the inverse matrix of $\|g_{ik}\|$, assuming that g, the determinant of $\|g_{ik}\| \neq 0$. (Since g_{ik} has signature -2, g is negative.) Thus we have

$$g_{ik} g^{kl} = \delta^l{}_i.$$

(2.25)

It can be shown that g^{ik} transforms as a contravariant tensor of rank 2. (See Exercise 6.)

Example 4

A physical example for tensors is found when one is discussing deformation of substances. Figure 2.5 illustrates the surface Σ of such a substance, which has normal n_i at a typical point P. If the surface is subjected to stress, the resulting force on an element of surface around P will be in the direction F_k, different from the normal, but related to it by the linear tensor relation

$$F_k = T^i{}_k n_i,$$

where $T^i{}_k$ is the stress tensor. If the stress is isotropic, then

$$T^i{}_k = p\delta^i{}_k$$

where p is the pressure which produces a force *normal* to the surface Σ.

2.2.5   Symmetric and antisymmetric tensors

If tensors S_{ik} and A_{ik} satisfy the relations

$$S_{ik} = S_{ki}, \qquad A_{ik} = -A_{ki}, \tag{2.26}$$

then they are, respectively, *symmetric* and *antisymmetric* tensors of rank 2. These ideas can be generalized to higher-rank tensors and we will encounter specific tensors having the properties of symmetry and antisymmetry with respect to some or all indices.

Example 1

g_{ik} and g^{ik} are symmetric tensors.

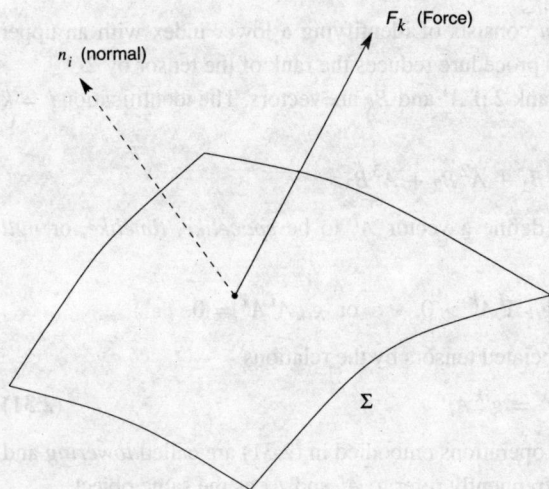

Figure 2.5 The stress-generated force F_k on the surface Σ of an elastic substance will not necessarily be along the normal n_i to the surface. The stress tensor relates the force vector to the normal vector.

Example 2

Consider the symbol ϵ_{ijkl} with the following properties:

$$\epsilon_{ijkl} = +1 \quad \text{if } (ijkl) \text{ is an even permutation of } (0123)$$
$$\epsilon_{ijkl} = -1 \quad \text{if } (ijkl) \text{ is an odd permutation of } (0123)$$
$$\epsilon_{ijkl} = 0 \quad \text{otherwise.} \tag{2.27}$$

We will now show that

$$e_{ijkl} = \sqrt{-g}\, \epsilon_{ijkl} \tag{2.28}$$

transforms as a tensor.

First take the determinant of (2.23). Let J denote the Jacobian $\left|\partial x^i/\partial x'^m\right|$. Then, using the rule that the determinant of a product of matrices is equal to the product of their determinants, we get

$$g' = J^2 g. \tag{2.29}$$

However, we have from the definition of a determinant

$$\epsilon_{mnpq} J = \epsilon_{ijkl} \frac{\partial x^i}{\partial x'^m} \frac{\partial x^j}{\partial x'^n} \frac{\partial x^k}{\partial x'^p} \frac{\partial x^l}{\partial x'^q}. \tag{2.30}$$

Using (2.28) and (2.29), the result follows: e_{ijkl} is a tensor that is totally antisymmetric. Strictly speaking, however, e_{ijkl} is a *pseudotensor*, since it changes sign under transformations involving reflection, such as $x'^0 = -x^0$, $x'^1 = x^1$, $x'^2 = x^2$ and $x'^3 = x^3$.

Exercises 3–10 at the end of this chapter will help you to understand the operation of vectors and tensors. We end this section with one important operation.

2.2.6   Contraction

The operation of *contraction* consists of identifying a lower index with an upper index in a mixed tensor. This procedure reduces the rank of the tensor by 2.

Thus $A^i B_k$ is a tensor of rank 2 if A^i and B_k are vectors. The identification $i = k$ gives a *scalar*:

$$A^i B_i = A^0 B_0 + A^1 B_1 + A^2 B_2 + A^3 B_3.$$

As in special relativity, we define a vector A^i to be *spacelike*, *timelike*, or *null* according to

$$g_{ik} A^i A^k < 0, \qquad g_{ik} A^i A^k > 0, \qquad \text{or } g_{ik} A^i A^k = 0.$$

It is convenient to define associated tensors by the relations

$$A_i = g_{ik} A^k, \qquad A^k = g^{ik} A_i. \tag{2.31}$$

Thus $g_{ik} A^i A^k = A_k A^k$. The operations embodied in (2.31) are called *lowering* and *raising* the indices. We may frequently refer to A^i and A_i as the same object.

From the above manipulations of tensors it is clear (and can be easily proved) that the product of two tensors is a tensor. A reverse result is sometimes useful in deducing that a certain quantity is a tensor. This result is known as the *quotient law*. It states that, if a relation such as

$$PQ = R$$

holds in all coordinate frames, where P is an *arbitrary* tensor of rank m and R a tensor of rank $m + n$, then Q is a tensor of rank n.

2.3 Covariant differentiation

A *vector field* is a vector function of position defined over a subspace of spacetime. Let $B_i(x^k)$ be a covariant vector field whose four components transform according to the rule in (2.19) at each point (x^k) where it is defined. Suppose that B_i is a differentiable function of (x^k). Do the derivatives $\partial B_i / \partial x^k$ transform as a tensor?

We have already seen that the derivatives $\partial \phi / \partial x^k$ of a scalar transform as a vector. So at first sight the answer to the above question might be 'yes'. Indeed, in special relativity we do encounter such results. For example, if A_i is the 4-potential of the electromagnetic field (described in the four-dimensional language of special relativity), then $\partial A_i / \partial x^k$, for Cartesian coordinates (x, y, z) and the time t of (2.1), do transform as a tensor. In our more general spacetime with an arbitrary coordinate system, however, the answer to the above question is in the negative.

This result is easily verified by differentiating (2.19) with respect to x'^m. We get

$$\frac{\partial B'_k}{\partial x'^m} = \frac{\partial x^i}{\partial x'^k} \frac{\partial x^n}{\partial x'^m} \frac{\partial B_i}{\partial x^n} + \frac{\partial^2 x^i}{\partial x'^m \, \partial x'^k} B_i. \qquad (2.32)$$

Thus, whereas the first term on the right-hand side does appear in the right form needed to make $\partial B_i / \partial x^n$ a tensor, the second term spoils the effect. It also gives a clue as to why this happens. The second derivative

$$\frac{\partial^2 x^i}{\partial x'^m \, \partial x'^k}$$

is in general non-zero and indicates that the transformation coefficients in equation (2.19) vary with position in spacetime. Now, when we seek to construct the derivative $\partial B_i / \partial x^n$, we have to define it as a limit:

$$\frac{\partial B_i}{\partial x^n} = \lim_{\delta x^n \to 0} \left(\frac{B^i(x^k + \delta x^k) - B^i(x^k)}{\delta x^n} \right).$$

However, the two terms in the numerator transform as vectors at two different points and, because of the variation of the transformation coefficients with position, their difference is not expected to be a vector. (The difference of two vectors is a vector, provided that both are so defined at the *same point*.)

This situation is illustrated in Figure 2.6. P and Q are the two neighbouring points (x^k) and $(x^k + \delta x^k)$, with the vectors B_i shown there with continuous arrows. In order to describe the change in the vector on going from P to Q, we must somehow measure the difference at *the same point*. How can this be achieved?

This is achieved by a device known as *parallel transport*. Assume that the vector B_i at P is moved from P to Q, parallel to itself, that is, as if its magnitude and direction did not change. In Figure 2.6 this is shown by a dotted vector at Q. The difference between the vector $B_i(x^k + \delta x^k)$ and this dotted vector is a vector at Q and this tells us the real physical difference in the vector on going from P to Q. So we may after all be able to define a process of differentiation of vectors, provided that we know what happens to B_i during parallel transportation from P to Q.

First we have to note that the dotted vector at Q need not have the same components as those of the undotted vector at P. It is only with Cartesian coordinates that the components are the same. Consider, for example, the Euclidean plane with a polar coordinate system. A vector **A** at a point P with coordinates (r, θ) has components A_r and A_θ in the radial and transverse directions. If we now move from P to a neighbouring point Q with polar coordinates $(r + \delta r, \theta + \delta \theta)$, as shown in Figure 2.7, the radial and transverse directions at Q will not necessarily be parallel to those at P. Hence, after parallel transportation of **A** from P to Q, its radial and transverse components at Q will be different from A_r and A_θ.

A simple calculation (see Exercise 11) shows that the components of **A** at Q are $A_r + \delta \theta \, A_\theta$ and $A_\theta - \delta \theta \, A_r$. Taking a cue from this example for our general case, we see that the changes in the components of B_i through parallel transportation will be proportional to the original components B_i and also to the displacement δx^k in position on going from P to Q. We may express the change as a linear function of both these quantities and the most general form that we can have is

$$\delta B_i = \Gamma^l_{ik} B_l \, \delta x^k, \tag{2.33}$$

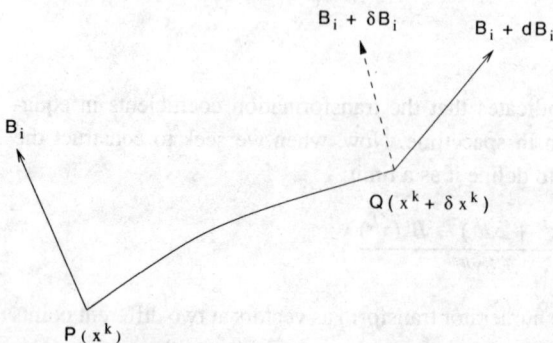

Figure 2.6 The vector field has the components B_i at P and $B_i + dB_i$ at Q. If B_i were transported parallel to itself along an infinitesimal curve connecting P to Q, its components at Q would be $B_i + \delta B_i$.

where the coefficients Γ^l_{ik} are, in general, functions of space and time. These quantities are called the *three index symbols* or the *Christoffel symbols*.

Notice that the introduction of (2.33) is something new in addition to the introduction of the metric. The metric tells us how to measure distances between neighbouring points, whereas (2.33) tells us how to define parallel vectors at neighbouring points. This property of connecting neighbouring vectors through the concept of local parallelism is often called the *affine connection* of spacetime. There is a practical way of describing parallel propagation in the following fashion.

We take the example of a sphere. Suppose that Γ is a curve drawn on the spherical surface connecting points P_1 and P_2. The arrow shown in Figure 2.8 represents in magnitude as well as direction a vector **A** at P_1. How do we transport it parallelly to P_2? Imagine a plane touching the sphere at P_1, with the vector **A** mapped (dotted arrow) at the corresponding point Q_1 on the plane. (By mapping, we mean that the magnitude and direction of the original vector on the sphere and the mapped vector on the tangent plane should match.) Now carefully roll the sphere on the plane so that it keeps touching it along the successive points of Γ. When you reach P_2, stop there. Let the corresponding point on the plane be Q_2. Draw a vector parallel to the starting vector *on the plane* at the point on the plane corresponding to P_2. Next map this vector onto the sphere. This will be the required transported vector at P_2.

Returning to (2.33), we see that the difference between the continuous and the dotted vectors at Q is given by

$$B_i(x^k + \delta x^k) - [B_i(x^k) + \delta B_i] = \left(\frac{\partial B_i}{\partial x^k} - \Gamma^l_{ik} B_l\right) \delta x^k. \tag{2.34}$$

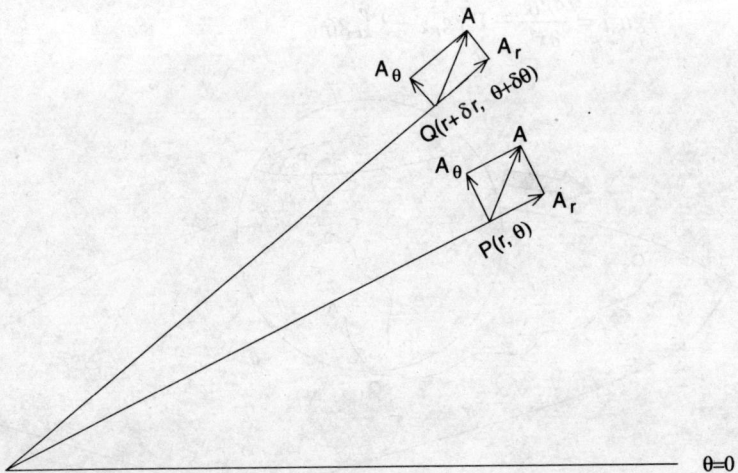

Figure 2.7 The radial and transverse directions at Q are *not* parallel to the radial and transverse directions at P. So equal and parallel vectors **A** at P and Q will have different radial and transverse components.

We may accordingly redefine the physically meaningful derivative of a vector by writing

$$B_{i;k} \equiv \frac{\partial B_i}{\partial x^k} - \Gamma^l_{ik} B_l \equiv B_{i,k} - \Gamma^l_{ik} B_l. \tag{2.35}$$

This derivative, by definition, must transform as a tensor. It is called the *covariant derivative* and will be denoted by a semicolon, in contrast to the ordinary derivative, which is denoted by a comma.

If $B_{i;k}$ must transform as a tensor, the coefficients Γ^i_{kl} have to transform according to the following law:

$$\Gamma'^i_{kl} = \frac{\partial x'^i}{\partial x^m} \frac{\partial x^n}{\partial x'^k} \frac{\partial x^p}{\partial x'^l} \Gamma^m_{np} + \frac{\partial^2 x^p}{\partial x'^k \partial x'^l} \frac{\partial x'^i}{\partial x^p}. \tag{2.36}$$

This result can be verified after some straightforward but tedious calculation.

A scalar, of course, does not change under parallel transportation, which is why $\partial \phi / \partial x^k$ transforms as a vector. If we use this result we see that, for a vector A_i, $(A_i A^i)_{,k}$ is a vector. This property allows us to construct the covariant derivative of a *contravariant* vector A^i:

$$A^i_{;k} \equiv \frac{\partial A^i}{\partial x^k} + \Gamma^i_{lk} A^l \equiv A^i_{,k} + \Gamma^i_{lk} A^l. \tag{2.37}$$

The rule of covariant differentiation of a tensor of arbitrary rank is easily obtained: we introduce a $(+\Gamma)$ term for each contravariant index and a $(-\Gamma)$ term for each covariant index. Thus, for the metric tensor, we have

$$g_{ik;l} = \frac{\partial g_{ik}}{\partial x^l} - \Gamma^p_{il} g_{pk} - \Gamma^p_{kl} g_{ip}. \tag{2.38}$$

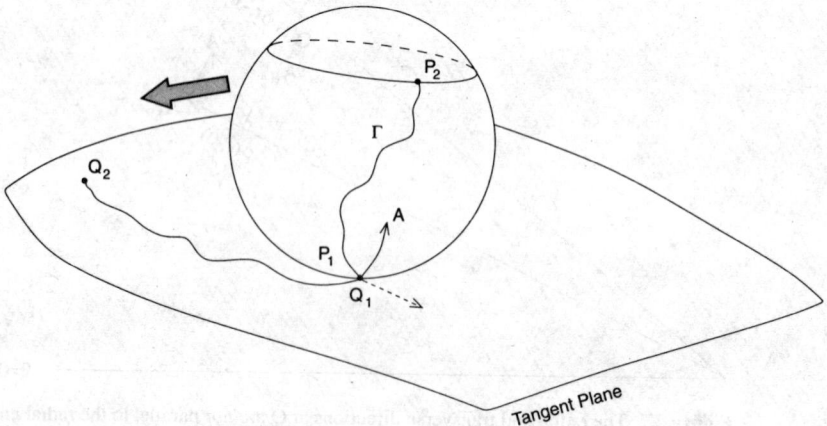

Figure 2.8 Parallel transport on a spherical surface. See the text for details.

2.4   Riemannian geometry

Einstein used the non-Euclidean geometry developed by Riemann to describe his theory of gravitation. The Riemannian geometry introduces the additional simplification that

$$\Gamma^i_{kl} = \Gamma^i_{lk}; \qquad g_{ik;l} \equiv 0. \tag{2.39}$$

Note that, as defined in the previous section, the affine connection need not satisfy these conditions. Indeed geometries for which the above relations are not satisfied also exist. For the theory of relativity, however, these conditions are *additionally assumed*.

Going back to (2.38), we see that $g_{ik;l} = 0$ gives us 40 linear equations for the 40 unknowns Γ^i_{kl}. These equations have a unique solution. For, from (2.38) and (2.39), we get

$$\Gamma_{k|il} + \Gamma_{i|kl} = g_{ik,l},$$

where

$$\Gamma_{k|il} = g_{pk}\Gamma^p_{il}.$$

Rotate the indices cyclically to obtain two more relations:

$$\Gamma_{l|ki} + \Gamma_{k|li} = g_{kl,i}, \qquad \Gamma_{i|lk} + \Gamma_{l|ik} = g_{li,k}.$$

Next use the symmetry condition (2.39) to eliminate $\Gamma_{l|ki} = \Gamma_{l|ik}$ and $\Gamma_{k|il} = \Gamma_{k|li}$ from the above three relations to get

$$2\Gamma_{i|kl} = g_{ik,l} + g_{li,k} - g_{kl,i}.$$

On raising the index i, we get the required solution:

$$\Gamma^i_{kl} = \frac{1}{2}g^{im}\left(\frac{\partial g_{mk}}{\partial x^l} + \frac{\partial g_{lm}}{\partial x^k} - \frac{\partial g_{kl}}{\partial x^m}\right). \tag{2.40}$$

We next consider some particular identities relating to these symbols, that are useful in various manipulations. If we differentiate the determinant of the metric tensor we get

$$dg = gg^{ik}\,dg_{ik}. \tag{2.41}$$

This relation is useful in expressing some combinations of Γ^i_{kl} and covariant derivatives in relatively simple forms. Thus, using (2.40) and (2.41), it is possible to prove the following relations:

$$\Gamma_{il}^l = \frac{1}{\sqrt{-g}} \frac{\partial}{\partial x^i} (\sqrt{-g})$$

$$\Gamma_{ik}^l g^{ik} = -\frac{1}{\sqrt{-g}} \frac{\partial}{\partial x^m} (\sqrt{-g} \, g^{ml})$$

$$A_{;i}^i = \frac{1}{\sqrt{-g}} \frac{\partial}{\partial x^i} (\sqrt{-g} \, A^i)$$

$$F_{;k}^{ik} = \frac{1}{\sqrt{-g}} \frac{\partial}{\partial x^k} (\sqrt{-g} \, F^{ik}) \quad \text{for} \quad F^{ik} = -F^{ki}. \tag{2.42}$$

(Here A^i and F^{ik} are, respectively, vector and tensor fields.) For example, to prove the first relation note that (2.40) gives, with $k \equiv i$,

$$\Gamma_{il}^i = \tfrac{1}{2} g^{im} (g_{mi,l} + g_{lm,i} - g_{il,m}).$$

Since $g_{lm,i} - g_{il,m}$ is antisymmetric in (i, m), its product with the symmetric g^{im} vanishes. The result then follows when we recall (2.41).

The symmetry condition (2.39) enables us to choose special coordinates in which the Christoffel symbols all vanish at any given point. Suppose that we start with $\Gamma_{np}^m \neq 0$ in the coordinate system (x^i) at point P. Let the coordinates of P be given x_P^i. Now define new coordinates in the neighbourhood of P by

$$x'^k = x^k - x_P^k - \tfrac{1}{2} \Gamma_{nm}^k (x^n - x_P^n)(x^m - x_P^m). \tag{2.43}$$

Then we have at P

$$x_P'^i = 0, \qquad \frac{\partial x'^i}{\partial x^m} = \delta_m^i, \qquad \frac{\partial^2 x'^i}{\partial x^n \partial x^m} = -\Gamma_{nm}^i,$$

with the result that, from (2.36),

$$\Gamma_{mn}'^i |_P = 0.$$

Furthermore, by a linear transformation we can arrange to have a coordinate system with

$$g_{ik} = \eta_{ik} = \mathrm{diag}(+1, -1, -1, -1), \qquad \Gamma_{kl}^i = 0 \tag{2.44}$$

at any chosen point P. Such a coordinate system is called a *locally inertial* coordinate system, for reasons that will become clear later. Apart from its physical implications in general relativity, the locally inertial coordinate system is often useful as a mathematical device for simplifying calculations. We also warn the reader that the operative word is 'local': the simplifications implied in (2.44) *cannot be achieved globally*. What prevents us from achieving a globally inertial coordinate system? In seeking an answer to this question we encounter the most crucial aspect in which a non-Euclidean geometry differs from its Euclidean counterpart.

2.5   Spacetime curvature

Figure 2.9(a) repeats the previous example of non-Euclidean geometry on the surface of a sphere. We have the triangle ABC of Figure 2.2(a) whose three angles are each 90°. Consider what happens to a vector (shown by a dotted arrow) as it is parallelly transported along the three sides of this triangle. As shown in Figure 2.9(a), this vector is originally perpendicular to AB when it starts its journey at A. When it reaches B it lies along CB. So it keeps pointing along this line as it moves from B to C. At C it is again perpendicular to AC. So, as it moves along CA from C to A, it maintains this perpendicularity with the result that, when it arrives at A, it is pointing along AB. In other words, one circuit around this triangle has resulted in a change of direction of the vector by 90°, although at each stage it was being moved parallel to itself!

A similar experiment with a triangle drawn on a flat piece of paper will tell us that there is no resulting change in the direction of the vector when it moves parallel to itself around the triangle. So our spherical triangle behaves differently from the flat Euclidean triangle.

The phenomenon illustrated in Figure 2.9(a) can also be described as follows. If we had moved our vector from A to C along two different routes – along AC and along AB followed by BC – we would have found it pointing in two different directions. In fact if we had taken any arbitrary curves from A to C we would have found that the outcome of parallel transport of a vector from A to C varies from curve to curve; that is, the outcome depends on the path of transport from A to C.

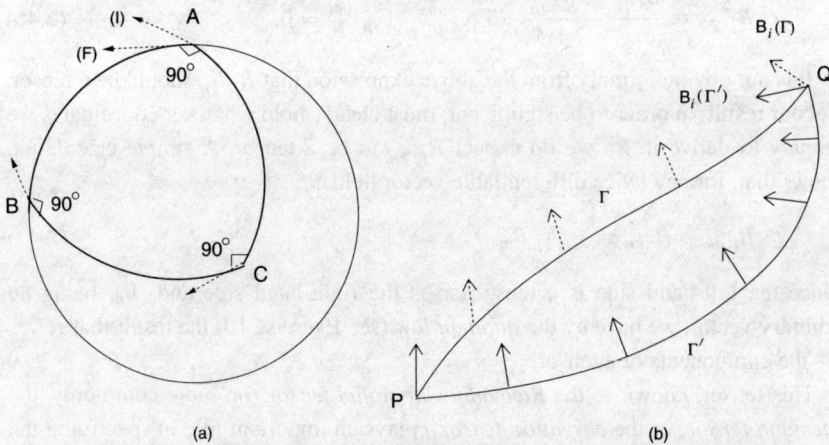

Figure 2.9 (a) Parallel transport of a vector around the triangle ABC alters its direction from I to F. (b) Parallel transport of a vector B_i from P to Q depends on the curve Γ connecting P to Q along which the vector is so transported. For the upper curve Γ, the final outcome is the dotted vector at Q, whereas for the lower curve Γ' the final outcome is shown by the full arrow.

This is one of the properties that distinguishes a curved space from a flat space. Let us consider it in more general terms for our four-dimensional spacetime. Let a vector B_i at P be transported parallelly to Q and let us ask for the condition that the answer should be *independent* of the curve joining P to Q (see Figure 2.9b). We have seen that, under parallel transport from a point $\{x^i\}$ to a neighbouring point $\{x^i + \delta x^i\}$, the components of the vector change according to (2.33). If it were possible to transport B_i from P to Q without the result depending on which path is taken, then we would be able to generate a vector field $B_i(x^k)$ satisfying the differential equation

$$\frac{\partial B_i}{\partial x^k} = \Gamma^l_{ik} B_l. \tag{2.45}$$

So the answer to our question depends on whether we can find a non-trivial solution to the system of four differential equations (2.45).

The necessary condition for the existence of a solution is easily derived. We differentiate (2.45) with respect to x^n to get

$$\frac{\partial^2 B_i}{\partial x^n \, \partial x^k} = \frac{\partial}{\partial x^n}(\Gamma^l_{ik} B_l) = \frac{\partial \Gamma^l_{ik}}{\partial x^n} B_l + \Gamma^l_{ik} \frac{\partial B_l}{\partial x^n}$$

$$= \left(\frac{\partial \Gamma^m_{ik}}{\partial x^n} + \Gamma^l_{ik} \Gamma^m_{ln}\right) B_m.$$

We now interchange the order of differentiation with respect to x^n and x^k and use the result $B_{i,nk} \equiv B_{i,kn}$. We then get the required necessary condition as

$$R_i{}^m{}_{kn} \equiv \frac{\partial \Gamma^m_{ik}}{\partial x^n} - \frac{\partial \Gamma^m_{in}}{\partial x^k} + \Gamma^l_{ik} \Gamma^m_{ln} - \Gamma^l_{in} \Gamma^m_{lk} = 0. \tag{2.46}$$

It is not obvious simply from the above expression that $R_i{}^m{}_{kn}$ should be a tensor. Yet our result, in order to be significant, must clearly hold whatever coordinates we employ to derive it. So we do expect $R_i{}^m{}_{kn}$ to be a tensor. A simple calculation shows that, for any twice differentiable vector field B_i,

$$B_{i;nk} - B_{i,kn} \equiv R_i{}^m{}_{kn} B_m. \tag{2.47}$$

Since the left-hand side is a tensor, so is the right-hand side and, B_m being an arbitrary vector, we have by the *quotient law* (see Exercise 10) the result that $R_i{}^m{}_{kn}$ are the components of a tensor.

This tensor, known as the *Riemann–Christoffel tensor* (or, more commonly, the *Riemann tensor*, or the *curvature tensor*), plays an important role in specifying the geometrical properties of spacetime. Although we have derived (2.46) as a necessary condition, a slightly more sophisticated technique shows that (2.46) is also the sufficient condition that a vector field $B_i(x^k)$ can be defined over the spacetime by parallel transport. We will not, however, go into the detailed mathematical proof here.

Spacetime is said to be *flat* if its Riemann tensor vanishes everywhere. Otherwise, it is said to be *curved*. Exercises 26 and 27 illustrate two other ways in which this tensor distinguishes the properties of a curved spacetime from those of a flat spacetime.

2.5.1 Symmetries of R_{iklm}

It is more convenient to lower the second index of the Riemann tensor to study its symmetry properties. Since the symmetry or antisymmetry of a tensor does not depend on what coordinates are used, it is more convenient to write (2.46) in the locally inertial coordinates (2.44). We then get

$$R_{iklm} = \tfrac{1}{2}(g_{kl,im} + g_{im,kl} - g_{km,il} - g_{il,km}).$$ (2.48)

From this expression the following symmetries are immediately obvious:

$$R_{iklm} = -R_{kilm} = -R_{ikml} = R_{lmik}.$$ (2.49)

We also get relations of the following type:

$$R_{iklm} + R_{imkl} + R_{ilmk} \equiv 0.$$ (2.50)

If we take all these symmetries into account, we find that, of the $4^4 = 256$ components of the Riemann tensor, only 20 at most are independent! Moreover, we will soon see that there are identities linking their derivatives too.

2.5.2 The Ricci and Einstein tensors

By the process of contraction we can construct lower-rank tensors from R_{iklm}. The tensor

$$R_{kl} = g^{im} R_{iklm} \equiv R^m{}_{klm}$$ (2.51)

is called the *Ricci tensor*. If we use the locally inertial coordinate system, we immediately see that

$$R_{kl} = R_{lk}.$$ (2.52)

Owing to the symmetries of (2.49) there are no other independent second-rank tensors that can be constructed out of R_{iklm}.

By further contraction we get a scalar:

$$R = g^{ik} R_{ik} \equiv R^k{}_k.$$ (2.53)

R is called the *scalar curvature*. The tensor

$$G_{ik} \equiv R_{ik} - \tfrac{1}{2}g_{ik}R$$ (2.54)

will turn out to have a special role to play in Einstein's general relativity. This tensor is called the *Einstein tensor*.

2.5.3   Bianchi identities

The expression (2.48) suggests another symmetry for the components of R_{iklm}. This symmetry is not algebraic but involves calculus. In covariant language we may express it as follows:

$$R_{iklm;n} + R_{iknl;m} + R_{ikmn;l} \equiv 0. \tag{2.55}$$

These relations are known as the *Bianchi identities*. Their proof is most easily given in the locally inertial system as in (2.48).

However, by multiplying (2.55) by $g^{im}g^{kn}$ and using (2.51)–(2.53), we can deduce from these identities another that is of importance to relativity:

$$\left(R^{ik} - \frac{1}{2}g^{ik}R \right)_{;k} \equiv 0. \tag{2.56}$$

In other words, *the Einstein tensor G^{ik} has zero divergence.*

2.6   Geodesics

So far we have talked about non-Euclidean geometries without mentioning whether, in general, they have the equivalents of straight lines in Euclidean geometry. We now show how equivalent concepts do exist in the Riemannian geometry under consideration here.

There are two properties of a straight line that can be generalized: the property of 'straightness' and the property of 'shortest distance'. Straightness means that, as we move along the line, its direction does not change. Let us see how we can generalize this concept first.

Let $x^i(\lambda)$ be the parametric representation of a curve in spacetime. Its tangent vector is given by

$$u^i = \frac{\mathrm{d}x^i}{\mathrm{d}\lambda}. \tag{2.57}$$

Our straightness criterion demands that u^i should not change along the curve (see Figure 2.10(a)). On going from λ to $\lambda + \delta\lambda$, the change in u^i is given by

$$\Delta u^i = \frac{\mathrm{d}u^i}{\mathrm{d}\lambda}\,\delta\lambda + \Gamma^i_{kl}u^k\,\delta x^l.$$

The second expression on the right-hand side arises from the change produced by parallel transport through a coordinate displacement δx^l. However, $\delta x^l = u^l\,\delta\lambda$. Therefore the condition of there being no change of direction u^i implies that $\Delta u^i = 0$; that is,

$$\frac{\mathrm{d}u^i}{\mathrm{d}\lambda} + \Gamma^i_{kl}u^k u^l = 0. \tag{2.58}$$

This is the condition that our curve must satisfy in order to be straight.

The second property of a straight line in Euclidean geometry is that it is the curve of shortest distance between two points. Let us generalize this property in the following way. Let the curve, parametrized by λ, connect two points P_1 and P_2 of spacetime, with parameters λ_1 and λ_2, respectively. Then the 'distance' of P_2 from P_1 is defined as

$$s(P_2, P_1) = \int_{\lambda_1}^{\lambda_2} \left(g_{ik} \frac{dx^i}{d\lambda} \frac{dx^k}{d\lambda} \right)^{1/2} d\lambda \equiv \int_{\lambda_1}^{\lambda_2} L \, d\lambda. \tag{2.59}$$

We now demand that $s(P_2, P_1)$ be 'stationary' for small displacements of the curve connecting P_1 and P_2, these displacements vanishing at P_1 and P_2 (see Figure 2.10(b)).

This is a standard problem in the calculus of variations and its solution leads to the familiar Euler–Lagrange equations

$$\frac{d}{d\lambda} \left(\frac{\partial L}{\partial x^i} \right) - \frac{\partial L}{\partial \dot{x}^i} = 0, \tag{2.60}$$

where $\dot{x}^i \equiv dx^i/d\lambda$ and $L \equiv [g_{ik}(dx^i/d\lambda)(dx^k/d\lambda)]^{1/2}$ is a function of x^i and $\dot{x}^i$. It is easy to see that (2.60) leads to

$$\frac{d}{d\lambda} \left(g_{ik} \frac{1}{L} \frac{dx^k}{d\lambda} \right) - \frac{1}{2} g_{mn,i} \frac{1}{L} \frac{dx^m}{d\lambda} \frac{dx^n}{d\lambda} = 0.$$

If we make the substitution

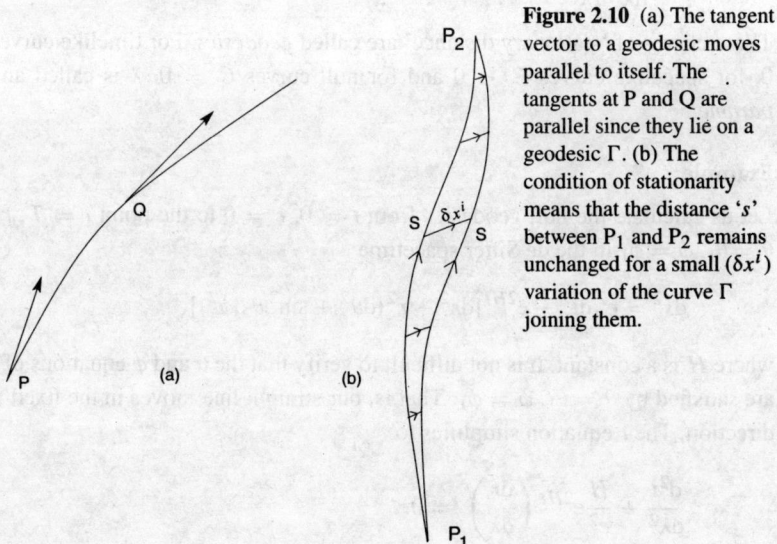

(a)

(b)

Figure 2.10 (a) The tangent vector to a geodesic moves parallel to itself. The tangents at P and Q are parallel since they lie on a geodesic Γ. (b) The condition of stationarity means that the distance 's' between P_1 and P_2 remains unchanged for a small (δx^i) variation of the curve Γ joining them.

$$ds = L \, d\lambda \tag{2.61}$$

and use (2.40), we get the above equation in the form

$$\frac{d^2 x^i}{ds^2} + \Gamma^i_{kl} \frac{dx^k}{ds} \frac{dx^l}{ds} = 0. \tag{2.62}$$

There are a few loose ends to be sorted out in the above derivation. First, L would be real only for timelike curves. Thus, if we want to use a real parameter along the curve, then, for spacelike curves, we must replace ds by

$$d\sigma = i \, ds, \qquad i = \sqrt{-1}. \tag{2.63}$$

For null curves, $L = 0$. The above treatment therefore breaks down. It is then more convenient to replace the integral (2.59) by another:

$$I = \int_{\lambda_1}^{\lambda_2} L^2 \, d\lambda \tag{2.64}$$

and consider $\delta I = 0$. We can always choose a new parameter $\lambda' = \lambda'(\lambda)$ such that the equation of the curve takes the same form as (2.62), with λ' replacing s.

It is easy to see that (2.62) is the same as (2.58). Although s in (2.62) has the special meaning 'length along the curve' while λ in (2.58) appears to be general, it is not difficult to see that, if (2.58) is satisfied, λ must be a constant multiple of s. This is because (2.58) has the first integral

$$g_{ik} \frac{dx^i}{d\lambda} \frac{dx^k}{d\lambda} = C, \qquad C = \text{constant.} \tag{2.65}$$

These curves of 'stationary distance' are called *geodesics*. For timelike curves $C > 0$, for spacelike curves $C < 0$ and for null curves $C = 0$. λ is called an *affine parameter*.

Example

Let us calculate the null geodesics from $t = 0, r = 0$ to the point $t = T, r = R$, $\theta = \theta_1, \phi = \phi_1$ in the de Sitter spacetime

$$ds^2 = c^2 \, dt^2 - e^{2Ht} [dr^2 + r^2 (d\theta^2 + \sin^2 \theta \, d\phi^2)]$$

where H is a constant. It is not difficult to verify that the θ and ϕ equations of (2.62) are satisfied by $\theta = \theta_1, \phi = \phi_1$. That is, our straight line moves in the fixed (θ, ϕ) direction. The t equation simplifies to

$$\frac{d^2 t}{d\lambda^2} + \frac{H}{c^2} e^{2Ht} \left(\frac{dr}{d\lambda} \right)^2 = 0.$$

The first integral (2.65) gives, on the other hand,

$$c^2 \left(\frac{dt}{d\lambda} \right)^2 = e^{2Ht} \left(\frac{dr}{d\lambda} \right)^2.$$

The two equations can be easily solved to give

$$t = \frac{1}{H} \ln \left(1 + \frac{\lambda}{\lambda_0} \right), \qquad r = \frac{c}{H} \frac{\lambda}{\lambda + \lambda_0},$$

where λ_0 is determined from the boundary condition that, when $r = R, t = T$. Note that a solution is possible only if R and T are related by the condition

$$R = \frac{c}{H} (1 - e^{-HT}).$$

We next consider the special role played by geodesics in general relativity.

2.7 The principle of equivalence

Having described the machinery of vectors and tensors and having outlined the salient features of Riemannian geometry, we now make our first contact with physics and introduce the so-called *principle of equivalence*, which has played the key role in general relativity.

Let us go back to the purely mathematical result embodied in the relations shown in (2.44) and attempt to describe their physical meaning. These relations tell us that special (locally inertial) coordinates exist in the neighbourhood of any point P in spacetime that behave like the coordinates (t, x, y, z) of special relativity. Physically, these coordinates imply a special frame of reference in which a momentary illusion that the geometry is of special relativity is created at P and in a small neighbourhood of P. The illusion is momentary and local to P because we have seen that the relations of (2.44) cannot be made to hold everywhere and at all times.

In view of the assertion made in §2.1 that gravitation manifests itself as non-Euclidean geometry, we would have to argue that, in the above-mentioned locally inertial frame, gravitation has been transformed away momentarily and in a small neighbourhood of P. How does this happen in practice? Consider Einstein's celebrated example of the freely falling lift. A person inside such a lift feels weightless. The accelerated frame of reference of the lift provides the locally inertial frame in the small neighbourhood of the falling person. Similarly, a spacecraft circling around the Earth is in fact freely falling in the Earth's gravity and the astronauts inside it feel weightless. (See Figure 2.11 describing floating astronauts in a space shuttle.)

It should be emphasized that this feeling of weightlessness in a falling lift or a spacecraft is limited to local regions: there is no universal frame that transforms away Earth's gravity everywhere, at all times. If we demand that the relations of (2.44) hold at all points of spacetime, we would need to have $\partial \Gamma^i_{kl} / \partial x^m = 0$ every-where, leading to $R^i{}_{klm} = 0$, that is, to a flat spacetime. Thus a curved spacetime with a non-vanishing Riemann tensor is necessary for describing genuine effects of gravitation.

The *weak principle of equivalence* states that effects of gravitation can be transformed away locally and over small intervals of time by using suitably accelerated frames of reference. Thus it is the physical statement of the mathematical relations given by (2.44). It is possible, however, to go from here to a much stronger statement, the so-called *strong principle of equivalence*, which states that any physical interaction (other than gravitation, which has now been identified with geometry) behaves in a locally inertial frame as if gravitation were absent. For example, Maxwell's equations will have their familiar forms (of special relativity) in a locally inertial frame. Thus an observer performing a local experiment in a freely falling lift would measure the speed of light to be c.

The strong principle of equivalence allows us to extend any physical law that is expressed in the covariant language of special relativity to the more general form it would have in the presence of gravitation. The law is usually expressed in vectors, tensors, or spinors in the Minkowski spacetime of special relativity. All we have to do is write it in terms of the corresponding entities in curved spacetime. Thus, in the flat spacetime of special relativity, the Maxwell electromagnetic field F^{ik} is related to the current vector j^k by

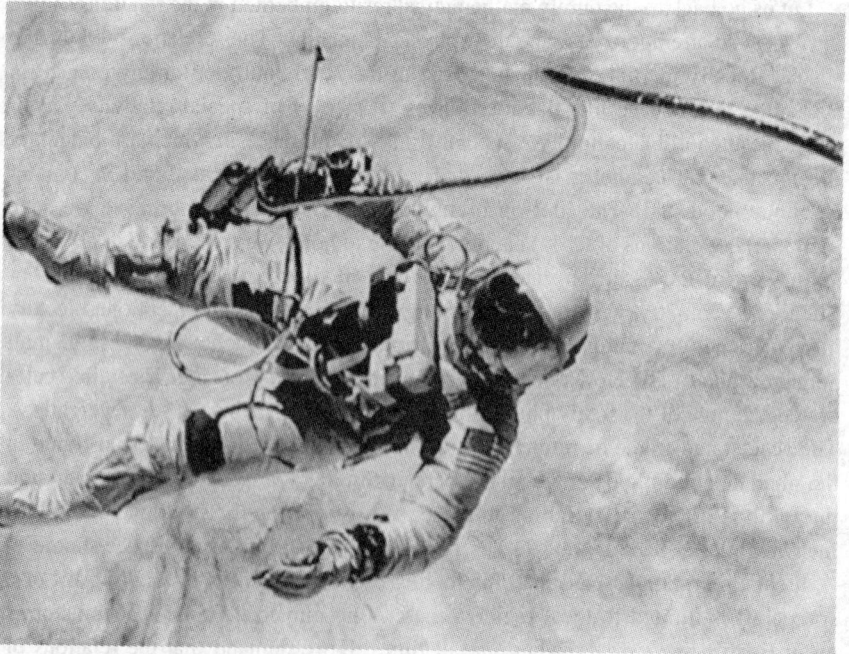

Figure 2.11 An astronaut floating in space. For him the gravity has been transformed away locally. Courtesy of NASA/Ames Research Center.

$$F^{ik}_{,i} = 4\pi j^k. \tag{2.66}$$

In curved spacetime the ordinary tensor derivative is replaced by the covariant derivative:

$$F^{ik}_{;i} = 4\pi j^k. \tag{2.67}$$

Notice that the effect of gravitation enters through the Γ^i_{kl} terms that are present in (2.67). This generalization of (2.66) to (2.67) is called the *minimal coupling* of the field with gravitation, since it is the simplest one possible.

So, in order to describe how other interactions behave in the presence of gravitation, we use the covariance under the general coordinate transformation as the criterion to be satisfied by their equations. It is immediately clear from the example of the electromagnetic field that a ray of light describes a null geodesic.

In the same vein we can now describe a moving object that is acted on by no other interaction except gravitation – for example, a probe moving in the gravitational field of the Earth. *In the absence of gravity*, this object would move in a straight line with uniform velocity; that is, with the equation of motion

$$\frac{du^i}{ds} = 0, \qquad u^i = \text{4-velocity}. \tag{2.68}$$

In the presence of gravity, (2.68) is modified to our geodesic equation (2.62).

We end this section with another example that provides a clue about how gravitational effects show up in spacetime geometry according to general relativity. Consider the Minkowski spacetime with the standard line element

$$ds^2 = c^2 \, dt^2 - dx^2 - dy^2 - dz^2. \tag{2.69}$$

If we make the coordinate transformation for a constant g,

$$x = \frac{c^2}{g}\left[\cosh\left(\frac{gt'}{c}\right) - 1\right] + x' \cosh\left(\frac{gt'}{c}\right),$$

$$y = y',$$

$$z = z',$$

$$t = \frac{c}{g}\sinh\left(\frac{gt'}{c}\right) + \frac{x'}{c}\sinh\left(\frac{gt'}{c}\right). \tag{2.70}$$

This leads to the line element

$$ds^2 = \left(1 + \frac{gx'}{c^2}\right)^2 dt'^2 - dx'^2 - dy'^2 - dz'^2. \tag{2.71}$$

What interpretation can we give to (2.71)? The origin of the (x', y', z') system has a world line whose parametric form in the old coordinates is given by

$$x = \frac{c^2}{g}\left[\cosh\left(\frac{gt'}{c}\right) - 1\right], \qquad y = 0, z = 0, \qquad t = \frac{c}{g}\sinh\left(\frac{gt'}{c}\right).$$

$$(2.72)$$

Using the kinematics of special relativity, it can easily be seen that (2.72) describes the motion of a point that has a uniform acceleration g in the x direction, a point that is momentarily at rest at the origin of (x, y, z) at $t = 0$. We may interpret the line element (2.71) and the new coordinate system as describing the spacetime in the rest frame of the uniformly accelerated observer.

Direct calculation shows that not all Γ^i_{kl} are zero in (2.71) at $x' = 0$, $y' = 0$, $z' = 0$. The frame is therefore non-inertial. For the neighbourhood of the origin, the metric component

$$g_{00} \cong 1 + \frac{2gx'}{c^2} = 1 + \frac{2\phi}{c^2},$$

$$(2.73)$$

where ϕ is the Newtonian gravitational potential for a uniform gravitational field that induces an acceleration due to gravity of $-g$. We have here the reverse situation to that of the falling lift: we seem to have generated a pseudo-gravitational field by choosing a suitably accelerated observer. The prefix 'pseudo-' is used because the gravitational field is not real – it is an illusory effect arising from the choice of co-ordinates. The Riemann tensor for the metric is zero. Nevertheless the relation (2.73) is also suggestive of the real gravitational field, as we shall see in the following example and later in §2.9.

Example

Consider a particle held at rest at the origin $x = 0$, $y = 0$, $z = 0$ in the manifestly Minkowski frame (2.69). What is its trajectory in the uniformly accelerated frame (2.71)?

Setting $x = 0$ in equation (2.70), we get

$$x' = \frac{c^2}{g}\left[\text{sech}\left(\frac{gt'}{c}\right) - 1\right],$$

which, for small t', i.e., for $t' \ll c/g$, approximates to

$$x' = -\tfrac{1}{2}gt'^2.$$

Thus, to an observer at rest in the accelerated frame, the particle will appear to have a 'free fall' in the negative x' direction; the observer will ascribe this to gravity in that direction.

2.8   The action principle and the energy tensors

Before examining relativity proper, let us see how we can write the laws of physics
in the covariant language in a Riemannian spacetime using the strong principle of
equivalence. We take the familiar example of charged particles interacting with an
electromagnetic field. The physical laws can be derived from an action principle.
First we write the action in Minkowski spacetime:

$$ \mathcal{A} = -\sum_a cm_a \int ds_a - \frac{1}{16\pi c} \int F_{ik}F^{ik}\, d^4x - \sum_a \frac{e_a}{c}\int A_i\, da^i. \quad (2.74) $$

Here A_i are the components of the 4-potential, which are related to the field tensor
F_{ik} by

$$ A_{k,i} - A_{i,k} = F_{ik}, \quad (2.75) $$

while e_a and m_a are the charge and rest mass of particle a, whose coordinates are
given by a^i and the proper time by s_a with

$$ ds_a^2 = \eta_{ik}\, da^i\, da^k. \quad (2.76) $$

How do we generalize (2.74) to Riemannian spacetime? First, we note that η_{ik} in
(2.76) are replaced by g_{ik}. Next, starting from the covariant vector A_i, we generate
F_{ik} by the covariant generalization of (2.75):

$$ A_{k;i} - A_{i;k} = F_{ik}. \quad (2.77) $$

However, since the expression (2.77) is antisymmetric in (i, k), the extra terms
involving the Christoffel symbols vanish and we are back to (2.75)! The volume
integral in (2.74) is modified to

$$ \int F_{ik}F^{ik}\sqrt{-g}\, d^4x. \quad (2.78) $$

The extra factor $\sqrt{-g}$ has crept in because the combination

$$ \sqrt{-g}\, dx^1\, dx^2\, dx^3\, dx^0 = \frac{1}{24}e_{ijkl}\, dx^i\, dx^j\, dx^k\, dx^l \quad (2.79) $$

acts as a scalar. We therefore have the generalized form of (2.74):

$$ \mathcal{A} = -\sum_a cm_a \int ds_a - \frac{1}{16\pi c} \int F_{ik}F^{ik}\sqrt{-g}\, d^4x - \sum_a \frac{e_a}{c}\int A_i\, da^i. \quad (2.80) $$

The variation of the world line of particle a gives its equation of motion

$$ \frac{d^2a^i}{ds_a^2} + \Gamma^i_{kl}\frac{da^k}{ds_a}\frac{da^l}{ds_a} = \frac{e_a}{m_a}F^i{}_l\frac{da^l}{ds_a}, \quad (2.81) $$

while the variation of A_i gives the field equations (2.67).

The transition from (2.74) to (2.80) has, however, introduced an additional independent feature into the action, besides the particle world lines and the potential
vector. The new feature is the spacetime geometry typified by the metric tensor g_{ik}.
What will happen if we demand that the g_{ik} be also dynamical variables and that the
action $\mathcal{A}$ remain stationary for small variations of the type

$$g_{ik} \to g_{ik} + \delta g_{ik}? \tag{2.82}$$

From the generalized action principle, should we not expect to get the equations that
determine the g_{ik} and through them the spacetime geometry? Let us investigate.

A glance at the action (2.80) shows that the last term does not contribute anything
under (2.82) if we keep the world lines and A_i fixed in spacetime. The first two
terms, however, do make contributions. Let us consider them in that order. First note
that

$$\delta(ds_a^2) = \delta g_{ik}\, da^i\, da^k,$$

that is,

$$\delta(ds_a) = \frac{1}{2}\delta g_{ik} \frac{da^i}{ds_a} \frac{da^k}{ds_a}\, ds_a.$$

Therefore,

$$\delta \sum_a cm_a \int ds_a = \frac{1}{2}\sum_a c \int m_a \frac{da^i}{ds_a} \frac{da^k}{ds_a}\, ds_a\, \delta g_{ik}. \tag{2.83}$$

Let us consider this variation in a small 4-volume $\mathcal{V}$ near a point P. If we consider
a locally inertial coordinate system near P, we can identify the above expression in
a more familiar form. Let us first identify

$$p_{(a)}^i = cm_a \frac{da^i}{ds_a}$$

as the 4-momentum of particle a. Then $cp_{(a)}^0 = E_a$ is the energy of the particle, and
we get

$$\frac{1}{2}cm_a \frac{da^i}{ds_a} \frac{da^k}{ds_a}\, ds_a = \frac{c^2}{2E_a} p_{(a)}^i p_{(a)}^k\, dt_a = \frac{c}{2E_a} p_{(a)}^i p_{(a)}^k\, dx_a^0.$$

Figure 2.12 shows the volume $\mathcal{V}$ as a shaded region in the neighbourhood of
P, t being the local time coordinate and x^μ ($\mu = 1, 2, 3$) the local rectangular
coordinates. We will briefly discuss the various cases described in Figure 2.12. The
expression (2.83) can then be looked upon as a volume integral over $\mathcal{V}$ of the form

$$\delta \sum_a cm_a \int ds_a = \frac{1}{2c} \int_{\mathcal{V}} \delta g_{ik} T_{(m)}^{ik}\, d^4x, \tag{2.84}$$

where $T^{ik}_{(m)}$ is the sum of the expressions

$$\frac{c^2}{E_a} p^i_{(a)} p^k_{(a)}$$

for each particle a that crosses a unit volume of the shaded region near P. We now interpret this sum.

2.8.1 The energy tensor of matter

This expression for T_{ik} is none other than the usual expression for the energy tensor of matter (also called the energy momentum tensor or the stress energy tensor). Since

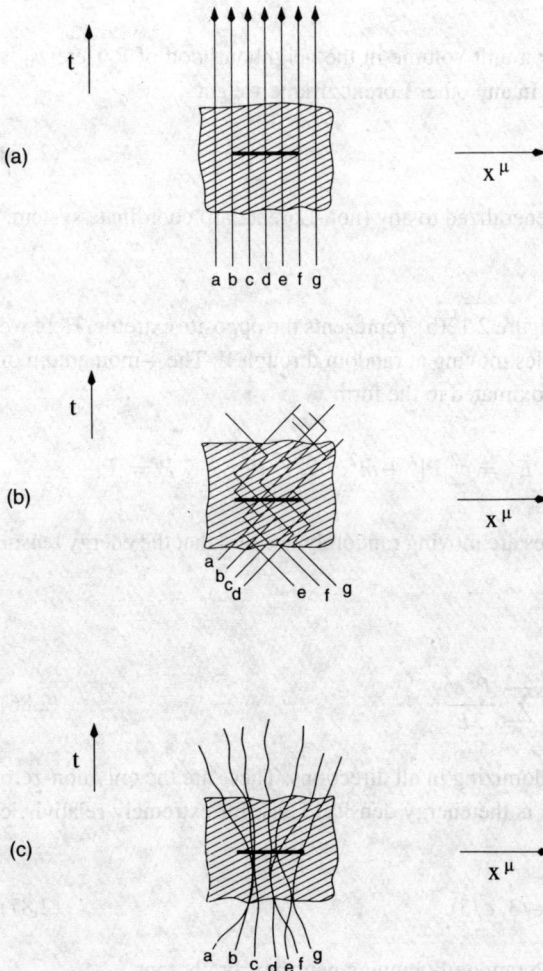

(a)

(b)

(c)

Figure 2.12 Three cases of particle motion in the locally inertial region $\mathcal{V}$ near a typical point P of spacetime. The thick line on the x^μ-axes in each case represents a unit 3-volume. All particles a, b, c, d, ... crossing this volume are counted for computing T^{ik}. (a) Particle world lines a, b, c, ... are nearly parallel. This is the dust approximation. (b) The particles move at random with speeds near the speed of light, frequently changing direction through collisions. This is the relativistic case. (c) The intermediate situation, in which the particles collide, change directions and generate pressures, but their motions are non-relativistic. This is the case of a fluid.

we will need this tensor frequently, it is derived below for three different types of matter.

Dust

This is the simplest situation, in which all the world lines going through the shaded region in Figure 2.12(a) are more or less parallel, indicating that the particles of matter are moving without any relative motion in the neighbourhood of P. Writing the typical 4-velocity as u^i and using a Lorentz transformation to make $u^i = (1, 0, 0, 0)$ (that is, transforming to the rest frame of the dust), the only non-zero component of the energy tensor is

$$T^{00} = \sum_a m_a c^2 = \rho_0 c^2,$$

where the summation is over a unit volume in the neighbourhood of P. Here ρ_0 is the rest mass density of dust. In any other Lorentz frame we get

$$T^{ik}_{(m)} = \rho_0 c^2 u^i u^k. \tag{2.85}$$

an expression that is easily generalized to any (non-Lorentzian) coordinate system.

Relativistic particles

This situation, described in Figure 2.12(b), represents the opposite extreme. Here we have highly relativistic particles moving at random through $\mathcal{V}$. The 4-momentum of a typical particle is then approximated to the form

$$p^i = \left(\frac{E}{c}, P \right), \qquad E^2 = c^2 |\mathbf{P}|^2 + m^2 c^4 \cong c^2 P^2, \qquad P = |\mathbf{P}|.$$

Using the fact that the particles are moving randomly, we find that the energy tensor has pressure components also:

$$T^{00} = \sum E = \epsilon,$$

$$T^{11} = T^{22} = T^{33} = \sum \frac{P^2 c^2}{3E}. \tag{2.86}$$

The factor $\frac{1}{3}$ comes from randomizing in all directions. These are the only non-zero pressure components. Here ϵ is the energy density. Thus, for extremely relativistic particles we get

$$T^{ik}_{(m)} = \text{diag}(\epsilon, \epsilon/3, \epsilon/3, \epsilon/3). \tag{2.87}$$

This form is also applicable to randomly moving neutrinos or photons.

Fluid

This situation is illustrated in Figure 2.12(c) and consists of a collection of particles with small (non-relativistic) random motions. If we choose the frame in which the fluid as a whole is at rest as the frame of reference, we can evaluate the components of $T_{(m)}^{ik}$ as follows. Let a typical particle have the momentum vector given by

$$p^0 = \frac{mc}{\sqrt{1 - v^2/c^2}}, \qquad p^\mu = \frac{m\mathbf{v}}{\sqrt{1 - v^2/c^2}} \qquad (\mu = 1, 2, 3). \qquad (2.88)$$

Then

$$T^{00} = \sum mc^2 \left(1 - \frac{v^2}{c^2}\right)^{-1/2} \cong \sum mc^2 \left(1 + \frac{v^2}{2c^2}\right) = \rho c^2,$$

$$T^{11} = T^{22} = T^{33} = \frac{1}{3} \sum mv^2 \left(1 - \frac{v^2}{c^2}\right)^{-1/2} \cong p. \qquad (2.89)$$

Here ρ and p are the density and pressure of the fluid. In a frame of reference in which the fluid as a whole has a 4-velocity u^i, the energy tensor becomes

$$T_{(m)}^{ik} = (p + \rho c^2)u^i u^k - p\eta^{ik}. \qquad (2.90)$$

The generally covariant form of (2.90) is obviously

$$T_{(m)}^{ik} = (p + \rho c^2)u^i u^k - pg^{ik}. \qquad (2.91)$$

Note that ρ is not just the rest-mass density, but also includes energy density of internal motion, as seen in (2.89).

We may now relax our restriction to the locally inertial coordinate system at P. The generalized form of (2.84) is then

$$\delta \sum_a cm_a \int ds_a = \frac{1}{2c} \int T_{(m)}^{ik} \sqrt{-g}\, \delta g_{ik}\, d^4 x. \qquad (2.92)$$

2.8.2 The energy tensor of the electromagnetic field

We next consider the variation of the second term of (2.74). If we keep A_i fixed, the F_{ik}, given by (2.77) or (2.75), remain unchanged under the variation of g_{ik}. Hence

$$\delta(F_{ik}F^{ik}\sqrt{-g}) = F_{ik}F_{lm}\,\delta(g^{il}g^{km}\sqrt{-g}).$$

From (2.25) we get

$$\delta g^{ik}g_{kl} = -g^{ik}\,\delta g_{kl},$$

that is,

$$\delta g^{ik} = -g^{im} g^{kn} \, \delta g_{mn}.$$ (2.93)

Also, from (2.41) we have

$$\delta \sqrt{-g} = \tfrac{1}{2} g^{ik} \sqrt{-g} \, \delta g_{ik}.$$ (2.94)

Substituting these expressions into the variation of the second term of the action gives

$$\delta \frac{1}{16\pi c} \int_{\mathcal{V}} F_{ik} F^{ik} \sqrt{-g} \, d^4 x = \frac{1}{2c} \int_{\mathcal{V}} T^{ik}_{(em)} \sqrt{-g} \, \delta g_{ik} \, d^4 x$$ (2.95)

with the electromagnetic energy tensor given by

$$T^{ik}_{(m)} = \frac{1}{4\pi} \left(\frac{1}{4} F_{mn} F^{mn} g^{ik} - F^i_{\ l} F^{lk} \right).$$ (2.96)

It is obvious from our two examples that the energy tensor of any term in the action of the form Λ is related to the variation of Λ by

$$\delta \Lambda = \frac{1}{2c} \int T^{ik}_{(\Lambda)} \sqrt{-g} \, \delta g_{ik} \, d^4 x.$$ (2.97)

In theories defined only in Minkowski spacetime the appearance of energy tensors is somewhat *ad hoc*. They do not enter explicitly into any dynamical or field equations. They appear only through their divergences, the typical conservation of energy and momentum being given by

$$T^{ik}_{\ ,k} = 0.$$ (2.98)

In our curved spacetime framework the T^{ik} find a natural expression through the variation of g_{ik}. It was this variation of the metric tensor that led Hilbert to derive the field equations of general relativity shortly after Einstein had proposed them from heuristic considerations. We now turn our attention to this topic.

2.9 Gravitational equations

The preceding section showed that the variation of the action $\mathcal{A}$ with respect to g_{ik} leads us to the energy tensors of various interactions. We still do not have dynamical equations that tell us how to determine the g_{ik} in terms of the distribution of matter and energy. Einstein's conjecture was that the energy tensors should act as the 'sources' of gravity. Following the general trend of nineteenth-century physics, especially the Maxwell equations, Einstein looked for an expression that would act like a wave equation for g_{ik}, with T_{ik} as the source. It is immediately clear that the standard wave equation in the covariant form

$$g^{mn} g_{ik;mn} = \kappa T_{ik},$$ (2.99)

where κ is a constant, will not do, for the left-hand size vanishes identically. Is there a second-rank tensor symmetric in its indices (like the T_{ik}) that involves second derivatives of g_{ik}? Clearly, if the tensor is to bring out the special feature of curvature of spacetime, it must be related to the Riemann tensor. Einstein first tried R_{ik}, before finally arriving at the tensor G_{ik} of (2.54). His field equations of general relativity, published in 1915, took the form

$$R_{ik} - \tfrac{1}{2}g_{ik}R \equiv G_{ik} = -\kappa T_{ik}. \tag{2.100}$$

These equations have the added advantage that, in view of the Bianchi identities in (2.56), we must have

$$T^{ik}_{\;;k} \equiv 0. \tag{2.101}$$

That is, the law of conservation of energy and momentum follows naturally from (2.100).

Although there are ten Einstein equations for ten unknown g_{ik}, the divergence condition of (2.101) reduces the number of independent equations to six. This underdeterminacy of the problem is due to the general covariance of the theory: if g_{ik} is a solution, then so is any tensor transform of g_{ik} obtained through a change of coordinates.

The expression (2.101) follows for any T^{ik} obtained from an action principle by the variation of g_{ik} (see Exercise 33). This result is in fact an example of Noether's theorem, which relates a conservation law to a basic symmetry. In this particular case the symmetry is that of coordinate invariance. It is therefore pertinent to ask whether the Einstein tensor can also be derived from an action principle. This problem was solved by Hilbert soon after Einstein proposed his equations of gravitation. Hilbert's problem can be posed as follows. Consider the variation of the term

$$\int_{\mathcal{V}} R\sqrt{-g}\, \mathrm{d}^4x$$

for $g^{ik} \to g^{ik} + \delta g^{ik}$ with the restriction that δg^{ik} and $\delta g^{ik}_{\;,l}$ vanish on the boundary of $\mathcal{V}$. It can be shown (see Exercises 34 and 35) that

$$\delta \int_{\mathcal{V}} R\sqrt{-g}\, \mathrm{d}^4x = \int_{\mathcal{V}} \delta g^{ik}\left(R_{ik} - \frac{1}{2}g_{ik}R\right)\sqrt{-g}\, \mathrm{d}^4x$$

$$= -\int_{\mathcal{V}} \delta g_{ik}\left(R^{ik} - \frac{1}{2}g^{ik}R\right)\sqrt{-g}\, \mathrm{d}^4x. \tag{2.102}$$

Thus it follows that Einstein's equations can be derived from an action principle if we add to $\mathcal{A}$ the term

$$\mathcal{A}_{\text{gravitation}} = \frac{1}{2\kappa c}\int_{\mathcal{V}} R\sqrt{-g}\, \mathrm{d}^4x. \tag{2.103}$$

Furthermore, if to the scalar R we add a constant (2λ, say) that is trivially a scalar, we get a modified set of field equations:

$$R_{ik} - \tfrac{1}{2} g_{ik} R + \lambda g_{ik} = -\kappa T_{ik}. \tag{2.104}$$

We may consider this equation as representing the variation of action in a spacetime region of prescribed volume, with λ playing the role of a Lagrangian undetermined multiplier. We will consider these equations only when we discuss cosmology, since the extra term (the λ-term) has cosmological significance. For the time being we return to (2.100) and relate κ to known physical constants.

2.9.1 The Newtonian approximation

We now come to the important question of the magnitude of κ and the relationship between general relativity and Newtonian gravitation. The first hint of the connection between Newtonian gravitation and the present theory was provided by (2.73), from which we saw that, provided that g_{00} did not differ significantly from unity, the difference $g_{00} - 1$ is proportional to the Newtonian gravitational potential. We now seek to formalize this relationship and thereby determine κ. We will show that, in the so-called slow-motion-plus-weak-field approximation, general relativity reduces to Newtonian gravitation.

This approximation is specified by the following assumptions:

1. The motions of particles are non-relativistic: $v \ll c$. In this case we are back to Newtonian mechanics.

2. The gravitational fields are weak in the sense that

$$g_{ik} = \eta_{ik} + h_{ik}, \qquad |h_{ik}| \ll 1. \tag{2.105}$$

   The inequality suggests that we ignore powers of $|h_{ik}|$ higher than 2 in the action principle and higher than 1 in the field equations.

3. The fields change slowly with time. This means that we ignore time derivatives in comparison with space derivatives.

Let us now see how the action is simplified under these approximations. First note that, with $x^0 = ct$,

$$ds^2 = (\eta_{ik} + h_{ik})\, dx^i\, dx^k \approx (1 + h_{00}) c^2\, dt^2 - v^2\, dt^2, \tag{2.106}$$

that is,

$$ds \approx \left(\sqrt{1 + h_{00} - \frac{v^2}{c^2}} \right) c\, dt \approx \left(1 + \frac{1}{2} h_{00} - \frac{v^2}{2c^2} \right) c\, dt. \tag{2.107}$$

We next look at the term involving the scalar curvature. The linearized expression for the Riemann tensor (see (2.48)) is

$$R_{iklm} \approx \tfrac{1}{2}(h_{kl,im} + h_{im,kl} - h_{km,il} - h_{il,km}). \tag{2.108}$$

The corresponding values of R can also be calculated. However, care is needed if we are to look at the action principle rather than the field equations in this approximation, for we anticipate quadratic expressions in the h_{ik} to appear in the geometrical term (2.103).

Item 3 above eliminates time derivatives altogether. Furthermore, the ratios of typical space and time displacements are $\delta x^\mu / \delta x^0 = v^\mu / c$, where v^μ are typical Newtonian velocities. Thus h_{00} is more important than any other h_{ik}, at least by the factor c/v. We will henceforth ignore all other h_{ik} in comparison with h_{00}. We then get

$$g^{00} \approx 1 - h_{00} \tag{2.109}$$

$$\sqrt{-g} \approx 1 + \tfrac{1}{2} h_{00}, \tag{2.110}$$

and

$$R\sqrt{-g} \approx -\left(1 - \frac{1}{2} h_{00}\right) \nabla^2 h_{00}. \tag{2.111}$$

Using these relations, we finally get the approximate action as

$$\mathcal{A} \approx -\frac{1}{2\kappa} \iint \left(1 - \frac{1}{2} h_{00}\right) \nabla^2 h_{00} \, d^3\mathbf{x} \, dt - \sum \frac{1}{2} mc^2 \int h_{00} \, dt$$
$$+ \sum \frac{1}{2} m \int v^2 \, dt + \text{constant}. \tag{2.112}$$

The constant represents path-independent terms that can be ignored in a variational problem. Here we have dropped particle labels $a, b, \ldots$ and used the 3-vector $\mathbf{x}$ to denote x^μ ($\mu = 1, 2, 3$). We can use Green's theorem and ignore surface terms. Thus in the three-dimensional spatial volume, we get

$$\int\limits_{\text{3-volume}} \left(1 - \frac{1}{2} h_{00}\right) \nabla^2 h_{00} \cdot d^3\mathbf{x} = \int\limits_{\text{2-surface}} \left(1 - \frac{1}{2} h_{00}\right) \nabla h_{00} \, d\mathbf{S}$$
$$+ \frac{1}{2} \int\limits_{\text{3-volume}} (\nabla h_{00})^2 \, d^3\mathbf{x}.$$

We can ignore the surface term. Hence

$$\mathcal{A} \approx -\frac{1}{4\kappa} \iint (\nabla h_{00})^2 \, d^3\mathbf{x} \, dt - \sum \frac{1}{2} mc^2 \int h_{00} \, dt + \sum \frac{1}{2} m \int v^2 \, dt. \tag{2.113}$$

Now compare this with the Newtonian action

$$\mathcal{A}_N \approx -\frac{1}{8\pi G} \iint (\nabla \phi)^2 \, d^3\mathbf{x} \, dt - \sum m \int \phi \, dt + \sum \frac{1}{2} m \int v^2 \, dt, \tag{2.114}$$

with ϕ as the gravitational potential. Clearly, (2.113) becomes the same as (2.114) if we put

$$\phi = \tfrac{1}{2}c^2 h_{00}, \qquad \kappa = 8\pi G/c^4. \tag{2.115}$$

Thus we have completed our project of evaluating κ and relating the relativistic framework to Newtonian gravitation. Assumptions 1–3 above are known collectively as the *Newtonian approximation*. It leads to the linear gravitation theory of Newton, which has wide applications ranging from the tidal phenomenon of the Earth's oceans to motions of planets of the Solar System and motions of stars and galaxies in clusters. Provided that these three assumptions hold, general relativity does not add anything new. If assumptions 1 and 3 are dropped but assumption 2 is retained, we are in the domain of the weak-field theory of *gravitational radiation*. For, in the weak-field limit, it is seen that effects of spacetime curvature propagate as waves with the speed of light. If this text were devoted primarily to general relativity, we would have discussed this intriguing phenomenon in detail. A few properties of gravitational radiation are outlined in Exercises 37–40. To get the full effects of general relativity, however, we must drop all three assumptions and face the non-linear equations of (2.100) in their most general form. Naturally this is a complicated task; after more than six decades of this theory there are only a handful of exact solutions of direct physical relevance. We will end this chapter with a discussion of the earliest, simplest and most important of these solutions.

2.10　The Schwarzschild solution

Shortly after Einstein published his equations of general relativity, Karl Schwarzschild (see Figure 2.13) solved them to find the spacetime geometry outside a spherical distribution of matter of mass M. The corresponding problem in Newtonian gravitation yields the solution for the gravitational potential as

$$\phi = -GM/r, \tag{2.116}$$

r being the distance from the centre of the spherical distribution.

At a large distance from the centre, we expect the gravitational field to be weak. So, under the Newtonian approximation, we expect

$$g_{00} \sim 1 - \frac{2GM}{c^2 r}. \tag{2.117}$$

We will now show how the Schwarzschild solution is obtained and how this exact solution relates to the above form.

The problem is simplified by making use of symmetry arguments. If the spacetime outside such a spherical distribution is empty, then its geometry should be spherically symmetric about the centre O of the distribution. So we start with the most general form of the line element that fulfils this requirement of spherical symmetry.

It can be shown that the most general form of such a line element is

$$ds^2 = e^\nu c^2\, dt^2 - e^\lambda\, dr^2 - r^2(d\theta^2 + \sin^2\theta\, d\phi^2), \tag{2.118}$$

where ν and λ are functions of r and t. If $\nu = \lambda = 0$, we get the Minkowski line element in spherical polar spatial coordinates. The non-Euclidean effects are therefore contained in the functions λ and ν. Although in this case r ceases to measure the radial distance from O, it still has the meaning that the spherical surface $r = \text{constant} = r_0$ (for example) has the surface area $4\pi r_0^2$. The arguments leading to (2.118) are group-theoretical ones, involving the invariance of spacetime under rotations about the point O. The techniques describing these arguments are beyond the scope of this text: see the classic book by Eisenhart listed in the bibliography for these details.

Given the line element (2.118), the next step is to calculate g^{ik}, $\sqrt{-g}$ and Γ_{kl}^i. We then calculate R_{kl}, which are given by (2.51) and are expressible in the form:

$$R_{kl} = -\frac{\partial \Gamma_{kl}^i}{\partial x^i} + \frac{\partial^2(\ln\sqrt{-g})}{\partial x^k\,\partial x^l} + \Gamma_{kn}^m\Gamma_{lm}^n - \frac{\partial}{\partial x^n}(\ln\sqrt{-g})\,\Gamma_{kl}^n. \tag{2.119}$$

Figure 2.13 Karl Schwarzschild (1873–1916).

Since, the space outside the distribution is empty, it has $T_{kl} = 0$. Therefore the contraction of the field equations (2.100) gives $R = 0$ and these equations reduce to

$$R_{ik} = 0. \tag{2.120}$$

The (00) and (11) components $(r = x^1)$ give, after some manipulation, the following equations:

$$e^{-\lambda}\left(\frac{\lambda'}{r} - \frac{1}{r^2}\right) + \frac{1}{r^2} = 0, \tag{2.121}$$

$$-e^{-\lambda}\left(\frac{\nu'}{r} + \frac{1}{r^2}\right) + \frac{1}{r^2} = 0. \tag{2.122}$$

From these we get

$$\nu' + \lambda' = 0;$$

that is,

$$\nu + \lambda = f(t)$$

(here a prime denotes differentiation with respect to r, whereas an overdot denotes differentiation with respect to t). The arbitrary function $f(t)$ can, however, be set to equal zero since we still have an arbitrary time transformation

$$t = g(\bar{t})$$

at our disposal, which changes ν to

$$\bar{\nu} = \nu + 2\ln\left(\frac{dg}{d\bar{t}}\right)$$

and preserves the form of the line element (2.118). Therefore we can take, without loss of generality,

$$\nu + \lambda = 0. \tag{2.123}$$

However, we also have, from $R_{01} = 0$,

$$\dot{\lambda} = 0. \tag{2.124}$$

Thus both λ and ν $(= -\lambda)$ are functions of r only. The equations (2.121) and (2.122) then yield the solution

$$e^{\nu} = e^{-\lambda} = 1 - A/r, \qquad A = \text{constant}.$$

However, if we are given the mass of the object M, we may use the boundary condition (2.117) to set $A = 2GM/c^2$. Thus we get our required solution as the line element

$$ds^2 = \left(1 - \frac{2GM}{c^2 r}\right) c^2 \, dt^2 - \left(1 - \frac{2GM}{c^2 r}\right)^{-1} dr^2 - r^2 (d\theta^2 + \sin^2 \theta \, d\phi^2).$$

$$(2.125)$$

This is known as the *Schwarzschild line element*. It turns out that, because of the symmetries of the problem, the other field equations are automatically satisfied: we need only the (11), (00) and (01) components in order to arrive at the solution. Also, the solution (2.125) is manifestly static. Thus there is no scope for a dynamical solution such as one involving gravitational radiation, even if our spherical source is expanding, contracting, or oscillating. This remarkable result is known as *Birkhoff's theorem*. We now consider a few observable implications of this solution.

2.11 Experimental tests of general relativity

Most of the present tests of general relativity are based on the Schwarzschild solution and seek to measure the fine differences between the predictions of Newtonian gravitation and those of general relativity. These are described briefly below.

Before confronting the experimental situation, however, it is necessary to clarify how to attach meanings to measurements in a spacetime that is non-Euclidean. We have already seen that coordinates have no absolute status, hence relying on them blindly might lead to incorrect results. The Schwarzschild metric (2.125) can be used to illustrate the concept of measurement as will be seen next.

Example

Suppose that an observer is located at a point with r = constant, θ = constant and ϕ = constant. How does he relate the time kept by his watch to the coordinate t? From the principle of equivalence we know that since $d\tau = ds/c$ measures the observer's proper time in a locally inertial frame, being a scalar, it does so in all frames. For our observer, $dr = 0$, $d\theta = 0$ and $d\phi = 0$; so from (2.125) we get

$$d\tau = \left(1 - \frac{2GM}{c^2 r}\right)^{1/2} dt.$$

This gives the required answer.

The experimental tests mostly revolve around application of the Schwarzschild line element to objects in the Solar System. However, beyond comparing the relativistic predictions with the corresponding Newtonian ones, there has also been interest in *other* theories of gravitation. Some of these (such as the Brans–Dicke theory to be discussed in Chapter 8) use the spacetime metric as in relativity, but come up with line elements different from Schwarzschild's. All these can be simultaneously looked at in their weak-field limits and by comparing their predictions in the various experiments. A series of parameters can be used to specify the various components of the metric with reference to these tests. Since we are looking at a level of approximation one step beyond the Newtonian limit, the procedure is called *parametrized post-Newtonian* approximation or simply the PPN approximation. The parameters are denoted by γ, β, ξ, α_1, α_2 and α_3 and ζ_1, ζ_2, ζ_3 and ζ_4. We will not discuss the details of how these parameters are derived in a particular theory (see a review by C. M. Will referred to at the end), except to identify the first two which have values of unity in general relativity and occur explicitly in the classical tests of this theory. The rest have value zero in relativity.

To identify β and γ, we express the Schwarzschild line element in the isotropic form, in which the spatial part of the metric is the Euclidean one multiplied by a radial function:

$$ds^2 = e^\mu c^2 \, dt^2 - e^\eta [dR^2 + R^2(d\theta^2 + \sin^2\theta \, d\phi^2)], \tag{2.126}$$

where μ and η are functions of the new radial coordinate R. (See Exercise 42.) Expanding these in powers of M/R, we get

$$e^\mu = 1 - 2\frac{M}{R} + 2\beta\left(\frac{M}{R}\right)^2, \qquad e^\eta = 1 + 2\gamma\frac{M}{R}, \tag{2.127}$$

where, as mentioned earlier for general relativity, both β and γ are unity. For some other theories they may have different values. We will later summarize the present status of the measured values of these parameters.

The gravitational redshift

Consider any static line element – that is, one in which g_{ik} do not depend on $x^0 \equiv ct$. Suppose that we have two observers A and B with world lines

$$x^\mu = \text{constant} = a^\mu, \ b^\mu, \tag{2.128}$$

respectively. Let Γ be a null geodesic from A and B, with parametric equations given by

$$x^i = x^i(\lambda), \tag{2.129}$$

with $x^\mu(0) = a^\mu$, $x^\mu(1) = b^\mu$, $x^0(0) = ct_A$ and $x^0(1) = ct_B$. To what does our geodesic correspond in physical terms?

It describes a ray of light leaving observer A at time t_A and reaching observer B at time t_B. Because of the static nature of the line element, we also have another null geodesic solution given by

$$x^\mu = x^\mu(\lambda), \qquad \mu = 1, 2, 3,$$
$$x^0 = x^0(\lambda) + \Delta, \quad \Delta = \text{constant.} \tag{2.130}$$

This describes a light ray leaving A at $t_A + \Delta/c$ and reaching B at $t_B + \Delta/c$. Figure 2.14 illustrates this result.

Now, in the rest frame of A, the time interval Δ/c corresponds to a proper time interval (measured by A) of

$$\delta\tau_A = \frac{\Delta}{c}[g_{00}(a^\mu)]^{1/2}.$$

If n light waves have left A in this time interval, then the frequency of these waves measured by A is

$$\nu_A = \frac{cn}{\Delta}[g_{00}(a^\mu)]^{-1/2}.$$

Since the same *number* of waves is received by B in the corresponding proper time interval $\delta\tau_B$, we get the ratio of frequencies measured by B and A as

$$\frac{\nu_B}{\nu_A} = \left(\frac{g_{00}(a^\mu)}{g_{00}(b^\mu)}\right)^{1/2}. \tag{2.131}$$

This is also the ratio of the wavelengths λ_A and λ_B measured by A and B, respectively.

If, in the Schwarzschild solution, A is an observer located on the surface of a star, at $r = r_s$, say, and B is a distant observer with $r \gg 2GM/c^2$, we get

$$\frac{\lambda_B}{\lambda_A} \cong \left(1 - \frac{2GM}{c^2 r_s}\right)^{-1/2}. \tag{2.132}$$

Thus spectral lines from a massive compact star should be redshifted. For $2GM/(c^2 r_s)$ small relative to unity, the redshift

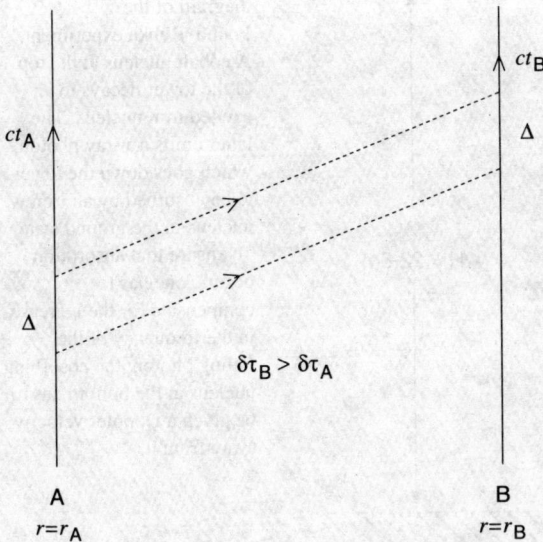

Figure 2.14 The proper time of an observer at $r =$ constant in Schwarzschild's spacetime runs more slowly than the coordinate time. The effect is more pronounced as r decreases. The figure shows that this effect is detectable by exchanges of light signals (shown by dotted lines) between two observers. Although signals are sent by A (located near the mass) at interval $\delta\tau_A$ of proper time, they are received by the remote observer B at interval $\delta\tau_B > \delta\tau_A$. The coordinate time difference between the signals at A and B remains the same at Δ. .

$$z = \frac{\lambda_B - \lambda_A}{\lambda_A} \approx \frac{GM}{c^2 r_s}. \qquad (2.133)$$

White dwarf stars like Sirius B and 40 Eridani B do exhibit redshifts in the range of 10^{-5}–10^{-4}, which are of the right order of magnitude. More reliable and quantitatively accurate measurements, however, are possible only in a terrestrial experiment. For example, in 1960 Pound and Rebka measured the change in the frequency of a γ-ray photon emitted by an excited iron nucleus as it fell from a height of 60–70 feet (see Figure 2.15). As such a photon falls through a height H, the Newtonian potential decreases by gH, where g is the acceleration due to gravity on the Earth's surface. From (2.131) we see that the photon should undergo a blueshift; that is, its frequency increases by a fraction gH/c^2. Although this fraction is as small as 10^{-15}, it can be measured by modern laboratory techniques. The Pound–Rebka experiment and later work have confirmed the gravitational-redshift effect to a high level of accuracy.

The perihelion precession of Mercury

If we treat the Sun as the mass M in the Schwarzschild solution and the planets as probes moving in the curved spacetime around the Sun, then, to a first approximation, each planet will move along a timelike geodesic. The equations of motion of a planet are therefore easily obtained (see Exercises 46 and 47). In the Newtonian approximation, the planet describes an ellipse given by its polar equation

$$l/r = 1 + e\cos(\phi - \phi_0). \qquad (2.134)$$

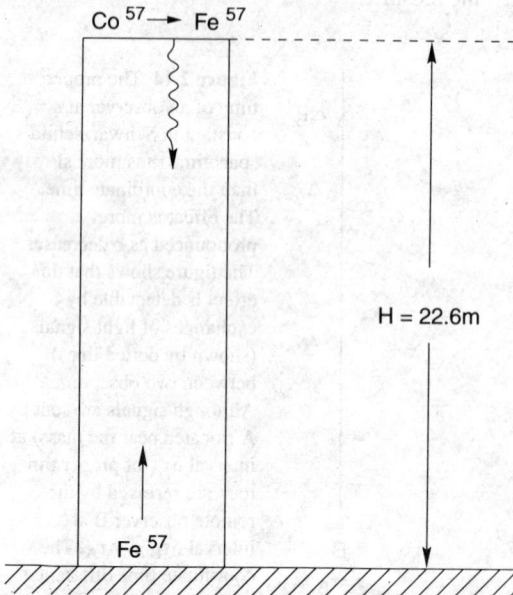

Figure 2.15 A schematic diagram of the Pound–Rebka experiment. A cobalt nucleus at the top of the tower decays to an excited iron nucleus. The latter emits a γ-ray photon, which goes down the tower to be absorbed by an iron nucleus in the ground state. To ensure that absorption occurs, one has to compensate for the increase in the frequency of the falling photon: the absorbing nucleus at the bottom has to be given a Doppler velocity away from it.

Here l is the *semi latus retum*, e the eccentricity and ϕ_0 the direction in which its perihelion (point of closest approach to the Sun) lies.

Observations of the orbit of planet Mercury had revealed that ϕ_0 is not a constant. Rather, the perihelion precesses steadily at a small but perceptible rate of 575 arc-seconds per century. Of this, all but 43 arcseconds per century could be accounted for by the perturbation of Mercury's orbit by the Newtonian gravitational effect of other planets.

A more careful solution of the equations (see Exercise 48) shows, however, that ϕ_0 is not a constant, but changes its magnitude at a steady rate, as illustrated in Figure 2.16. This precession of perihelion is at a rate

$$n = \frac{6\pi G M_\odot}{lTc^2}, \tag{2.135}$$

where $M_\odot$ is the mass of the Sun and T is the period of the planet. The value of n is largest for Mercury, which of all the planets has the most eccentric orbit and the closest orbit to the Sun. The rate for Mercury, $n \cong 43$ arcseconds per century, can explain exactly the rate of precession, which had long remained unaccounted for in the Newtonian theory.

In the late 1970s, a more dramatic example of such a precession was observed for the binary star system that houses the pulsar PSR 1913 + 16. Here the gravitational effects are stronger than those in the Sun–Mercury system and the rate of precession is as high as 4.23 degrees per year – about 3.6×10^4 times the value for Mercury. (We should caution the reader, however, that, unlike in the Sun–Mercury case, in which, because of the large disparity of their masses, the Sun could be considered at rest and Mercury moving around it as a 'test' particle, in the binary-pulsar case the two stars have comparable masses and hence the Schwarzschild solution is not strictly applicable. Ideally one should solve the relativistic two-body problem. This

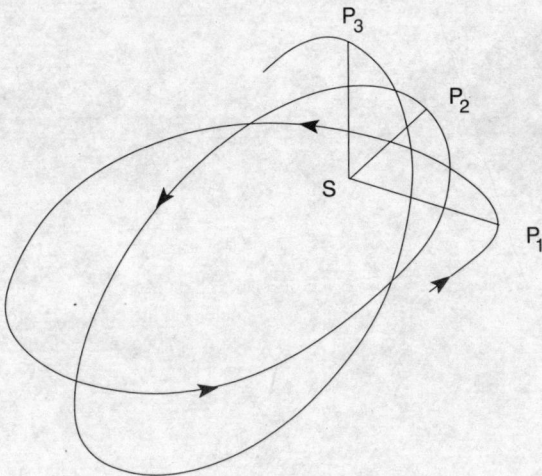

Figure 2.16 Perihelion is the point of closest approach to the Sun (S) in the orbit of Mercury. As successive orbits are completed, the perihelion advances steadily from P_1 to P_2 to P_3 and so on. (The rate of advance per orbit is actually much smaller than that shown here.)

has not been possible so far and hence only an approximate extrapolation of the Sun–Mercury problem is generally used for theoretical comparison.)

The bending of light

Just as timelike geodesics determine the tracks of planets, we can calculate the track of a ray of light by determining the equations of null geodesics. These equations are straightforward to write down (see Exercise 48) and integrate (see Exercise 49). The most dramatic effects arise when a null geodesic goes very close to the mass distribution.

For a ray light of grazing the solar limb (see Figure 2.17), the spatial direction of the ray changes by an angle

$$\alpha = \frac{4GM_\odot}{c^2 R_\odot} \simeq 1.75 \text{ arcseconds},\tag{2.136}$$

where $R_\odot$ is the radius of the Sun. The bending angle is indeed very small; its measurement was first attempted by Eddington (see Figure 2.18) and his colleagues in 1919 at the time of a solar eclipse. (The experiment involves measuring the apparent change in the direction of a star as its line of sight grazes the solar limb. For obvious reasons, optical astronomers have to wait for a total solar eclipse in

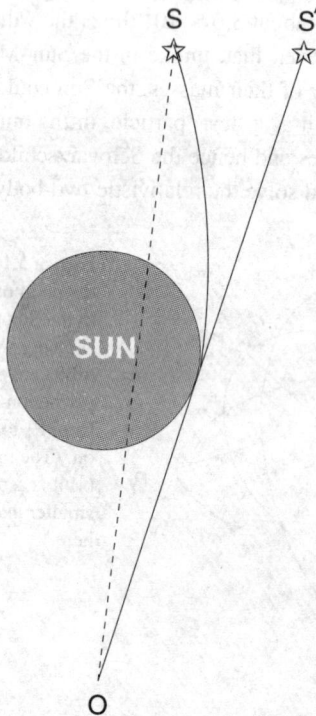

Figure 2.17 The 'bending' of rays of light from star S shifts its image to S′. This effect is noticeable when the ray of light grazes the solar surface, as shown here.

order to perform this experiment.) That measurement seemed to favour Einstein. However, it had large error bars. Subsequent attempts by optical astronomers yielded somewhat inconclusive results, largely because of the limited sensitivity of the measuring equipment and the uncertain nature of systematic errors. In the 1970s, however, measurements with microwaves confirmed the above-mentioned value of the bending angle much more precisely with only about 5% experimental error. This technology has subsequently been improved to reduce the error bars further, as can be seen in Table 2.1 later.

It is worth pointing out with regard to the gravitational redshift and the bending of light that, since the strict Newtonian theory did not predict any effects of gravity on light, their observation implies a disproof of Newtonian gravitation. It is, however, possible to enlarge the scope of the Newtonian framework and argue that light is made of particles (photons) that are also subject to the inverse-square law. (Indeed, Newton himself had speculated concerning this possibility.) We can then show that such an enlarged theory gives (2.131) for gravitational redshift (see Exercise 53), the same as in relativity, but half the relativistic value for the bending of light (see Exercise 52). The observed bending of microwaves therefore rules out such a theory.

Figure 2.18 Arthur Stanley Eddington (1882–1944).

Radar echo delay

Just as the direction of a ray of light is altered by the Sun's gravity, so is its apparent travelling time. This effect can also be calculated in a straightforward manner. In the 1970s measurements were made by bouncing radar signals emitted from the spacecrafts Mariner 6 and 7 off the surface of the Earth as the signals grazed the solar limb. The expected delays of the order of 200 μs were observed within 3% error bars. This test has also been made more accurate with time and Table 2.1 gives the latest information.

The equality of inertial and gravitational mass

An important consequence of the principle of equivalence is the equality of inertial and gravitational mass. A little thought will convince us that Galileo's experiment from the leaning tower of Pisa, which demonstrated that all bodies fall freely with equal rapidity, is an essential part of Einstein's thought experiment involving the freely falling lift. Both experiments are possible because the same quantity enters the law of motion as inertial mass and the law of gravitation as gravitational mass.

Recent experiments with lunar laser ranging have been successful in measuring the distance of the Moon from the Earth to within a few centimetres. Such experiments also demonstrate that the Moon moves around the Earth as predicted by the equations of general relativity. In particular, these experiments ruled out certain alternative theories of gravitation, like the Brans–Dicke theory, that allow for the variation of inertial mass with the distance from another mass.

Laboratory experiments of the torsion-balance type have been conducted very accurately with various materials to establish this equality with high sensitivity. Such experiments place stringent upper limits on the possibility of the presence of a 'fifth force' operating at a range of a few metres. For a review of the measured accuracy of the principle of equivalence, see the article by C. M. Will referred to in the bibliography.

The precession of a gyroscope

Although the Schwarzschild solution describes the gravitational effects of the Sun or the Earth with great accuracy, there is scope for further improvement. For instance, a rotating mass would introduce a $d\phi\, dt$ term into the metric. Although the effects of such terms are very small for the Earth and the Sun, modern technology can measure them.

A proposed experiment that can measure the effect of a rotating mass makes use of gyroscopes. The axis of a gyroscope sent on an equatorial orbit around the Earth will precess slowly. An estimated rate of precession of ~7 arcseconds per year can be detected with present technology; such an experiment has been on the drawing board for three decades, but has not yet been performed.

Table 2.1 gives the measured values of the PPN parameters, or rather the limits set on their deviation from the predictions of general relativity. Although the experiments described there go beyond what we have outlined above, it is clear that the theory comes out with flying colours.

Gravitational radiation

Calculations based on the weak-field theory show that the magnitude of gravitational radiation from terrestrial apparatus is extremely small and beyond the scope of present technology. Celestial objects, however, can and do emit appreciable quantities of gravitational radiation and attempts to devise detectors to measure them are being made. Typical sources of gravity waves are supernova explosions, coalescing binary stars and, possibly, primordial developments in the very early universe (see Chapter 6). The quasi-steady-state cosmology described in Chapter 8 also predicts gravitational waves from mini-creation events. Several detectors are being set up around the globe to catch these feeble signals (see Figure 2.19).

2.11.1 Black holes

All the effects discussed above are those of *weak* gravity. For the Sun the ratio $2GM_\odot/(c^2 R_\odot)$ is as low as 4×10^{-6}, and for the Earth it is even smaller. Can we visualize an object that is so compact that its mass M and radius R are related by

Table 2.1 Limits on the measured values of the PPN parameters (Based on data reviewed by C. M. Will)

Parameter	Effect	Limit	Remarks
$\gamma - 1$	Time delay	2×10^{-3}	Viking ranging
	Deflection of light	3×10^{-4}	VLBI
$\beta - 1$	Perihelion shift	3×10^{-3}	$J_2 = 10^{-7}$ from helioseismology
	Nordtvedt effect	6×10^{-4}	$\eta = 4\beta - \gamma - 3$ assumed
ξ	Tides on Earth	10^{-3}	Gravimeter data
α_1	Orbital polarization	4×10^{-4}	Lunar laser ranging
		2×10^{-4}	PSR J2317 + 1439
α_2	Spin precession	4×10^{-7}	Solar alignment with the ecliptic
α_3	Pulsar acceleration	2×10^{-20}	Pulsar $\dot{P}$ statistics
η^{a}	Nordtvedt effect	10^{-3}	Lunar laser ranging
ζ_1	–	2×10^{-2}	Combined PPN bounds
ζ_2	Binary acceleration	4×10^{-5}	$\ddot{P}_p$ for PSR 1913 + 16
ζ_3	Newton's third law	10^{-8}	Lunar acceleration
ζ_4	–	–	Not independent

[a] Here $\eta = 4\beta - \gamma - 3 - 10\xi/3 - \alpha_1 - 2\alpha_2/3 - 2\zeta_1/3 - \zeta_2/3$.

$$\frac{2GM}{c^2 R} \sim 1?$$

A glance at the Schwarzschild solution will show that, for such an object, the space-time geometry near the surface will be markedly non-Euclidean. Several unexpected results arise if such objects exist (see, for example, Exercises 54 and 55).

An object whose coordinate radius R satisfies the condition

$$R \le \frac{2GM}{c^2} \equiv R_S \qquad (2.137)$$

is called a *black hole*. As its name implies, such an object is dark because its strong gravity traps light and prevents it from getting away. A glance at (2.132) shows that the gravitational redshift of a black hole is infinite. Since a redshift, z, implies a decrease in the energy of a photon of light by the factor $(1 + z)^{-1}$, no photon with finite energy can come out of a black hole.

Astrophysicists find that a black hole may be the end state of a massive star which has exhausted all its nuclear fuel that normally provides it with thermal and radiation pressures to support it in equilibrium against its self-gravity. If the mass of such a star exceeds the critical mass (known as the *Chandrasekhar limit* in the case of white-dwarf stars, for example) below which the degenerate particle pressure can hold it in equilibrium, it continues to contract until its radius reaches or becomes less than the above-mentioned critical value. However, it should be stated that, in the proper time of external observers like ourselves, that stage is never reached in a finite interval. Thus all claims about the existence of a black hole (in a binary star or a galactic nucleus) are technically incorrect. At best they are approximate

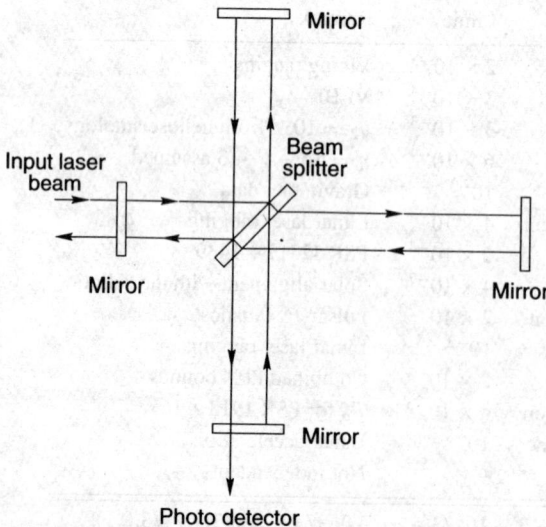

Figure 2.19 A schematic diagram of an interferometric detector of gravitational waves based on Michelson's interferometer, but using laser beams travelling to and fro several kilometres. With the detector one expects to measure minute fractional changes ($\sim 10^{-21}$) in lengths as gravitational waves move across it, causing the geometry to change.

statements describing an object with a Schwarzschild radius slightly in excess of the critical value $2GM/c^2$.

Important advances in general relativity were made between 1965 and 1975 through attempts to study black holes and their weird properties. Since the emphasis of this text is on cosmology, we must be content with the superficial introduction given here. Our purpose in laying the foundations of general relativity as the working theory for cosmological models has now been served. For those interested in learning about general relativity in greater depth we suggest other textbooks written specially for the purpose. It is to a discussion of cosmology that we must now proceed.

Exercises

1 Verify that a piece of string stretched across the spherical globe of the Earth lies along the arc of a great circle, which is a 'straight line' on the spherical surface. Check whether lines of latitude and longitude are straight.

2 One of the 'self-evident truths' on which Euclid's geometry is based is the so-called postulate of parallelism. This states that given a straight line l and a point P not on it, one and only one line parallel to l can be drawn through P (that is, a line that does not meet l if both lines are extended indefinitely). What happens to this postulate in the geometries on the surface of a sphere and on a saddle-shaped surface?

3 Calculate tangent vectors at typical points on the curves in spacetime given by the following relations ($x^1 = x$, $x^2 = y$, $x^3 = z$ and $x^0 = ct$):

(a) $x = ct_0 \cos(t/t_0)$, $y = ct_0 \sin(t/t_0)$, $z = ct$, $t_0 = $ constant;
(b) $x = 0$, $y = 0$, $z^2 - c^2t^2 = 0$;
(c) $x = ct \cos(t/t_0)$, $y = ct \sin(t/t_0)$, $z = $ constant, $t_0 = $ constant.

Determine whether these tangent vectors are spacelike, timelike, or null in Minkowski spacetime.

4 Calculate the components of the normals of the following surfaces and determine whether they are spacelike, timelike, or null in Minkowski spacetime ($x^1 = x$, $x^2 = y$, $x^3 = z$, $x^0 = ct$):

(a) $x^2 + y^2 + z^2 - \lambda^2 t^2 = $ constant, $\lambda = $ constant;
(b) $x^2 + y^2 - \lambda^2 t^2 = $ constant, $\lambda = $ constant;
(c) $x^2 - \lambda^2 t^2 = $ constant, $\lambda = $ constant.

5 Which of the following expressions are invalid with respect to the summation convention: (a) $A_{ij} B^{jk} A_{jl}$, (b) $g_{ik} g^{ik}$, (c) $R_{ik} g_{ik}$, (d) $e_{iklm} e^{iklm}$ and (e) $T^{ik} g^k_l$? Simplify those expressions that are valid.

6 A_{ik} is a tensor such that the matrix $\|A_{ik}\|$ is non-singular. Show that the components of the inverse matrix transform as a tensor. (An example of this result is the tensor g^{ik}.)

7 Show that the property of symmetry or antisymmetry with respect to indices of a tensor is invariant under coordinate transformations.

8 Construct, with the help of g_{ik} only, a fourth-rank tensor that is symmetric with respect to the interchange of any two of its indices.

9 Verify that, if F_{ik} is an antisymmetric tensor field, then

$$F_{ik,l} + F_{kl,i} + F_{li,k}$$

is a third-rank tensor.

10 Prove the *quotient law* in the following form: if $A_{ik}B^k$ is a vector for any arbitrary vector B^k, then A_{ik} must transform as a tensor.

11 Use two-dimensional polar coordinates (r, θ) on a Euclidean plane. Let A_r and A_θ be the radial and transverse components of a vector $\mathbf{A}$ at a typical point P chosen with respect to locally Cartesian axes with directions coinciding with $\theta = $ constant and $r = $ constant, respectively. Show that parallel transportation of the vector at P to a neighbouring point $Q(r + \delta r, \theta + \delta\theta)$ gives the two components of the vector at Q as

$$A_r + \delta\theta\, A_\theta, \qquad A_\theta - \delta\theta\, A_r.$$

12 By using the requirement that $B_{i;k}$ transforms as a tensor, deduce the transformation relation (2.36) for Γ^i_{kl}.

13 Deduce the form (2.37) for $A^i{}_{;k}$, using (2.35) for $B_{l;k}$ and assuming that the covariant derivative of a scalar equals its ordinary derivative.

14 Suppose that two metrics are defined on the same spacetime and let Γ^i_{kl} and $\bar{\Gamma}^i_{kl}$ be the two corresponding Riemannian affine connections. Deduce that the quantities

$$Q^i_{kl} = \Gamma^i_{kl} - \bar{\Gamma}^i_{kl}$$

transform as a tensor. (The coordinates are the same in the two cases.)

15 Show that, to arrive at a locally inertial system, it is necessary to have $\Gamma^i_{kl} = \Gamma^i_{lk}$.

16 Deduce the relations shown in (2.42) from first principles.

17 Show that, for a scalar field ϕ, the wave operator takes the form

$$\Box\phi = g^{ik}\phi_{;ik} = \frac{1}{\sqrt{-g}}\frac{\partial}{\partial x^k}\left(\sqrt{-g}\,g^{ik}\frac{\partial\phi}{\partial x^i}\right).$$

18 The line element on the surface of a sphere of radius a in the Euclidean space is given by

$$ds^2 = a^2(d\theta^2 + \sin^2\theta\, d\phi^2).$$

For this space calculate Γ^i_{kl}; $i, k, l = 1, 2$ (with $\theta = x^1$ and $\phi = x^2$) and verify the result discussed in the text regarding the change of direction of a vector under parallel displacement around the three-right-angled triangle ABC.

19 Prove from first principles that $B_{i;nk} - B_{i;kn} = R^m_{i\,kn} B_m$.

20 For an antisymmetric tensor field F_{ik}, show that $F^{ik}_{;ik} = 0$.

21 A_i is a vector field satisfying $A^i_{\;;i} = 0$. If $F_{ik} = A_{k;i} - A_{i;k}$, show that

$$F^{ik}_{;i} = g^{mn} A^k_{\;;mn} + R^k_m A^m.$$

22 Deduce the form (2.48) taken by R_{iklm} in the locally inertial coordinate system. Use the same coordinates to deduce the symmetric nature of R_{ik}.

23 Show by direct enumeration that the number of algebraically independent components of R_{iklm} is 20.

24 Using a locally inertial coordinate system, deduce the Bianchi identities. From these identities in their covariant form, show that

$$R^l_{k;l} = \tfrac{1}{2} R_{,k}.$$

25 Show that the first integral of (2.58) is

$$g_{ik} \frac{dx^i}{d\lambda} \frac{dx^k}{d\lambda} = \text{constant}.$$

26 Show that the change in the direction of a vector under a parallel displacement around a closed infinitesimal curve can be expressed in terms of the Riemann tensor and the area spanned by the curve.

27 Let a bundle of geodesics be specified by a parameter μ, so that a typical point on the μ geodesic has the coordinates $x^k(\lambda, \mu)$, λ being the affine parameter. The vector $v^k = \partial x^k / \partial \mu$ denotes the rate of deviation from one geodesic to another across the bundle. Deduce the following relations:

(a) $v^k_{\;;l} u^l = u^k_{\;;l} v^l$, where $u^k = \partial x^k / \partial \lambda$;

(b) $d^2 v^k / d\lambda^2 + R^k_{lmn} u^l v^m u^n = 0$.

The latter is the equation of *geodetic deviation*.

28 Construct a Newtonian analogue of geodetic deviation by comparing the deviations of two test particles falling on the (spherical) Earth along two neighbouring radial trajectories.

29 Verify the existence of the factor $\tfrac{1}{3}$ in (2.86).

30 Show that the results of (2.89) are based on standard kinetic theory.

31 Calculate the form of the energy tensor for a plane electromagnetic wave.

32 Show that, if the Lagrangian density of a physical interaction in curved spacetime is L, so that its contribution to action is

$$\int L \sqrt{-g}\, \mathrm{d}^4 x,$$

then, provided that L depends on the geometry only through g_{ik} and $g_{ik,l}$, the energy tensor of the interaction is given by

$$T^{ik} = -\frac{2c}{\sqrt{-g}} \left[\frac{\partial L \sqrt{-g}}{\partial g_{ik}} - \left(\frac{\partial L \sqrt{-g}}{\partial g_{ik,l}} \right)_{,l} \right].$$

33 Show that, from the scalar nature of L in Exercise 32, it is possible to deduce that

$$T^{ik}_{\ \ ;k} = 0.$$

Hint: use (2.97) and the fact that an infinitesimal change in the coordinates $x^i \to x^i + \xi^i$ gives $\delta g_{ik} = -(\xi_{i;k} + \xi_{k;i})$.

34 Show that under $g_{ik} \to g_{ik} + \delta g_{ik}$, the variation $\delta \Gamma^i_{\ kl}$ transforms as a tensor.

35 Show that

$$\delta \int_V R \sqrt{-g}\, \mathrm{d}^4 x = \int_V \left(R_{ik} - \frac{1}{2} g_{ik} R \right) \delta g^{ik} \sqrt{-g}\, \mathrm{d}^4 x$$

for variations of the metric that vanish together with their derivatives on the boundary of V. Hint: write $R = R_{ik} g^{ik}$ so that $\delta R = \delta R_{ik} + R_{ik}\, \delta g^{ik}$. Use a locally inertial coordinate system to deduce that

$$\sqrt{-g} g^{ik}\, \delta R_{ik} = -\sqrt{-g}[(g^{ik}\, \delta \Gamma^l_{\ ik})_{;l} - (g^{il}\, \delta \Gamma^k_{\ ik})_{;l}] = \sqrt{-g} w^k_{\ ;k},$$

where w^k (from Exercise 34) is a vector. Then use Green's theorem.

36 From the Newtonian approximation of Einstein's field equations and the geodesic equations, deduce Poisson's equation and the Newtonian equations of motion in a gravitational field.

37 Show that, in the weak-field approximation for gravitational radiation, it is possible to make a coordinate transformation to ensure that one has a gauge condition:

$$\psi^k_{i,k} = 0,$$

where

$$\psi^k_i = h^k_i - \tfrac{1}{2} h^l_l \delta^k_i.$$

Also, show that the ψ^k_i satisfy the wave equation in flat space

$$\Box \psi^k_i = -\frac{16\pi G}{c^4} T^k_i.$$

38 Compare the linearized theory of gravitational waves mentioned above with the electromagnetic theory of Maxwell. Construct plane-wave solutions in the case $T_i^k = 0$.

39 In a plane-wave solution of the gravitational wave equation, estimate the components of the Riemann tensor. Show that, in principle, a gravitational wave can be detected by the measurement of the components of the Riemann tensor with the help of the equation of geodetic deviation.

40 Just as the second time derivative of a changing electric dipole moment acts as the elementary source of electromagnetic radiation, the third time derivative of a changing quadrupole moment (of mass) acts as the simplest source of gravitational radiation. Use this result and dimensional arguments to show why the emission of gravitational waves is energetically very weak under laboratory conditions.

41 Show that, if we apply the line element (2.118) to the interior ($\dot{r} \le r_s$) of a spherically symmetric distribution of matter, we get from the (00) component of the field equations

$$e^{-\lambda} = 1 - 2GM(r)/c^2,$$

where

$$M(r) = \int_0^r 4\pi\rho^2 T_0^0 \, d\rho.$$

The quantity $M(r_s)$ may be identified with the gravitational mass M that appears in the exterior solution ($r > r_s$).

42 Show that, by a radial coordinate transformation, of the form $r = f(R)$, the Schwarzschild line element can be rewritten in isotropic form as follows:

$$ds^2 = e^\mu \, dt^2 - e^\eta[dR^2 + R^2(d\theta^2 + \sin^2\theta \, d\phi^2)],$$

where

$$e^\mu = \frac{[1 - MG/(2R)]^2}{[1 + MG/(2R)]^2}, \qquad e^\eta = \left(1 + \frac{MG}{2R}\right)^4.$$

(*Isotropic* here means that the radial and transverse directions are treated in the same way.)

43 Calculate the ratio $GM/(c^2 r_s)$ in order that the entire visible spectrum (4000–8000 Å) in the light from the surface of the spherical object is just about redshifted out.

44 Calculate the proportionate increase in the frequency of a γ-ray photon as it descends from a height of 100 m along a vertical path to the surface of the Earth.

45 Show that, if instead of being stationary, the observer in the Schwarzschild spacetime is moving radially and his radial coordinate at time t is given by $r = f(t)$, his proper time interval corresponding to dt is

$$d\tau = \left(1 - \frac{2GM}{c^2 r}\right)^{1/2} dt \left[1 - \left(\frac{df(t)/dt}{1 - 2GM/(c^2 f(t))}\right)^2\right]^{1/2}.$$

46 Show that the equations of a timelike geodesic in the Schwarzschild spacetime are given by

$$\frac{d^2t}{ds^2} + \frac{dv}{dr}\frac{dr}{ds}\frac{dt}{ds} = 0, \qquad \frac{d^2\theta}{ds^2} + \frac{2}{r}\frac{dr}{ds}\frac{d\theta}{ds} - \sin\theta\cos\theta\left(\frac{d\phi}{ds}\right)^2 = 0.$$

$$\frac{d^2\phi}{ds^2} + \frac{2}{r}\frac{dr}{ds}\frac{d\phi}{ds} + 2\cot\theta\frac{d\phi}{ds}\frac{d\theta}{ds} = 0,$$

$$\frac{d^2r}{ds^2} - \frac{1}{2}\frac{dv}{dr}\left(\frac{dr}{ds}\right)^2 - r e^v\left(\frac{d\theta}{ds}\right)^2 - r^2\sin^2\theta\, e^v\left(\frac{d\phi}{ds}\right)^2 + \frac{1}{2}e^{2v}\frac{dv}{dr}\left(\frac{dt}{ds}\right)^2 c^2 = 0,$$

with $e^v = 1 - \dfrac{2GM}{c^2 r}$.

47 Show that the equations in Exercise 46 may be integrated as follows without loss of generality:

$$\theta = \frac{\pi}{2}, \qquad \frac{dt}{ds} = \left(1 - \frac{2GM}{c^2 r}\right)^{-1} E, \qquad r^2\frac{d\phi}{ds} = h,$$

where E and h are constants of motion. What other integral of these equations is known?

48 Show that, for the Sun–Mercury system an approximate solution of the equations of Exercise 47 is provided by

$$r = l[1 + e\cos(\phi - \phi_0)]^{-1},$$

where ϕ_0 is a slowly increasing function of time. Evaluate $d\phi_0/dt$ and relate the result to the observed precession of the perihelion of Mercury.

49 Show that null geodesics (describing rays of light, for example) in the Schwarzschild spacetime are given by the following equations in terms of the affine parameter λ:

$$\frac{d^2t}{d\lambda^2} + \frac{dv}{dr}\frac{dr}{d\lambda}\frac{dt}{d\lambda} = 0, \qquad \frac{d^2\theta}{d\lambda^2} + \frac{2}{r}\frac{dr}{d\lambda}\frac{d\theta}{d\lambda} - \sin\theta\cos\theta\left(\frac{d\phi}{d\lambda}\right)^2 = 0,$$

$$\frac{d^2\phi}{d\lambda^2} + \frac{2}{r}\frac{dr}{d\lambda}\frac{d\phi}{d\lambda} + 2\cot\theta\frac{d\phi}{d\lambda}\frac{d\theta}{d\lambda} = 0,$$

$$c^2\left(\frac{dt}{d\lambda}\right)^2 e^v = e^{-v}\left(\frac{dr}{d\lambda}\right)^2 + r^2\left(\frac{d\theta}{d\lambda}\right)^2 + r^2\sin^2\theta\left(\frac{d\phi}{d\lambda}\right)^2.$$

50 Show how to obtain first integrals of the equations in Exercise 48 analogous to those of Exercise 46.

51 Show that the following is an approximate solution of the null geodesic equations:

$$r \cos \phi = r_s - \frac{GM}{c^2 r_s}(r \cos^2 \phi + 2r \sin^2 \phi).$$

Interpret this solution as describing the bending of light by a massive object.

52 Considering the photon of light as a projectile moving under the Newtonian inverse-square law, calculate the bending of light produced by a massive object. Show that the nett bending is half that given by general relativity.

53 Show how Newtonian gravitation can be adapted suitably to describe the phenomenon of gravitational redshifting. (Ascribe a gravitational mass $h\nu/c^2$ to a photon of frequency ν.)

54 A star of solar mass slowly contracts from initial radius $R_\odot$. Show how its gravitational redshift increases as a function of its radial coordinate r_s. What happens when $r_s < 2GM_\odot/c^2$?

55 A ray of light describes a circular trajectory around a black hole. Show how this is possible and calculate the size of the orbit.

Chapter 3

From relativity to cosmology

3.1   The historical background

In 1915 Einstein (Figure 3.1) put the finishing touches to the general theory of relativity. The Schwarzschild solution described in Chapter 2 was the first physically significant solution of the field equations of general relativity. It showed how spacetime is curved around a spherically symmetric distribution of matter. The problem solved by Schwarzschild was basically a local problem, in the sense that the deviations of spacetime geometry from the Minkowski geometry of special relativity gradually diminish to zero as we move further and further away from the gravitating sphere. This result can be easily verified from the line element (2.125) by letting the radial coordinate r go to infinity. In technical jargon a spacetime satisfying this property is called *asymptotically flat*. In general any spacetime geometry generated by a local distribution of matter is expected to have this property. Even from Newtonian gravity we expect an analogous result: that the gravitational field of a local distribution of matter will die away at a large distance from the distribution. Can the universe be approximated by a local distribution of matter?

Einstein felt that the answer to the above question would be in the negative. Rather, he expected the universe to be filled with matter, howsoever far we are able to probe it. A Schwarzschild-type solution cannot therefore provide the correct spacetime geometry of such a distribution of matter. Since we can never get away from gravitating matter, the concept of asymptotic flatness must break down. A new type of solution is therefore needed in order to describe a universe filled everywhere with matter. Einstein published such a solution in 1917.

Before we consider Einstein's solution, it is worth noting that, more than two centuries earlier, Newton also had attempted a solution describing a matter-filled universe of infinite extent. A highly symmetric distribution of matter does lead to a solution in Newtonian gravity. Imagine, for example, a uniform distribution of matter filling the infinite Euclidean space. An observer viewing the universe from any vantage point will find that it looks the same in all directions and that it presents the same aspect from all vantage points. These two properties are known as *isotropy* and *homogeneity*; they will turn out to play simplifying roles in relativistic cosmology too. Newton found that such a universe would be static, for any particle of matter, being attracted equally in all directions, should stay put where it is.

On the other hand, homogeneity precludes any pressure gradients in the universe and we know that any finite distribution of pressure-free matter would tend to shrink under its own gravity. Stars are able to maintain a stationary shape only because they have large enough pressure gradients inside to withstand their own gravity. Clearly, in going from a finite to an infinite universe something new has entered

Figure 3.1 Albert Einstein (1879–1955).

the argument: the boundary conditions at infinity. Considerable ambiguity arises in Newtonian theory when we try to interpret these boundary conditions.

Newton also found his solution to be unstable: any local inhomogeneity would precipitate gravitational contraction that would tend to augment the local inhomogeneity. Newton compared the instability of the solution to that of a set of needles finely balanced on their points.

Nevertheless, in 1934 W. H. McCrea and E. A. Milne showed how some of the problems of Newtonian cosmology can be resolved. Before coming to the relativistic cosmology initiated by Einstein, let us take a brief look at this Newtonian version.

3.2   **Newtonian cosmology**

In Newtonian cosmology space is Euclidean and time has the meaning implicit in Newtonian dynamics. We will assume that the universe is homogeneous and isotropic (we will elaborate on these assumptions in the following section) and that there is no preferred position or direction in the universe. Also, at each point in space there is a preferred observer who sees the universe as isotropic. We will assume that a galaxy at such a point is at rest with respect to this observer, henceforth referred to as the *fundamental observer*.

In such a universe let $\mathbf{r} \equiv (x_1, x_2, x_3)$ denote the coordinates (Cartesian, of course!) of a typical galaxy G and let $\mathbf{v} = (v_1, v_2, v_3)$ denote its velocity relative to a galaxy G_0 located at the origin (see Figure 3.2). The origin of course has no special status: any fundamental observer may be taken to be at the origin of coordinates. An observer in G_0 observes a velocity–distance relation for galaxies like G of the following form:

$$\mathbf{v} = \mathbf{f}(\mathbf{r}). \tag{3.1}$$

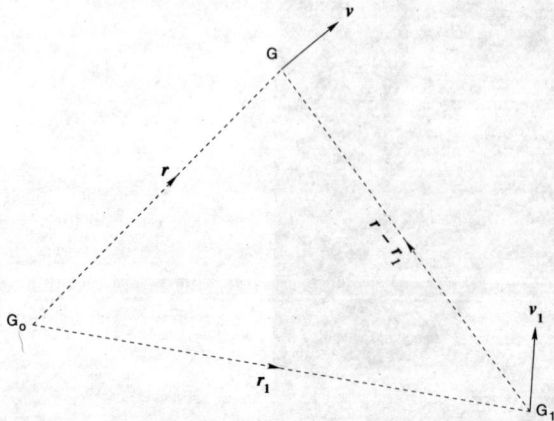

Figure 3.2 Three galaxies G_0, G_1 and G are shown above as three typical galaxies with the position vectors of G_1, and G being $\mathbf{r}_1$ and $\mathbf{r}$ relative to G_0. Each should observe the same velocity distance relation for the other two. For example, if G_0 observes the velocities of G_1 and G to be $\mathbf{v}_1$ and $\mathbf{v}$, respectively, then G_1 sees the velocity of G to be $\mathbf{v} - \mathbf{v}_1$ and its distance vector as $\mathbf{r} - \mathbf{r}_1$. Symmetry requires that $\mathbf{v} - \mathbf{v}_1$ must be the *same function* of $\mathbf{r} - \mathbf{r}_1$ as $\mathbf{v}$ is of $\mathbf{r}$.

If it is assumed that the same relation is observed for any other galaxy G_1 with co-ordinates $\mathbf{r}_1$ and velocity $\mathbf{v}_1$, then, because, all fundamental observers must observe the same general features of the universe, the function $\mathbf{f}$ must satisfy the condition

$$\mathbf{f}(\mathbf{r} - \mathbf{r}_1) = \mathbf{f}(\mathbf{r}) - \mathbf{f}(\mathbf{r}_1). \tag{3.2}$$

From this functional relation it is not difficult to deduce that f must be a linear function of r. That is, we can write

$$v_\mu = H_{\mu\nu} x_\nu, \tag{3.3}$$

where we are using the summation convention for the three space coordinates, but with all indices in lower (subscript) form. For the Euclidean geometry of Newtonian physics, the distinction between upper and lower indices is not needed for Cartesian coordinates. The tensorial coefficient $H_{\mu\nu}$ can be a function of time t, at most.

Next, under the assumption of isotropy during any epoch t, the tensor must have the form $H_{\mu\nu} = H(t)\delta_{\mu\nu}$ and the velocity–distance relation then takes the form

$$\mathbf{v} = H(t)\mathbf{r}. \tag{3.4}$$

This is nothing other than the velocity–distance relation obtained by Hubble described in Chapter 1. Thus Hubble's law is consistent with our postulate of homogeneity and isotropy: we do not enjoy any 'special status' by being at $\mathbf{r} = \mathbf{0}$, say.

We can complete the integration of the differential equation (3.4) by writing

$$\mathbf{r} = S(t)\mathbf{r}_0, \tag{3.5}$$

with

$$\dot{S}/S = H(t). \tag{3.6}$$

(The overdot denotes differentiation of the quantity with respect to t.)

3.2.1 The redshift

To relate the velocity–distance relation into the redshift (found in Hubble's law), we need to do some more work, however. Consider a galaxy at $\mathbf{r}_0 = \mathbf{a}_0$ sending light to us. Now we will work out the propagation of light from it to the observer at the origin, by using the assumption that all velocities add as per Newtonian kinematics and that velocity of light is seen to be c by every fundamental observer. As illustrated in Figure 3.3, let light leave $\mathbf{r}_0 = \mathbf{a}_0$ at $t = t_a$ and reach $\mathbf{r}_0 = \mathbf{0}$ at $t = t_0$.

The ray of light propagating from $\mathbf{r}_0 = \mathbf{a}_0$ to $\mathbf{r}_0 = \mathbf{0}$ will pass at time t in the range $t_a < t < t_0$ intermediate observers at $\mathbf{r}_0 = \eta \mathbf{a}_0$, $0 < \eta < 1$. Since the velocity of such a typical en-route observer is $HS(t)\mathbf{r}_0$ away from us, the light has an effective velocity

$$\frac{dr}{dt} = -c + \eta a_0 \dot{S}(t), \tag{3.7}$$

where $r = r_0 S(t) = \eta a_0 S(t)$.

Since $dr/dt = \dot{\eta} a_0 S + \eta a_0 \dot{S}$, we get

$$\frac{d\eta}{dt} = -\frac{c}{a_0 S}, \tag{3.8}$$

i.e.,

$$a_0 = \int_{t_a}^{t_0} \frac{c\, dt}{S(t)}, \tag{3.9}$$

since $\eta = 1$ at $t = t_a$ and $\eta = 0$ at $t = t_0$.

In deriving (3.9) we have added the velocity of light to the velocity of the intermediate observer as per the Newtonian formula for vectorial addition of velocities. Although our operation is inconsistent with special relativity, it is fully consistent within the Newtonian framework.

Consider now two light-wave crests of wavelengths λ_a emitted by the above galaxy. The first crest leaves at t_a and arrives at t_0. The second one leaves at $t_a + \Delta t_a$, say, and arrives at $t_0 + \Delta t_0$. Then a relation similar to (3.9) holds, viz.,

$$a_0 = \int_{t_a + \Delta t_a}^{t_0 + \Delta t_0} \frac{c\, dt}{S(t)}. \tag{3.10}$$

Subtracting (3.9) from (3.10) and using the approximation that Δt_0 and Δt_a are small enough intervals for us to treat $S(t)$ as constant over them, we get

$$\frac{c\, \Delta t_0}{S(t_0)} = \frac{c\, \Delta t_a}{S(t_a)}. \tag{3.11}$$

However, if λ_0 is the wavelength received by us, then $c\, \Delta t_0 = \lambda_0$ while $c\, \Delta t_a = \lambda_a$. Therefore

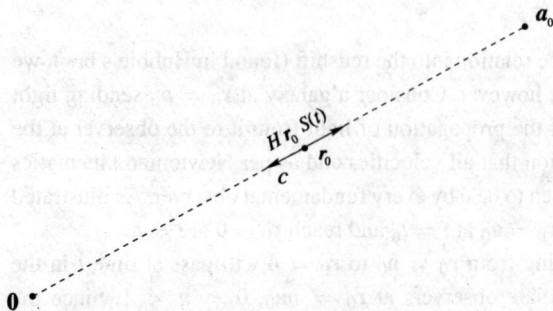

Figure 3.3 A ray of light from the galaxy at $\mathbf{r}_0 = \mathbf{a}_0$, coming towards $\mathbf{r}_0 = \mathbf{0}$, is retarded by the moving medium expanding away from $\mathbf{r}_0 = \mathbf{0}$. At an intermediate value of $\mathbf{r}_0$ (shown above) the velocity c of the ray towards 0 is *reduced* by the expansion velocity $H\mathbf{r}_0 S(t)$.

$$1 + z \equiv \frac{\lambda_0}{\lambda_\mathbf{a}} = \frac{S(t_0)}{S(t_\mathbf{a})}. \tag{3.12}$$

This is the relationship between z, the redshift and the scale factor $S(t)$. Since $z > 0$, $S(t_\mathbf{a}) < S(t_0)$ for $t_\mathbf{a} < t_0$. In other words, the scale factor increases with time, implying that *the universe is expanding*.

It is possible to deduce the linear redshift–distance relation for small distances from the above derivation of the redshift and show that Hubble's constant is given by $\dot{S}/S$, evaluated at $t = t_0$.

For small distances $t_\mathbf{a} \approx t_0$ and Taylor expansion near $t = t_0$ gives

$$S(t_\mathbf{a}) \cong S(t_0) - (t_0 - t_\mathbf{a})\dot{S}(t_0).$$

Hence,

$$1 + z = \frac{S(t_0)}{S(t_\mathbf{a})} = \left(1 - (t_0 - t_\mathbf{a})\frac{\dot{S}(t_0)}{S(t_0)}\right)^{-1}$$

$$\approx 1 + (t_0 - t_\mathbf{a})\frac{\dot{S}}{S}\bigg|_{t_0}$$

However, from (3.9), under the same approximation

$$a_0 \approx \frac{c(t_0 - t_\mathbf{a})}{S(t_0)}.$$

The distance of the galaxy from the observer at $t = t_0$ is $D = a_0 S(t_0) \cong c(t_0 - t_\mathbf{a})$. From the above relations, the result follows.

In this chapter we will be largely concerned with the kinematical aspects of cosmology and so will leave the Newtonian discussion here. As we will discover soon, the relativistic models give similar results, although they are more securely based by virtue of being consistent with the extrapolation from special to general relativity. (We will not be required to add the speed of light to the speed of the medium, for example, in deriving the redshift formula.) We shall return to Newtonian cosmology, however, in the following chapter to show that, in terms of the dynamical aspects also, it resembles the relativistic version. We will now begin our discussion of relativistic cosmology with Einstein's classic solution of 1917.

3.3 The Einstein universe

It is evident from the field equations of general relativity derived in Chapter 2 that their solution in the most general form – the solution of an interlinked set of non-linear partial differential equations – is beyond the present range of techniques available in applied mathematics. It is necessary to impose simplifying assumptions regarding symmetry in order to make any progress towards a solution. Just as

Schwarzschild assumed spherical symmetry in his local solution, Einstein assumed homogeneity and isotropy in his cosmological problem. He further assumed, like Schwarzschild, that spacetime is static. This allowed him to choose a time coordinate t such that the line element of spacetime could be described by

$$ds^2 = c^2 \, dt^2 - \alpha_{\mu\nu} \, dx^\mu \, dx^\nu, \tag{3.13}$$

where $\alpha_{\mu\nu}$ are functions of space coordinates x^μ ($\mu, \nu = 1, 2, 3$) only.

Note that the constraint of homogeneity implies that the coefficient of dt^2 can only be a constant, which we have normalized to c^2. Similarly, the condition of isotropy tells us that there should be no terms of the form $dt \, dx^\mu$ in the line element. This can be seen easily in the following way. If we had terms like $g_{0\mu} \, dt \, dx^\mu$ in the line element, then spatial displacements dx^μ and $-dx^\mu$ would contribute oppositely to ds^2 over a small time interval dt and such directional variation would be observable and inconsistent with isotropy. Can we say anything more about $\alpha_{\mu\nu}$?

Einstein believed that the universe has so much matter as to 'close' the space. This assumption led him to a specific form for $\alpha_{\mu\nu}$. We will now elaborate a little on the notion of closed space and on how to arrive at $\alpha_{\mu\nu}$. Let us begin with examples from lower-dimensional spaces.

Since the simplest example of an open space is the Euclidean straight line extending indefinitely in both directions, we can use a real variable r to denote a typical point on the line with $-\infty < r < \infty$. Figure 3.4(a) shows such a straight line.

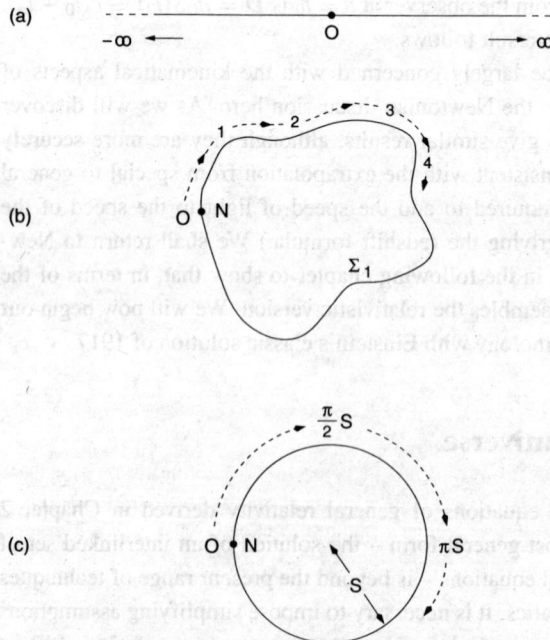

Figure 3.4 Curves in one-dimensional space. (a) A straight line extending from $-\infty$ to ∞. This is an example of open space. (b) A closed curve Σ_1. Starting from a point N on it as the origin, we can use the length r along the curve to label points on it. If the length of the curve is L, when $r = L$ we come back to the starting point. This is a closed space. (c) A closed space S_1 that is homogeneous: it is a circle. If it has radius S, $L = 2\pi S$.

Figure 3.4(b) shows an example of a closed curve Σ_1. It has no boundary, but, if we use a real variable r to denote points on the curve, we will find that a finite range of r will suffice. If we go beyond this range we will begin to go over the curve again and again. A familiar simple example of this is the circle S_1 of radius S shown in Figure 3.4(c). If we use the Euclidean measure of distance to locate a point on the circle and denote by r the distance of this point from a fixed point N on the circle, we find that the range $0 \leq r < 2\pi S$ describes all the points on the circle.

While both the curves in Figures 3.4(b) and (c) are closed, the circle evidently has more symmetries than does the curve Σ_1. This can be demonstrated as follows. If we take a small section (an arc) of the circle and slide it along the circle, it will always lie flush to it. We cannot do the same for the curve Σ_1. We can express this by saying that the circle S_1 describes homogeneous space, whereas the curve Σ_1 does not.

Figure 3.5 illustrates the corresponding situation in two dimensions. Two co-ordinates r and $\phi\,(0 \leq r < \infty, 0 \leq \phi < 2\pi)$ are needed in order to locate a point on the Euclidean plane of Figure 3.5(a). The surface Σ_2 shown in Figure 3.5(b) and the sphere S_2 of radius S shown in Figure 3.5(c) are closed surfaces, of which S_2 is

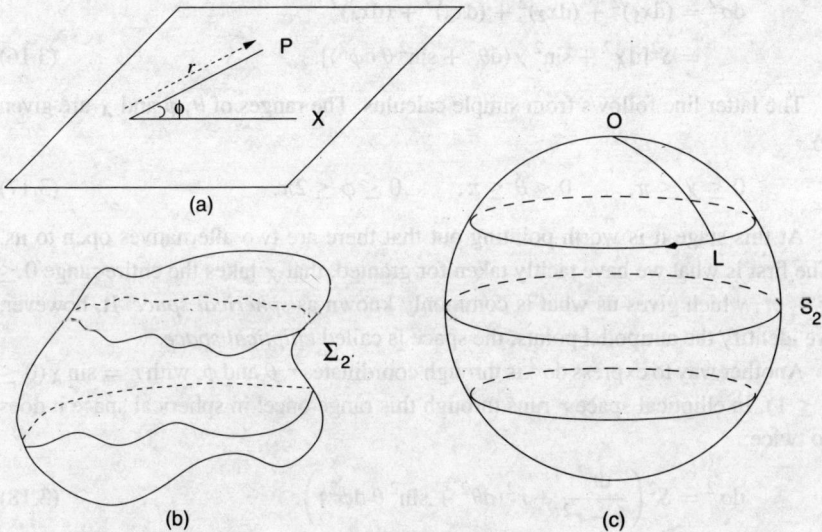

Figure 3.5 (a) The plane is an open two-dimensional space. From any point O on it draw the straight line OX in any direction in the plane. The coordinates (r, ϕ) in the illustration show how to specify any point P on the plane. (b) An arbitrary closed surface Σ_2. (c) A closed surface S_2 that is homogeneous and isotropic. It is a sphere. Take any point O on S_2 and draw a small arc of a great circle OL lying on S_2. As OL is rotated around O, the point L moves along a small circle on S_2 and the arc always stays on S_2. This is an example of isotropy: as seen from O, the surface S_2 exhibits no preferential direction.

homogeneous but Σ_2 is not. This property can easily be verified by our technique of sliding a small section of each surface along itself.

There is another symmetry inherent in the spherical surface, which can be demonstrated as follows. At any point O on it draw a small arc lying on the surface and then rotate this arc around the point O, trying all the while to keep the arc lying flush to the surface. Again the spherical surface S_2 allows you to do this, but Σ_2 does not. This means that the surface S_2 exhibits isotropy about O.

We can now see how to construct the homogeneous and isotropic closed space of three dimensions that Einstein wanted for his model of the universe. It is S_3, the 3-surface of a four-dimensional hypersphere of radius S. The equation of such a 3-surface is given in Cartesian coordinates x_1, x_2, x_3 and x_4 by

$$(x_1)^2 + (x_2)^2 + (x_3)^2 + (x_4)^2 = S^2. \tag{3.14}$$

To use coordinates *intrinsic* to the surface, we define

$$x_4 = S\cos\chi, \qquad x_1 = S\sin\chi\cos\theta, \qquad x_2 = S\sin\chi\sin\theta\cos\phi,$$
$$x_3 = S\sin\chi\sin\theta\sin\phi. \tag{3.15}$$

The spatial line element *on* the surface S_3 is therefore given by

$$d\sigma^2 = (dx_1)^2 + (dx_2)^2 + (dx_3)^2 + (dx_4)^2$$
$$= S^2[d\chi^2 + \sin^2\chi(d\theta^2 + \sin^2\theta\, d\phi^2)]. \tag{3.16}$$

The latter line follows from simple calculus. The ranges of θ, ϕ and χ are given by

$$0 \le \chi \le \pi, \qquad 0 \le \theta \le \pi, \qquad 0 \le \phi \le 2\pi. \tag{3.17}$$

At this stage it is worth pointing out that there are two alternatives open to us. The first is what we have tacitly taken for granted, that χ takes the entire range $0 \le \chi \le \pi$, which gives us what is commonly known as *spherical space*. If, however, we identify the antipodal points, the space is called *elliptical space*.

Another way to express $d\sigma^2$ is through coordinates r, θ and ϕ, with $r = \sin\chi$ ($0 \le r \le 1$). In elliptical space r runs through this range once: in spherical space it does so twice:

$$d\sigma^2 = S^2\left(\frac{dr^2}{1 - r^2} + r^2(d\theta^2 + \sin^2\theta\, d\phi^2)\right). \tag{3.18}$$

The constant S is called the 'radius' of the universe. The line element for the Einstein universe is therefore given by

$$ds^2 = c^2\, dt^2 - d\sigma^2$$
$$= c^2\, dt^2 - S^2[d\chi^2 + \sin^2\chi(d\theta^2 + \sin^2\theta\, d\phi^2)]$$
$$= c^2\, dt^2 - S^2\left(\frac{dr^2}{1 - r^2} + r^2(d\theta^2 + \sin^2\theta\, d\phi^2)\right). \tag{3.19}$$

Note that we have derived the line element (3.19) entirely from the various assumptions of symmetry. The field equations have not yet been used. We will now see what happens when we use the above line element to compute the left-hand sides of Einstein's equations.

This is easily done with the machinery developed in Chapter 2. We write $x^0 = ct$, $x^1 = r$, $x^2 = \theta$ and $x^3 = \phi$, so that

$$g_{00} = 1, \qquad g_{11} = -\frac{S^2}{1 - r^2}, \qquad g_{22} = -S^2 r^2, \qquad g_{33} = -S^2 r^2 \sin^2 \theta,$$

$$g^{00} = 1, \qquad g^{11} = -\frac{1 - r^2}{S^2}, \qquad g^{22} = -\frac{1}{S^2 r^2}, \qquad g^{33} = -\frac{1}{S^2 r^2 \sin^2 \theta}.$$

Elementary calculus then tells us that the only non-zero components of Γ^i_{kl} are the following:

$$\Gamma^1_{11} = \frac{r}{1 - r^2}, \qquad\qquad \Gamma^2_{12} = \Gamma^3_{13} = \frac{1}{r}, \qquad\qquad \Gamma^1_{22} = -r(1 - r^2),$$

$$\Gamma^1_{33} = -r(1 - r^2) \sin^2 \theta, \qquad \Gamma^2_{33} = -\sin\theta\cos\theta, \qquad \Gamma^3_{23} = \cot\theta.$$

Next, using the formulae given in the last chapter, we find the following non-zero components of the Einstein tensor:

$$R^0_0 - \frac{1}{2} R = -\frac{3}{S^2}, \tag{3.20}$$

$$R^1_1 - \frac{1}{2} R = R^2_2 - \frac{1}{2} R = R^3_3 - \frac{1}{2} R = -\frac{1}{S^2}. \tag{3.21}$$

To complete the field equations, Einstein used the energy tensor for dust derived in (2.85). For dust at rest in the above frame of reference u^i has only one non-zero component, the time component. We therefore get

$$T^0_0 = \rho_0 c^2,$$
$$T^1_1 = T^2_2 = T^3_3 = 0. \tag{3.22}$$

Thus the two equations (3.20) and (3.21) lead to two independent equations:

$$-\frac{3}{S^2} = -\frac{8\pi G}{c^2} \rho_0, \qquad -\frac{1}{S^2} = 0. \tag{3.23}$$

Clearly no sensible solution is possible from these equations, thus suggesting that no static homogeneous isotropic and dense model of the universe is possible under the regime of Einstein equations stated in (2.100).

It was his inability to generate such a model that led Einstein to modify his equations from (2.100) to (2.104), thus introducing the now famous (or infamous) λ-term. If we introduce this additional constant into the picture, our equations in (3.23) are modified to

$$\lambda - \frac{3}{S^2} = -\frac{8\pi G}{c^2}\rho_0 \tag{3.24}$$

and

$$\lambda - \frac{1}{S^2} = 0. \tag{3.25}$$

We now do have a sensible solution. We get

$$S = \sqrt{\frac{1}{\lambda}} = \frac{c}{2\sqrt{\pi G \rho_0}}. \tag{3.26}$$

Einstein considered this solution as justifying his conjecture that, with sufficiently high density, it should be possible to 'close' the universe. In (3.26) we have the radius S of the universe given by the density of matter ρ_0, with the result that the larger the value of ρ_0 the smaller the value of S. However, if λ is a given universal constant like G, both ρ_0 and S are determined in terms of λ (as well as G and c). How big is λ?

In 1917 very little information was available about ρ_0, from which λ could be determined. The value of

$$S \approx 10^{26}\text{--}10^{27} \text{ cm}$$

quoted in those days is therefore only of historical interest. If we take ρ_0 as $\sim 10^{-31}$ g cm^{-3} as the rough estimate of mass density in the form of galaxies (see Chapter 9), we get $S \approx 10^{29}$ cm and $\lambda \approx 10^{-58}$ cm^{-2}.

The λ-term introduces a force of repulsion between two bodies that increases in proportion to the distance between them. The above value of λ is too small to make any detectable difference from the prediction of standard general relativity (that is, with $\lambda = 0$) in any of the Solar-System tests mentioned in Chapter 2. Thus the Einstein universe faced no threat from the local tests of gravity. The model, however, did not survive much longer than a decade, for reasons discussed next.

3.4　　The expanding universe

In the late nineteenth century the philosopher and scientist Ernst Mach raised certain conceptual objections to Newton's laws of motion. Mach critically examined the role of a background against which motion is to be measured and argued that, unless there is a material background, it is not possible to attach any meaning to the concepts of rest and motion. Einstein was greatly influenced by Mach's discussion. The Einstein universe described above includes matter-filled space and thus a background of distant matter against which a local observer can measure motion and formulate laws of mechanics. In fact, as we have just seen, the density of matter determines the precise geometrical nature of spacetime in the Einstein model.

Einstein believed this to be a unique feature of general relativity. He argued that the presence of matter was essential for a meaningful spacetime geometry. However, his expectation that general relativity can yield only such matter-filled spacetimes as solutions of the field equations was proved wrong shortly after the publication of his paper in 1917. For, a few months later in the same year, W. de Sitter (Figure 3.6) published another solution of the field equations in (2.104) with the line element given by

$$ds^2 = c^2\left(1 - \frac{H^2 R^2}{c^2}\right)dT^2 - \frac{dR^2}{1 - \frac{H^2 R^2}{c^2}} - R^2(d\theta^2 + \sin^2\theta\, d\phi^2)$$

(3.27)

where H is a constant related to λ by

$$\lambda = 3H^2/c^2.$$

(3.28)

The remarkable feature of the de Sitter universe is that *it is empty*. Moreover, although the above coordinates give the impression that the universe is static, it is possible to find a new set of coordinates (t, r, θ, ϕ) in terms of which the line element (3.27) takes the dynamical form

Figure 3.6 W. de Sitter (1872–1934).

$$ds^2 = c^2\,dt^2 - e^{2Ht}[dr^2 + r^2(d\theta^2 + \sin^2\theta\,d\phi^2)]. \tag{3.29}$$

It is easy to verify that test particles with constant values of (r, θ, ϕ) follow timelike geodesics in this model. Thus the proper separation between any two particles measured at a given time t increases with time as e^{Ht}. That is, these particles are all moving apart from one another.

However, these particles have no material status. They have no masses and they do not influence the geometry of spacetime. In the dynamical sense the universe is empty, although in the kinematical sense it is expanding. As Eddington once put it, the de Sitter universe has motion without matter, in contrast to the Einstein universe, which has matter without motion.

The de Sitter universe showed, however, that empty spacetimes could be obtained as solutions of general relativity. For reasons discussed above, a universe of this type fails to satisfy Mach's criterion that there should be a background of distant matter against which local motion can be measured. Although the property of emptiness of the de Sitter universe was embarrassing, its property of expansion turned out to contain the germ of the truth. For, by the end of the third decade of this century, the observations of Hubble and Humason indicated that the universe is not static but is indeed expanding.

Chapter 1 summarized these observations. The phenomenon of nebular redshift observed by Hubble and Humason in the 1920s has now been observed for practically all extragalactic objects. As mentioned in §1.8, a Newtonian interpretation of such redshifts involves the Doppler effect. How can we express this phenomenon in the language of general relativity? Can we generate models of the universe that combine de Sitter's notion of expansion with Einstein's notion of non-emptiness? The Friedmann models to be discussed in Chapter 4 do just that and were in fact obtained by Alexander Friedmann between 1922 and 1924, seven years *before* Hubble's data became well known. Later Abbé Lemaître in 1927 independently obtained models similar to Friedmann's. However, until the impact of Hubble's observations of 1929, these ideas remained largely unrecognized.

The rest of this chapter outlines the kinematical features of the expanding models of the universe. We will first describe how to generalize the arguments that led Einstein to the static line element (3.19). This generalization will lead us to a non-static line element that preserves the properties of homogeneity and isotropy assumed by Einstein, but that is potentially capable of explaining Hubble's data.

3.5 Simplifying assumptions of cosmology

Once we decide to generalize from a static to a non-static model of the universe, our task becomes more complicated. Figure 3.7(a) shows a spacetime diagram with a swarm of world lines representing particles moving in arbitrary ways. There is no order in this picture and, where two world lines intersect, we have colliding particles. It would indeed be very difficult to solve the Einstein field equations for such a mess of gravitating matter. Fortunately, the real universe does not appear to be so messy.

Hubble's observations indicate that the universe is (or at least seems to be) an orderly structure in which the galaxies, considered as basic units, are moving apart from one another. Figure 3.7(b) represents a typical spacetime section of the universe in which the world lines represent the histories of galaxies. These world lines, unlike those of Figure 3.7(a), are not intersecting and form a funnel-like structure in which the separation between any two world lines is steadily increasing. One may compare Figure 3.7(b) with that of the disciplined march of an army unit; and Figure 3.7(a) with a jostling mob after a rowdy football match.

3.5.1   Weyl's postulate

This intuitive picture of regularity is often expressed formally as the *Weyl postulate*, after the early work of the mathematician Hermann Weyl. The postulate states that

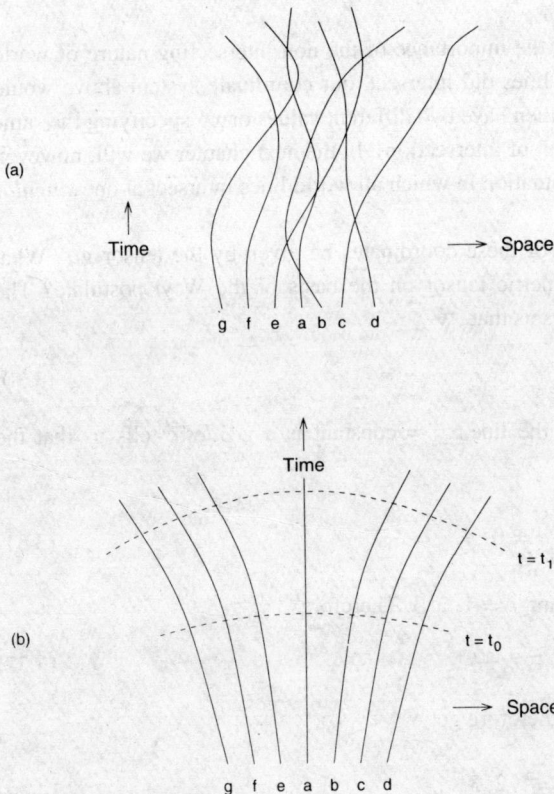

Figure 3.7 (a) An arbitrary bundle of world lines a, b, c, ... describes particles moving haphazardly. Intersecting world lines denote interparticle collisions. (b) Particles move along non-intersecting world lines a, b, c, ... that have no wobbles or irregularities. This is the regularity expressed formally by the Weyl postulate. Note that this regularity allows us to construct a sequence of spacelike hypersurfaces orthogonal to the world lines of the bundle. These are hypersurfaces of constant cosmic time t. Thus the cosmologist can talk of cosmic epochs $t = t_0$, $t = t_1$ and so on in an unambiguous fashion.

the world lines of galaxies form a 3-bundle of non-intersecting geodesics orthogonal to a series of spacelike hypersurfaces.

To appreciate the full significance of Weyl's postulate, let us try to express it in terms of the coordinates and metric of spacetime. Accordingly we use three spacelike coordinates $x^{\mu}(\mu = 1, 2, 3)$ to label a typical world line in the 3-bundle of galaxy world lines. Furthermore, let the coordinate x^0 label a typical member of the series of spacelike hypersurfaces mentioned above. Thus

$$x^0 = \text{constant}$$

is a typical spacelike hypersurface orthogonal to the typical world line given by

$$x^{\mu} = \text{constant}.$$

Although in practice the galaxies form a discrete set, we can extend the discrete set (x^{μ}) to a continuum by the *smooth-fluid approximation*. This approximation is none other than the widely used device of going over from a discrete distribution of particles to a continuum density distribution. In this case we can treat the quantities x^{μ} as forming a continuum together with x^0 and use them as the four coordinates x^i to describe space and time.

It is worth emphasizing the importance of the non-intersecting nature of world lines. If two galaxy world lines did intersect, our coordinate system above would break down, for we would then have two different values of x^{μ} specifying the same point in spacetime (the point of intersection). In the next chapter we will, however, encounter an exceptional situation in which all world lines intersect at one *singular* point!

Let the metric in terms of these coordinates be given by the tensor g_{ik}. What can we assert about this metric tensor on the basis of the Weyl postulate? The orthogonality condition tells us that

$$g_{0\mu} = 0. \tag{3.30}$$

Furthermore, the fact that the line $x^{\mu} = \text{constant}$ is a *geodesic* tells us that the geodesic equations

$$\frac{d^2 x^i}{dx^2} + \Gamma^i_{kl} \frac{dx^k}{ds} \frac{dx^l}{ds} = 0 \tag{3.31}$$

are satisfied for $x^i = \text{constant}$, $i = 1, 2, 3$. Therefore

$$\Gamma^{\mu}_{00} = 0, \qquad \mu = 1, 2, 3. \tag{3.32}$$

From (3.30) and (3.32) we therefore get

$$\frac{\partial g_{00}}{\partial x^{\mu}} = 0, \qquad \mu = 1, 2, 3. \tag{3.33}$$

Thus g_{00} depends on x^0 only. We can therefore replace x^0 by a suitable function of x^0 to make g_{00} constant. Hence we take, without loss of generality,

$$g_{00} = 1. \tag{3.34}$$

The line element therefore becomes

$$
\begin{aligned}
\mathrm{d}s^2 &= (\mathrm{d}x^0)^2 + g_{\mu\nu}\,\mathrm{d}x^\mu\,\mathrm{d}x^\nu \\
&= c^2\,\mathrm{d}t^2 + g_{\mu\nu}\,\mathrm{d}x^\mu\,\mathrm{d}x^\nu,
\end{aligned} \tag{3.35}
$$

where we have put $ct = x^0$. This time coordinate is called the *cosmic time*. It is easily seen that the spacelike hypersurfaces in Weyl's postulate are the surfaces of simultaneity with respect to the cosmic time. Moreover, t is the proper time kept by any galaxy.

3.5.2 The cosmological principle

The second important assumption of cosmology is embodied in the *cosmological principle*. This principle states that, at any given cosmic time, the universe is homogeneous and isotropic. That is, the surfaces $t = $ constant exhibit the properties discussed earlier in connection with the Einstein universe. There we saw that the three-dimensional surface S_3 of a hypersphere has the requisite properties of homogeneity and isotropy. Is this the only alternative available?

Einstein, as we saw earlier, selected this alternative because he believed space to be closed. However, if we do not insist on closed space, two more alternatives are available to us, which can be seen in the following way. First let us consider an analogy in lower dimensions.

Figure 3.8 shows three surfaces. Figure 3.8(a) shows a section of the Euclidean plane, Figure 3.8(b) a spherical surface and Figure 3.8(c) a saddle-shaped surface. Suppose that we try to cover these surfaces with a plain sheet of paper. We will find that our sheet fits exactly and smoothly on the plane surface. If we try to cover the spherical surface, the sheet of paper develops wrinkles, indicating that the sheet of paper has area in excess of that needed to cover the surface. Similarly, in trying to cover the saddle our paper will be torn, being short of the necessary covering area. These differences can be expressed in differential geometry by the notion of curvature. The plain surface has zero curvature, the spherical surface has positive curvature and the saddle has negative curvature. Our paper-covering experiment tells us in general whether a given surface has a zero, positive, or negative curvature. These ideas can be extended to higher dimensions as well.

In the Einstein universe the space sections were the 3-surfaces of hyperspheres and hence they had a constant *positive curvature*. The constancy of curvature is necessary in order to ensure that the properties of homogeneity and isotropy hold;

for, if the curvature of space differs from place to place, physical measurements to detect the differences could be devised. We can similarly get other homogeneous and isotropic spaces by considering them as 3-surfaces of *constant negative curvature* or of *zero curvature*.

In terms of the Cartesian coordinates x_1, x_2, x_3 and x_4 used earlier, a 3-surface of constant negative curvature is given by an equation of the form

$$x_1^2 + x_2^2 + x_3^2 - x_4^2 = -S^2,$$

(3.36)

where S is a constant. The substitution

$$x_1 = S \sinh \chi \cos \theta, \qquad x_2 = S \sinh \chi \sin \theta \cos \phi,$$

(3.37)

$$x_3 = S \sinh \chi \sin \theta \sin \phi, \quad x_4 = S \cosh \chi$$

(3.38)

gives

$$dx_1^2 + dx_2^2 + dx_3^2 - dx_4^2 = S^2[d\chi^2 + \sinh^2 \chi (d\theta^2 + \sin^2 \theta \, d\phi^2)].$$

(3.39)

Notice the minus sign in front of dx_4^2. It means that we are embedding our 3-surface not in a Euclidean space but in a pseudo-Euclidean space. For example, Minkowski space is pseudo-Euclidean.[1] If we further substitute

$$r = \sinh \chi,$$

(3.40)

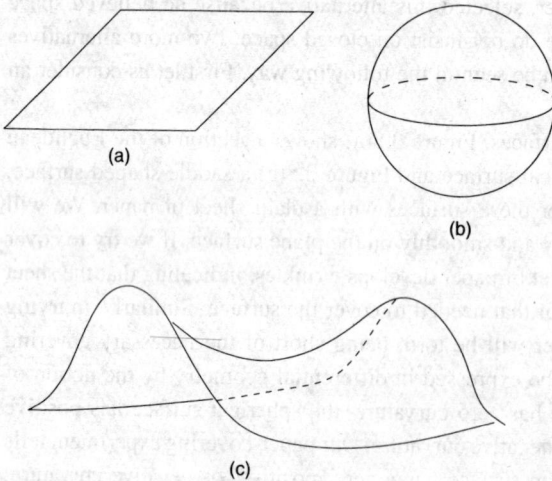

Figure 3.8 Examples of surfaces of (a) zero curvature (b) positive curvature and (c) negative curvature.

(a)

(b)

(c)

---

[1] In Euclidean space the Pythagoras theorem holds with the line element given by $ds^2 = dx_1^2 + dx_2^2 + dx_3^2 + \cdots$. If some of the + signs on the right-hand side are changed to − signs, the resulting space is called a pseudo-Euclidean space.

(3.39) becomes

$$d\sigma^2 = S^2\left(\frac{dr^2}{1+r^2} + r^2(d\theta^2 + \sin^2\theta\, d\phi^2)\right). \tag{3.41}$$

Compare this with the expression (3.18) for the space of positive curvature:

$$d\sigma^2 = S^2\left(\frac{dr^2}{1-r^2} + r^2(d\theta^2 + \sin^2\theta\, d\phi^2)\right). \tag{3.42}$$

These two expressions can be combined into a single expression by introducing a parameter k that takes the values ± 1:

$$d\sigma^2 = S^2\left(\frac{dr^2}{1-kr^2} + r^2(d\theta^2 + \sin^2\theta\, d\phi^2)\right). \tag{3.43}$$

Notice that, if we set $k = 0$, we also get the third alternative – the 3-surface of zero curvature:

$$d\sigma^2 = S^2[dr^2 + r^2(d\theta^2 + \sin^2\theta\, d\phi^2)]. \tag{3.44}$$

The right-hand side of (3.44) is simply the Euclidean line element scaled by the constant factor S.

The constant S can, however, depend on cosmic time, since we were considering a typical $t = $ constant hypersurface in the above argument. Thus the most general line element satisfying the Weyl postulate and the cosmological principle is given by

$$ds^2 = c^2\,dt^2 - S^2(t)\left(\frac{dr^2}{1-kr^2} + r^2(d\theta^2 + \sin^2\theta\, d\phi^2)\right) \tag{3.45}$$

where the 3-spaces $t = $ constant are Euclidean for $k = 0$, closed with positive curvature for $k = +1$ and open with negative curvature for $k = -1$. For reasons that will become clearer later, the scale factor $S(t)$ is often called the *expansion factor*.

The line element (3.45) that we have obtained using partly intuitive and partly heuristic arguments was rigorously derived in the 1930s by H. P. Robertson and A. G. Walker (independently). It is often referred to as the *Robertson–Walker line element*.

The Robertson–Walker line element is sometimes expressed in a slightly different form with the help of the following radial coordinate transformation:

$$\bar{r} = \frac{2r}{1 + \sqrt{1 - kr^2}}. \tag{3.46}$$

We then get the line element as

$$ds^2 = c^2\,dt^2 - \frac{S^2(t)}{\left(1 + k\bar{r}^2/4\right)^2}[d\bar{r}^2 + \bar{r}^2(d\theta^2 + \sin^2\theta\, d\phi^2)]. \tag{3.47}$$

This line element is manifestly isotropic in $\bar{r}$, θ and ϕ. We will, however, continue to use (3.45).

Notice how the simplifying postulates of cosmology have reduced the number of unknowns in the metric tensor from 10 to the single function $S(t)$ and the discrete parameter k that characterize the Robertson–Walker metric. The task of the relativist is now simplified to solving an ordinary differential equation in the independent variable t. We will defer the solution of this problem to the next chapter.

We next consider some of the important observational features of a typical Robertson–Walker spacetime. These features show how a non-Euclidean geometry can substantially alter conclusions based on naive Euclidean concepts. Perhaps this is the right place to alert the reader against the common but inexact practice of referring to the $k = 0$ model as a 'flat model'. The Robertson–Walker spacetime for such a model is *not flat*; only its subspaces $t = $ constant are flat.

3.6   The redshift

Let us first try to understand how the nebular redshift found by Hubble and Humason is accounted for by the Robertson–Walker model. We begin by recalling that the basic units of Weyl's postulate are galaxies with constant coordinates x^μ. We can easily identify the x^μ with the (r, θ, ϕ) of Robertson–Walker spacetime. Thus each galaxy has a constant set of coordinates (r, θ, ϕ). This coordinate frame is often referred to as the *cosmological rest frame*. As observers we are located in our Galaxy, which also has constant (r, θ, ϕ) coordinates.[2] Without loss of generality we can take $r = 0$ for our Galaxy. Although this assumption suggests that we are placing ourselves at the centre of the universe, it does not confer any special status on us. Because of the assumption of homogeneity, *any* galaxy could be chosen to have $r = 0$. Our particular choice is simply dictated by convenience.

Consider a galaxy G_1 at (r_1, θ_1, ϕ_1) emitting light waves towards us. Let us denote by t_0 the present epoch of observation. At what time should a light wave leave G_1 in order to arrive at $r = 0$ at the present time $t = t_0$? To find the answer to this question we need to know the path of the wave from G_1 to us. Since light travels along null geodesics, as described in Chapter 2, we need to calculate the null geodesic from G_1 to us.

From the symmetry of a spacetime we can guess that a null geodesic from $r = 0$ to $r = r_1$ will maintain a constant spatial direction. That is, we expect to have $\theta = \theta_1$ and $\phi = \phi_1$ all along the null geodesic. This guess proves to be correct when we substitute these values into the geodesic equations. Accordingly we will assume that only r and t change along the null geodesic. Next we recall that a first integral of the null geodesic equation is simply $ds = 0$. For the Robertson–Walker line element this gives us

[2] Later on, in Chapter 10, we will show that this remark is only approximately correct, because our Galaxy has a small motion relative to this cosmological frame.

$$c \, dt = \pm \frac{S \, dr}{\sqrt{1 - kr^2}}. \tag{3.48}$$

Since r decreases as t increases along this null geodesic, we should take the minus sign in the above relation. Suppose that the null geodesic left G_1 at time t_1. Then we get from the above relation

$$\int_{t_1}^{t_0} \frac{c \, dt}{S(t)} = \int_0^{r_1} \frac{dr}{\sqrt{1 - kr^2}}. \tag{3.49}$$

Thus, if we know $S(t)$ and k, we know the answer to our question.

However, consider what happens to successive wave crests emitted by G_1. Suppose that the wave crests were emitted at t_1 and $t_1 + \Delta t_1$ and received by us at t_0 and $t_0 + \Delta t_0$, respectively. Then, similarly to (3.49), we have

$$\int_{t_1 + \Delta t_1}^{t_0 + \Delta t_0} \frac{c \, dt}{S(t)} = \int_0^{r_1} \frac{dr}{\sqrt{1 - kr^2}}. \tag{3.50}$$

If $S(t)$ is a slowly varying function so that it effectively remains unchanged over the small intervals Δt_0 and Δt_1, we get by subtraction of (3.49) from (3.50)

$$\frac{c \, \Delta t_0}{S(t_0)} - \frac{c \, \Delta_1}{S(t_1)} = 0;$$

that is,

$$\frac{c \, \Delta t_0}{c \, \Delta t_1} = \frac{S(t_0)}{S(t_1)} \equiv 1 + z. \tag{3.51}$$

It is not difficult to see that the quantity z defined above is the redshift. The term $c \, \Delta t_1$ is the wavelength λ_1 measured by an observer at rest in the galaxy G_1, while $c \, \Delta t_0$ is the wavelength λ_0 measured by an observer at rest in our Galaxy, since, in the Robertson–Walker spacetime, the cosmic time measures the proper time kept by any galaxy. Thus the wavelength of the light wave increases by a fraction z in the transmission from G_1 to us, provided that $S(t_0) > S(t_1)$. In other words, Hubble's observations of the redshift are explained if we assume $S(t)$ to be an increasing function of time.

It is worth comparing the way in which this redshift has been obtained with the Newtonian way we used earlier. Except for the curvature parameter k, which has the value of unity in the Newtonian case, the formula (3.9) of Newtonian cosmology is the same as the formula (3.49) derived above. The subsequent argument leading to the redshift is also the same. However, the rationales in the two cases are different. In the Newtonian case we had considered the passage of a ray of light through an expanding medium with pre-relativistic notions of the change in its velocity if the medium were moving relative to Newtonian absolute space. In a sense it is a version of the Doppler effect.

Our derivation above shows that the effect in the relativistic case arises from the passage of light through a non-Euclidean spacetime. It does *not* arise from the Doppler effect, since in our coordinate frame all galaxies have constant (r, θ, ϕ) coordinates. In a non-Euclidean spacetime it is not possible to attach an unambiguous meaning to the relative velocity of two objects separated by a great distance. People are often tempted to relate z to velocity by the special relativistic relation

$$1 + z = \sqrt{\frac{1 + v/c}{1 - v/c}}. \tag{3.52}$$

Such an interpretation is not valid in our present framework because, as we saw in Chapter 2, special relativity applies only in a local region of spacetime.

It is also necessary to contrast (3.51) to the gravitational redshift described in Chapter 2. The gravitational redshift is characterized by the fact that, if light travelling from object B to object A is redshifted, the light travelling from A to B is blueshifted. In the present case, if light travelling from galaxy A to galaxy B is redshifted, that from B to A will also be redshifted provided that $S(t)$ is increasing during the transmission of light.

We will refer to the present redshift as the *cosmological redshift*. There is a way of unifying all three redshifts under a single banner, however, which is indicated by Exercise 18.

3.7   **Apparent brightness**

The redshift discussed above shows up in the spectrum of a galaxy. The astronomer measures another quantity associated with the galaxy – its apparent brightness. Let us now see how the apparent brightness is related to the luminosity of the galaxy and its distance from us in the expanding universe described by the Robertson–Walker spacetime.

Let L be the total energy emitted by the galaxy G_1 in unit time during the epoch t_1 when light left it in order to reach us in the present epoch t_0. The redshift z of the galaxy is therefore given by (3.51). It is now necessary to specify the wavelength range of observation. To fix our ideas, suppose that the intensity distribution of light from G_1 over wavelengths λ is given by the normalized function $I(\lambda)$. Thus

$$\mathrm{d}L = LI(\lambda)\,\mathrm{d}\lambda \tag{3.53}$$

is the energy emitted by G_1 per unit time over the bandwidth $(\lambda, \lambda + \mathrm{d}\lambda)$. If, instead of wavelengths, we wanted to use frequencies, the corresponding intensity function $J(v)$ is related to $I(\lambda)$ by

$$cJ(v) = \lambda^2 I(\lambda). \tag{3.54}$$

Both $J(\nu)$ and $I(\lambda)$ are used by the astronomer, the choice depending on convenience.

In the case of isotropic emission of light by G_1, by the time its light reaches us it is distributed uniformly across a sphere of coordinate radius r_1 centred on G_1 (see Figure 3.9). What is the proper surface area of this sphere?

In the Robertson–Walker line element, put $t = $ constant and also $r = $ constant to get

$$ds^2 = -r^2 S^2 (d\theta^2 + \sin^2 \theta \, d\phi^2).$$

This is the line element on the surface of a Euclidean sphere of radius rS. Hence the answer to the above question is that light from G_1 is distributed over a total surface area of $4\pi r_1^2 S^2(t_0)$ at time t_0. We may occasionally refer to $rS(t_0)$ as the *proper distance* of a source with coordinate r, during the epoch t_0. We now need to know how much light is received per unit time by us across unit proper area held perpendicular to the line of sight to G_1, over a bandwidth $(\lambda_0, \lambda_0 + \Delta\lambda_0)$. Denote this quantity by $\mathcal{F}(\lambda_0) \, \Delta\lambda_0$.

Note first that, because of the redshift, the light arriving with wavelengths in the range $(\lambda_0, \lambda_0 + \Delta\lambda_0)$ left G_1 in the wavelength range

$$\left(\frac{\lambda_0}{1+z}, \frac{\lambda_0 + \Delta\lambda_0}{1+z} \right).$$

Now the total amount of energy that leaves G_1 between the epochs t_1 and $t_1 + \Delta t_1$ in the above frequency range is

$$LI\left(\frac{\lambda_0}{1+z} \right) \frac{\Delta\lambda_0}{1+z} \Delta t_1.$$

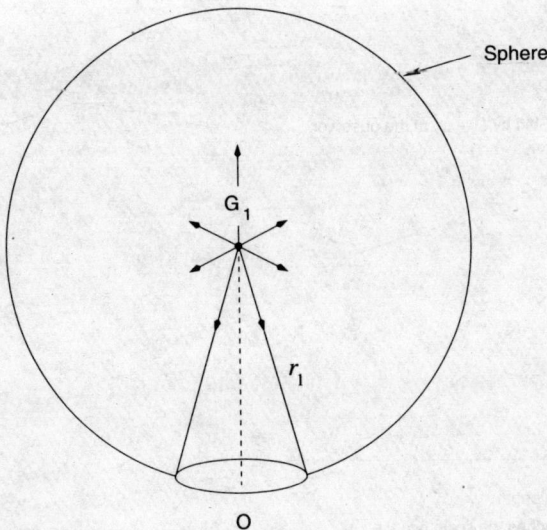

Figure 3.9 The radiation emitted by galaxy G_1 is distributed uniformly across a sphere of coordinate radius r_1 with G_1 as the centre. The observer O (that is, ourselves) located on this sphere would expect to receive a proportionate quantity of this radiation across a unit area held normal to the direction $G_1 O$.

Figure 3.10 illustrates this effect.

How many photons carry the above quantity of energy? For a small enough bandwidth, we may assume that a typical photon had, at emission, the wavelength $\lambda_0/(1+z)$, a frequency $(1+z)c/\lambda_0$ and hence an energy equal to $(1+z)ch/\lambda_0$, where h is Planck's constant. Therefore the required number of photons is

$$\delta\mathcal{N} = LI\left(\frac{\lambda_0}{1+z}\right)\frac{\Delta\lambda_0}{1+z}\frac{\Delta t_1}{(1+z)ch/\lambda_0}$$

$$= \frac{L\lambda_0}{ch}\frac{1}{(1+z)^2}I\left(\frac{\lambda_0}{1+z}\right)\Delta\lambda_0\,\Delta t_1.$$

During the epoch of reception, these photons are distributed across a surface area of $4\pi r_1^2 S^2(t_0)$ and are received over a time interval $(t_0, t_0 + \Delta t_0)$. Thus the number of photons received by us per unit area held normal to the line of sight and per unit time is given by

$$\frac{L\lambda_0}{ch}\frac{1}{(1+z)^2}I\left(\frac{\lambda_0}{1+z}\right)\Delta\lambda_0\frac{\Delta t_1}{\Delta t_0}\frac{1}{4\pi r_1^2 S^2(t_0)}.$$

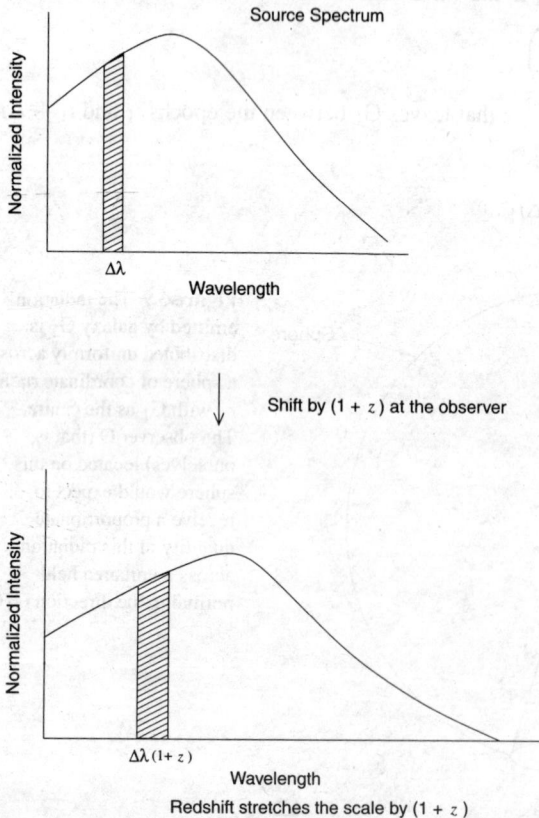

Source Spectrum

Normalized Intensity

$\Delta\lambda$

Wavelength

Shift by $(1 + z)$ at the observer

Normalized Intensity

$\Delta\lambda\,(1+z)$

Wavelength

Redshift stretches the scale by $(1 + z)$

Figure 3.10 The intensity distribution of a galaxy over various wavelengths is modified by the redshift. The effect is like stretching the λ-axis by the factor $1 + z$. To preserve the area under the curve, its height decreases by the same factor.

During this epoch, because of a scaling down of its frequency by redshifting, each photon has been degraded in energy by the factor $(1 + z)^{-1}$. Thus each photon now has the energy ch/λ_0. If we multiply the above expression by this factor, we get the quantity we were after:

$$\mathcal{F}(\lambda_0)\,\Delta\lambda_0 = L\frac{1}{(1+z)^2}\frac{\Delta t_1}{\Delta t_0}I\left(\frac{\lambda_0}{1+z}\right)\frac{1}{4\pi r_1^2 S^2(t_0)}\,\Delta\lambda_0.$$

However, we note from (3.51) that $\Delta t_1/\Delta t_0$ gives us another factor $(1 + z)^{-1}$ in the denominator. Thus finally we get

$$\mathcal{F}(\lambda_0) = \frac{LI(\lambda_0/1+z)}{(1+z)^3 4\pi r_1^2 S^2(t_0)}. \tag{3.55}$$

In terms of frequencies the result is quoted as the *flux density*

$$S(\nu_0) = \frac{LJ[\nu_0(1+z)]}{(1+z)4\pi r_1^2 S^2(t_0)}. \tag{3.56}$$

Here $S(\nu_0)\,\Delta\nu_0$ is the amount of radiation received perpendicular to unit area in unit time across a frequency range $(\nu_0, \nu_0 + \Delta\nu_0)$.

The optical astronomer uses this result in the form (3.55), while the radio-astronomer uses it in the form (3.56). The X-ray astronomer uses energies instead of frequencies, so that (3.56) is scaled by h. We will have occasion to use these expressions when we look at the various observational tests of cosmology. We will end this section by deriving a few results of interest to optical astronomy.

The expression (3.55) integrated over all wavelengths gives

$$\mathcal{F}_{\text{bol}} = \frac{L_{\text{bol}}}{4\pi r_1^2 S^2(t_0)(1+z)^2}, \tag{3.57}$$

where L_{bol} ($=L$) is the absolute *bolometric* luminosity of G_1. $\mathcal{F}_{\text{bol}}$ is correspondingly the apparent bolometric luminosity of G_1. On the logarithmic scale of magnitudes familiar to the optical astronomer, (3.57) becomes

$$m_{\text{bol}} = -2.5 \log(\mathcal{F}_{\text{bol}}/\mathcal{F}_0),$$
$$M_{\text{bol}} = -2.5 \log(L_{\text{bol}}/L_\odot) + 4.75,$$
$$m_{\text{bol}} - M_{\text{bol}} = 5 \log D_1 - 5, \tag{3.58}$$

where

$$\mathcal{F}_0 = 2.48 \times 10^{-5} \text{ erg cm}^{-2} \text{ s}^{-1},$$
$$L_\odot = \text{solar luminosity} = 2 \times 10^{33} \text{ erg s}^{-1},$$
$$D_1 = r_1 S(t_0)(1+z). \tag{3.59}$$

D_1 is called the *luminosity distance* of G_1. If we are interested in a magnitude defined for a particular waveband around λ_0, say, we may similarly use (3.55) in the logarithmic form with the apparent magnitude defined by

$$m(\lambda_0) = -2.5 \log \mathcal{F}(\lambda_0) + \text{constant},$$

the constant depending on the filter used to select that waveband. It is customary to indicate the filter by a suffix attached to m. Thus m_{pg} stands for photographic magnitude, m_v for visual magnitude, m_b for blue magnitude and so on.

Note, however, that, because of the redshift, the astronomer has to apply a correction to include the effect of the term $I(\lambda_0/1 + z)$. Thus an astronomer using a red filter may be actually receiving the photons that originated in the blue part of the spectrum of G_1 if $z \approx 1$. This correction, which is crucial to many cosmological observations, is called the *K correction*.

Example

Suppose that the spectrum of a class of sources in the optical part has the form $J(\nu)\,\Delta\nu = K\nu^2\,\Delta\nu$. Let a source P be in our local neighbourhood while source Q has redshift $z = 1$. When we compare the apparent magnitudes of the two sources in the wavelength band of 700 nm, say, then the source Q will have an extra brightness because its spectrum is redshifted from the wavelength 350 nm at which the spectral factor $\nu^2\,\Delta\nu$ will be multiplied by $(1 + z)^3 = 8$. If this spectral effect is not taken into consideration and formula (3.57) is used, there will be an error in estimating the luminosity of the source.

3.8   Hubble's law

Hubble's law was derived for galaxies with low redshifts. The largest redshift in Hubble's 1929 paper was $z \cong 0.003$. At these small redshifts we can use the Taylor expansion to derive a simple linear relation between D_1 and z, the relation arrived at by Hubble from his early observations. The calculation is quite analogous to that which we used in Newtonian cosmology:

$$D_1 = r_1 S(t_0). \tag{3.60}$$

We also get, by the Taylor expansion of (3.49),

$$\int_0^{r_1} \frac{dr}{\sqrt{1 - kr^2}} \approx r_1, \tag{3.61}$$

$$\int_{t_1}^{t_0} \frac{c\,dt}{S(t)} \approx \frac{c(t_0 - t_1)}{S(t_0)}, \tag{3.62}$$

$$S(t_1) \approx S(t_0) - (t_0 - t_1)\left(\frac{\dot{S}}{S}\right)_{t_0} S(t_0), \tag{3.63}$$

$$S(t_1) = \frac{S(t_0)}{1+z} \approx S(t_0)(1-z). \tag{3.64}$$

From these relations and from (3.60) we get

$$D_1 \approx r_1 S(t_0) \approx c(t_0 - t_1)$$

$$\approx \left[\left(\frac{\dot{S}}{S}\right)_{t_0}\right]^{-1} cz, \tag{3.65}$$

which can be expressed in the form

$$cz = H_0 D_1, \tag{3.66}$$

with H_0, *the Hubble constant*, given by

$$H_0 = \left(\frac{\dot{S}}{S}\right)_{t=t_0}. \tag{3.67}$$

From a Doppler-shift point of view, cz may be identified with the velocity of recession at small z. In this form (3.66) is sometimes called the *velocity–distance relation*. Expressed as part of the velocity–distance relation, the Hubble constant has the unit of velocity per unit distance, the most common unit in usage being kilometres per second per megaparsec. In many calculations of observational and physical cosmology we shall use

$$H_0 = h_0 \times 100 \, \text{km s}^{-1} \, \text{Mpc}^{-1}. \tag{3.68}$$

Although Hubble originally obtained $h_0 \sim 5.3$, the present estimate of h_0 is much lower. It is still uncertain and, until recently, was believed to lie in the range $0.5 \leq h_0 \leq 1$. Observations with the Hubble Space Telescope (HST) and some ground-based telescopes have narrowed this range down to around [0.55–0.75]. We will discuss in Chapter 10 how modern techniques arrive at the above result.

Another useful way of expressing H_0 is in units of reciprocal time; that is, by expressing

$$\tau_0 = H_0^{-1} \tag{3.69}$$

in units of time. A good time unit for τ_0 is the gigayear (Gyr). The present estimate of τ_0 is in the range of approximately 9–18 Gyrs.

3.9   Angular size

Figure 3.11 illustrates a somewhat unusual effect of the non-Euclidean geometry of the Robertson–Walker spacetime. We consider our Galaxy G_1 to have a linear extent d, as shown in Figure 3.11. What angle does this length d subtend at our location?

To decide the answer to this question, consider two neighbouring null geodesics (representing rays of light) from the two points A and B at the two extremities of G_1 directed towards our Solar System. Without loss of generality we can choose our angular coordinates such that A has the coordinates (θ_1, ϕ_1), while B has the coordinates $(\theta_1 + \Delta\theta_1, \phi_1)$. (Although we have used homogeneity to take $r = 0$ at our location, we can also use isotropy to choose any particular direction as the polar axis $\theta = 0, \theta = \pi$.)

According to the Robertson–Walker line element, the proper distance between A and B is obtained by putting $t = t_1 = $ constant, $r = r_1 = $ constant, $\phi = \phi_1 = $ constant and $d\theta = \Delta\theta_1$ in (3.45). We then get

$$ds^2 = -r_1^2 S^2(t_1)(\Delta\theta_1)^2 = -d^2,$$

since, in the rest frame of G_1 the spacelike separation AB $= d$. Thus

$$\Delta\theta_1 = \frac{d}{r_1 S(t_1)} = \frac{d(1+z)}{r_1 S(t_0)} \tag{3.70}$$

gives the answer to our question.

Notice that, as r_1 increases, we are looking at more and more remote galaxies, which must therefore be seen during earlier and earlier epochs t_1. However, in an expanding universe $S(t_1)$ was smaller during earlier epochs t_1, so it is not obvious that $r_1 S(t_1)$ should get progressively larger as we look at more and more remote galaxies. The effect can be ascribed to 'gravitational bending or lensing' of light as it passes through curved spacetime. Clearly, we need to know how fast $S(t_1)$ decreases as r_1 increases. Although (3.49) provides the answer in an implicit form, we still need to know $S(t)$ in order to be able to perform these integrations. All that we can say at present is that the observed angular size of a galaxy need not be a monotonically decreasing function of its distance away from us, as it would be in a Euclidean universe.

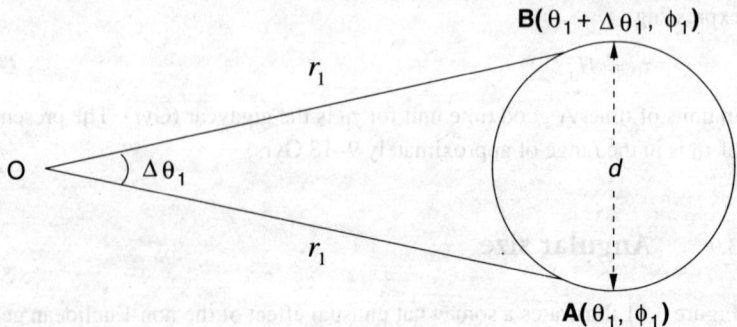

Figure 3.11 The angle subtended by galaxy G_1 at the observer O.

3.10   Surface brightness

A related result is the variation of apparent surface brightness with redshift. We may look upon surface brightness as the average apparent brightness of the source per unit angular surface area of its image. Although we need to take note of the spectrum of the source (for K correction), let us see the effect in terms of the apparent bolometric brightness.

Formula (3.57) gives us the apparent bolometric luminosity, whereas formula (3.70) gives us its angular diameter. On relating the surface brightness to

$$B = \frac{\mathcal{F}_{bol}}{(\pi/4)\,\Delta\theta^2} \qquad (3.71)$$

we get

$$B = \text{constant} \times (1+z)^{-4}. \qquad (3.72)$$

Notice that B only depends on the redshift in a rather simple way and this can be taken as a signature of the expanding-universe hypothesis. For there exist other explanations of the redshift, not all of which predict a dependence of surface brightness as per formula (3.72). Hence one may look for an observationally determined B–z relation to test the expanding-universe hypothesis.

3.11   Source counts

The distribution of discrete luminous sources out to great distances may give indications that spacetime geometry is non-Euclidean. How does the number of galaxies out to coordinate distance r_1 (that is, out to the distance of galaxy G_1) increase with r_1? Let us suppose that, during any epoch t, there are $n(t)$ galaxies in a *unit comoving coordinate volume* (using the r, θ, ϕ coordinates). The word 'comoving' indicates that, although the galaxies individually retain the same coordinates (r, θ, ϕ), the proper separation between any two of them during any epoch increases with epoch according to the scale factor $S(t_1)$. Thus the proper volume of any region bounded by such galaxies increases as S^3.

When we observe galaxies at radial coordinates between r and $r + dr$, we see them at times in the range $t, t + dt$, where, from (3.49),

$$\int_t^{t_0} \frac{c\,dt'}{S(t')} = \int_0^r \frac{dr'}{\sqrt{1 - kr'^2}}. \qquad (3.73)$$

The number of galaxies seen in this shell is therefore

$$dN = \frac{4\pi r^2\,dr}{\sqrt{1 - kr^2}} n(t), \qquad (3.74)$$

where t is related to r through (3.73). Thus the required number of galaxies out to $r = r_1$ is given by

$$N(r_1) = \int_0^{r_1} \frac{4\pi r^2 n(t)\, dr}{\sqrt{1 - kr^2}}. \tag{3.75}$$

If no galaxies are created or destroyed between $r = 0$ and $r = r_1$, we may take $n(t) = $ constant and the integral can be evaluated explicitly. Clearly, the answer must depend on the parameter k. The function $N(r_1)$ apparently increases faster than the Euclidean form $(\propto r_1^3)$ in closed universes $(k = +1)$ and slower than this form in open universes of negative curvature. In physical terms, however, it means the following. If we draw a sphere whose surface lies at a proper distance R from the centre in the $k = 0$ (Euclidean) space, its volume will be $4\pi R^3/3$. However, a similar sphere drawn in the $k = +1$ (closed) space will have a volume *less* than $4\pi R^3/3$, whereas a sphere drawn in the $k = -1$ (open) space will have a volume exceeding this value. In Chapter 11 we will cast this idea in a somewhat different form to make it suitable for observations of galaxies, radio sources and quasars.

Having discussed some of the general properties of the Robertson–Walker universes, it is now appropriate for us to turn to specific models – the models first considered by Friedmann, which are described in Chapter 4. These models will provide us with information missing so far, viz., the value of k and the function $S(t)$.

Exercises

1 Express the velocity–distance tensor $H_{\mu\nu}$ in the Newtonian cosmology as the sum of an isotropic tensor, a symmetric tensor with zero trace and an antisymmetric tensor. Relate these three tensors to (i) expansion, (ii) shear and (iii) spin.

2 In a Newtonian cosmological model, $S(t) \propto t^{2/3}$ and H_0 is the present measured value of Hubble's constant. Show that, at a time $2/(3H_0)$ ago, the entire framework of fundamental observers was concentrated in a point.

3 Verify by direct substitution that (3.16) follows from (3.15).

4 Show that the volume of the Einstein universe is $2\pi^2 S^3$. Comment on the statement that this universe is 'finite but unbounded'. Does this volume refer to spherical or to elliptical space?

5 A ray of light is emitted in a given direction in the Einstein universe. How long will the ray take to make one circuit about the universe and return to its starting point?

6 Using the metric components and the Christoffel symbols given in the text, verify the relations (3.20) and (3.21).

7 Taking $\rho_0 = 10^{-31}$ g cm^{-3}, calculate the radius of the Einstein universe and its total mass in spherical space.

8 With the density given in Exercise 7, calculate the λ-term and estimate the fraction by which the Sun's attraction for the Earth is reduced because of the λ repulsion. Comment on the effect of this force on the experimental tests of general relativity.

9 Given that (3.29) is the coordinate transform of (3.27), find the transformation law between (R, T) and (r, t).

10 Comment on why we cannot look upon the de Sitter universe as a static universe in spite of its apparently static line element (3.27).

11 The de Sitter universe has an *event horizon* in the following sense. If test particles with constant r, θ and ϕ emit light signals towards the origin $r = 0$, then, at a given time t, there is a critical value r_0 such that signals from all particles with $r \geq r_0$ emitted at t will never reach their destination. Calculate r_0.

12 Suppose that we drop the condition of orthogonality of the surfaces $t =$ constant with respect to the geodesics $x^\mu =$ constant from the Weyl postulate, so that (3.30) does not hold. Show that $g_{0\mu}$ must be independent of t.

13 By calculating the 3-volume of space within the coordinate region $r =$ constant in the spaces with the spatial line element

$$d\sigma^2 = S^2 \left(\frac{dr^2}{1 - kr^2} + r^2(d\theta^2 + \sin^2\theta \, d\phi^2) \right), \qquad k = 0, 1, -1$$

develop the three-dimensional analogue of the experiment for covering the surfaces of zero, positive and negative curvature described in the text.

14 Derive from first principles the coordinate transformation $r = f(\bar{r})$ that takes the Robertson–Walker line element from the form (3.45) to the form (3.47).

15 Determine the affine parameter for the radial null geodesic from galaxy G_1 to the origin $r = 0$ in Robertson–Walker spacetime.

16 A particle of mass m is fired from a galaxy at $t = t_0$ with a linear momentum P_0. Show that the momentum of the particle when it reaches another galaxy during a later epoch t (measured in the rest frame of that galaxy) is given by

$$P = P_0 \frac{S(t_0)}{S(t)}.$$

Compare this result with the cosmological redshift for photons.

17 A galaxy, instead of strictly following a typical Weyl geodesic, has a small random velocity relative to it. Use the non-relativistic version of Exercise 16 to find out how this velocity decreases with time.

18 Take a galaxy G_1 at (r_1, θ, ϕ) as a fundamental observer and write u_1^k as its velocity vector in the Robertson–Walker frame. Consider parallel propagation of this vector (see Chapter 2) along the null ray connecting the galaxy to the observer O at the origin during the present epoch t_0 of observation. Let this vector be v_1^k at O. This represents

a radial velocity relative to the cosmological rest frame at O. Use the Doppler effect to work out the redshift for this motion and show that it is none other than z given by the formula (3.51). You can do this exercise for the Schwarzschild line element of Chapter 2 and you can show that the gravitational redshift can also be understood as a Doppler effect for the parallelly transported velocity vector of the source along the null geodesic to the observer.

19 In a universe with $S(t) \propto t^{2/3}$ and $k = 0$, a galaxy is observed to have a redshift $z = 1.25$. How long has light taken to travel from that galaxy to us? Express your answer in units of τ_0.

20 How will the forms (3.55) and (3.56) look if the spectrum of the emitting source is given by $J(\nu) \propto \nu^{-\alpha}$, $\alpha =$ constant in the relevant range of observations?

21 For $S \propto \exp(H_0 t)$ and $k = 0$, $H_0 =$ constant, show that (3.57) takes the form

$$\mathcal{F}_{\text{bol}} = \frac{L_{\text{bol}}}{4\pi (c/H_0)^2 z^2 (1 + z)^2}.$$

22 Calculate the redshift–magnitude relation for bolometric magnitudes in the universe of Exercise 21.

23 Work out the formula (3.57) for the universe with $S \propto t^{2/3}$ and $k = 0$ and compare your result with the result of Exercise 21. In which model is the galaxy apparently brighter?

24 If the Hubble constant is given by h_0 in the units of 100 km s^{-1} Mpc^{-1}, show that $\tau_0 \simeq 9.8 h_0^{-1}$ billion years.

25 Show that, if $k = 0$, $S \propto t^{2/3}$, the apparent angular sizes of distant objects of the same linear size have a minimum at $z = 1.25$.

26 Repeat Exercise 25 for the universe with $k = 0$, $S \propto \exp(H_0 t)$. At what redshift does the minimum apparent angular size lie?

27 If in (3.75) $k = 1$ and $n(t)$ is constant and equal to n_0, show that the number of galaxies in the entire universe is given by $2\pi^2 (c/H_0)^3 n_0$. Clarify whether this answer refers to spherical space or to elliptical space.

28 On a unit sphere mark out the circle of distance r from a point O, r being measured along the arcs of great circles from O. Show that the area of this circle is $4\pi \sin^2 r/2$, which is less than πr^2, the Euclidean circle. Find the three-dimensional analogues of this result for the $k = +1$ and -1 universes.

29 In (3.75), put $k = -1$ and $n(t) = n_0$ (constant) and show that, for $S(t) = ct$, the number of galaxies with redshifts less than z is given by

$$N(z) = \pi \left(\frac{c}{H_0}\right)^3 n_0 \left(\frac{(z^2 + 2z)(z^2 + 2z + 2)}{2(1 + z)^2} - 2\ln(1 + z)\right).$$

Chapter 4

The Friedmann models

4.1   Introduction

The work covered in Chapter 3 did not tell us two important items of information about the universe: (1) the rate at which it expands as given by the function $S(t)$, and (2) whether its spatial sections $t = $ constant are open or closed as indicated by the parameter k. To find answers to these questions it is necessary to go beyond the Weyl postulate and the cosmological principle. We require a dynamical theory that tells us how the scale factor is determined by the matter/radiation contents of the universe. Although we have already developed a relativistic framework in which to express these ideas, it is worthwhile first to see the problem from a Newtonian perspective. The Newtonian problem is intuitively easy to understand and, as we will see thereafter, it gives results very similar to those of the more elaborate and rigorously sounder relativistic models.

4.1.1   Newtonian cosmology

Continuing from the second section of Chapter 3, we will expand the Newtonian framework in order to put in dynamical effects. We will assume that matter in the universe is in the form of 'dust'. By dust we mean pressureless fluid, an idealization that we will justify when we consider the relativistic models. Thus we have a typical fluid element containing density ρ of matter with a bulk velocity $\mathbf{v}$, given by the Hubble law

$$\mathbf{v} = H(t)\mathbf{r}, \qquad H(t) = \dot{S}/S. \tag{4.1}$$

The continuity equation of fluid mechanics then gives

$$\frac{\partial \rho}{\partial t} + \text{div}(\rho \mathbf{v}) = 0, \tag{4.2}$$

i.e.,

$$\frac{\partial \rho}{\partial t} + \rho \, \text{div} \, \mathbf{v} = 0. \tag{4.3}$$

However, from (4.1), div $\mathbf{v} = 3H(t)$, which leads to

$$\frac{\partial \rho}{\partial t} + 3\frac{\dot{S}}{S}\rho = 0,$$

i.e.,

$$\rho S^3 = \text{constant} = \rho_0 S_0^3, \text{ say.} \tag{4.4}$$

This is the dilution of density during adiabatic expansion.

Next we consider the Navier–Stokes equation for fluid dynamics:

$$\rho\left(\frac{\partial \mathbf{v}}{\partial t} + (\mathbf{v} \cdot \nabla)\mathbf{v}\right) = -\nabla p + \rho \mathbf{F}, \tag{4.5}$$

where p is the pressure and $\mathbf{F}$ is the external force per unit mass on the fluid element. In our case it is gravitational and satisfies the relation

$$\nabla \cdot \mathbf{F} = -4\pi G \rho. \tag{4.6}$$

On substituting (4.1) into (4.5) with $p = 0$, we get

$$\dot{H}\mathbf{r} + H^2 \mathbf{r} = \mathbf{F}. \tag{4.7}$$

Taking the divergence of this relation and using the fact that $\nabla \cdot \mathbf{r} = 3$, we get

$$\dot{H} + H^2 = -\frac{4\pi G \rho}{3}. \tag{4.8}$$

With $H = \dot{S}/S$ and ρ given by (4.4) we finally get the following differential equation for $S(t)$:

$$\frac{\ddot{S}}{S} = -\frac{4\pi G \rho_0 S_0^3}{3}\frac{1}{S^3},$$

i.e.,

$$\ddot{S} = -\frac{4\pi G \rho_0 S_0^3}{3S^2}. \tag{4.9}$$

This equation can be easily integrated after multiplying by $2\dot{S}$, to give

$$\dot{S}^2 = \frac{8\pi G \rho_0 S_0^3}{3}\frac{1}{S} - kc^2. \tag{4.10}$$

This gives us a one-parameter family of differential equations, the parameter k being a positive, zero, or negative number. We may relate it to the energy of the initial expression: the universe is assumed to have exploded at a small value of S. If $k > 0$, we find that the universe has finite kinetic energy at $S \to \infty$. For $k < 0$ the universe comes to a halt at a finite S and then collapses towards $S \to 0$. The $k = 0$ case represents the limiting case in which the universe comes to rest as $S \to \infty$.

Although we have obtained a set of working models of the universe in this way, we are left with some uncomfortable feelings. Equation (4.6), for example, seems to indicate a force per unit volume acting towards the centre of our chosen coordinate system. Why should there be such a preferred direction, when we have assumed that the universe is homogeneous and isotropic? Closer examination will show that the result arises because of our dealing with an infinitely extended system, in which the limit towards infinity has been taken with the origin of coordinates as the centre. In the relativistic version of these models there is no such difficulty of interpretation.

Although we can carry on discussing physical properties of these models within the Newtonian framework, we will not do so. Rather, we will now return to relativistic cosmology and see *how closely* the models there resemble the Newtonian ones. We will then compare the equations (4.4), (4.9) and (4.10) with their relativistic counterparts.

4.2 The Einstein field equations simplified for cosmology

As discussed in Chapter 1, Einstein's general relativity is *prima facie* the most suitable theory for discussing cosmology. In Chapters 8 and 9 we will consider alternative approaches to cosmology but for the present we will rely on the standard relativity theory.

Once we decide to use relativity, our procedure is cut and dried. Although the theory has a rather forbidding set of field equations (ten non-linear partial differential equations in four independent variables), these are simplified in the cosmological context because of the simplifying symmetries we introduced in Chapter 3, viz. the Weyl postulate and the cosmological principle. Because of these, we already have the line element considerably simplified to start with:

$$ds^2 = c^2 \, dt^2 - S^2(t) \left(\frac{dr^2}{1 - kr^2} + r^2 (d\theta^2 + \sin^2 \theta \, d\phi^2) \right). \tag{4.11}$$

We use it first to compute the Einstein tensor and thereby formulate the general relativistic field equations. To solve them we will next require the energy tensor of the material contents of the universe.

Accordingly, we set

$$x^0 = ct, \qquad x^1 = r, \qquad x^2 = \theta, \qquad x^3 = \phi \tag{4.12}$$

so that the non-zero components of g_{ik} and g^{ik} are

$$g_{00} = 1, \qquad g_{11} = -\frac{S^2}{1 - kr^2}, \qquad g_{22} = -S^2 r^2, \qquad g_{33} = -S^2 r^2 \sin^2 \theta$$

$$g^{00} = 1, \qquad g^{11} = -\frac{1 - kr^2}{S^2}, \qquad g^{22} = -\frac{1}{S^2 r^2}, \qquad g^{33} = -\frac{1}{S^2 r^2 \sin^2 \theta}$$

$$\sqrt{-g} = \frac{S^3 r^2 \sin \theta}{\sqrt{1 - kr^2}}. \tag{4.13}$$

The non-zero components of Γ^i_{kl} are then as follows:

$$\Gamma^1_{01} = \Gamma^2_{02} = \Gamma^3_{03} = \frac{1}{c} \frac{\dot{S}}{S},$$

$$\Gamma^0_{11} = \frac{S \dot{S}}{c(1 - kr^2)}, \qquad \Gamma^0_{22} = \frac{S \dot{S} r^2}{c}, \qquad \Gamma^0_{33} = \frac{S \dot{S} r^2 \sin^2 \theta}{c},$$

$$\Gamma^1_{11} = \frac{kr}{1 - kr^2}, \qquad \Gamma^2_{12} = \Gamma^3_{13} = \frac{1}{r},$$

$$\Gamma^1_{22} = -r(1 - kr^2), \qquad \Gamma^1_{33} = -r(1 - kr^2) \sin^2 \theta,$$

$$\Gamma^2_{33} = -\sin \theta \cos \theta, \qquad \Gamma^3_{23} = \cot \theta.$$

Now we use the expression for the Ricci tensor (see Chapter 2), which may be put in the following form:

$$R_{ik} = \frac{\partial^2 \ln \sqrt{-g}}{\partial x^i \, \partial x^k} - \frac{\partial \Gamma^l_{ik}}{\partial x^l} + \Gamma^m_{in} \Gamma^n_{km} - \Gamma^l_{ik} \frac{\partial \ln \sqrt{-g}}{\partial x^l}. \tag{4.14}$$

Straightforward but tedious calculation then gives the following non-zero components of R^i_k:

$$R^0_0 = \frac{3}{c^2} \frac{\ddot{S}}{S}, \tag{4.15}$$

$$R^1_1 = R^2_2 = R^3_3 = \frac{1}{c^2} \left(\frac{\ddot{S}}{S} + \frac{2 \dot{S}^2 + 2kc^2}{S^2} \right). \tag{4.16}$$

From these we get

$$R = \frac{6}{c^2} \left(\frac{\ddot{S}}{S} + \frac{\dot{S}^2 + kc^2}{S^2} \right), \tag{4.17}$$

and hence

$$G^1_1 \equiv R^1_1 - \frac{1}{2} R = -\frac{1}{c^2} \left(2\frac{\ddot{S}}{S} + \frac{\dot{S}^2 + kc^2}{S^2} \right) = G^2_2 = G^3_3, \tag{4.18}$$

$$G^0_0 \equiv R^0_0 - \frac{1}{2} R = -\frac{3}{c^2} \left(\frac{\dot{S}^2 + kc^2}{S^2} \right). \tag{4.19}$$

We have gone through the details of the calculation to illustrate how techniques of general relativity developed in Chapter 2 can be applied to the problem of cosmology. The reader may check that putting $S = \text{constant} = S_0$ and $k = +1$ gives us the formulae (3.20) and (3.21) obtained for the Einstein universe in Chapter 3. As a general comment we remark that because we have spatial homogeneity, the tensor components above (equations (4.15)–(4.19)) do not contain any space coordinates. Furthermore, because of isotropy, we have the three space–space components of the Einstein tensor equal. Recalling now the Einstein equations, we get from (4.18) and (4.19) the only non-trivial equations of the set as

$$2\frac{\ddot{S}}{S} + \frac{\dot{S}^2 + kc^2}{S^2} = \frac{8\pi G}{c^2}T_1^1 = \frac{8\pi G}{c^2}T_2^2 = \frac{8\pi G}{c^2}T_3^3, \tag{4.20}$$

$$\frac{\dot{S}^2 + kc^2}{S^2} = \frac{8\pi G}{3c^2}T_0^0. \tag{4.21}$$

We next consider the energy tensor.

4.3   Energy tensors of the universe

Before we consider specific forms of T_k^i, it is worth noting that two properties must be satisfied by any energy tensor in the present framework of cosmology. The first is obvious from (4.20):

$$T_1^1 = T_2^2 = T_3^3 = -p, \quad \text{say.} \tag{4.22}$$

The fact that these three components of T_k^i are equal is hardly surprising since we have already emphasized the condition of isotropy imposed on the universe. In the light of our discussion in Chapter 2, we identify the quantity p with pressure. We further define the energy density by

$$T_0^0 = \epsilon. \tag{4.23}$$

The second property is not quite so obvious, but is derivable from (4.20) and (4.21). It relates the pressure to the energy density. We note that, if we differentiate (4.21) with respect to t, we can express the resulting answer as a linear combination of (4.20) and (4.21). The result is in fact equivalent to the following identity:

$$\frac{d}{dt}\left[S(\dot{S}^2 + kc^2)\right] \equiv \dot{S}\left(2S\ddot{S} + \dot{S}^2 + kc^2\right);$$

that is,

$$\frac{d}{dS}(\epsilon S^3) + 3pS^2 = 0. \tag{4.24}$$

It is not necessary, however, to write down the full field equations (4.20) and (4.21) in order to arrive at (4.24). The above result is a direct consequence of the conservation law implicit in the Einstein equations:

$$T_{k;i}^i = 0. \tag{4.25}$$

Recall that, from the Bianchi identities, (4.25) follows *identically*. We now turn our attention to the specific forms of the energy tensor.

4.3.1   Random motions in an expanding universe

Present observations suggest that galaxies are the major constituents of the universe. If galaxies followed the Weyl postulate strictly, we would have the typical velocity vector of a galaxy as

$$u^i = (1, 0, 0, 0). \tag{4.26}$$

In our smooth-fluid approximation a velocity field like (4.26) represents an orderly motion with no pressure. Thus we have in this case the system of galaxies behaving like *dust*, with

$$p = 0, \qquad \epsilon = \rho c^2, \tag{4.27}$$

ρ being the rest-mass density of galaxies.

Considering the details discussed in Chapter 1, we are making an approximation here, with the typical galaxy following the track of a fundamental observer and the density ρ representing the 'smoothed-out' density of matter in the form of galaxies. This smooth-fluid approximation is valid, if the typical volume under consideration has a large number of galaxies. If we believe that homogeneity applies over a scale of $\sim$200 Mpc, say, then the number of galaxies in this volume is over a million. For comparison, the so-called *Hubble radius* of the universe, that is, the length scale associated with the expansion of the universe, is $\sim 3000 h_0^{-1}$. Thus the assumption of homogeneity in the distribution of matter is justified, but *only just so!* This was the assumption we made in the Newtonian cosmology also, by assuming the model of uniformly distributed dust for the universe.

In practice galaxies do not follow the Weyl postulate strictly and their velocity vectors depart from (4.26). Such departures in velocity are measurable for galaxies in clusters and are of the order of $\leq$1000 km s^{-1}. If we take a typical departure in velocity of $v \approx 1000$ km s^{-1}, then, from our discussions in Chapter 2, we would have a non-zero value of p in (4.27) of the order of

$$p \approx \frac{v^2}{c^2}\epsilon \sim 10^{-5}\epsilon. \tag{4.28}$$

Therefore we would be justified in ignoring the p term at the present epoch, in comparison with the ϵ term as in the idealized situation of (4.27).

What about the role of pressure in the future and past epochs? To assess the importance of the pressure term we have to investigate how the random motions of galaxies change in an expanding universe. We may express the 4-velocity of a galaxy as

$$u^i \equiv [1, u^\mu], \qquad |u^\mu| \ll 1.$$

The $\ll$ sign implies that the squares of u^μ are to be neglected in comparison with unity. Therefore the requirement $u_i u^i = 1$ is satisfied and we also have the built-in assumption that the random motions are small. In the absence of any external forces, therefore, the velocity u^i satisfies the geodesic equation:

$$\frac{du^i}{ds} + \Gamma^i_{kl}u^k u^l = 0.$$

Substitution of the Γ^i_{kl} for the Robertson–Walker line element then gives the result

$$u^\mu S^2 = \text{constant}.$$

However, u^μ measures the velocity in the comoving (r, θ, ϕ) coordinates. These motions are not, however, physical motions, since they refer to comoving coordinates. The proper distances and speeds are obtained from the coordinate distances and speeds, by multiplication by the scale factor S. Thus proper random velocities v change with S as

$$v = S(t)(u_\mu u^\mu)^{1/2} \propto S^{-1}. \tag{4.29}$$

(Also refer to Exercises 16 and 17 at the end of Chapter 3, where this result is seen as the non-relativistic approximation for how the momentum decreases in the expanding universe.)

Hence, so long as S goes on increasing beyond the present epoch the approximation $p \ll \epsilon$ will continue to apply. If, however, we turn towards the past epoch, we should find the motions of galaxies becoming more and more turbulent, since, according to (4.29), v was larger in the past. Thus, if we use $S \approx 10^{-2}S_0$ (S_0 being the value of S during the present epoch), (4.29) would give $p \sim 10^{-1}\epsilon$. Clearly the p term would no longer be negligible during this epoch and prior to it.

For such epochs we have to abandon our simplified picture of cosmology and ask whether galaxies existed as single units then. This question leads us to *cosmogony*, the subject of the origin of large-scale structure of the universe. Obviously, galaxies were formed at some stage in the past and, in a proper theory of cosmology and cosmogony, we have to say how and when they were formed. We will return to this question in Chapter 7. At present we simply state that the present cosmological framework of galaxies receding from one another breaks down, as does the dust approximation (4.27), for epochs like these.

If, however, we simply extrapolate $v \propto S^{-1}$ to very low values of S, v becomes comparable to c and our non-relativistic approximation that led us to $v \propto S^{-1}$ breaks down. The correct formula (see Exercise 3.16) then tells us that the 3-momentum P goes as S^{-1}. In this relativistic domain we have to use the formula (2.87) and we set

$$p = \tfrac{1}{3}\epsilon. \tag{4.30}$$

Thus, we may look upon a typical volume of these early epochs as containing matter particles moving at random relativistically, but any such spherical volume would have a centre of mass at rest in the Robertson–Walker frame. In this case the Weyl postulate is not satisfied for a typical particle, but it may still be applied to the centre of mass of a typical spherical volume.

4.3.2   Matter versus radiation domination

So, if S continues to increase from very small values, then (4.30) would hold for the early epochs, just as (4.27) holds in the present and relatively recent epochs. The transition between the two epochs is through a rather messy phase for which neither (4.27) nor (4.30) holds.

If (4.27) holds, then from (4.24) we get

$$\frac{d}{dS}(\rho S^3) = 0, \tag{4.31}$$

which integrates to

$$\rho = \rho_0 \frac{S_0^3}{S^3}, \tag{4.32}$$

ρ_0 and S_0 being the values of ρ and S for the present epoch. Similarly, substitution of (4.30) into (4.24) leads to

$$\frac{d}{dS}(\epsilon S^4) = 0, \tag{4.33}$$

giving

$$\epsilon \propto S^{-4}. \tag{4.34}$$

We therefore have the following picture. For a distribution of matter (4.34) holds when S was very small relative to S_0, whereas (4.32) holds for the more recent epochs. If, however, on top of matter we also have electromagnetic radiation present in the universe, it will also contribute to T_k^i. For small S, (4.30) holds uniformly for matter (moving relativistically) and for radiation. However, as S increases we have to be more careful in distinguishing between the contributions of matter and radiation to T_k^i. For, as we shall see later, although matter and radiation were in close interaction at small S, during later epochs they became effectively decoupled from each other. We will go into these details more fully in Chapter 5.

For the present discussion let us assume that, beyond a certain epoch $t = t_{\text{dec}}$ when S was given by $S = S_{\text{dec}}$, radiation and matter decoupled from each other, each going its own way. Thus we can write

$$T_k^i = T_{k|\text{matter}}^i + T_{k|\text{radiation}}^i \tag{4.35}$$

and assume that the divergence of each energy tensor separately vanishes. Since for the radiation energy tensor we have (for $\mu = 1, 2, 3$), say,

$$-T_{\mu|\text{radiation}}^{\mu} = \frac{1}{3} T_{0|\text{radiation}}^0 = \frac{1}{3}\epsilon, \tag{4.36}$$

we get for $S > S_{\text{dec}}$

$$\epsilon = \epsilon_0 \frac{S_0^4}{S^4}. \tag{4.37}$$

What is t_{dec}? Why, if at all, should matter decouple from radiation? What happened prior to $t = t_{\text{dec}}$? We defer a discussion of these questions to Chapter 5. There is, however, another important epoch in the past history of the universe, when the densities of matter and radiation were equal. We will denote it by $t = t_{eq}$, when S was equal to S_{eq}, say. It is easy to estimate this scale as follows.

The present estimates of $\epsilon_0 \approx 4 \times 10^{-13}$ erg cm^{-3} and of $\rho_0 c^2 \geq 3 \times 10^{-10}$ erg cm^{-3} mean that the density of matter is more than 10^3 times the density of radiation. Thus $\epsilon_0 \ll \rho_0 c^2$ and we may ignore the contribution of radiation (in comparison with the contribution of matter) to the field equations (4.20) and (4.21) during the present epoch and for $S > S_0$. However, for the past epochs with $S < S_0$, we have from (4.32) and (4.37)

$$\frac{\epsilon}{\rho c^2} = \frac{\epsilon_0}{\rho_0 c^2} \frac{S_0}{S} \tag{4.38}$$

and we cannot ignore the contribution of radiation for, say, $S_0/S \sim 10^3$. This is the epoch t_{eq}. Indeed, prior to this epoch, that is for $S < S_{eq}$, the relative importance of radiation and matter was inverted: the radiation was the more dominant factor in deciding how S should vary with t.

From the above discussion we see that, at $S = S_{eq} \approx 10^{-3} S_0$, we have a transition from a *radiation-dominated* universe to a *matter-dominated* one. The radiation-dominated phase will be discussed in the following chapter; here we will limit ourselves to the matter-dominated epochs. The equations (4.20) and (4.21) are therefore to be solved with

$$T_1^1 = 0, \qquad T_0^0 = \rho_0 c^2 \frac{S_0^3}{S^3}. \tag{4.39}$$

This simplification leads us to the classic models first considered by A. Friedmann (see Figure 4.1) between 1922 and 1924. Basically, these models ignore any contributions of electromagnetic radiation to T_k^i and suppose that the matter in the universe can be approximated by dust.

Example

If $\rho_0 c^2 / \epsilon_0 = \xi$, then estimate the redshift of the epoch prior to which the universe was radiation-dominated.

During the epoch t, in relation to the present epoch t_0, the redshift z is given by

$$1 + z = \frac{S(t_0)}{S(t)}$$

(see Chapter 3). From (4.38), during this epoch the ratio of the matter and radiation energy densities was $(1 + z)^{-1}$ of its present value ξ. Therefore, the epoch t_{eq}, when the two energy densities were equal, is given by the redshift $z_{eq} = \xi - 1$. For epochs prior to t_{eq}, i.e., for $z > z_{eq}$, the universe was radiation-dominated. Figure 4.2 illustrates this switchover of dominance. Notice that, if there is a substantial amount of dark matter present, then the present-day estimate of ρ_0 goes up and so does the value of z_{eq}.

Of course, we assume here that the scale factor has been monotonically growing during the past epochs. This is what the simplest models of the universe deduce, as we shall see next.

4.4   The Friedmann models

We will assume that the universe is (as at present) dust-dominated. For dust models, (4.20) and (4.21) become

$$2\frac{\ddot{S}}{S} + \frac{\dot{S}^2 + kc^2}{S^2} = 0, \tag{4.40}$$

$$\frac{\dot{S}^2 + kc^2}{S^2} = \frac{8\pi G\rho_0}{3}\frac{S_0^3}{S^3}. \tag{4.41}$$

In view of the conservation law given in (4.31), the above two differential equations are not independent and just one of them suffices to determine $S(t)$. Since it is of lower order, we will choose (4.41) for our solution and consider the three cases $k = 0, 1$ and -1 separately. However, before proceeding further we note the similarity between these equations and those of Newtonian cosmology obtained earlier in this chapter. See, for example, equation (4.10), which is analogous to (4.41) above. That equation also had a constant k, which was related to the initial total kinetic energy of explosion that led to the expanding models. Here the constant is simply related to the curvature of spatial sections. The dynamical behaviour of these

Figure 4.1 Alexander Friedmann (1888–1925).

models is, however, similar to that of the Newtonian models. We will now look at the three cases $k = 0$, 1 and -1 in some detail.

4.4.1 Euclidean sections ($k = 0$)

This is the simplest case and is also known as the *Einstein–de Sitter model*, since it was given by Einstein and de Sitter in a joint paper in 1932. Equation (4.41) becomes

$$\dot{S}^2 = \frac{8\pi G\rho_0}{3} \frac{S_0^3}{S}. \tag{4.42}$$

We now recall from Chapter 3 that the present value of Hubble's constant is given by

$$\left.\frac{\dot{S}}{S}\right|_{t_0} = H_0. \tag{4.43}$$

Hence, applying (4.43) to the present epoch, we get

$$\rho_0 = \frac{3H_0^2}{8\pi G} \equiv \rho_c. \tag{4.44}$$

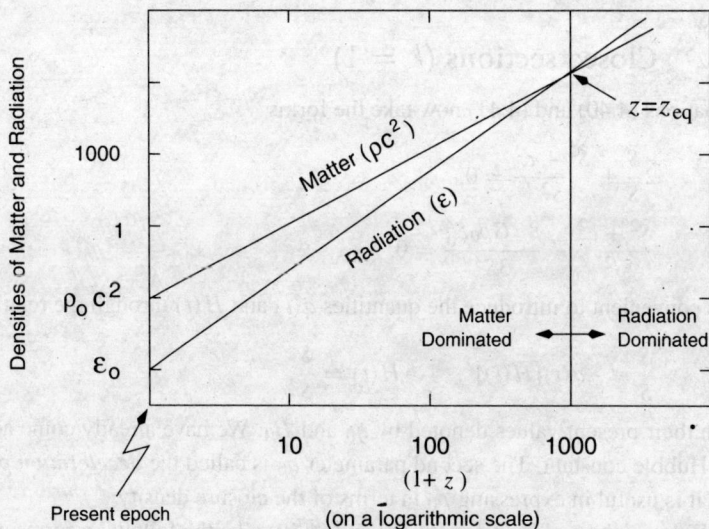

Figure 4.2 A plot of densities of radiation and matter in the universe against the redshift on a log–log scale. Note that the density of radiation drops more steeply than does the density of matter as the universe expands. The lines have been adjusted to show that the present-day density of matter is about a thousand times the density of radiation. Thus the two were comparable when the universe was $\sim 10^3$ times smaller in size. This is indicated by the point of intersection of the two lines. The redshift z_{eq} corresponds to the epoch when the two densities were equal.

For reasons that will become clear later, ρ_c is often called the *closure density*. With the range of values of H_0 quoted in Chapter 3, we have

$$\rho_c = 2 \times 10^{-29} h_0^2 \text{ g cm}^{-3}. \tag{4.45}$$

These values are considerably higher than the density of matter actually observed at present; we will take up this point in detail in later chapters.

Returning to (4.42), it is easy to verify that it has the solution

$$S = S_0 \left(\frac{t}{t_0} \right)^{2/3}. \tag{4.46}$$

An arbitrary constant that arises from the integration of the differential equation can be set equal to zero by assuming that $S = 0$ at $t = 0$. We also get from equation (4.43) the *age of the universe* as the present value of t:

$$t_0 = \frac{2}{3H_0}. \tag{4.47}$$

The constant S_0, the value of the scale factor during the present epoch, is not determined. It has the dimensions of length and it can be absorbed into the unit of length chosen. Figure 4.3 illustrates this solution. If we drop the suffix 0, the results (4.43), (4.44) and (4.47) hold for any arbitrary epoch t.

4.4.2   Closed sections ($k = 1$)

Equations (4.40) and (4.41) now take the forms

$$2\frac{\ddot{S}}{S} + \frac{\dot{S}^2 + c^2}{S^2} = 0, \tag{4.48}$$

$$\frac{\dot{S}^2 + c^2}{S^2} - \frac{8\pi G\rho_0 S_0^3}{3S^3} = 0. \tag{4.49}$$

It is convenient to introduce the quantities $q(t)$ and $H(t)$ through the relations

$$\frac{\ddot{S}}{S} - -q(t)[H(t)]^2, \qquad H(t) = \frac{\dot{S}}{S}, \tag{4.50}$$

with their present values denoted by q_0 and H_0. We have already come across H_0, the Hubble constant. The second parameter q_0 is called the *deceleration parameter* and it is useful in expressing ρ_0 in terms of the closure density.

With the above definitions, (4.48) and (4.49) take the following forms when they are applied to the present epoch:

$$\frac{c^2}{S_0^2} = (2q_0 - 1)H_0^2, \tag{4.51}$$

$$\rho_0 = \frac{3}{8\pi G} \left(H_0^2 + \frac{c^2}{S_0^2} \right) = \frac{3H_0^2}{4\pi G} q_0. \tag{4.52}$$

The density ρ_0 is often expressed in the following form:

$$\rho_0 = \rho_c \Omega_0, \tag{4.53}$$

so that, from (4.52), (4.53) and (4.44), we get the *density parameter*

$$\Omega_0 = 2q_0. \tag{4.54}$$

Since the left-hand side of (4.51) is positive, we must have

$$q_0 > \tfrac{1}{2}, \qquad \Omega_0 > 1. \tag{4.55}$$

Thus our closed model has a density *exceeding* the so-called closure density ρ_c. This explains the name 'closure density'. It is the value of the universal density that must be exceeded if the model is to describe a closed universe. We mention at this stage the result (to be proved shortly) that, for the open models ($k = -1$), the inequalities of (4.55) are reversed.

Using (4.51) and (4.52) to eliminate S_0 and ρ_0 we get the following differential equation from (4.49):

$$\dot{S}^2 = c^2\left(\frac{\alpha}{S} - 1\right), \tag{4.56}$$

with α given by

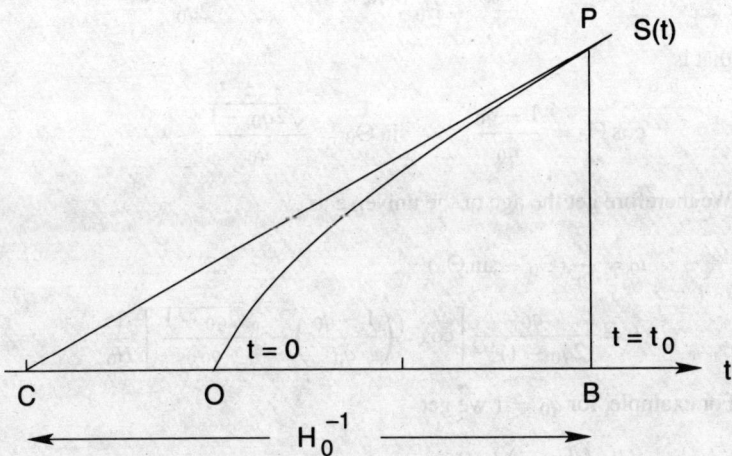

Figure 4.3 A schematic graph of $S(t)$ as a function of t for the Einstein–de Sitter model. The present epoch t_0 is denoted by the point B on the t-axis. The ordinate at B , PB $= S_0$, the present-day value of the scale factor. Taking PB as a unit of length, the present-day value of the Hubble constant is given by the ratio 1/BC, where C is the point common to the t-line and the tangent to the $S(t)$ curve at P. Thus BC $= H_0^{-1}$. As can be verified from the figure, the age of the universe, represented by OB, is two thirds of the time intercept BC.

$$\alpha = \frac{2q_0}{(2q_0 - 1)^{3/2}} \frac{c}{H_0}. \tag{4.57}$$

The parameter α has the dimensions of length. Thus the model is characterized by the parameters H_0 and q_0 (or, alternatively, Ω_0).

Equation (4.56) can be integrated as follows. We get

$$ct = \int \frac{\sqrt{S}\, dS}{\sqrt{\alpha - S}}.$$

Make the substitution

$$S = \alpha \sin^2 \left(\frac{\Theta}{2}\right) = \frac{1}{2}\alpha(1 - \cos\Theta). \tag{4.58}$$

Then the integral becomes

$$ct = \int \alpha \sin^2 \left(\frac{\Theta}{2}\right) d\Theta = \frac{1}{2}\alpha(\Theta - \sin\Theta). \tag{4.59}$$

Again, as in the case $k = 0$ we have taken $S = 0$ at $t = 0$ ($\Theta = 0$). We therefore get $t = t_0$ by requiring that $S = S_0$. From (4.51) and (4.57) we see that $S = S_0$ at $\Theta = \Theta_0$, where

$$\frac{1}{2}\alpha(1 - \cos\Theta_0) = \frac{c}{H_0}(2q_0 - 1)^{-1/2} = \frac{2q_0 - 1}{2q_0}\alpha;$$

that is,

$$\cos\Theta_0 = \frac{1 - q_0}{q_0}, \qquad \sin\Theta_0 = \frac{\sqrt{2q_0 - 1}}{q_0}. \tag{4.60}$$

We therefore get the age of the universe as

$$t_0 = \frac{\alpha}{2c}(\Theta_0 - \sin\Theta_0)$$

$$= \frac{q_0}{(2q_0 - 1)^{3/2}}\left[\cos^{-1}\left(\frac{1 - q_0}{q_0}\right) - \frac{\sqrt{2q_0 - 1}}{q_0}\right]\frac{1}{H_0}. \tag{4.61}$$

For example, for $q_0 = 1$ we get

$$t_0 = \left(\frac{\pi}{2} - 1\right)H_0^{-1}. \tag{4.62}$$

Note that S reaches a maximum value at $\Theta = \pi$, when

$$S = S_{\max} = \alpha = \frac{2q_0}{(2q_0 - 1)^{3/2}} \frac{c}{H_0}. \tag{4.63}$$

Thus, for $q_0 = 1$, the universe expands to twice its present size.

In closed models, therefore, expansion is followed by contraction and S decreases to zero. The value $S = 0$ is reached when $\Theta = 2\pi$; that is, when

$$t = t_L = \frac{\pi\alpha}{c} = \frac{2\pi q_0}{(2q_0 - 1)^{3/2}} \frac{1}{H_0}. \tag{4.64}$$

The quantity t_L may be termed the *lifespan* of this universe. For $q_0 = 1$, $t_L = 2\pi H_0^{-1} = 2\pi\tau_0$, where τ_0 is defined by the relation (3.69).

Figure 4.4 illustrates the function $S(t)$ for the closed models for a number of parameter values q_0. All curves have been adjusted to have the same value of H_0 at point P. Notice that the value $S = 0$ is reached sooner in the past as q_0 is increased from just over $\frac{1}{2}$.

4.4.3 Open sections ($k = -1$)

Equations (4.40) and (4.41) become in this case

$$2\frac{\ddot{S}}{S} + \frac{\dot{S}^2 - c^2}{S^2} = 0, \tag{4.65}$$

$$\frac{\dot{S}^2 - c^2}{S^2} - \frac{8\pi G\rho_0 S_0^3}{3S^3} = 0. \tag{4.66}$$

We again use the definitions of (4.50) and apply them for the present epoch to get

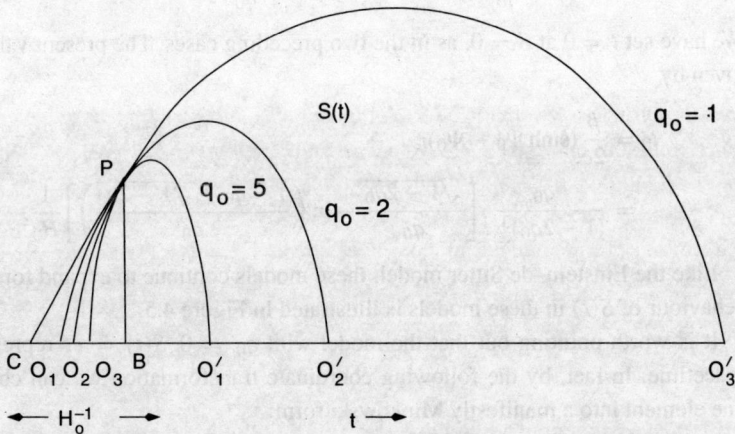

Figure 4.4 The $S(t)$ curves for $q_0 = 1, 2$ and 5. All curves have been scaled to touch at P, the present point, and they all have the common tangent PC. The intercept BC $= H_0^{-1}$. Notice that, as q_0 increases, the curves for $S(t)$ intersect the past section of the t-line at points $O_1, O_2, O_3, \ldots$ lying closer and closer to B, implying that the age of the universe is reduced if q_0 is increased. The points $O_1', O_2', O_3', \ldots$ in the future section of the t-line show the singularities at which these models end their existence.

$$\frac{c^2}{S_0^2} = (1 - 2q_0)H_0^2,$$ (4.67)

$$\rho_0 = \frac{3H_0^2}{4\pi G}q_0.$$ (4.68)

Thus, instead of (4.55), we now have

$$0 \le q_0 \le \tfrac{1}{2}, \qquad 0 \le \Omega_0 < 1.$$ (4.69)

In place of (4.56) we get

$$\dot{S}^2 = c^2\left(\frac{\beta}{S} + 1\right)$$ (4.70)

with

$$\beta = \frac{2q_0}{(1 - 2q_0)^{3/2}}\frac{c}{H_0}.$$ (4.71)

As in the $k = +1$ case, the solution of (4.70) may be parametrized by an angle Ψ with

$$S = \tfrac{1}{2}\beta(\cosh \Psi - 1), \qquad ct = \tfrac{1}{2}\beta(\sinh \Psi - \Psi).$$ (4.72)

The present value of Ψ is given by

$$\cosh \Psi_0 = \frac{1 - q_0}{q_0}, \qquad \sinh \Psi_0 = \frac{\sqrt{1 - 2q_0}}{q_0}.$$ (4.73)

We have set $t = 0$ at $S = 0$, as in the two preceding cases. The present value of t is given by

$$
\begin{aligned}
t_0 &= \frac{\beta}{2c}(\sinh \Psi_0 - \Psi_0) \\
&= \frac{q_0}{(1 - 2q_0)^{3/2}}\left[\frac{\sqrt{1 - 2q_0}}{q_0} - \ln\left(\frac{1 - q_0 + \sqrt{1 - 2q_0}}{q_0}\right)\right]\frac{1}{H_0}.
\end{aligned}
$$ (4.74)

Like the Einstein–de Sitter model, these models continue to expand forever. The behaviour of $S(t)$ in these models is illustrated in Figure 4.5.

It is worth pointing out that the model with $q_0 = 0$, $S(t) = ct$ represents flat spacetime. In fact, by the following coordinate transformation we can change the line element into a manifestly Minkowski form:

$$R = ctr, \qquad T = t\sqrt{1 + r^2},$$
$$ds^2 = c^2\,dT^2 - dR^2 - R^2(d\theta^2 + \sin^2\theta\,d\phi^2).$$ (4.75)

This model arose naturally in Milne's kinematical relativity, which was a cosmological theory with foundations different from those of general relativity. For this reason the above model is sometimes referred to as the *Milne model*.

For a comparison, the three types of Friedmann models ($k = 0, \pm 1$) are shown together on the same plot in Figure 4.6. There is a unique 'flat' model (the $k = 0$ case is often so referred; but this can be misleading as the 'flatness' refers to the

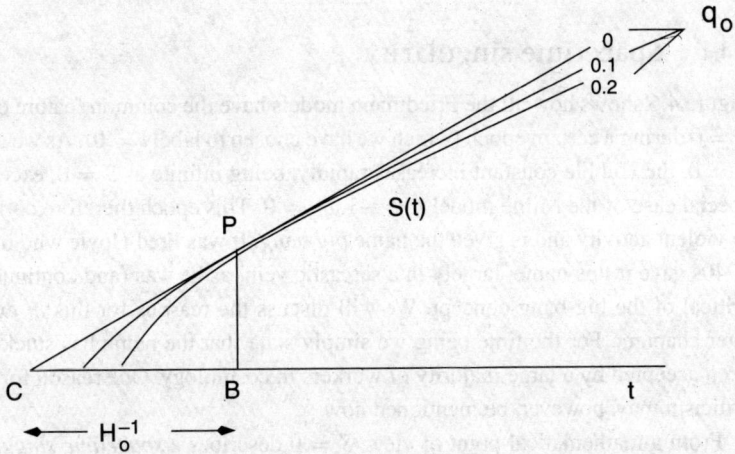

Figure 4.5 The $S(t)$ curves for $q_0 = 0, 0.1$ and 0.2. As in Figure 4.4, all curves have the same value of H_0 at P. The age of the universe is seen to increase as q_0 decreases, being maximum $(=H_0^{-1})$ for $q_0 = 0$.

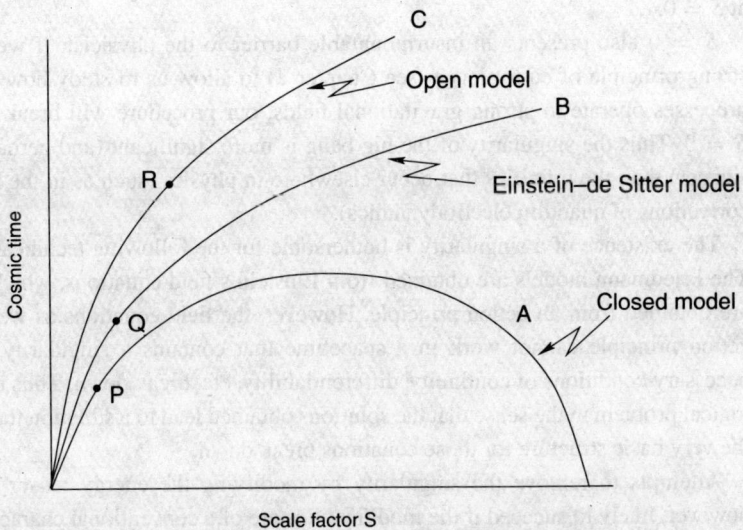

Figure 4.6 The three types of Friedmann models are shown together in this plot of $S(t)$ versus t. Points P, Q and R indicate where the present epoch occurs on the various curves. As we go from open to closed models, this epoch falls closer to the initial singular epoch.

spatial sections t = constant, not to spacetime as a whole), but a continuous range of $k = \pm 1$ models, of which two representative ones are shown. The dot on a curve shows a typical $H(t)$ = constant epoch. Thus, for the same Hubble constant, the open models give a larger age. We will return to this property later in this chapter.

4.4.4   Spacetime singularity

Figure 4.6 shows how all the Friedmann models have the common feature of having $S = 0$ during a certain epoch (which we have chosen to label $t = 0$). As we approach $S = 0$, the Hubble constant increases rapidly, being infinite at $S = 0$, except in the special case of the Milne model $k = -1$, $q_0 = 0$. This epoch therefore corresponds to violent activity and is given the name *big bang*. It was Fred Hoyle who in the late 1940s gave it this name, largely in a sarcastic vein, as he was (and continues to be) critical of the big-bang concept. We will discuss the reasons for this *in extenso* in later chapters. For the time being we simply state that the name has stuck and has been accepted by a large majority of workers in cosmology. One reason for Hoyle's criticism may, however, be mentioned now.

From a mathematical point of view, $S = 0$ describes a *spacetime singularity*. If we compute the components of R_{iklm} and construct invariants out of these, such as

$$R, \ R_{ik}R^{ik}, \ R_{iklm}R^{iklm}, \ldots ,$$

these invariants diverge. It is therefore meaningless to talk of a spacetime geometry at $S = 0$.

$S = 0$ also presents an insurmountable barrier to the physicist. If we use the strong principle of equivalence (see Chapter 2) to allow us to study how physical processes operate in strong gravitational fields, our procedure will break down at $S = 0$. Thus the singularity of the big bang is more significant (and perhaps more sinister) than the infinities that occur elsewhere in physics (such as in the radiative corrections of quantum electrodynamics).

The existence of a singularity is bothersome for the following technical reason. The Friedmann models are obtained from Einstein's field equations, which in turn are obtained from an action principle. However, the field equations as well as the action principle cannot work in a spacetime that contains a singularity, for the necessary conditions of continuity, differentiability, etc. break down. Thus there is a logical problem in the sense that the solutions obtained lead to a situation that makes the very basic structure for those equations break down.

Attempts to remove the singularity by modifying the energy tensor are not, however, likely to succeed if the modifications are of a conventional character. This was demonstrated in a general manner in the late 1960s by Roger Penrose, Stephen Hawking, Robert Geroch and others. For the time being we will accept the existence of this singularity as a fact of life under the regime of general relativity and learn to live with it.

4.5 The luminosity distance

Since the Friedmann models are frequently used to interpret cosmological observations, we will now derive some of the observable quantities in these models, starting with the luminosity distance defined in Chapter 3. Our aim is to express the final answer in terms of the two parameters that characterize a Friedmann model: H_0 and q_0.

4.5.1 The Einstein–de Sitter model

We use (3.49) to relate r_1, the radial coordinates of the galaxy G_1, to the time t_1 and to its redshift z:

$$
\begin{aligned}
r_1 &= \int_{t_1}^{t_0} \frac{c\,dt}{S(t)} = \frac{c}{S_0}\int_{t_1}^{t_0} t_0^{2/3} t^{-2/3}\,dt \\
&= \frac{c}{S_0} t_0^{2/3} \times 3\big(t_0^{1/3} - t_1^{1/3}\big) \\
&= \frac{3c}{S_0} t_0 \left[1 - \left(\frac{t_1}{t_0}\right)^{1/3}\right].
\end{aligned}
$$

We now use (3.51) to note that

$$
1 + z = \frac{S(t_0)}{S(t_1)} = \left(\frac{t_0}{t_1}\right)^{2/3},
$$

so that with the help of (4.47),

$$
\begin{aligned}
r_1 &= \frac{3ct_0}{S_0}\big[1 - (1+z)^{-1/2}\big] \\
&= \frac{2c}{S_0 H_0}\big[1 - (1+z)^{-1/2}\big].
\end{aligned}
\tag{4.76}
$$

The luminosity distance is therefore given by

$$
\begin{aligned}
D_1 &= r_1 S_0 (1+z) \\
&= \frac{2c}{H_0}\big[(1+z) - (1+z)^{1/2}\big].
\end{aligned}
\tag{4.77}
$$

4.5.2 The closed model

This calculation is more involved. Equation (3.49) becomes

$$
\int_0^{r_1} \frac{dr}{\sqrt{1 - r^2}} = \int_{t_1}^{t_0} \frac{c\,dt}{S(t)}.
$$

The left-hand side can be easily integrated. To integrate the right-hand side we use (4.56) to get

$$\int_{t_1}^{t_0} \frac{c \, dt}{S(t)} = \int_{S_1}^{S_0} \frac{dS}{\sqrt{S(\alpha - S)}}.$$

Now we use the parametric form (4.58): $S = \alpha \sin^2(\Theta/2)$. We then get

$$\int_{S_1}^{S_0} \frac{dS}{\sqrt{S(\alpha - S)}} = \int_{\Theta_1}^{\Theta_0} d\Theta = \Theta_0 - \Theta_1.$$

Remembering that the integral on the left-hand side of (3.49) gives $\sin^{-1} r_1$, we have

$$r_1 = \sin(\Theta_0 - \Theta_1). \tag{4.78}$$

We must now relate this answer to z. We have

$$1 + z = \frac{S(t_0)}{S(t_1)} = \frac{\sin^2(\Theta_0/2)}{\sin^2(\Theta_1/2)},$$

giving

$$\sin\Theta_1 = \frac{2}{1+z} \sin\left(\frac{\Theta_0}{2}\right)\left[z + \cos^2\left(\frac{\Theta_0}{2}\right)\right]^{1/2},$$

$$\cos\Theta_1 = \frac{z + \cos\Theta_0}{1+z}. \tag{4.79}$$

Also, from (4.60), we have

$$\sin\left(\frac{\Theta_0}{2}\right) = \sqrt{\frac{2q_0 - 1}{2q_0}}, \qquad \cos\left(\frac{\Theta_0}{2}\right) = \sqrt{\frac{1}{2q_0}}.$$

Putting all these together and performing algebraic simplification, we get

$$r_1 = \frac{\sqrt{2q_0 - 1}}{q_0^2(1+z)}[q_0 z + (1 - q_0)(1 - \sqrt{1 + 2zq_0})]. \tag{4.80}$$

The luminosity distance is therefore given by

$$D_1 = r_1 S_0(1 + z)$$
$$= \frac{c}{H_0} \frac{1}{q_0^2}[q_0 z + (q_0 - 1)(\sqrt{1 + 2zq_0} - 1)]. \tag{4.81}$$

This formula was first derived by Mattig in 1958.

4.5.3 The open model

The calculation in this case is similar to that for the closed model, with the difference that the trigonometric functions are replaced by hyperbolic ones. We will not go through the intermediate steps, but simply quote the final results:

$$r_1 = \frac{\sqrt{1-2q_0}}{q_0^2(1+z)}[q_0 z + (1-q_0)(1-\sqrt{(1+2zq_0)})], \tag{4.82}$$

$$D_1 = \frac{c}{H_0}\frac{1}{q_0^2}[q_0 z + (q_0-1)(\sqrt{1+2zq_0}-1)]. \tag{4.83}$$

It is interesting to note that the final expressions for D_1 are the same for $k = \pm 1$, $|q_0 - \frac{1}{2}| > 0$. If we let $q_0 \to \frac{1}{2}$, it is easy to see that the result (4.77) for the Einstein–de Sitter model also follows from the same formula.

Figure 4.7 plots $D_1(q_0, z)$ as a function of z for various parametric values of q_0. Note that all curves start off with the linear Hubble law (3.66) for small z, but then fan out, with only the curve for $q_0 = 1$ staying linear all the way. As a rule we notice that, for the same redshift, the luminosity distance is larger for lower values of q_0. Thus, for $q_0 = 1$, we have

$$D_1 = \frac{c}{H_0}z, \tag{4.84}$$

whereas for $q_0 = 0$ we get

$$D_1 = \frac{c}{H_0}z\left(1+\frac{z}{2}\right). \tag{4.85}$$

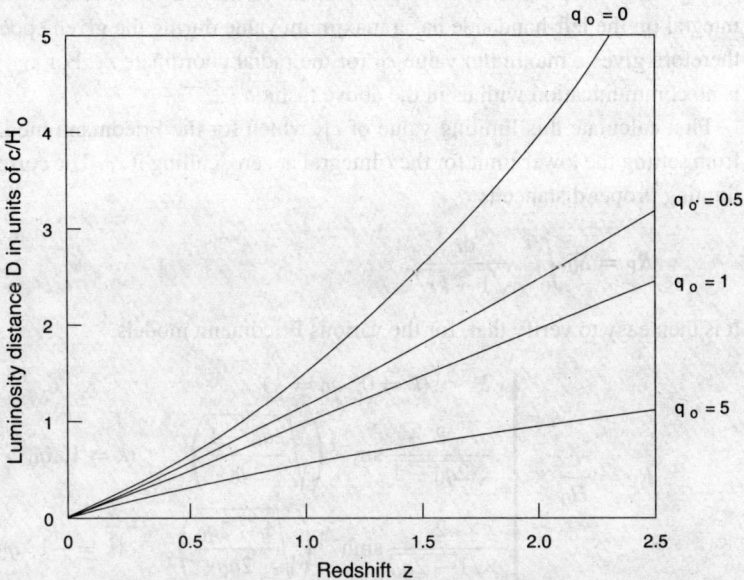

Figure 4.7 The luminosity distance expressed (in units of c/H_0) as a function of the redshift z for $q_0 = 0, \frac{1}{2}, 1$ and 5. The relationship is linear, as predicted by Hubble's law for $q_0 = 1$. For $q_0 < 1$, D_1 increases with z faster than predicted by Hubble's linear law; for $q_0 > 1$, D_1 increases more slowly with z. All curves merge for small z.

For S_0 at $z = 1$, (4.85) exceeds (4.84) by as much as 50%. In Chapter 11 we will discuss the feasibility of determining q_0 from Hubble-type observations of remote galaxies. We end this discussion by restating the formula (4.81) in terms of H_0 and Ω_0:

$$D_1 = \frac{2c}{H_0\Omega_0^2}\{[\Omega_0 z + (\Omega_0 - 2)[(1 + z\Omega_0)^{1/2} - 1]\}. \tag{4.86}$$

4.6   Horizons and the Hubble radius

In cosmological discussions two kinds of horizons often crop up. Of these the *particle horizon* relates to limits on communication in the past whereas the *event horizon* relates to limits on communication in the future. We will deal with these two concepts in that order.

4.6.1   The particle horizon

It is pertinent to ask the following question. What is the limit on the proper distance up to which we are able to see sources of light? This question is answered as follows. Going back to equation (3.49) of Chapter 3, we may have a situation wherein the integral on the left-hand side has a maximum value during the given epoch t_0. This therefore gives a maximum value r_P for the radial coordinate r_1. For $r_1 > r_P$ there is no communication with us in the above fashion.

First calculate this limiting value of r_1, which for the Friedmann models comes from setting the lower limit for the t integral at zero, calling it r_P. The corresponding limiting proper distance is

$$R_P = S_0 \int_0^{r_P} \frac{dr}{\sqrt{1 - kr^2}}.$$

It is then easy to verify that, for the various Friedmann models,

$$R_P = \frac{c}{H_0} \times \begin{cases} 2 & (k = 0,\ q_0 = \tfrac{1}{2}) \\ \dfrac{2}{\sqrt{2q_0 - 1}} \sin^{-1}\left(\sqrt{\dfrac{2q_0 - 1}{2q_0}}\right) & (k = 1,\ q_0 > \tfrac{1}{2}) \\ \dfrac{2}{\sqrt{1 - 2q_0}} \sinh^{-1}\left(\sqrt{\dfrac{1 - 2q_0}{2q_0}}\right) & (k = -1,\ q_0 < \tfrac{1}{2}). \end{cases} \tag{4.87}$$

The existence of a finite value of R_P means that the universe has a *particle horizon*. Particles with $S(t_0)r_1 > R_P$ are not visible to us at present, no matter how good our techniques of observation are.

Example

Consider the Einstein–de Sitter model. The result (4.87) gives, in this case, $R_P = 2c/H_0$. This means that, at present, we are able to see only those galaxies whose proper distances from us happen to be less than $2c/H_0$. See Figure 4.8.

4.6.2 The event horizon

The particle horizon sets a limit on communications from the past. Let us now see how the event horizon sets a limit on communications to the future. Let us ask the following question. A light source at $r = r_1$, $t = t_0$ sends a light signal to an observer at $r = 0$. Will the signal ever reach its destination? Suppose that it does and let t_1 be the time of arrival. Then, from (3.49), we get

$$\int_{t_0}^{t_1} \frac{c\,dt}{S(t)} = \int_0^{r_1} \frac{dr}{\sqrt{1 - kr^2}}.$$

This relation determines t_1 for any given r_1, provided that the integral on the left-hand side is large enough to match that on the right. Now it may happen that, as $t_1 \to \infty$, the integral on the left-hand side converges to a finite value that corresponds to a value of the integral on the right-hand side for $r_1 = r_E$, say. In that case it is not possible to satisfy the above relation for $r_1 > r_E$. In other words the signal from the

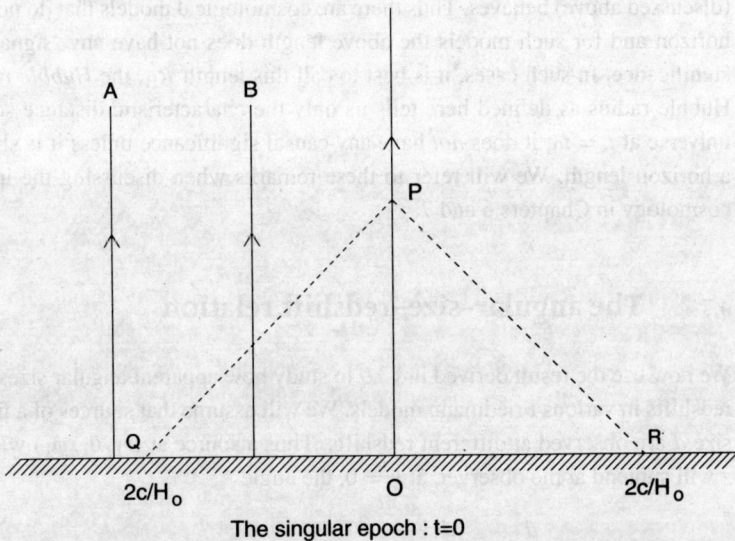

The singular epoch : t=0

Figure 4.8 The past light cone drawn from P the present epoch at $r = 0$ in the Einstein–de Sitter universe meets the $t = 0$ epoch in the section QR. World lines of a fundamental observer like B within this section intersect the cone. The light signal from B therefore reaches P. The line A, on the other hand, lies outside the particle horizon of P represented by QR.

light source at $r_1 > r_E$ will *never* reach the observer at r_0. Thus no observer beyond a proper distance

$$R_E = S_0 \int_{t_0}^{\infty} \frac{c\,dt}{S(t)} \tag{4.88}$$

at $t = t_0$ can communicate with another observer.

This limit is called the *event horizon*. It does not exist for Friedmann models but has the value c/H_0 for the de Sitter model as seen in the following calculation.

Example

Consider the de Sitter model described in Chapter 3. Here we have $k = 0$ and $S = e^{Ht}$. Then we get

$$R_E = e^{Ht_0} \int_{t_0}^{\infty} c e^{-Ht}\,dt = \frac{c}{H_0}.$$

That is, if any light source sends a ray of light from beyond this range at time t_0 towards the observer at $r = 0$, it will *never* reach the observer. See Figure 4.9.

Notice that both the event horizon and the particle horizon have radii comparable to c/H_0, which has led to the erroneous conclusion that the length $R_H = c/H_0$ is of the size of the horizon in *any* cosmology. Whether a horizon (particle or event) exists in a cosmological model depends on the scale factor and how the relevant integral (discussed above) behaves. Thus there are cosmological models that do not have any horizon and for such models the above length does not have any 'signal-limiting' significance. In such cases, it is best to call this length R_H, the *Hubble radius*. The Hubble radius as defined here tells us only the characteristic distance scale of the universe at $t = t_0$; it does *not* have any causal significance unless it is shown to be a horizon length. We will refer to these remarks when discussing the inflationary cosmology in Chapters 6 and 7.

4.7   The angular-size–redshift relation

We now use the result derived in §3.9 to study how apparent angular sizes vary with redshifts in various Friedmann models. We will assume that sources of a fixed linear size d are observed at different redshifts. Thus a source at (r_1, θ_1, ϕ_1) with redshift z will subtend at the observer, at $r = 0$, the angle

$$\Delta\theta_1 = \frac{d}{r_1 S(t_1)} = \frac{d(1+z)^2}{D_1}. \tag{4.89}$$

Since we know D_1 from (4.83), $\Delta\theta_1$ is determined as a function of z and q_0. The interesting fact that emerges is that $\Delta\theta_1$ does not steadily decrease as z increases, but has a minimum at a certain value of z that depends on q_0.

It is easy to derive this result for $q_0 = \frac{1}{2}$. From (4.77) we get

$$\Delta\theta_1 = \frac{dH_0}{2c} \frac{(1+z)^{3/2}}{(1+z)^{1/2} - 1}. \tag{4.90}$$

Straightforward differentiation gives us the result that the minimum value of $\Delta\theta_1$ ($= \theta_{min}$, say) and the redshift $z = z_m$ at which it occurs are given by

$$\theta_{min} = 3.375 \frac{dH_0}{c}$$

and

$$z_m = 1.25. \tag{4.91}$$

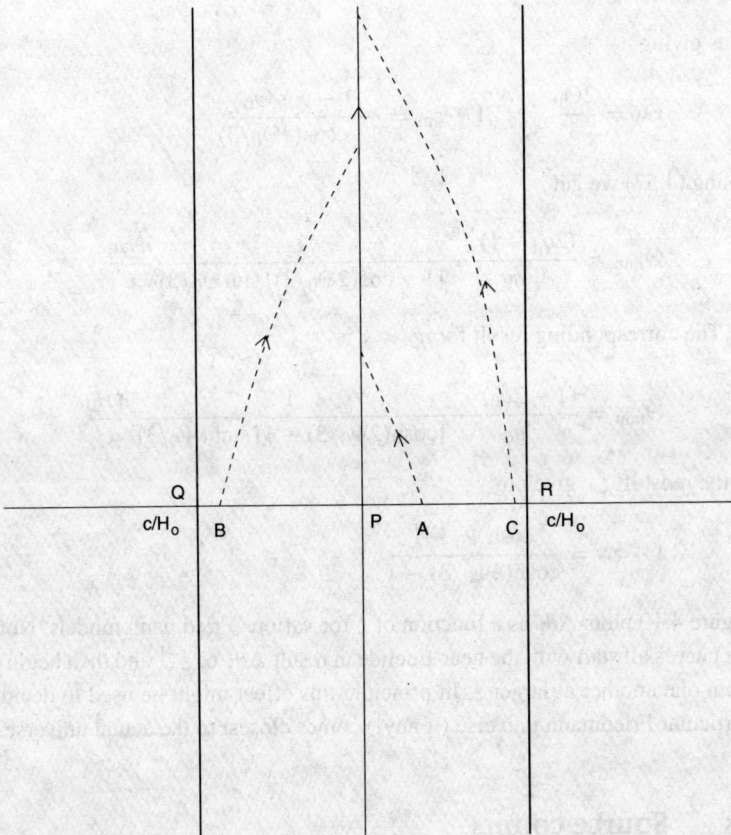

Figure 4.9 The event horizon in the de Sitter universe is shown by the section QR. Particles like A, B, C, etc. in this section can communicate with the observer P at the centre. No signal from an observer lying outside this section can ever reach P.

The cases $q_0 \geq \frac{1}{2}$ are more involved. We illustrate the case $q_0 > \frac{1}{2}$. Instead of using D_1 as given by (4.81), it is more convenient to use the parameter Θ introduced in (4.58) and (4.59) and the relation (4.78). We then get

$$\Delta\theta_1 = \frac{d}{r_1 S(t_1)} = \frac{2d}{\alpha}[(1 - \cos\Theta_1)\sin(\Theta_0 - \Theta_1)]^{-1}. \tag{4.92}$$

The constant α is defined by equation (4.57). Differentiation with respect to Θ_1 tells us that the minimum occurs when

$$\sin\Theta_1 \sin(\Theta_0 - \Theta_1) - (1 - \cos\Theta_1)\cos(\Theta_0 - \Theta_1) = 0;$$

that is,

$$\sin\left(\Theta_0 - \frac{3\Theta_1}{2}\right) = 0,$$

thus giving

$$\Theta_1 = \frac{2\Theta_0}{3}, \qquad 1 + z_m = \frac{1 - \cos\Theta_0}{1 - \cos(2\Theta_0/3)}. \tag{4.93}$$

Using (4.57) we get

$$\Theta_{min} = \frac{(2q_0 - 1)^{3/2}}{q_0} \frac{1}{[1 - \cos(2\Theta_0/3)]\sin(\Theta_0/3)} \frac{dH_0}{c}. \tag{4.94}$$

The corresponding result for $q_0 < \frac{1}{2}$ is

$$\Theta_{min} = \frac{(1 - 2q_0)^{3/2}}{q_0} \frac{1}{[\cosh(2\Psi_0/3) - 1]\sinh(\Psi_0/3)} \frac{dH_0}{c} \tag{4.95}$$

at the redshift z_m given by

$$1 + z_m = \frac{\cosh\Psi_0 - 1}{\cosh(2\Psi_0/3) - 1}. \tag{4.96}$$

Figure 4.10 plots $\Delta\theta_1$ as a function of z for various Friedmann models. Notice how the curves all start with the near-Euclidean result $\Delta\theta_1 \propto z^{-1}$ and then begin to differ from one another at larger z. In principle this effect might be used to decide which particular Friedmann universe (if any!) comes closest to the actual universe.

4.8   Source counts

We now return to the general formula (3.75) and apply it to Friedmann models. It is more convenient to use the redshift as the distance parameter instead of r or t.

As before, we will work with the case $k = +1$. From (4.58) and the relations that follow it we have

$$r = \sin(\Theta_0 - \Theta_1).$$

$$\left| \frac{dr}{\sqrt{1 - r^2}} \right| = |d\Theta_1|, \qquad 1 + z = \frac{\sin^2(\Theta_0/2)}{\sin^2(\Theta_1/2)},$$

$$\left| \frac{dz}{1 + z} \right| = \cot\left(\frac{\Theta_1}{2}\right) |d\Theta_1| = \sqrt{\frac{1 + 2q_0 z}{2q_0 - 1}} \, |d\Theta_1|.$$

Therefore the number of astronomical sources with redshifts in the range $(z, z + dz)$ is given by

$$dN = 4\pi \sin^2(\Theta_0 - \Theta_1) \, n(t) \left| \frac{d\Theta_1}{dz} \right| dz.$$

Let us suppose that $n(t)$ is specified as a function $n(z)$ of z. Using (4.80) and some algebraic manipulation, we get

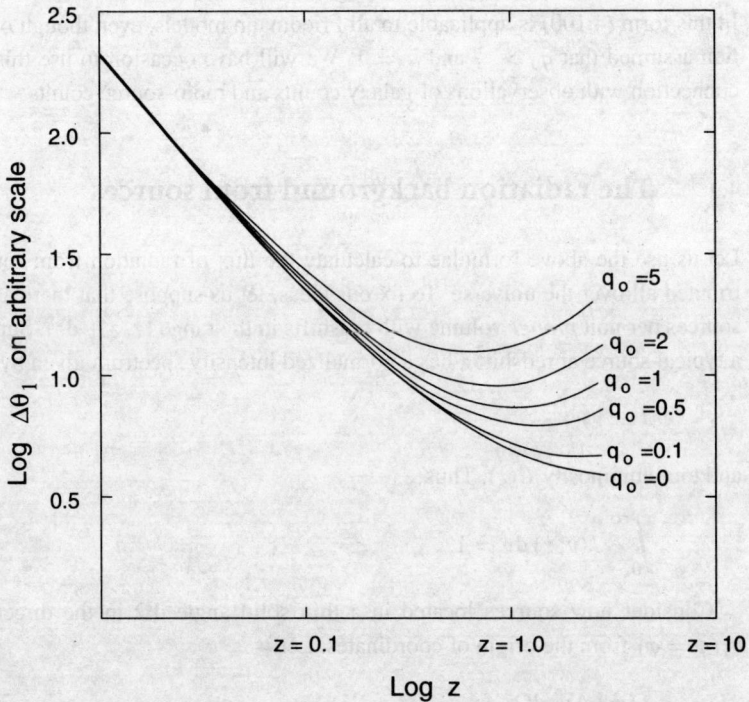

Figure 4.10 This graph plots $\log \Delta\theta_1$ against $\log z$ for the Friedmann models with $q = 0, 0.1, 0.5, 1, 2$ and 5. All curves merge at small z into a straight line that describes the variation of $\Delta\theta_1$ with distance in a Euclidean universe.

$$dN = 4\pi n(z)\frac{(2q_0 - 1)^{3/2}}{q_0^4}\frac{[q_0 z + (q_0 - 1)(\sqrt{1 + 2zq_0} - 1)]^2\, dz}{\sqrt{1 + 2q_0 z}(1 + z)^3}. \quad (4.97)$$

Suppose that $n(z)$ is expressed in a slightly different form. We recall that n was specified as the number of sources per unit *coordinate* volume, in terms of the comoving (r, θ, ϕ) coordinates. What is the relationship between n and the number of sources per unit *proper* volume? Denoting the latter by $\bar{n}$, we have

$$n = \bar{n}S^3 = \frac{\bar{n}S_0^3}{(1 + z)^3}. \quad (4.98)$$

From (4.51) we get

$$\frac{\bar{n}}{(1 + z)^3} = (2q_0 - 1)^{3/2}\left(\frac{H_0}{c}\right)^3 n. \quad (4.99)$$

Substitution into (4.97) gives

$$dN = 4\pi\left(\frac{c}{H_0}\right)^3\frac{[q_0 z + (q_0 - 1)(\sqrt{1 + 2zq_0} - 1)]^2 \bar{n}\, dz}{q_0^4(1 + z)^6\sqrt{1 + 2q_0 z}}. \quad (4.100)$$

In this form (4.100) is applicable to all Friedmann models, even though our derivation assumed that $q_0 > \frac{1}{2}$ and $k = 1$. We will have occasion to use this result in connection with observations of galaxy counts and radio-source counts.

4.9   The radiation background from sources

Let us use the above formulae to calculate the flux of radiation from sources distributed all over the universe. To fix our ideas, let us suppose that there are $\bar{n}(z)\, dz$ sources per unit *proper* volume with redshifts in the range $(z, z + dz)$. Suppose that a typical source at redshift z has a normalized intensity spectrum given by

$$J(\nu; z)$$

and total luminosity $L(z)$. Thus

$$\int_0^\infty J(\nu; z)\, d\nu = 1. \quad (4.101)$$

Consider now sources located in a thin solid angle $d\Omega$ in the direction $\theta = \theta_1, \phi = \phi_1$ from the origin of coordinates. Let

$$f(\nu_0).\Delta\nu_0\, d\Omega$$

denote the total flux of radiation received at $r = 0$ in the frequency range $(\nu_0, \nu_0 + \Delta\nu_0)$ from all the sources located within our solid angle.

Now the number of sources in a typical redshift range $(z, z + dz)$ is given by multiplying dN by $d\Omega/(4\pi)$, and the flux of radiation from a source in this range is given by the application of (3.56). Putting the two results together, we get

$$f(\nu_0) = \frac{c}{H_0}\frac{1}{4\pi}\int_0^\infty \frac{\bar{n}(z)L(z)J[\nu_0(1+z); z]\,dz}{(1+z)^5\sqrt{1+2q_0z}}. \quad (4.102)$$

This formula is useful for estimating the contributions of sources to the cosmic radiation background. Note that the flux density $\mathcal{S}(\nu_0, z)$ from a typical source in the above calculation is related to the quantity dN/dz by the relation

$$\frac{dN}{dz}[\tilde{\mathcal{S}}(\nu_0, z)] = \frac{c}{H_0}\frac{L(z)\bar{n}(z)J[\nu_0(1+z); z]}{(1+z)^5\sqrt{1+2q_0z}}. \quad (4.103)$$

Now, in a Euclidean space with a uniform distribution of sources, the number of sources out to a Euclidean distance R would be given by

$$N = \frac{4\pi}{3}R^3\bar{n}_0,$$

$\bar{n}_0$ being the number density of sources, which is assumed constant.

Furthermore, a typical source at a distance R and with a luminosity L would produce a flux at the origin given by

$$S = \frac{L}{4\pi R^2}.$$

We therefore get

$$\frac{dN}{dR}S = \bar{n}_0L = \text{constant}. \quad (4.104)$$

To discover the analogue of this result in a Friedmann universe, we assume that $\bar{n}(z) = \bar{n}_0(1+z)^3$, corresponding to a constant number of sources in the unit coordinate volume. We also assume that $L(z) = $ constant and integrate (4.103) over all ν_0. Then, using (4.101), we get

$$\frac{dN}{dz}S = \left(\frac{c}{H_0}\right)\frac{L\bar{n}_0}{(1+z)^3\sqrt{1+2q_0z}}. \quad (4.105)$$

Thus the product on the left-hand side steadily decreases with increasing z in all Friedmann models. The redshift factors in the denominator see to it that the product of the differential number count and the flux is less for remote sources than it is for nearby ones.

We also see this effect in the contribution of sources to the overall radiation background in (4.105). The contribution of remote shells is steadily reduced by the redshift effect. This was therefore considered one way of resolving a long-standing paradox known as the *Olbers paradox*. In 1826 Heinrich Olbers from Germany had computed the background from a uniform distribution of sources in a Euclidean universe of infinite extent in space and time. Using (4.104), Olbers concluded that the nett flux in infinite! The Olbers paradox is often phrased as the question 'Why is the sky dark at night?'. By using (4.105) instead of (4.104) we see that attenuation at large redshifts results in $f(\nu_0)$ being finite. Various aspects of the Olbers paradox are discussed in Exercises 25–29, which demonstrate that the expanding universe is not the only way of arriving at a finite answer.

4.10   Cosmological models with the λ-term

Although our concern in this chapter is with the Friedmann models, we now discuss briefly another class of models that have a close relationship with the Friedmann models. These are the models given by the modified Einstein equations of (2.104) – the equations containing the cosmological constant λ. We have already discussed two special cases of this class of solutions in the last chapter, the static Einstein model and the empty de Sitter model. When Hubble's observations established the expanding-universe picture, Einstein conceded that there was no special need for the λ-term in his equations. In the post-Hubble-law era, he dropped this term from his equations; the Einstein–de Sitter model discussed in this chapter was the outcome of Einstein's collaboration with de Sitter after abandoning the λ-term.

Nevertheless, in the 1930s eminent cosmologists such as A. S. Eddington and Abbé Lemaître felt that the λ-term introduced certain attractive features into cosmology and that models based on it should also be discussed at length. In modern cosmology the reception given to the λ-term has varied from the hostile to the ecstatic. The term is quietly forgotten if the observational situation does not demand models based on it. It is resurrected if it is found that the standard Friedmann models without this term are being severely constrained by observations. The present compulsion for this term comes partly because of the observational constraints and partly because inputs of particle physics in the very early stages of the universe have provided a new interpretation for the λ-term, which we shall discuss in Chapter 6.

Putting $\lambda \neq 0$, (4.20) and (4.21) are modified to the following:

$$2\frac{\ddot{S}}{S} + \frac{\dot{S}^2 + kc^2}{S^2} - \lambda c^2 = \frac{8\pi G}{c^2}T_1^1, \tag{4.106}$$

$$\frac{\dot{S}^2 + kc^2}{S^2} - \frac{1}{3}\lambda c^2 = \frac{8\pi G}{3c^2}T_0^0. \tag{4.107}$$

The conservation laws discussed in §4.3 are not affected by the λ-term. If we restrict ourselves to dust only, (4.107) gives us the following differential equation in place of (4.41):

$$\frac{\dot{S}^2 + kc^2}{S^2} - \frac{1}{3}\lambda c^2 = \frac{8\pi G\rho_0}{3}\frac{S_0^3}{S^3}. \tag{4.108}$$

Similarly, (4.106) becomes

$$2\frac{\ddot{S}}{S} + \frac{\dot{S}^2 + kc^2}{S^2} - \lambda c^2 = 0. \tag{4.109}$$

Let us first recover the static model of Einstein. By setting $S = S_0$, $\dot{S} = 0$ and $\ddot{S} = 0$ in (4.108) and (4.109), we get

$$\frac{kc^2}{S_0^2} - \frac{1}{3}\lambda c^2 = \frac{8\pi G\rho_0}{3}; \qquad \frac{kc^2}{S_0^2} = \lambda c^2.$$

From these relations it is not difficult to verify that $k = +1$ and we recover the relations obtained in §3.3:

$$\lambda = \frac{1}{S_0^2} \equiv \lambda_c, \tag{4.110}$$

$$\rho_0 = \frac{\lambda_c c^2}{4\pi G}. \tag{4.111}$$

We shall denote by λ_c the critical value of λ for which a static solution is possible. It was pointed out by Eddington that the Einstein universe is unstable. A slight perturbation destroying the equilibrium conditions (4.110) and (4.111) leads to either a collapse to singularity ($S \to 0$) or an expansion to infinity ($S \to \infty$). Eddington and Lemaître instead proposed a model in which λ exceeds λ_c by a small amount. In this case the universe erupts from $S = 0$ (the big bang) and slows down near $S = S_0$, staying thereabouts for a long time and then expanding away to infinity. It was argued that the quasi-stationary phase of the universe would be suitable for the formation of galaxies. This model is illustrated in Figure 4.11, which plots $S(t)$ for a range of values of λ for $k = +1$. The initial (explosive) phase of the Eddington–Lemaître model is shown along the section OP of the curve OPQR, with PQ the quasi-stationary phase and QR the final accelerated expansion. Notice that, for $\lambda < \lambda_c$ the universe contracts (as in the Friedmann case), whereas for $\lambda > \lambda_c$ it ultimately disperses to infinity, resembling the de Sitter universe.

Figure 4.11 also shows by a dotted line another series of models that contract from infinity to a minimum value of $S > 0$ and then expand back to $S \to \infty$. These models are sometimes called *oscillating models of the second kind*, to distinguish them from the models that shrink back to $S = 0$ and are called *oscillating models of the first kind*. This terminology is, however, not quite apt, since there is no repetition of phases in these models as implied by the word 'oscillating'.

The models with $k = 0$ or $k = -1$ do not exhibit these different types of behaviour for $\lambda > 0$. We get from (4.108) a relation of the following type:

$$\dot{S}^2 = -kc^2 + \frac{1}{3}\lambda c^2 S^2 + \frac{8\pi G\rho_0 S_0^3}{3S}, \tag{4.112}$$

wherein each term on the right-hand side is non-negative. Thus $\dot{S}$ does not change sign and we get ever-expanding models. For $\lambda < 0$, however, we can get universes that expand and then recontract as in the $k = 1$ case for $\lambda < \lambda_c$.

This concludes our discussion of the general dynamical behaviour of the λ cosmologies. We end this section by writing (4.108) and (4.109) for the present epoch in terms of H_0 and q_0. Thus, in place of earlier relations, we have

$$H_0^2 + \frac{kc^2}{S_0^2} - \frac{1}{3}\lambda c^2 = H_0^2\Omega_0,$$

$$(1 - 2q_0)H_0^2 + \frac{kc^2}{S_0^2} - \lambda c^2 = 0.$$

From these we get

$$\Omega_0 = 2q_0 + \frac{2}{3}\lambda\frac{c^2}{H_0^2}. \tag{4.113}$$

Now there is no unique relationship between q_0 and Ω_0: we have an additional parameter entering the relation. *Note also that it is possible to have negative q_0, that is, an accelerating expansion, if $\lambda > 0$.* This is because the λ-term introduces a force of cosmic *repulsion*.

Finally, if the universe is spatially flat, i.e., $k = 0$, then the following relation can confirm the fact:

$$\Omega_0 + \frac{1}{3}\frac{\lambda c^2}{H_0^2} = 1.$$

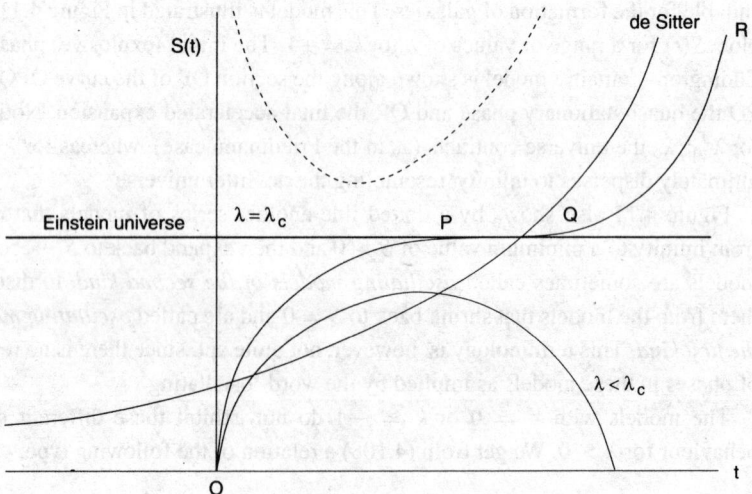

Figure 4.11 The λ cosmologies for $k = 1$. For details see the text.

By writing

$$\frac{1}{3}\frac{\lambda c^2}{H_0^2} = \Omega_\Lambda \tag{4.114}$$

the above relation is expressed in the form

$$\Omega_0 + \Omega_\Lambda = 1. \tag{4.115}$$

See Figure 4.12 showing these relationships in the $(\Omega_0, \Omega_\Lambda)$ plane.

A few words are needed here to explain to the reader one reason why the λ models are preferred these days. The measurements of Hubble's constant and the estimates of ages of stars in globular clusters suggest that the ages of the $\lambda = 0$ models are inadequate to accommodate the stellar ages. As Figure 4.13 shows, by having a

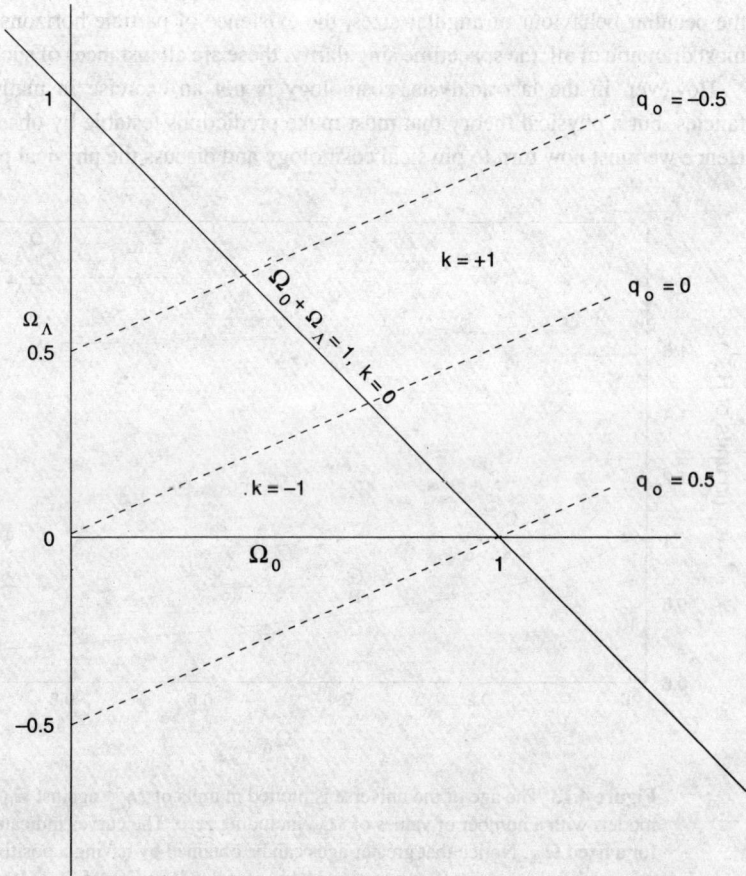

Figure 4.12 The $(\Omega_0, \Omega_\Lambda)$ plane with the line $\Omega_0 + \Omega_\Lambda = 1$ drawn, to indicate $k = 0$. Above this line $k = 1$; below it, $k = -1$.

positive cosmological constant, the age of the universe can be increased. Therefore, the age constraint is relaxed if $\lambda > 0$.

4.11   Concluding remarks

Our discussion of the dynamical and geometrical properties of the expanding universe is now complete. We started with Newtonian cosmology and derived simple models, which turn out to be very similar to those of general relativity – the theory that introduced the unique idea that gravitational effects are intimately connected with the non-Euclidean geometry of spacetime. Indeed, relativity provides a better perspective on cosmology than does the Newtonian theory, as we have discussed before. Nowhere except in cosmology do we see examples of the large-scale effects of non-Euclidean geometry. The redshift, the dimming of light from distant sources, the peculiar behaviour of angular sizes, the existence of particle horizons and, the most dramatic of all, the spacetime singularity: these are all instances of such effects.

However, in the last analysis, cosmology is not an exercise in mathematical fancies, but a physical theory that must make predictions testable by observations. Hence we must now turn to physical cosmology and discuss the physical properties

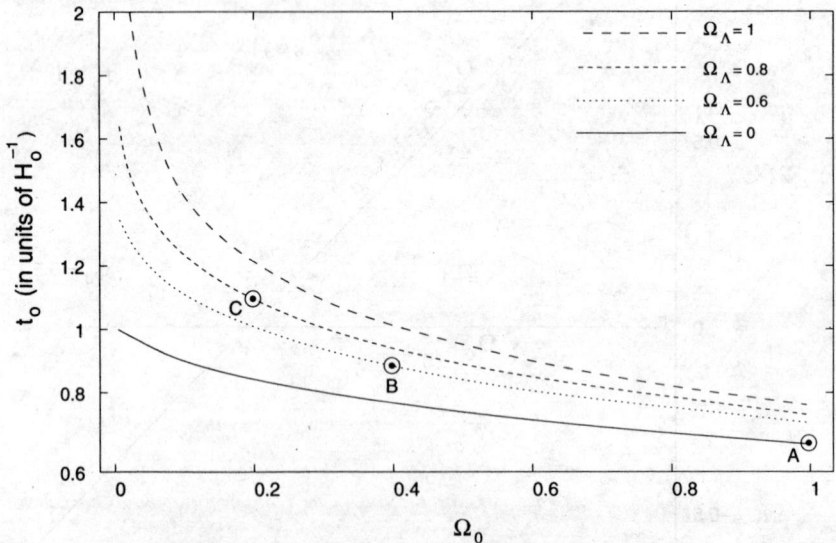

Figure 4.13 The age of the universe is plotted in units of H_0^{-1} against Ω_0 for models with a number of values of Ω_Λ, including zero. The curves indicate the ages for a fixed Ω_Λ. Notice that greater ages can be obtained by having a positive cosmological constant. (See equation (4.114) for the definition of Ω_Λ.) It is also interesting to see that points A, B, C, ... correspond to progressively greater ages as the proportion of Ω_Λ in the relation $\Omega_\Lambda + \Omega_0 = 1$ is increased.

of the expanding universe. Do we have any relics of the early epochs just after the big bang? How did galaxies, which we have taken as the basic units of the universe, form in the first place? How did matter itself come into existence in its elementary state? The next three chapters will deal with such issues.

Exercises

1 Verify the expressions for the Ricci tensor and the Einstein tensor for the Robertson–Walker line element.

2 Show how the assumptions of the Weyl postulate and the cosmological principle reduce the number of independent Einstein equations from ten to two. What more can be deduced about these equations with the help of the conservation law?

3 Deduce (4.24) from (4.25).

4 Using the Einstein–de Sitter model, estimate the epoch during which the matter and radiation densities in the universe were equal. For this calculation take $\rho_0 = 10^{-29}$ g cm^{-3} and $\epsilon_0 = 10^{-13}$ erg cm^{-3} and express your answer as the fraction of the age of the universe.

5 What is the significance of the closure density? Show that there is a unique relationship between the deceleration parameter q_0 and the density parameter Ω_0 in a Friedmann universe. How is this relation modified by the λ-term?

6 A galaxy is observed with redshift 0.69. How long did light take to travel from the galaxy to us if we assume that we live in the Einstein–de Sitter universe with Hubble's constant $H_0 = 70$ km s^{-1} Mpc^{-1}?

7 Calculate from first principles the age of a Friedmann universe with $q_0 = 2$.

8 In the Friedmann universe with $q_0 = 1$, a galaxy is seen with redshift $z = 1$. How old was the universe at the time this galaxy emitted the light received today? (Take $H_0 = 100$ km s^{-1} Mpc^{-1}.)

9 A ray of light is emitted during the present epoch in the closed Friedmann universe. Discuss the possibility of this ray making a round trip of the universe and coming back to its starting point.

10 Derive the formulae for r_1 and D_1 for the open Friedmann model with given q_0 and H_0.

11 Show that the expressions for D_1 in the cases $q_0 = \frac{1}{2}$ and $q_0 = 0$ can be obtained from (4.81) by a suitable limiting process.

12 Show that there is a unique value of q_0 for which the linear Hubble law holds exactly.

13 Invert the formula (4.81) to express z as a function of $x \equiv D_1 H_0/c$. Show that

$$z = q_0 x - (q_0 - 1)\left(\sqrt{1 + 2x} - 1\right).$$

Use this relation to show how the linear Hubble velocity–distance relation begins to fan out for cosmological models with different values of q_0.

14 Show by computing R_{iklm} that the Friedmann model with $q_0 = 0$ describes a flat spacetime.

15 Given that the Friedmann model with $q_0 = 0$ describes a flat spacetime, find coordinates in which its line element is manifestly that of Minkowski spacetime.

16 What is meant by a particle horizon? How does the size of the particle horizon depend on the epoch of observation in a given Friedmann model?

17 Show why the Friedmann models with $\lambda = 0$ do not have event horizons.

18 In the Einstein–de Sitter model there are n sources in a unit comoving coordinate volume. Calculate the number of sources in principle visible to us during the present epoch, for n = constant.

19 In a Friedmann model the minimum of angular size occurs at $z = 1$. Deduce from this the value of q_0.

20 The *surface brightness* of an astronomical object is defined by the flux received from the object divided by the angular area subtended by the object at the observation point. How does the surface brightness vary with the redshift?

21 Show from first principles that the angular sizes of astronomical objects of fixed linear size will have a minimum at $z = 1.25$ in the Einstein–de Sitter model.

22 Derive (4.93) by direct differentiation of $D_1(z)$ with respect to z.

23 Derive (4.96) from first principles and use it to show that $z_m \to \infty$ as $q_0 \to 0$.

24 If in a Friedmann universe we have a fixed number of sources in a unit comoving coordinate volume and each source emits a line radiation of fixed total intensity L_0 at frequency $\bar{\nu}$, show that the radiation background produced by such sources during the present epoch will have the frequency spectrum $S(\nu)\, d\nu$ where $S(\nu) = 0$ for $\nu > \bar{\nu}$, whereas for $\nu < \bar{\nu}$

$$S(\nu) = \frac{c}{H_0} n_0 L_0 \frac{\nu^{3/2}}{\bar{\nu}^2 \sqrt{2q_0\bar{\nu} - (2q_0 - 1)\nu}},$$

where n_0 is the proper number density of sources during the present epoch.

25 Discuss Olbers' calculation on the darkness of the night sky in the Euclidean universe.

26 Show that a finite answer can be obtained in an Olbers-type calculation if the universe is finite in extent or finite in age.

27 Show that a finite answer can be obtained in an Olbers-type calculation by assuming that the sources are finite in size and therefore that nearby sources tend to block radiation from the more distant sources.

28 Show that the Olbers paradox can be resolved by assuming that a typical source can radiate only for a finite interval of time because of its finite reservoir of energy.

29 Review all the possible means of resolving the Olbers paradox and state your own preference for any particular resolution.

30 Derive (4.106) and (4.107) for the λ cosmologies and deduce the conservation law from them.

31 Given that objects during the quasi-stationary phase of the Eddington–Lemaître cosmology are now seen with the redshift $z = 2$, what can you say about the value of λ?

32 Deduce that the scale factor in the λ cosmology with $k = 1$ satisfies the differential equation

$$\dot{S}^2 = c^2\left(\frac{1}{3}\lambda S^2 - 1 + \frac{\gamma}{S}\right),$$

where

$$\gamma = \frac{2q_0 + \dfrac{2}{3}\dfrac{\lambda c^2}{H_0^2}}{\left(2q_0 - 1 + \dfrac{\lambda c^2}{H_0^2}\right)^{3/2}}\left(\frac{c}{H_0}\right).$$

33 Write down an integral that gives the age of a big-bang universe for $\lambda \neq 0$. Discuss qualitatively how the λ-term may be used to increase the age of the universe.

34 In λ cosmology, what is the lower limit on the value of λ given the value of q_0?

35 Compute the invariants R, $R_{ik}R^{ik}$ and $R_{iklm}R^{iklm}$ for the Friedmann models and show that they all diverge as $S \to 0$. Is there an exceptional case?

36 Repeat Exercise 35 for λ cosmologies and show that the same conclusions follow for the models with $S \to 0$.

37 Give a general argument to show that, for sufficiently small S, the λ force is ineffective at preventing the spacetime singularity.

Chapter 5

Relics of the big bang

5.1   The radiation-dominated universe

In Chapter 4 we saw that all Friedmann models have an epoch in the past when the scale factor S was zero. We referred to this epoch as the big-bang epoch. To mathematicians the big bang implies a breakdown of the concept of spacetime geometry and they have come to recognize it as an inevitable feature of Einstein's general relativity. It is a feature that prevents the physicist from investigating what happened at $S = 0$ or prior to it. To some physicists this abrupt termination of the past signifies an incompleteness of the theory of relativity. To them a more complete theory of the future may show a way of avoiding the catastrophic nature of the $S = 0$ epoch. A universe that has been expanding forever or that has been oscillating between maximum and minimum (but finite) values of S might result from such a theory. We shall look at models of this kind in Chapter 9.

Here we will follow the so-called *standard* approach in cosmology and continue to put our faith in the validity of general relativity while pushing our investigations into the past of the universe as close as possible to the $S = 0$ epoch. The purpose of such investigations will be to find out whether we can point to any present-day evidence that the universe indeed had a past epoch when S was close to zero. In short, we will be looking for *relics* of the big bang.

Pioneering work in this field was done by George Gamow in the mid-1940s. Gamow (see Figure 5.1) was concerned with the problem of the origin of the elements. Starting from the (then available) basic building blocks of neutrons and protons, Gamow attempted to describe the formation of nuclei of deuterium, helium and so on. The process envisaged by him involved nuclear fusion, that is, a process in which nuclei are formed by bringing together neutrons and protons. Astrophysicists were already sure by the 1940s that such processes operate inside stars, where the necessary conditions of high temperature and density were known to exist. Gamow pointed out that similar conditions must have existed in a typical Friedmann universe soon after the big bang.

We know from (4.32) that the density ρ was very high at small values of S. However, we also saw in Chapter 4 that sufficiently early on the universe would have been radiation-dominated. That is, the expansion of such a universe was controlled largely by radiation rather than by matter. Figure 5.2 reproduces Figure 4.2 illustrating how the present ratio of matter to radiation energy density determines the epoch when the universe switched over from radiation domination to matter domination. The redshift of such an epoch *vis-à-vis* the present epoch would have been at least $\sim 10^3$.

What about temperature? In the radiation-dominated era, the temperature was determined by radiation and a simple calculation shows how the temperature also might have been high. This calculation requires the assumption that at present we have a radiation density u_0 that is a relic of an early hot era. With this assumption, the radiation energy density for a past epoch S is given by (4.37):

$$u = u_0 \frac{S_0^4}{S^4}. \tag{5.1}$$

Figure 5.1 George Gamow (1904–1968).

Gamow therefore assumed that the dynamics of expansion during the early epochs was determined by radiant energy rather than by matter in the form of dust and that these were high-temperature epochs.

We illustrate the above ideas with a simplified calculation by assuming that the radiation was in blackbody form with temperature T, so that

$$u = aT^4, \tag{5.2}$$

where a is the radiation constant. This means that, in the early stages of the big-bang universe,

$$T_0^0 = aT^4, \qquad T_1^1 = T_2^2 = T_3^3 = -\tfrac{1}{3}aT^4. \tag{5.3}$$

We also expect that the space-curvature parameter k will not affect the dynamics of the early universe significantly and set it equal to zero. This assumption, known as the *assumption of flatness*, is non-trivial, but we will await the next chapter before discussing it in detail. For the time being we accept it uncritically. Thus, from (4.21),

$$\frac{\dot{S}^2}{S^2} = \frac{8\pi Ga}{3c^2}T^4. \tag{5.4}$$

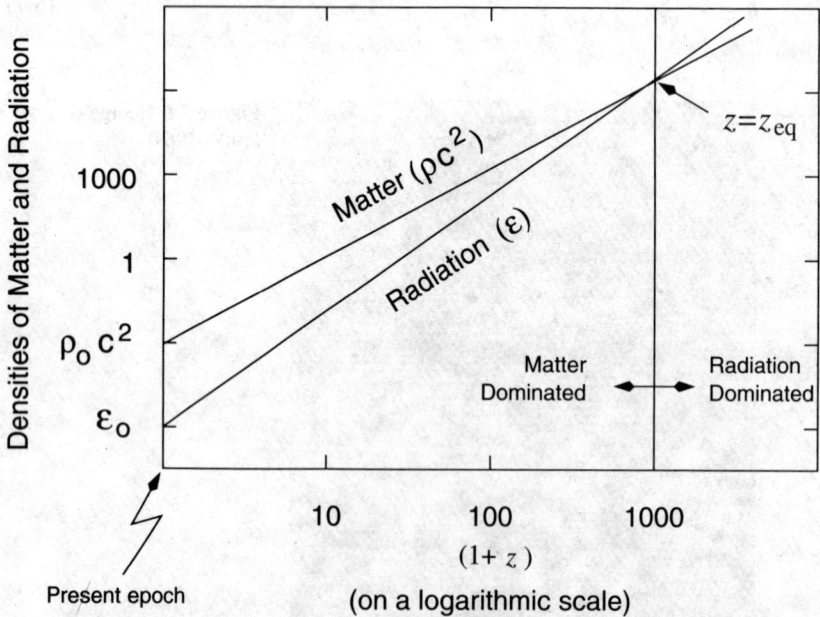

Figure 5.2 Figure 4.2 from the previous chapter has been reproduced here for ready reference. As explained there, the energy density of radiation drops as S^{-4}, whereas the energy density of matter drops as S^{-3}. Since at present the latter is at least 10^3 times the former, the epoch of equality of the two occurred when the scale factor was at least that much smaller.

Furthermore, from (5.1) and (5.2) we get

$$T = A/S, \qquad A = \text{constant}. \tag{5.5}$$

Substituting (5.5) into (5.4) gives a differential equation for S that can be easily solved. Setting $t = 0$ at $S = 0$ we get

$$S = A \left(\frac{3c^2}{32\pi Ga} \right)^{-1/4} t^{1/2} \tag{5.6}$$

and, more importantly,

$$T = \left(\frac{3c^2}{32\pi Ga} \right)^{-1/4} t^{1/2}. \tag{5.7}$$

Notice that all the quantities inside the parentheses on the right-hand side of the above equation are known physical quantities. Thus, by substituting their values into (5.7), we can express the above result in the following form:

$$T_{\text{kelvin}} = 1.52 \times 10^{10} t_{\text{second}}^{-1/2}. \tag{5.8}$$

In other words, about 1 s after the big bang the radiation temperature of the universe was 1.52×10^{10} K. The universe at this stage was certainly hot enough to have free neutrons and protons around, which, as the universe expanded, cooled down to facilitate nucleosynthesis, as Gamow supposed.

The idea of a *hot big bang*, as the above picture is called, depends therefore on the assumption that there is relic radiation present today. Later in this chapter we will present the argument that the microwave background discovered in 1965 by Arno Penzias and Robert Wilson is that relic radiation. For the present we will accept this evidence as confirming Gamow's notion of the hot big bang and proceed further.

5.2 Thermodynamics of the early universe

It was in 1946 that George Gamow first wrote a paper on primordial nucleosynthesis. The first detailed paper on this subject appeared in 1948 under the authorship of Gamow's student Ralph Alpher, the nuclear physicist Hans Bethe (who had worked out the solar models of Eddington with nuclear energy generation explicitly put in) and Gamow himself. It is said that Gamow persuaded Bethe to add his name to the paper to make the list of authors read as 'Alpha, Beta and Gamma'! In any case, the idea became popular as the *α–β–γ theory*.

Considerable progress has been made in our understanding of the properties of particles and their basic interactions since the days when Gamow and his students Alpher and Robert Herman did their calculations of primordial nucleosynthesis. Gamow's original programme of building up *essentially all elements* during the early moments of the universe did not work out. For reasons that we shall elaborate on later in this chapter, only the light nuclei could be synthesized in the primordial fusion processes. In the following pages we will briefly outline the basic principles on which the modern calculations are based.

First it is necessary to specify the building blocks from which nuclei were constructed in the early epochs. The physicist would naturally like to imagine that the universe started with the simplest possible composition (whatever that may be!) and that more complex structures were built out of simpler ones by physical interactions. Thus the cosmologist is encouraged to take stock of the knowledge of particle physics. Although Gamow and his colleagues took the existence of particles like protons, neutrons, electrons and so on for granted, modern particle physicists believe that a more basic framework should account for the creation or existence of these particles. In Chapter 6 we will consider the more speculative earlier epochs and discuss how these particles came into existence.

Here we take up the story from the stage when baryons (neutrons and protons), leptons (electrons, muons, neutrinos and their antiparticles) and photons (the particles of light) were already in existence and were in thermodynamic equilibrium as particles of an ideal gas. Before proceeding with calculations we must clarify what is meant by 'thermodynamic equilibrium' and 'ideal gas'. We have already mentioned that the dominant form of energy during these early epochs was in particles moving relativistically. The question that arises therefore is that of whether these particles were interacting with one another or instead were mostly moving freely. Such particles would interact and collide, of course, but these instances are assumed to have occupied very brief time spans, so that their effects on motions may be otherwise neglected. We will shortly express this idea in a quantitive manner.

The collisions and scatterings of the particles would, however, have helped to redistribute their energies and momenta. If these redistributions occurred frequently enough, the system of particles as a whole would have reached a state of thermodynamic equilibrium. In this case, for each species of particles there is a definite rule governing the number of particles in a given range of momentum. For thermodynamic equilibrium to be reached, the time scales between successive scatterings should be small relative to the time scale for expansion of the universe. Again, we will express this idea quantitively in a short while.

5.2.1   Distribution functions

Assuming that the ideal gas approximation and thermodynamic equilibrium hold, it is possible to write down the distribution functions of any given species of particles. Let us use the symbol A to denote typical species ($A = 1, 2, \ldots$). Thus $n_A(P) \, dP$ denotes the number density of species in the momentum range $(P, P + dP)$, where

$$n_A(P) = \frac{g_A}{2\pi^2 \hbar^3} P^2 \left[\exp\left(\frac{E_A(P) - \mu_A}{kT}\right) \pm 1 \right]^{-1}. \tag{5.9}$$

In the above formula T is the temperature of the distribution, g_A is the number of spin states of the species and k is Boltzmann's constant, while

$$E_A^2 = c^2 P^2 + m_A^2 c^4 \tag{5.10}$$

is the energy corresponding to the rest mass m_A of a typical particle. Thus for the electron $g_A = 2$; for the neutrino $g_A = 1$ and $m_A = 0$; and so on. The $+$ sign in (5.9) applies to particles obeying the Fermi–Dirac statistics (these particles are called *fermions*), whereas the $-$ sign applies to particles obeying the Bose–Einstein statistics (particles known as *bosons*). For example, electrons and neutrinos are fermions, whereas photons are bosons.

The quantity μ_A is the chemical potential of the species A. For a detailed discussion of chemical potentials, see any standard text on thermodynamics and statistical mechanics. We note here that, in any reaction involving these particles, the μ_A are conserved (just as electric charge, energy, spin and so on are conserved). Because photons can be absorbed or emitted in any number in a typical reaction, we set $\mu_A = 0$ for photons. From thermodynamic considerations we can also deduce this result by arguing that, for blackbody radiation, the number of photons is not fixed but determined from conditions of thermal equilibrium. Thus we require that the Helmholtz free energy $U - TS$ should be a minimum, which leads to the result $\mu_A = 0$ for photons. Since particles and antiparticles (such as electrons and positrons) annihilate in pairs and produce photons, their chemical potentials are equal and opposite.

Apart from the dynamical quantities as well as the electric charge, several other quantities are found to be conserved in the interactions of particles. These are the baryon number, the muon lepton number and the electron lepton number. In computing these numbers, a value of $+1$ is assigned to a particle and -1 to its antiparticle. The electron lepton number counts electrons (e^-) and their neutrinos (ν_e), while the muon lepton number counts muons (μ^-) and their neutrinos (ν_μ). Under these conservation rules reactions such as

$$n \to p + e^- + \bar{\nu}_e, \qquad p + \bar{\nu}_\mu \to \mu^+ + n$$

are permitted, whereas a reaction like the following is not:

$$n \to p + e^- + \nu_e.$$

(In Chapter 6 we will consider the situation in which the baryon number is not conserved. For the epochs with which we are concerned here, however, we may safely assume the conservation of baryon number to apply.)

Hence, if we assume that electric charge, the baryon number, the electron lepton number and the muon lepton number are conserved in any reaction, then we have only four independent chemical potentials – those corresponding to protons, electrons, electron neutrinos and muon neutrinos. (In Chapter 6 we will consider the possibility of more species of leptons/neutrinos being present.) From (5.9) we see that the total number of particles per unit volume of each of these species is needed in order to determine the corresponding μ_A and that the number densities will be large for large $\mu_A > 0$. These number densities are not known with any degree of accuracy, except that (as we shall see later in this chapter) the ratio

$$\frac{N_B}{N_\gamma} = \frac{\text{Number density of baryons}}{\text{Number density of photons}} \sim 10^{-8}\text{--}10^{-10}$$

is small relative to unity.

The smallness of the baryon number density suggests that the number densities of leptons may also be small relative to N_γ, and it is usually assumed that this hypothesis provides a good justification for taking $\mu_A = 0$ for all species. We will assume that $\mu_A = 0$ for all species in our calculations to follow.

We then get the following integrals for the number density (N_A), the energy density (ϵ_A), pressure (p_A) and entropy density (s_A) of particle A:

$$N_A = \frac{g_A}{2\pi^2\hbar^3} \int_0^\infty \frac{P^2\,dP}{\exp[E_A(P)/(kT)] \pm 1}, \tag{5.11}$$

$$\epsilon_A = \frac{g_A}{2\pi^2\hbar^3} \int_0^\infty \frac{P^2 E_A(P)\,dP}{\exp[E_A(P)/(kT)] \pm 1}, \tag{5.12}$$

$$p_A = \frac{g_A}{6\pi^2\hbar^3} \int_0^\infty \frac{c^2 P^4 [E_A(P)]^{-1}\,dP}{\exp[E_A(P)/(kT)] \pm 1}. \tag{5.13}$$

$$s_A = (p_A + \epsilon_A)T. \tag{5.14}$$

We can deduce a simple relation from these formulae to show that the entropy in a given comoving volume remains constant as the universe expands. Differentiate p_A with respect to T to get

$$\frac{dp_A}{dT} = \frac{g_A}{6\pi^2\hbar^3} \int_0^\infty \frac{c^2 P^4 \exp[E_A(P)/(kT)]\,dP}{\{\exp[E_A(P)/(kT)] \pm 1\}^2 kT^2}.$$

Now integrate by parts to get for the above integral

$$\frac{g_A}{g\pi^2\hbar^3 T} \int_0^\infty \frac{(3P^2 E_A + c^2 P^4 E_A^{-1})\,dP}{\exp[E_A(P)/(kT)] \pm 1} = \frac{p_A + \epsilon_A}{T}.$$

Defining the pressure, energy density and entropy density for a mixture of such gases in thermodynamic equilibrium by

$$p = \sum_A p_A, \qquad \epsilon = \sum_A \epsilon_A, \qquad s = \sum_A s_A, \tag{5.15}$$

we have the following relation:

$$\frac{dp}{dT} = \frac{p + \epsilon}{T}. \tag{5.16}$$

We will use this relation next in the expanding universe. For we shall see that, as the universe expands, it cools adiabatically.

5.2.2 The behaviour of entropy

We first recall the conservation law satisfied by ϵ and p during the early stages of the expanding universe, the law given by (4.24),

$$\frac{d}{dS}(\epsilon S^3) + 3pS^2 = 0, \tag{5.17}$$

and use it in conjunction with the second law of thermodynamics. This law tells us that the entropy in a given volume S^3 stays constant as the volume expands adiabatically. From (5.14) and (5.15) we therefore get

$$\frac{d}{dt}(S^3 s) = \frac{d}{dt}\left(\frac{S^3}{T}(p + \epsilon)\right) = 0, \tag{5.18}$$

where, as defined in (5.15), $s = \Sigma_A s_A$ is the total entropy of all the particles in the expanding volume.

Rewriting (5.18) with the help of (5.17) we get

$$0 = \frac{d}{dt}\left(\frac{S^3 p}{T}\right) + \frac{1}{T}\frac{d}{dt}(S^3 \epsilon) + S^3 \epsilon \frac{d}{dt}\left(\frac{1}{T}\right) \tag{5.19}$$

$$= \frac{d}{dt}\left(\frac{S^3 p}{T}\right) - \frac{3pS^2}{T}\dot{S} + S^3 \epsilon \frac{d}{dt}\left(\frac{1}{T}\right),$$

that is,

$$\frac{dp}{dT} = \frac{1}{T}(p + \epsilon). \tag{5.20}$$

Notice that we directly derived the above result in (5.16) from (5.12) and (5.13) by a simple manipulation of the integrals in the last subsection. Thus, starting from (5.19) we can derive (5.18). We will use the constancy of

$$\sigma = \frac{S^3}{T}(p + \epsilon) \tag{5.21}$$

in our later calculations.

In the high-temperature approximation we get $p = \epsilon/3 \propto S^{-4}$ from (5.18). Hence, from the constancy of σ, we recover the relation (5.5):

$$T \propto 1/S. \tag{5.22}$$

A simple relation like this does not hold if the high-temperature approximation is not valid.

5.2.3   High- and low-temperature approximations

The above expressions become simplified for particles moving relativistically. In this case, the mean kinetic energy per particle far exceeds the rest-mass energy of the particle, an inequality expressed by

$$T \gg \frac{m_A c^2}{k} \equiv T_A. \tag{5.23}$$

This is called the *high-temperature approximation*, or the *relativistic limit*.

The thermodynamic details for the various species of interest are given in Table 5.1. The numbers are expressed in units of the quantities for the photon ($g_A = 2$, the symbol for photon is γ):

$$N_\gamma = \frac{2.404}{\pi^2} \left(\frac{kT}{c\hbar} \right)^3, \qquad \epsilon_\gamma = \frac{\pi^2 (kT)^4}{15\hbar^3 c^3} = 3p_\gamma, \qquad s_\gamma = \frac{4\pi^2 k}{45} \left(\frac{kT}{c\hbar} \right)^3. \tag{5.24}$$

In this approximation consider the electric potential energy of any two electrons separated by distance r. This is given by

$$U = e^2/r.$$

Table 5.1 Thermodynamic quantities for various particle species at $T \gg T_A$

Particle species A	Symbol	T_A (K)	g_A	N_A/N_γ	$\epsilon_A/\epsilon_\gamma$	S_A/S_γ
Photon	γ	0	2	1	1	1
Electron	e^-	5.93×10^9	2	3/4	7/8	7/8
Positron	e^+		2	3/4	7/8	7/8
Muon	μ^-	1.22×10^{12}	2	3/4	7/8	7/8
Antimuon	μ^+		2	3/4	7/8	7/84
Muon and electron	ν_μ, ν_e	0	1	3/8	7/16	7/16
neutrinos and their	$\bar{\nu}_\mu, \bar{\nu}_e$		1	3/8	7/16	7/16
antineutrinos						
Pions	π^+		1	1/2	1/2	1/2
	π^-	1.6×10^{12}	1	1/2	1/2	1/2
	π^0		1	1/2	1/2	1/2
Proton	p	10^{13}	2	3/4	7/8	7/8
Neutron	n	$T_n - T_p$	2	3/4	7/8	7/8
		$\sim 1.5 \times 10^{10}$				

Now the average inter-electron distance is given by $N_e^{-1/3} \sim c\hbar/(kT)$. Thus the average interaction energy is

$$(U) \sim \frac{e^2}{\hbar c} kT.$$

However, kT measures the energy of motion of electrons. Thus the interaction energy is $e^2/(\hbar c) \sim 1/137$ of the energy of motion. Since the fraction is small, we are justified in treating the electrons as a free gas.

In contrast, at *low temperatures* $T \leq T_A$ we have for all species with $m_A \neq 0$

$$N_A = \frac{g_A}{\hbar^3}\left(\frac{m_A kT}{2\pi}\right)^3 \exp\left(-\frac{T_A}{T}\right),$$

$$\epsilon_A = m_A N_A, \qquad p_A = N_A kT, \qquad s_A = \frac{m_A N_A}{T} c^2. \qquad (5.25)$$

Notice that, with falling temperature all these quantities drop off rapidly. We will often refer to this limit as the *non-relativistic approximation*. (For the photon and a zero-rest-mass neutrino $T_A = 0$ and this approximation never applies.)

When one is applying these results to cosmology, the following considerations usually count. First, the expansion of the universe is controlled by the species that are in the relativistic limit, for these are the particles that are present in abundance. Heavier species are fewer in number because of the exponential damping of equation (5.25). Thus, as the temperature of the universe drops with expansion, the heavier species progressively diminish in dynamical importance.

Example

What will the energy–temperature relationship for a cosmic brew containing particles of Table 5.1 be when the temperature is 10^{10} K? We note that the relativistic approximation applies to the electrons and positrons, whereas the photons and the two neutrino–antineutrino pairs will always be relativistic. Adding contributions to the energy density as per Table 5.1 for these species in the above order and using (5.24), we get

$$\epsilon = \left(\frac{7}{8} + \frac{7}{8} + 1 + 4 \times \frac{7}{16}\right)\epsilon_\gamma = \frac{9}{2}\epsilon_\gamma = \frac{9}{2}aT^4.$$

Notice that, if the same calculation were carried out at the temperature of 10^8 K, only the photons and the four neutrino species would be relativistic and at that stage we would have the energy-density–temperature relation as $\epsilon = (11/4)aT^4$. Thus the coefficient in the ϵ–T relationship changes as per the relativistic component of the cosmic brew.

Next, thermodynamic equilibrium is required in general for ascribing a common temperature to the various species. It may happen that some species do not interact strongly enough to maintain the equilibrium. Whether this is actually the case can be determined by comparing the rate of reaction for the specific species (with other particles) with the rate of expansion of the universe given by $H(t)$. This procedure is illustrated by our description next of how neutrinos interact during the expansion of the universe.

5.3   Primordial neutrinos

From Table 5.1 we see that for $T < 1.5 \times 10^{12}$ K, the only particles that could be present with appreciable number densities in thermal equilibrium were $\mu^{\pm}, e^{\pm}, \nu_e, \bar{\nu}_e, \nu_\mu, \bar{\nu}_\mu$ and γ. The baryons (p and n) and pions ($\pi^{\pm}$ and π^0) would have cooled below their critical temperatures T_A and so were subject to the low-temperature approximation. The photons, $e^{\pm}$, and $\mu^{\pm}$ follow their respective distributions of the type (5.9). The neutrinos, however, require some attention, since this phase happens to be crucial in determining the extent of their survival.

Apart from gravitation, the neutrinos take part only in the weak interaction and are absorbed, emitted, or scattered in reactions such as the following:

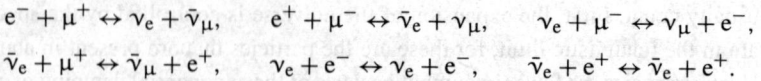

$$e^- + \mu^+ \leftrightarrow \nu_e + \bar{\nu}_\mu, \qquad e^+ + \mu^- \leftrightarrow \bar{\nu}_e + \nu_\mu, \qquad \nu_e + \mu^- \leftrightarrow \nu_\mu + e^-,$$

$$\bar{\nu}_e + \mu^+ \leftrightarrow \bar{\nu}_\mu + e^+, \qquad \nu_e + e^- \leftrightarrow \nu_e + e^-, \qquad \bar{\nu}_e + e^+ \leftrightarrow \bar{\nu}_e + e^+.$$

The rates of these are all determined by the coupling strength of weak interactions. For $T \leq T_\mu$ the cross section of a typical reaction is of the order

$$\Sigma = \mathcal{G}^2 \hbar^{-4} (kT)^2 c^{-4}, \tag{5.26}$$

where $\mathcal{G} = 1.4 \times 10^{-49}$ erg cm^{-3} is the weak-interaction coupling constant. From (5.24) and Table 5.1 we see that the number density of participating particles $e^{\pm}$ is of the order

$$[kT/(c\hbar)]^3,$$

whereas for muons we should take account of (5.25) and introduce an exponential damping factor of

$$\exp(-T_\mu/T).$$

Thus the typical reaction rate for neutrinos is

$$\eta = c\Sigma \left(\frac{kT}{c\hbar}\right)^3 \exp\left(-\frac{T_\mu}{T}\right) = \mathcal{G}^2 \hbar^{-7} c^{-6} (kT)^5 \exp\left(-\frac{T_\mu}{T}\right). \tag{5.27}$$

We must now take note of the other rate that is relevant to the maintenance of equilibrium of neutrinos – the rate at which a typical volume enclosing them expands. From Einstein's equations we get

$$H^2 = \frac{\dot{S}^2}{S^2} = \frac{8\pi G}{3c^2}\epsilon \approx \frac{16\pi^3 G}{90\hbar^3 c^5}(kT)^4. \tag{5.28}$$

H, the Hubble constant for the particular epoch, measures the rate of expansion of the volume in question. Thus the ratio of the reaction rate to the expansion rate is given by

$$\frac{\eta}{H} \sim G^{-1/2}\hbar^{-11/2}\mathcal{G}^2 c^{-7/2}(kT)^3 \exp\left(-\frac{T_\mu}{T}\right) \tag{5.29}$$

$$\sim \left(\frac{T}{10^{10}\,\text{K}}\right)^3 \exp\left(-\frac{10^{12}\,\text{K}}{T}\right)$$

$$= T_{10}^3 \exp\left(-\frac{1}{T_{12}}\right). \tag{5.30}$$

Here we have substituted the values of $G, \hbar, \mathcal{G}, c, k$ and T_μ and arrived at the above numerical expression. Furthermore, we have written the temperatures using the notation that T_n indicates temperature expressed in units of 10^n K, i.e.,

$$T_n = \frac{T}{10^n\,\text{K}}.$$

5.3.1 The neutrino-decoupling epoch

What does (5.30) tell us? As the temperature drops below 10^{12} K, the exponential decreases rapidly. This means that the reactions involving neutrinos run at a slower rate than the rate of expansion of the universe. The neutrinos then cease to interact with the rest of the matter and therefore drop out of thermal equilibrium as temperatures fall appreciably below $T_{12} = 1$. How far below?

The original theory of weak interactions suggested that this temperature may be around $T_{11} = 1.3$. In the late 1960s and early 1970s successful attempts to unify the weak interaction with the electromagnetic interaction led to additional (neutral-current) reactions that keep neutrinos interacting with other matter at even lower temperatures. The outcome of these investigations is that the neutrinos can remain in thermal equilibrium down to temperatures of the order of $T_{10} = 1$.

However, even though neutrinos decouple themselves from the rest of the matter, their distribution function still retains its original form with the temperature dropping as $T \propto S^{-1}$. This is because as the universe expands the momentum and energy of each neutrino fall as S^{-1} and the number density of neutrinos falls as S^{-3} (see Chapter 3, Exercise 16). Since the temperature of the rest of the mixture also drops as S^{-1} and since the two temperatures were equal when the neutrinos were coupled with the rest of the matter, they continue to remain equal even though neutrinos and the rest of the matter are no longer interacting with one another. These remarks about neutrinos are meant to apply to all four species $\nu_e, \bar{\nu}_e, \nu_\mu$ and $\bar{\nu}_\mu$.

5.3.2   The era of e^- and e^+ annihilation

There is, however, another (later) epoch when the neutrino temperature begins to differ from the temperature of the rest of the matter. First consider the universe in the temperature range $T_{12} = 1$ to $T_{10} = 1$. In this phase we have the neutrinos, the electron–positron pairs and the photons, each with distribution functions of type (5.9) in the high-temperature approximation (see Table 5.1). Thus, referring back to the example at the end of the previous section, we get

$$\epsilon = \tfrac{9}{2}aT^4. \tag{5.31}$$

Thus, for this period the expansion equation is modified from our simplified formula (5.4) to

$$\frac{\dot{S}^2}{S^2} = \frac{12\pi Ga}{c^2}T^4 \tag{5.32}$$

and the relation (5.7) is changed to

$$T = \left(\frac{c^2}{48\pi Ga}\right)^{1/4} t^{-1/2}, \tag{5.33}$$

which we may rewrite as

$$T_{10} = 1.04 t_{\text{seconds}}^{-1/2}. \tag{5.34}$$

However, in the next phase the situation becomes complicated, since the electron–positron pairs are no longer relativistic. Thus the high-temperature approximation is no longer valid for them and we have to use the full formulae (5.12) and (5.13) to determine ϵ and p and the rate of expansion of the universe. We will not go into details of this phase but instead jump across to its end, when the pairs have annihilated, leaving only photons:

$$e^- + e^+ \rightarrow \gamma + \gamma$$

Thus the energy, originally in $e^\pm$ and photons, has now been vested only in photons, raising their number and temperature. How can we evaluate this change? It is here that (5.21), telling us of the constancy of σ, comes to our help.

In the relativistic phase ($T_9 > 5$) of $e^\pm$ we have

$$\sigma = \frac{4S^3}{3T}\left(\epsilon_{e^-} + \epsilon_{e^+} + \epsilon_\gamma\right) = \frac{11}{3}a(ST)^3. \tag{5.35}$$

When the $e^\pm$ have annihilated and left only photons, we have the photon temperature T_γ given by

$$\sigma = \frac{4}{3}\frac{S^3}{T_\gamma}\epsilon_\gamma = \frac{4}{3}a(ST_\gamma)^3. \tag{5.36}$$

We now use the result that the neutrino temperature always changes as S^{-1}. Let us write it as

$$T_\nu = \frac{B}{S}, \qquad B = \text{constant}. \tag{5.37}$$

Then (5.35) gives

$$\sigma = \frac{11}{3} a B^3 \left(\frac{T}{T_\nu} \right)^3. \tag{5.38}$$

Similarly (5.36) gives

$$\sigma = \frac{4}{3} a B^3 \left(\frac{T_\gamma}{T_\nu} \right)^3. \tag{5.39}$$

Now, in the pre-annihilation era $T = T_\nu$, so that (5.38) tells us that $\sigma = (11/3) a B^3$. After annihilation σ must have the same value, so we may equate it to the value given by (5.39). Thus we arrive at the conclusion that the photon temperature at the end of $e^\pm$ annihilation has risen *above* the neutrino temperature by the factor

$$\frac{T_\gamma}{T_\nu} = \left(\frac{11}{4} \right)^{1/3} \cong 1.4. \tag{5.40}$$

So the present-day neutrino temperature is *lower* than the photon temperature by the factor $(1.4)^{-1}$. If we take the latter ~ 2.7 K, the former is ~ 1.9 K.

5.4 The neutron-to-proton ratio

We have so far developed a picture of the early universe that is best expressed in the form of a time–temperature table of events, as shown in Table 5.2 (see also Figure 5.3). We will now be interested in the last entry of Table 5.2.

In our discussion so far we have not paid much attention to baryons – the protons and neutrons that are also present in the mixture. In our approximation of setting the chemical potentials to zero we took the baryon number to be zero. The validity of the approximation depended on the baryon number density being several (8–10) orders of magnitude smaller than the photon density. Nevertheless, we must now take note of the existence of baryons, howsoever small their number density; for we need them in order to consider Gamow's idea of nucleosynthesis in the hot universe. We also emphasize that the baryons at this stage of the universe (when nucleosynthesis could occur) are not playing any significant role in determining the expansion of the universe.

Insofar as chemical potentials are concerned, we will take explicit note of them in the following section. However, first notice that the critical temperatures T_n and

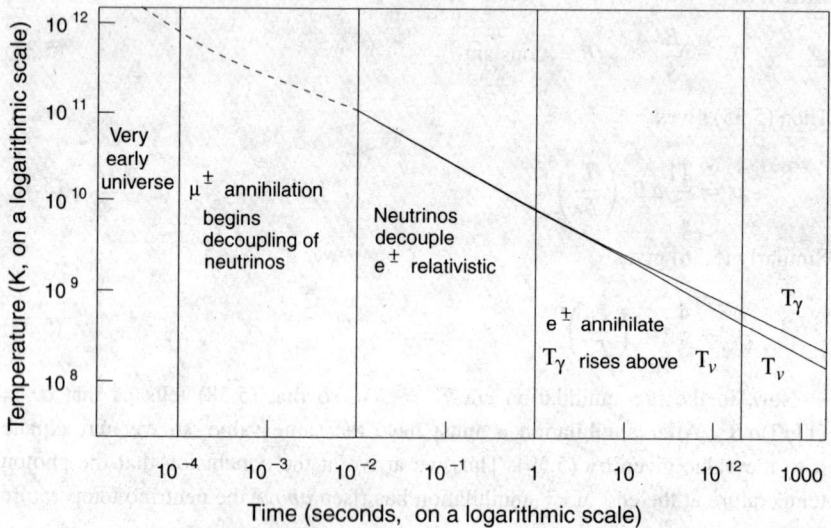

Figure 5.3 The time–temperature plot of the early universe. The dotted portion does not describe the (t, T) relationship accurately, since the particles (especially pions and μ^+) interact and are not really free. A more reliable picture emerges for $t \geq 10^{-2}$ s. Notice the difference between the temperatures of neutrinos and photons for $t \geq 10$ s. See the text for details.

Table 5.2 A time–temperature table of events preceding nucleosynthesis in the early universe

Time since big bang (s)	Temperature (K)	Events
$\leq 10^{-4}$	$> 10^{12}$	Baryons, mesons, leptons and photons are in thermal equilibrium.
10^{-4}–10^{-2}	10^{12}–10^{11}	$\mu^{\pm}$ begin to annihilate and disappear from the mixture. Neutrinos begin to decouple from the rest of the matter.
10^{-2}–1	10^{11}–10^{10}	Neutrinos decouple completely. $e^{\pm}$ pairs are still relativistic.
1–180	10^{10}–10^{9}	The pairs of $e^{\pm}$ annihilate and disappear, raising the photon-gas temperature to ~ 1.4 times the temperature of the neutrinos.

T_p of Table 5.1 are very high, so that the neutron and proton distribution functions follow the non-relativistic approximations of (5.25). Thus we get

$$N_p = \frac{2}{\hbar^3}\left(\frac{m_p kT}{2\pi}\right)^{3/2}\exp\left(-\frac{T_p}{T}\right),$$
$$N_n = \frac{2}{\hbar^3}\left(\frac{m_n kT}{2\pi}\right)^{3/2}\exp\left(-\frac{T_n}{T}\right). \tag{5.41}$$

In this approximation the neutron to proton member ratio is given by

$$\frac{N_n}{N_p} \cong \exp\left(\frac{T_p - T_n}{T}\right) = \exp\left(-\frac{1.5}{T_{10}}\right). \tag{5.42}$$

The ratio therefore drops with temperature, from near $1:1$ at $T \geq 10^{12}$ K to about $5:6$ at $T = 10^{11}$ K and to 3.5 at 3×10^{10} K. Figure 5.4 illustrates this change with temperature.

For thermodynamic equilibrium to be maintained, the reactions that convert neutrons into protons and *vice versa* have to be rapid enough relative to the rate at which the universe expands. These interactions are none other than the weak interactions considered earlier when we discussed the decoupling of neutrinos from the rest of the primordial brew (see §5.3). There is one difference, however. In discussing the decoupling of neutrinos we were concerned mainly with the reaction of a neutrino

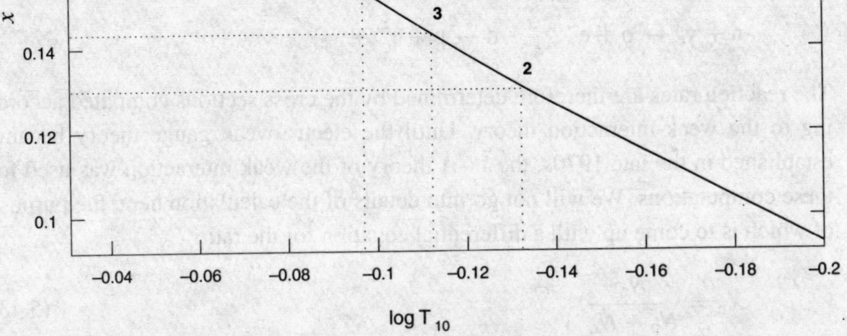

Figure 5.4 The neutron-to-proton number ratio x is plotted against temperature. Three points are identified on the curve to mark the stage at which the ratio effectively determines the abundance of helium: the three cases shown correspond to $l = 2, 3$ and 4 species of neutrino.

with leptons like $e^{\pm}$ and $\mu^{\pm}$ and the cross section Σ given by (5.26) was determined for such interactions. Similarly the reaction rate η given by (5.27) was obtained by multiplying by the number densities of participating leptons.

In the present case the cross section for a typical reaction like

$$\nu_e + n \leftrightarrow e^- + p$$

is larger than that for a purely leptonic reaction like

$$\nu_\mu + e \leftrightarrow \nu_e + \mu.$$

Also, the lepton densities used in (5.27) were considerably higher than the nucleon densities we are considering now. So the probability of a given nucleon interacting with any neutrino is higher than the probability of a given neutrino interacting with any nucleon. The result is that the effective temperature at which n and p cease to be in thermodynamic equilibrium is lower than the effective temperature for decoupling of neutrinos determined earlier.

Quantitatively, instead of $\Sigma \propto T^2$ as in (5.26), the cross section in the present case goes as $\propto T$ and the effective decoupling temperature T_* at which the rate of reaction is just about equal to H is $<10^{10}$ K. Note that, if the universe were expanding faster, T would be higher and the ratio N_n/N_p at decoupling given by (5.42) would be higher. We will recall this point in §5.5.1 when we relate the abundance of helium to the number of neutrino species.

Once the thermodynamic equilibrium has ceased to be maintained, the ratio N_n/N_p is given not by (5.42) but rather by detailed consideration of specific reactions involving the nucleons.

As the universe cooled further, this ratio was therefore determined by the reactions that change protons into neutrons and *vice versa*. These are essentially weak interactions of the type

$$n + \nu_e \leftrightarrow p + e^-, \qquad n \leftrightarrow p + e^- + \bar{\nu}_e.$$

The reaction rates are therefore determined by the cross sections computed according to the weak-interaction theory. Until the electro-weak gauge theory became established in the late 1970s, the $V-A$ theory of the weak interaction was used for these computations. We will not go into details of the calculation here, the purpose of which is to come up with a differential equation for the ratio

$$X_n = \frac{N_n}{N_p + N_n}. \tag{5.43}$$

If $\lambda(n \to p)$ denotes the rate at which neutrons are converted to protons and $\lambda(p \to n)$ the corresponding rate for protons changing the neutrons then clearly X_n satisfies the equation

$$\frac{dX_n}{dt} = (1 - X_n)\lambda(p \to n) - X_n\lambda(n \to p). \tag{5.44}$$

The rates λ depend on distribution functions of leptons, which in turn depend on the temperature, which is related to the scale factor of the expanding universe. The integration of (5.44) has to be done numerically and it is continued until all $e^{\pm}$ pairs have dropped out of the mixture – which happens at $T \geq 10^9$ K.

When all $e^{\pm}$ have disappeared it is still possible for the neutron to decay via the reaction

$$n \to p + e^- + \bar{\nu}_e$$

with a characteristic time $\tau = 1013$ s. So, from the time the pairs disappear to the onset of nucleosynthesis the neutron ratio X_n will decrease by the exponential factor $\exp(-t/\tau)$.

Thus the ratio of neutrons to protons is uniquely determined at the time nucleosynthesis begins, once we know all the parameters of the weak-interaction process. This is one good aspect of primordial nucleosynthesis theory, which was first pointed out by Chushiro Hayashi in 1950. We now proceed to discuss its outcome.

5.5   The synthesis of light nuclei

A typical nucleus Q is described by two quantities, the atomic mass A and the atomic number Z, and is written

$$_Z^A Q.$$

This nucleus has Z protons and $A - Z$ neutrons. If m_Q is the mass of the nucleus, its binding energy is given by

$$B_Q = [Zm_p + (A - Z)m_n - m_Q]c^2. \tag{5.45}$$

Let us now consider a unit volume of cosmological medium containing N_N nucleons, bound or free. Since the masses of protons and neutrons are nearly equal, we may denote the typical nucleon mass by m. Thus $m_n \approx m_p = m$. If there are N_n free neutrons and N_p free protons in the mixture, the ratios

$$X_n = \frac{N_n}{N_N}, \qquad X_p = \frac{N_p}{N_N} \tag{5.46}$$

will denote the fractions by weight of free neutrons and free protons. If a typical bound nucleus Q has atomic mass A and there are N_Q of them in our unit volume, we may similarly denote the weight fraction of Q by

$$X_Q = \frac{N_Q A}{N_N}. \tag{5.47}$$

Now, at very high temperatures ($T \gg 10^{10}$ K), the nuclei are expected to be in thermal equilibrium. However, even at these temperatures $T \ll T_Q$ and the formula (5.25) holds. Furthermore, since we are now concerned with relative number densities, we can no longer ignore the chemical potentials. Thus

$$N_Q = g_Q \left(\frac{m_Q kT}{2\pi \hbar^2} \right)^{3/2} \exp\left(\frac{\mu_Q - m_Q c^2}{kT} \right), \tag{5.48}$$

where we have reinstated the chemical potentials μ_Q. Since chemical potentials are conserved in nuclear reactions,

$$\mu_Q = Z\mu_p + (A - Z)\mu_n, \tag{5.49}$$

assuming that the nuclei were built out of neutrons and protons by nuclear reactions.

Using equation (5.49) the unknown chemical potentials can be eliminated between (5.48) and similar relations for N_p and N_n. The result is expressed in this form:

$$X_Q = \frac{1}{2} g_Q A^{5/2} X_p^Z X_n^{A-Z} \xi^{A-1} \exp\left(\frac{B_Q}{kT} \right) \tag{5.50}$$

where

$$\xi = \frac{1}{2} N_N \left(\frac{mkT}{2\pi \hbar^2} \right)^{-3/2}. \tag{5.51}$$

For an appreciable build-up of complex nuclei, T must drop to a low enough value to make $\exp[B_Q/(kT)]$ large enough to compensate for the smallness of ξ^{A-1}. This happens for nucleus Q when T has dropped to

$$T_Q \sim \frac{B_Q}{k(A-1)|\ln \xi|}. \tag{5.52}$$

Let us consider what happens when we apply the above formula to the nucleus of ^{4}He. The binding energy of this nucleus is given approximately by 4.3×10^{-5} erg. If we substitute this value into (5.50) and estimate N_N from the nucleon density of around 10^{-6} cm^{-3} observed at present, we find that T_Q is as low as $\sim 3 \times 10^9$ K (see Exercise 23). However, at this low temperature the number densities of participating nucleons are so low that four-body encounters leading to the formation of ^{4}He are extremely rare. Thus we need to proceed in a less ambitious fashion in order to describe the build-up of complex nuclei.

Hence we try using two-body collisions (which are not so rare) to describe the build-up of heavier nuclei. Thus deuterium (D), tritium (^{3}H), and helium (^{3}He and ^{4}He) are formed via reactions like

$$\begin{aligned} p + n &\leftrightarrow D + \gamma, \\ D + D &\leftrightarrow {}^3\text{He} + n \leftrightarrow {}^3\text{H} + p, \\ {}^3\text{H} + D &\leftrightarrow {}^4\text{He} + n. \end{aligned} \tag{5.53}$$

Since the formation of deuterium involves only two-body collisions, it quickly reaches its equilibrium abundance given by

$$X_D = \frac{3}{\sqrt{2}} X_p X_n \xi \exp\left(\frac{B_D}{kT}\right). \tag{5.54}$$

However, the binding energy B_D of deuterium is low so that, unless T drops to less than 10^9 K, X_D is not high enough to start further reactions leading to ^{3}H, ^{3}He, and ^{4}He. In fact, with the exception of the first one, the reactions given in (5.53) do not proceed fast enough until the temperature has dropped to $\sim 8 \times 10^8$ K.

Although at such temperatures nucleosynthesis does proceed rapidly enough, it cannot go beyond ^{4}He. This is because there are no stable nuclei with $A = 5$ or 8 and nuclei heavier than ^{4}He break up as soon as they are made. Their primordial abundances are extremely small. So the process effectively terminates there. Detailed calculations by several authors have now established this result quite firmly.

So, starting with primordial neutrons and protons, we end up finally with ^{4}He nuclei and free protons. All neutrons have been gobbled up by helium nuclei. Thus, if we consider the fraction by weight of primordial helium, it is very simply related to the quantity X_n – the concentration of neutrons before nucleosynthesis began. Denoting this fraction by weight by the symbol Y, we get

$$Y = 2X_n. \tag{5.55}$$

In Figure 5.5 the cosmic weight fractions of ^{4}He, ^{3}He, deuterium (^{2}H) and so on are plotted against a parameter η defined by

$$\eta = \left(\frac{\rho_0}{1.97 \times 10^{-26} \text{ g cm}^{-3}}\right)\left(\frac{2.7}{T_0}\right)^3. \tag{5.56}$$

Thus η essentially measures the nucleon density in the early universe through the formula

$$\rho = \eta T_9^3, \qquad T_9 < 3. \tag{5.57}$$

We will now discuss the implications of this parameter for primordial production of light nuclei.

5.5.1 Primordial helium and neutrino species

Note that the weight fraction of ^{4}He is insensitive to the parameter η. This is because, as we saw just now, it depends only on X_n, which in turn depends more critically on the epoch when the rate of weak interactions fell below the rate of expansion. If we go back to (5.42) we see that, in the very early stages, the neutron-to-proton ratio depends on the temperature T_*. A faster rate of expansion implies that the ratio becomes frozen at a higher temperature and so is higher, thus leading to a higher abundance of ^{4}He.

To see the effect quantitatively, recall from (5.42) that there was a 'last epoch' of temperature T_* for which the neutron-to-proton ratio was determined from considerations of thermodynamic equilibrium:

$$x = \frac{N_n}{N_p} = \exp\left(-\frac{1.5}{T_{*10}}\right). \tag{5.58}$$

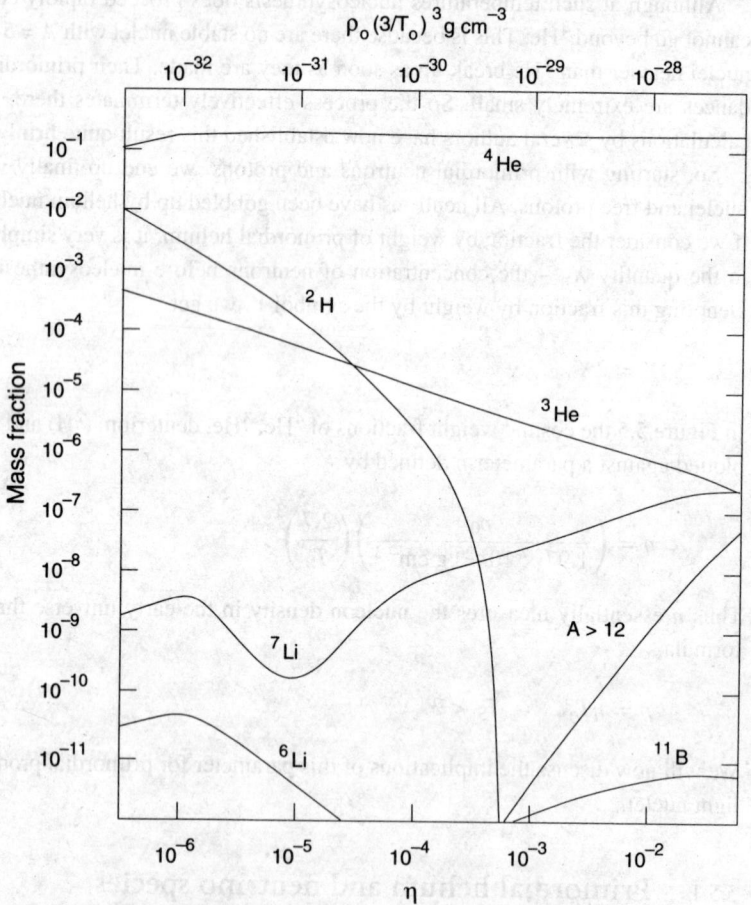

Figure 5.5 Primordial abundances of light nuclei as functions of the present-day density of matter in the universe. The number of types of neutrino is assumed to be three. The relation between ρ_0 and η is given by (5.56). Adapted from R. V. Wagoner, 1979, 'The early universe', in R. Balian, J. Audouze, and D. N. Schramm, eds. *Physical Cosmology*, Les Houches Lectures Session XXXII, p. 395 (Amsterdam: North Holland).

The temperature T_* was determined by equating the Hubble constant H to the reaction rate η for n $\leftrightarrow$ p conversions. Now

$$H \propto g^{1/2} T_*^2 \qquad \text{and} \qquad \eta \propto T_*^4$$

so that

$$T_*^2 \propto g^{1/2}. \tag{5.59}$$

Writing $Y = 2N_n/(N_n + N_p)$ for the weight fraction of helium, we can estimate the change in Y due to a change in g (from increasing the number of neutrino flavours) as follows. As $g \to g + \delta g$, the above relations imply that $x \to x + \delta x$ and $Y \to Y + \delta Y$, where

$$\delta Y = -\frac{x \ln x}{(1+x)^2} \frac{\delta g}{g}. \tag{5.60}$$

If there are l lepton families,

$$g = \tfrac{7}{8}(4 + 2l) + 2. \tag{5.61}$$

Hence an increase from $l = 2$ to $l = 3$ gives $\delta g/g \sim 1/5$. For $Y = 1/4$ and $x = 1/7$ we get $\delta Y \cong 0.02$.

This result is relevant to the question of how many different types of neutrinos exist primordially. In the GUT formalism described in Chapter 6 there are three types of neutrino, ν_e, ν_μ and ν_γ. Other formalisms may permit even more types of neutrinos to exist, thereby forcing the value of Y upwards. When we look at observations we will discover that the present estimates of the abundance of helium rule out the existence of more than three types of neutrino.

It is also interesting that the particle-accelerator experiments appear to lead to the same conclusion. A series of experiments carried out in 1990 with the Large Electron–Positron Collider (LEP) at CERN (see Figure 5.6) produced the intermediate Z^0 boson in large numbers. The presence of these particles (which mediate in electro-weak interactions) could be inferred by detecting resonance peaks in the energy-dependent cross sections for producing hadrons and leptons. The width of the peak measures the lifetime of the Z^0 boson, which in turn can be linked to the number of neutrino species present. The estimate is very close to three, which is consistent with the above cosmological considerations. This circumstance is considered a notable success of the enterprise of bringing together cosmology and particle physics.

5.5.2 Deuterium and other light nuclei

In contrast to the behaviour of Y, which does not sensitively depend on the parameter η, the abundances of other nuclei do depend on η. These abundances are very small relative to Y. Only nuclei heavier than ^{4}He eventually survive; ^{3}H (tritium) decays to ^{3}He. Of nuclei heavier than ^{4}He, only ^{7}Li (lithium) appears in any appreciable quantity, although the amount is smaller than that of ^{3}He. The most interesting situation exists for deuterium, whose abundance sharply drops as η rises above 10^{-4}. The present estimate of the mass fraction of deuterium is $\sim 2 \times 10^{-5}$. From Figure 5.5, we have $\eta \cong 2 \times 10^{-5}$. For $T_0 = 2.7$ K, this corresponds to a present-day density of

$$\rho_0 \sim 4 \times 10^{-31} \text{ g cm}^{-3}. \tag{5.62}$$

On comparing this with (4.45), we see that $h_0\Omega_0^2 \lesssim 0.02$ and hence $q_0 \lesssim 0.01$. Therefore, if even such a small amount of deuterium believed to be primordial in origin were found, Friedmann models of the closed variety would be ruled out. There is, however, a loophole in this argument: we can still accommodate non-baryonic matter in the universe. Such matter does not affect the abundance of deuterium, but does contribute to Ω_0. Matter of this kind will have to be dark. We will discuss the observational situation and the dark matter option in Chapter 10.

Figure 5.5 shows that the primordial production of heavy nuclei ($A \geq 12$) is very small and that it cannot account for their observed abundances. The main reason for this is that there are no stable nuclei with $A = 5$ and 8. Thus any attempt to synthesize

Figure 5.6 The Large Electron–Positron Collider (LEP) at CERN, Geneva, where the experiments have placed limits close to three on the number of species of neutrino. The particles are accelerated to very high energy as they are moved in a gigantic ring spread over several kilometres seen in this photograph. Photograph by courtesy of CERN, Geneva.

heavier nuclei by adding to ^{4}He fails, whether we add a proton or another ^{4}He. In fact, to cross this gap and reach stable heavy nuclei like ^{12}C, ^{16}O, etc., we need an altogether different scenario. Deep interiors of stars on their way to becoming red giants are suitable sites for making such nuclei. In fact the classic paper by Margaret and Geoffrey Burbidge, William Fowler and Fred Hoyle in 1957 (these authors are jointly referred to as B^{2}FH) demonstrated how all such nuclei can be made during the various stages of stellar evolution.

We can sum up by saying that Gamow's expectation that the early hot universe would synthesize all types of nuclei has only partially been fulfilled. The idea works for light nuclei like deuterium, ^{4}He, etc. To obtain complex nuclei heavier than ^{4}He (and possibly ^{7}Li), astrophysicists have to look to other sources: the stars.

5.6   The microwave background

In 1948, Gamow's colleagues Alpher and Herman (see Figure 5.7) made another prediction, however, that appears to have received confirmation. This is the prediction that the photons of the early hot era would have cooled down to provide a thermal radiation background in the microwaves at present. Alpher and Gamow estimated that the temperature of this radiation today would be around 5 K. In the absence of a firm quantitative basis, this estimate was more of a guess. Gamow himself in a later paper guessed the temperature at 7 K and later on increased this value in further stages to as much as 50 K. In the 1950s, these ideas on primordial nucleosynthesis did not catch on, largely because of the failure of the theory to explain the origin of heavier nuclei. With the stellar nucleosynthesis scoring a major success on this front, interest in the primordial scenario dwindled and the prediction of a relic radiation background was more or less forgotten.

By the early sixties, however, the interest in the primordial phase of the universe had been rekindled by the realization that not all helium found in the universe could have been made in stars. Work by Fred Hoyle and Roger Tayler in Cambridge, UK, by Robert Dicke and Jim Peebles at Princeton in the USA and by Yakob Zel'dovich in Moscow in the USSR during 1964–65 showed that a revised version of Gamow's nucleosynthesis programme does succeed in yielding the correct abundance of helium. Hence the idea of a relic radiation became interesting enough for Dicke to plan a radiometer experiment to detect it.

As mentioned earlier, such radiation was first detected in 1965 by Arno Penzias and Robert Wilson (see Figure 5.8), more or less serendipitously. They had planned using a 20-foot horn-shaped reflector antenna to study radiation in the microwaves in the Milky Way. While testing the antenna, they pointed it in various directions and used the wavelength 7.35 cm since it did not attract much galactic noise. It was these test measurements that contained an unaccounted-for component that was isotropic, i.e., one that could not be ascribed to any specific galactic or extragalactic source. It was only when they compared notes with the Princeton group that they could identify this radiation with the relic background.

The Penzias–Wilson measurement at one wavelength, if it were interpreted as blackbody radiation, had a temperature of 3.5 K. In Chapter 10 we will discuss the details of subsequent observations of this radiation. In Figure 5.9 we show the spectrum of the radiation measured by the Cosmic Background Explorer (COBE) satellite in 1990, with a temperature of 2.735 ± 0.06 K. The blackbody nature of the intensity–frequency curve has gone a long way towards confirming in most cosmologists' minds that the picture of an early hot universe discussed by Gamow is correct. To see how this background formed we have to follow our history of the early universe to stages subsequent to nucleosynthesis.

Figure 5.7 Ralph Alpher (left) and Robert Herman (right), who in 1948 predicted the existence of a relic background radiation from their calculations (with George Gamow) of primordial nucleosynthesis. Photographs by courtesy of Ralph Alpher.

Figure 5.8 Robert Wilson and Arno Penzias in front of their horn-shaped antenna, which was used in the discovery of the microwave background. Photograph by courtesy of Bell Labs/Lucent Technologies.

The era of nucleosynthesis took place when the temperature was around 10^9 K. The universe in subsequent phases continued to cool as it expanded with the radiation temperature dropping as S^{-1}. The presence of nuclei, free protons and electrons did not have much effect on the dynamics of the universe, which was still radiation-dominated. However, these particles, especially the lightest of them, the electrons, acted as scattering centres for the ambient radiation and kept it thermalized. The universe was therefore quite opaque to start with.

However, as the universe cooled, the electron–proton electrostatic attraction began to assert itself. In detailed calculations performed by P. J. E. Peebles, the mixture of electrons and protons and of hydrogen atoms was studied at varying temperatures. Because of Coulomb attraction between the electron and the proton, the hydrogen atom has a certain binding energy B. The problem of determining the relative number densities of free electrons, free protons (that is, ions) and neutral H atoms in thermal equilibrium is therefore analogous to that we considered earlier in deriving (5.50) in §5.5 for the mixture of free and bound nucleons. The only difference is that the binding to be considered now is electrostatic rather than nuclear. Following the same method, we arrive at the formula relating the number densities of electrons (N_e), protons ($N_p = N_e$) and H atoms (N_H) at a given temperature T:

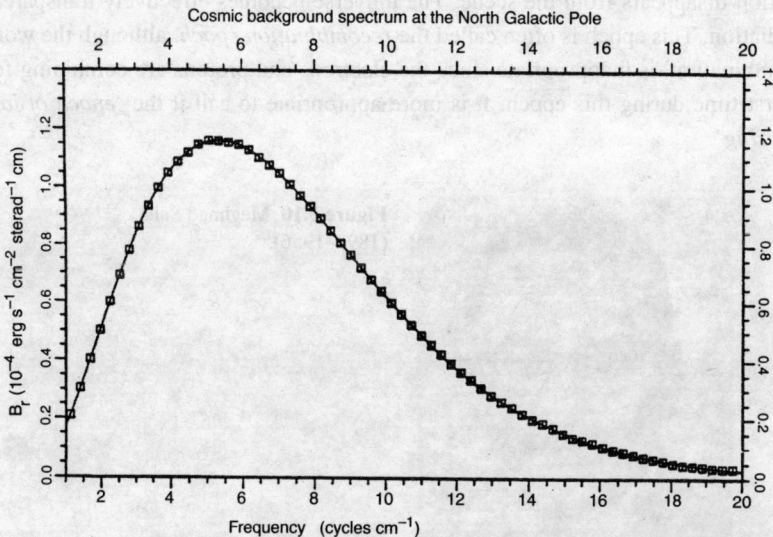

Cosmic background spectrum at the North Galactic Pole

B_r (10^{-4} erg s^{-1} cm^{-2} sterad^{-1} cm)

Frequency (cycles cm^{-1})

Figure 5.9 The spectrum of the cosmic microwave background determined by the COBE satellite with the continuous curve through the error rectangles representing the best fit blackbody curve at temperature $\sim$2.73 K. Courtesy of NASA Goddard Space Flight Center and the COBE Science Working Group.

$$\frac{N_e^2}{N_H} = \left(\frac{m_e kT}{2\pi\hbar^2}\right)^{3/2} \exp\left(-\frac{B}{kT}\right),$$ (5.63)

where m_e is the electron mass. This equation is a particular case of *Saha's ionization equation*. Around 1920 Meghnad Saha (see Figure 5.10) had looked at the problem of ionization in the context of stellar atmospheres and had derived this equation.

Writing N_B for the total baryon number density, we may express the fraction of ionization by the ratio

$$x = N_e/N_B.$$ (5.64)

Then, since $N_H = N_B - N_e$, we get from (5.63)

$$\frac{x^2}{1-x} = \frac{1}{N_B}\left(\frac{m_e kT}{2\pi\hbar^2}\right)^{3/2} \exp\left(-\frac{B}{kT}\right).$$ (5.65)

For the H atom, $B = 13.59$ eV. By substituting for various quantities on the right-hand side of (5.65), we can solve for x as a function of T. The results (see Figure 5.11) show that x drops sharply from unity to near zero in the temperature range of around 5000–2500 K, depending on the value of N_B, that is, on the parameter $\Omega_0 h_0^2$ (see Chapter 4). For example, for $\Omega_0 h_0^2 = 0.01$, $x = 0.003$ at $T = 3000$ K.

Thus, by this time most of the free electrons have been removed from the cosmological brew and as a result the main agent responsible for the scattering of radiation disappears from the scene. The universe becomes effectively transparent to radiation. This epoch is often called the *recombination epoch*, although the word 'recombination' is inappropriate since the electrons and protons are combining for the first time during this epoch. It is more appropriate to call it the *'epoch of last scattering'*.

Figure 5.10 Meghnad Saha (1893–1956).

The transparency of the universe means that a photon of light can go a long way ($\sim c/H$) without being absorbed or scattered. Therefore this epoch signifies the beginning of the new phase when matter and radiation become decoupled. This phase has lasted up to the present epoch. During this phase, the frequency of each photon is redshifted according to the rule

$$\nu \propto 1/S \tag{5.66}$$

while the number density of photons falls as

$$N_\gamma \propto 1/S^3. \tag{5.67}$$

It is easy to see that, under these conditions, the photon distribution function preserves the Planckian form with the temperature dropping as

$$T \propto 1/S. \tag{5.68}$$

A present-day background temperature of ~ 2.7 K therefore means that the epoch when matter decoupled from radiation corresponds to a redshift of $\sim 10^3$. However, in §4.3 we also saw that the universe changed over from being radiation-dominated to being matter-dominated around the same epoch. Why the transition from opaqueness to transparency and from radiation domination to matter domination should take place around the same time is at present unexplained and must be considered a coincidence.

Another result as yet unexplained by early universe physics is the observed ratio of photons to baryons:

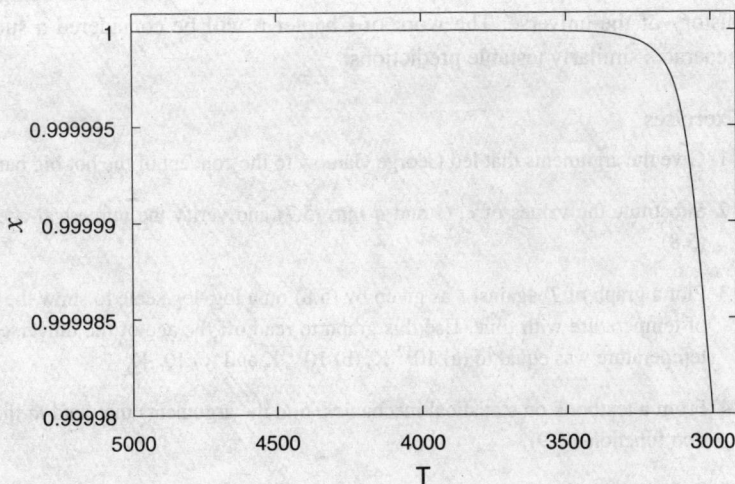

Figure 5.11 The plot of x against T shows how the universe becomes progressively transparent as the temperature falls and the number of free electrons declines.

$$\frac{N_\gamma}{N_B} = 3.33 \times 10^7 (\Omega_0 h_0^2)^{-1} \left(\frac{T_0}{2.7}\right)^3. \tag{5.69}$$

This ratio has been conserved since the time the universe became essentially transparent, although both N_γ and N_B can be studied theoretically for even earlier epochs. Why the above ratio and no other? Many physicists feel that deeper ideas from particle physics are needed in order to throw light on this mystery.

The important signature of the relic radiation is, however, its spectrum. There may be small perturbations of the radiation background caused by formation of discrete structures, but, these apart, we should find the background spectrum to be very close to the Planckian form. This has indeed been confirmed. We will recall this prediction when taking stock of the observations of the microwave background in Chapter 10.

5.7   Concluding remarks

Our investigations of the early universe in this chapter started with the epoch when the universe was very hot and barely 10^{-4} s old. They concluded with the epoch of redshift $\sim 10^3$ when the universe became transparent. We now have two ways to go: backwards from 10^{-4} s or forwards from the epoch with a of redshift 10^3. In Chapter 6 we go backwards; in Chapter 7 we will go forwards.

The opaqueness of the universe prevents us from 'seeing' directly beyond the redshifts of $\sim 10^3$. Thus any evidence of the big bang or hot universe must come indirectly. In this sense the abundances of light nuclei and the detailed observations of the microwave background provide us with the only means of checking the early history of the universe. The work of Chapter 6 will be considered a success if it generates similarly testable predictions.

Exercises

1 Give the arguments that led George Gamow to the concept of the hot big bang.

2 Substitute the values of c, G and a into (5.7) and verify the numerical coefficient in (5.8).

3 Plot a graph of T against t as given by (5.8) on a log–log scale to show the variations of temperature with time. Use this graph to read off the age of the universe when the temperature was equal to (a) 10^{12} K, (b) 10^{11} K and (c) 10^9 K.

4 From a textbook on statistical mechanics, find the arguments that lead to the distribution functions (5.9).

5 From the reactions

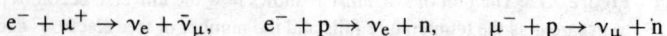

$$e^- + \mu^+ \to \nu_e + \bar{\nu}_\mu, \qquad e^- + p \to \nu_e + n, \qquad \mu^- + p \to \nu_\mu + n$$

deduce that the corresponding chemical potentials satisfy the relations

$$\mu_{e^-} - \mu_{\nu_e} = \mu_{\mu^-} - \mu_{\nu_\mu} = \mu_n - \mu_p.$$

6 Give arguments to show that there are just four independent conserved quantum numbers in nuclear reactions of the type shown in Exercise 5.

7 Derive the relations (5.11)–(5.14) from the formula (5.9) in the approximation in which the chemical potentials are neglected.

8 Obtain (5.16) from (5.11)–(5.14) in the high-temperature approximation.

9 Using the table of constants at the end of this book, compute T_A for the species given in Table 5.1 and verify the numbers given in that table.

10 Deduce (5.17) from (5.11)–(5.14) in the low-temperature approximation.

11 Deduce (5.20) directly from (5.13) and (5.12).

12 Show that the McCrea–Milne description of Newtonian cosmology can be extended to the early universe, provided that we replace the density ρ in relation (4.6) by $\rho + 3p$ for radiation. Compare the results with those of relativistic cosmology.

13 Write down all possible reactions involving the electron, the muon, their respective neutrinos and the antiparticles of all of them.

14 Why does the neutrino have the degeneracy factor $g_\nu = 1$?

15 Use the table of constants at the end of this book to derive (5.24).

16 Use the Einstein equations and equation (5.42) to deduce that, if the universe were to expand faster, the neutron-to-proton ratio would be frozen at a higher temperature.

17 Give arguments to show that the neutrino temperature drops as S^{-1} after neutrinos decouple from the rest of the matter.

18 Why is the present neutrino temperature expected to be lower than the photon temperature? Derive the ratio of the two temperatures from considerations of the early universe.

19 Using the formulae (5.12)–(5.14), deduce that, during the phase in which the pairs of e^+ and e^- are annihilating and producing photons, the constancy of σ tells us that the photon temperature T changes with S according to the law

$$ST\xi(T) = \text{constant},$$

where

$$[\xi(x)]^3 = 1 + \frac{15}{2\pi^4} \int_0^\infty \frac{y^2(3x^2 + 4y^2)\,dy}{\sqrt{x^2 + y^2}[\exp(\sqrt{x^2 + y^2}) + 1]}.$$

20 A primordial mixture of relativistic bosons and fermions in the early universe of temperature T has the total energy density given by the formula

$$\epsilon = \frac{\pi^2}{30\hbar^3 c^3} g_*(kT)^4.$$

Show that $g_* = g_b + \frac{7}{8}g_f$, where g_b is the total spin degeneracy of all bosons and g_f is the total spin degeneracy of all fermions.

21 Discuss, with the help of Exercise 20, how the rate of expansion of the universe is affected by the number of species of relativistic particles that are present in it. How is the time–temperature relation affected?

22 From relations of the type (5.48), express μ_Q, μ_n, and μ_p in terms of the remaining quantities. Then use (5.49) to derive (5.50).

23 The binding energy of the ^{4}He nucleus is $B \cong 4.3 \times 10^{-5}$ erg. Show that $B/k(A-1) \cong 10^{11}$ K. Next, assume that the present value of the radiation temperature is 3 K and that the nucleon density is 10^{-6} cm^{-3}. Using the result that $N_N T^3 =$ constant, show that (5.52) gives T_Q for ^{4}He as $\sim 3.2 \times 10^9$ K.

24 Give arguments to show why the primordial abundance of helium is insensitive to the number density of baryons in the universe.

25 The abundances of which nuclei are likely to provide a sensitive test of the baryonic density of the universe?

26 Can you think of a loophole that would allow $\Omega_0 = 1$ and still permit deuterium to be formed primordially in a standard hot-universe model?

27 If m is the mass of a nucleon and Ω_0 is the density parameter, show that the present-day number density of baryons is $3H_0^2 \Omega_0/(8\pi Gm)$. Use this formula and the present-day microwave background temperature $T_0 = 2.7$ K to estimate N_B in (5.65). Solve the Saha equation for $\Omega_0 = 0.1$ and $h_0 = 1$ to show that $x = 0.003$ at 3000 K.

28 Derive (5.69) using the blackbody spectrum and the Friedmann cosmology. What is the corresponding ratio N_ν/N_B for neutrinos?

29 Show that the form of the blackbody spectrum is preserved as the universe expands, with the effective temperature declining as S^{-1}.

30 Why is it not possible to observe the past of the universe beyond the redshift of $\sim 10^3$?

31 What could be considered possible candidates for relics of the big bang?

32 If the spectrum of the microwave background had turned out to be markedly different from the Planckian form, what implication would it have had for the hot big bang?

33 Show that the space-curvature parameter k and the cosmological constant are unlikely to affect the calculations for the early universe.

34 Using the Thomson-scattering cross section for the electrons, show that the optical depth of the universe during the present epoch would be given by $0.08\Omega_0 h_0$ if all electrons in the universe were free and equal in number to the baryons and there were no non-baryonic matter.

35 Assuming that in the past the electron number density increased as $(1 + z)^3$, use the analysis of Exercise 34 to estimate the smallest redshift at which the Einstein–de Sitter universe was opaque to radiation. (Take $h_0 = 1$.) Comment on the fact that your answer comes out very much lower than $z \sim 1000$.

Chapter 6

The very early universe

6.1   Cosmology and particle physics

In Chapter 5 we discussed the properties of the big-bang universe starting from the epoch in which it was $\sim 10^{-4}$ s old, when a mixture of baryons, mesons, leptons and photons was in thermodynamic equilibrium with a temperature of $\sim 10^{12}$ K. We discussed how this hot primordial gas evolved as the universe expanded and cooled down. We ended our story with the formation of the helium nucleus, by which time the universe was ~ 3 min old.

In the 1960s the above range of epochs would have been considered as describing the early universe. Indeed, in the late forties, when George Gamow first opened this subject, the physics community considered it too speculative to be taken seriously. The abundances of light nuclei found in the universe, however, provided persuasive evidence in favour of the scenario. In 1967, Robert Wagoner, William Fowler and Fred Hoyle extensively reworked the picture of primordial nucleosynthesis with the then-available nuclear data and established the subject on a firm footing. The discovery of the microwave background added further credibility to the early-hot-universe scenario.

Today, therefore, cosmologists have become bolder in extending their interest to the 'very early universe': the era preceding the above phase, when matter was in an even more elementary form than that considered above. The reason for this shift lies less in any development in cosmology than in particle physics. Theoretical developments in particle physics, which signify progress towards a unification of the basic interactions of physics, have remarkably found their echoes in cosmology. In fact, because particle physicists and big-bang cosmologists have found a joint venture into studies of the very early epochs mutually beneficial, a new subject

194

known as *astroparticle physics* has emerged since the late 1970s. The rationale for it is as follows.

So far particle physicists have relied on the use of powerful accelerators to study the interactions of particles at high energy. From elementary quantum theory it follows that, to be able to probe smaller and smaller distances, higher and higher momenta must be achieved. Thus high-energy accelerators are required in order to probe the structures of particles like the proton and the pion. Present-day accelerators achieve energies of the order of a few tens or hundreds of giga-electron-volts $(1 \text{ GeV} = 10^9 \text{ eV} \approx 1.6 \times 10^{-3} \text{ erg})$. The LEP in the high-energy labs at CERN, Geneva, can produce particle energies of the order of 10^3 GeV. These values may be compared with the energies of $\sim 10^{16}$ GeV at which interesting unification phenomena are predicted by particle physicists. Energy ranging as high as this value is far beyond what could be achieved by human technology of the foreseeable future. Thus, for lack of testability, ideas of unification of all basic interactions would remain mere speculations.

It is against this background that particle physicists have turned to cosmology in the realization that the very early hot universe is the poor man's high-energy accelerator, for the typical particle energy in the universe only 10^{-37} s old would have been as high as the claimed 'grand-unification energy' of $\sim 10^{16}$ GeV. Going even closer to the big bang, within 10^{-43} s of it, we arrive at particle energies of the order of 10^{19} GeV, when the dynamics in the universe would have been controlled by a quantum theory of gravitation. In short, the sufficiently early universe provides a locale, and apparently the only one, for studying the interplay of various basic interactions at very high energy. Which is why the physics community has turned around from its attitude of caution and scepticism in the late forties to one of speculation and adventure today.

This is not the first time that physicists have turned to astronomy in order to study the behaviours of physical processes under conditions unattainable in a terrestrial laboratory. Even before thermal fusion could be achieved on the Earth, physicists were studying the process inside stars. To go even further back in history, it was the astronomy of the Solar System that provided the real testing ground for the laws of gravitation, the first of the fundamental interactions of nature to be studied by physics.

Naturally, the interplay of cosmology and particle physics that we plan to discuss in this chapter is highly speculative on both fronts. It depends on the validity of the cosmological model and on the viability of ideas of particle physics that as yet remain fluid. The best that can be claimed is consistency between the two. Furthermore, two matching speculations cannot take the place of fact. The reader should bear this in mind throughout the various calculations given in this chapter.

6.1.1   The particle composition of the very early universe

Let us first consider what particles might exist in the early universe, particularly those out of which the baryons and mesons are formed. This information is supplied by particle physics and is listed in Table 6.1. Note that the building blocks are called 'quarks' and they are listed by their six 'flavours': up, down, strange, charmed, truth and beauty (or 'top' and 'bottom'). Each quark comes in three types, classified by abstract labels called 'colours': red, white (sometimes called green), and blue, while their antiparticles, the *antiquarks*, come in three anti-colours, cyan, yellow and magenta. These are the constituents of baryons and mesons, three quarks making a baryon and a quark–antiquark pair making a meson. The quarks interact with each other by exchanging bosonic particles called 'gluons', just as electrons interact with each other by exchanging photons.

Table 6.1 also lists six leptons, which come in pairs. Two pairs, e and ν_e and μ and ν_μ, we have already encountered in Chapter 5. A third pair, τ and ν_τ, is now known. The list of bosons includes the graviton, the photon and the eight gluons, as well as the charged particles $W^\pm$ and the neutral particle Z^0. Do these numbers have any special significance? Why six quarks? Why six leptons? Why eight gluons? Particle physicists have found it useful to describe the framework of all these particles in the abstract language of group theory (see §6.3).

The masses in Table 6.1 are listed in mega-electron-volts (1 MeV $\equiv 10^6$ eV). It is convenient to spend some time relating this unit to others used in macrophysics, since we shall be expressing many ideas from particle physics, in which this unit is commonly used. Thus, for each mass m expressed in grammes, mc^2 is energy expressed in ergs. We then use the following conversion scale:

$$1 \text{ MeV} = 1.602\,191\,7 \times 10^{-6} \text{ erg.}$$

Furthermore, since we are going to describe the hot universe, it is also convenient to express the temperature in the same unit. Thus, for T expressed in kelvins, kT is energy expressed in ergs, which can be written in units of mega- or giga-electron-volts. We therefore have

$$1 \text{ g} \sim 5.618 \times 10^{28} \text{ MeV} = 5.618 \times 10^{25} \text{ GeV,}$$

$$1 \text{ K} \sim 8.617 \times 10^{-11} \text{ MeV} = 8.617 \times 10^{-14} \text{ GeV.}$$

Although these conversion factors involve many powers of ten, they show why these are good units for the early universe. For example, a temperature of the order of 10^{12} K is a few mega-electron-volts. Similarly, Table 6.1 shows that the masses of the listed particles are given by moderate numbers when they are expressed in mega-electron-volts. For higher energies we may use giga-electron-volts.

We now recall from Chapter 5 the result that relates the temperature of the universe to its rate of expansion given by the Einstein equation

$$\frac{\dot{S}^2}{S^2} = \frac{8\pi G}{3}\rho. \tag{6.1}$$

If there are bosons with a total g_b of g-factors and fermions with a total g_f of g-factors, then the above equation has the solution

$$\rho c^2 = \tfrac{1}{2}gaT^4 \tag{6.2}$$

with

$$g = g_b + \frac{7}{8}g_f. \tag{6.3}$$

Table 6.1 Elementary particles in the early universe

Particle		Mass (MeV)[a]	Spin (h)	Electric charge (e)	Interaction[b]
Quarks	u	$? + 4$		$\frac{2}{3}$	
	d	$? + 8$		$-\frac{1}{3}$	G, W
	c	$? + 1150$	$-\frac{1}{2}$	$\frac{2}{3}$	E, C
	s	$? + 150$		$-\frac{1}{3}$	
	t	$? + \geq 5000$		$\frac{2}{3}$	
	b	$? + 4500$		$-\frac{1}{3}$	
Leptons	ν_e	$< 6 \times 10^{-4}?$		0	G, W
	e	0.5110		-1	G, W, E
	ν_μ	< 0.65	$\frac{1}{2}$	0	G, W
	μ^-	105.66		-1	G, W, E
	ν_τ	< 250		0	G, W
	τ^-	< 1780		-1	G, W, E
Bosons	Graviton	$\leq 10^{-36}$	2	0	G
	γ	$\leq 7 \times 10^{-22}$	1	0	G, E
	Gluons (8)	≤ 100	1	0	G, C
	$W^\pm$	$\sim 8 \times 10^4$	1	± 1	G, W, E
	Z^0	$\sim 9 \times 10^4$	1	0	G, W

[a] Quark masses are not uniquely determined, since free quarks have not yet been found. There is some indication that the mass of ν_e may exceed the value given here.

[b] G, gravitation; W, weak interaction; E, electromagnetism; and C, chromodynamics.

Source: Based on R. V. Wagoner, 1979, 'The Early Universe', in R. Balian, J. Audouze and D. N. Schramm, eds., *Physical Cosmology*, Les Houches Lectures Session XXXII, p. 395 (Amsterdam: North-Holland).

Thus we have for $g =$ constant

$$S \propto t^{1/2} \tag{6.4}$$

with

$$t = \left(\frac{3c^2}{16\pi Ga}\right)^{1/2} g^{-1/2} T^{-2}. \tag{6.5}$$

This relation can be expressed as

$$t_{\text{second}} = 2.4 g^{-1/2} T_{\text{MeV}}^{-2} = 2.4 \times 10^{-6} g^{-1/2} T_{\text{GeV}}^{-2}, \tag{6.6}$$

where we have used the above conversion factors to write the temperature in mega- and giga-electron-volts.

6.2 The survival of massive particles

We will begin with a simple extrapolation of the approach adopted in Chapter 5. We will assume in this section that quarks have combined to form particles (and antiparticles) and investigate the criteria that determine the survival of a particular species of particles. In the ideal-gas approximation, we will assume the distribution functions to be those given by (5.9). In the relativistic (high-temperature) approximation of §5.2, we have the following formula for the number density of particles of species A:

$$N_A = \eta g_A N_\gamma = \eta g_A \frac{2.4}{\pi^2} \left(\frac{kT}{c\hbar}\right)^3. \tag{6.7}$$

where N_γ is the number density of photons and $\eta = \frac{1}{2}$ for bosons and $\frac{3}{8}$ for fermions. In the non-relativistic approximation we get

$$N_A = \frac{g_A}{\hbar^3} \left(\frac{m_A kT}{2\pi}\right)^{3/2} \exp\left(-\frac{m_A c^2}{kT}\right). \tag{6.8}$$

The assumption leading to (6.7) or (6.8) is that the species is in thermodynamic equilibrium with the rest of the particles. For (6.7) to hold, we require $T \gg T_A \equiv m_A c^2/k$, whereas for (6.8) to hold we should have $T \ll T_A$. Exactly similar results must hold if the species A has antiparticles $\bar{A}$. To fix our ideas (since we are eventually going to use these formulae for baryons – protons and neutrons) we will assume A to be a fermion. Thus $\eta = \frac{3}{8}$.

In general A and $\bar{A}$ may annihilate if they are brought together. In a typical reaction two photons will be produced:

$$A + \bar{A} \rightarrow \gamma + \gamma. \tag{6.9}$$

In the reverse reaction pairs $(A, \bar{A})$ are produced. The question we wish to answer is this: how does the interchange affect the number density N_A or $N_{\bar{A}}$?

To start with, suppose that $N_A = N_{\bar{A}}$ and consider the particles (and antiparticles) in a comoving volume V_0. The corresponding proper volume is $V_0 S^3(t)$. Define

$$\mathcal{N}_A = N_A V_0 S^3(t), \qquad \mathcal{N}_{\bar{A}} = N_{\bar{A}} V_0 S^3(t). \tag{6.10}$$

Let $\psi(T)$ denote the production rate per unit volume and $\beta(T)$ the annihilation rate coefficient. Both ψ and β will depend on the temperature T and

$$\beta = \langle v\sigma \rangle, \tag{6.11}$$

where σ is the annihilation cross section and v the velocity of particles. Accelerator experiments on nucleon–antinucleon cross sections give us $\beta \sim 10^{-15}$ cm^3 s^{-1}, in the energy range 0.4–7 GeV. Thus it is convenient to write

$$\beta = 10^{-15}\bar{\beta} \text{ cm}^3 \text{ s}^{-1} \tag{6.12}$$

and expect that $\bar{\beta} \sim 1$. It is also worth noting that, if we consider the Compton wavelength for a particle of mass m and define $\sigma = \pi(\hbar/mc)^2$, then, for a proton or a neutron with $v = c$, (6.11) gives $\beta \sim 4 \times 10^{-17}$ cm^3 s^{-1}. Thus we may set

$$\beta = \zeta \frac{\pi \hbar^2}{m^2 c} \tag{6.13}$$

and expect $\zeta \sim 100$ for a proton or neutron.

The rate of change of $\mathcal{N}_A$ ($\mathcal{N}_{\bar{A}}$) is then given by

$$\frac{d\mathcal{N}_A}{dt} = \frac{d\mathcal{N}_{\bar{A}}}{dt} = [\psi(T) - \beta(T)N_{\bar{A}}^2(T)]V_0 S^3(t). \tag{6.14}$$

Frequent collisions are necessary to establish equilibrium. The collision rate is given by

$$\Gamma(T) = N_A(T)\beta(T) \propto T^3 \beta(T). \tag{6.15}$$

In general $\beta(T)$ does not decrease as T increases. Hence, in the very early stages, $\Gamma(T)$ was so large that it exceeded the rate of expansion of the volume, given by

$$3H(t) = \frac{3}{2t} \propto T^2. \tag{6.16}$$

Thus initially

$$t\Gamma(T) \gg 1, \tag{6.17}$$

guaranteeing that collisions occur frequently. Under such circumstances an equilibrium with detailed balancing between the processes of creation and annihilation is reached. In equilibrium $\psi(T) = \beta(T)N_{A0}^2(T)$, where $N_{A0}(T)$ denotes the equilibrium value of $N_A(T)$. Thus (6.14) becomes, for either A or $\bar{A}$,

$$\frac{d\mathcal{N}}{dt} = \beta(N_0 + N)(\mathcal{N}_0 - \mathcal{N}). \tag{6.18}$$

If we now refer back to (6.7) we see that, in the relativistic approximation, $N \propto T^3 \propto S^{-3}$, so that $\mathcal{N} \propto NS^3$ = constant. Thus, if particles are relativistic, then $\mathcal{N}$ = constant = $\mathcal{N}_0$ is a solution of (6.18). If on top of this we suppose that the relativistic regime lasted long enough for $t\Gamma(T)$ to drop below unity, then we encounter the situation in which $\mathcal{N}_0$ *is preserved for subsequent epochs*. This is because the rarity of collisions makes it unlikely that production or annihilation will significantly alter $\mathcal{N}$ once $t\Gamma(T)$ has dropped well below unity. We will now follow the analysis of G. Steigman.

For massless particles the relativistic regime lasts forever. Hence for these particles the above result always holds. Indeed, we encountered an example of this reasoning in the context of massless neutrinos in Chapter 5. The present-day neutrino distribution could be traced back to the epoch when they decoupled from the rest of the matter – when the weak-interaction processes became slower than the rate of expansion of the universe.

It may, however, happen that the particles are massive and are in equilibrium even when they become non-relativistic. In that case (6.8) applies and the number $\mathcal{N}_A$ drops rapidly as T decreases. At some stage, with $\mathcal{N}_A \ll \mathcal{N}_\gamma$, the collision rate drops ($t\Gamma \ll 1$) so that further changes in $\mathcal{N}_A$ through creation and annihilation are not possible. Let us denote this epoch by t_* and the corresponding temperature by T_*. The value of $\mathcal{N}_A$ during this epoch then becomes frozen; that is, it remains unaltered for subsequent epochs. This number would survive as a relic of the hot universe.

Figure 6.1 shows how, for various particles, $\mathcal{N}_A$ depends on the mass of the species A. For massless particles such as the photon (and the neutrino if it has $m = 0$), $\mathcal{N}_A$ is unchanged. The neutral leptons become frozen at the next lower value. The charged leptons can interact longer through the electromagnetic force and hence they decouple later and at lower values of $\mathcal{N}_A$ than do the neutral leptons. The lowest are the hadrons (mesons, neutrons, protons and so on), which have strong interactions to hold them together and the largest masses.

Let us try to estimate this effect quantitatively. At t_* we have for species A in the non-relativistic regime

$$N_A = \frac{g_A}{\hbar^3} \left(\frac{m_A k T_*}{2\pi} \right)^{3/2} \exp\left(-\frac{m_A c^2}{k T_*} \right). \tag{6.19}$$

Applying the condition $t_*\Gamma(T_*) = 1$ and using (6.15) and (6.19), we get

$$t_*\beta \frac{g_A}{\hbar^3} \left(\frac{m_A k T_*}{2\pi} \right)^{3/2} \exp\left(-\frac{m_A c^2}{k T_*} \right) = 1. \tag{6.20}$$

Define

$$x_* = \frac{m_A c^2}{kT_*} \tag{6.21}$$

and express masses and temperatures in mega-electron-volts. In these units $x_* = m_A / T_*$. Then from (6.6) we have

$$2.4 g^{-1/2} T_*^{-2} \beta \frac{g_A}{\hbar^3 c^3} \left(\frac{m_A T_*}{2\pi} \right)^{3/2} e^{-x_*} = 1; \tag{6.22}$$

that is

$$2.4 g^{-1/2} \frac{g_A}{\hbar^3 c^3} \frac{m_A}{(2\pi)^{3/2}} x_*^{1/2} e^{-x_*} = 1. \tag{6.23}$$

We may express this relation in the following form:

$$x_*^{-1/2} e^{x_*} = \Lambda g_A Z, \tag{6.24}$$

where

$$Z = m_A \bar{\beta} g^{-1/2}. \tag{6.25}$$

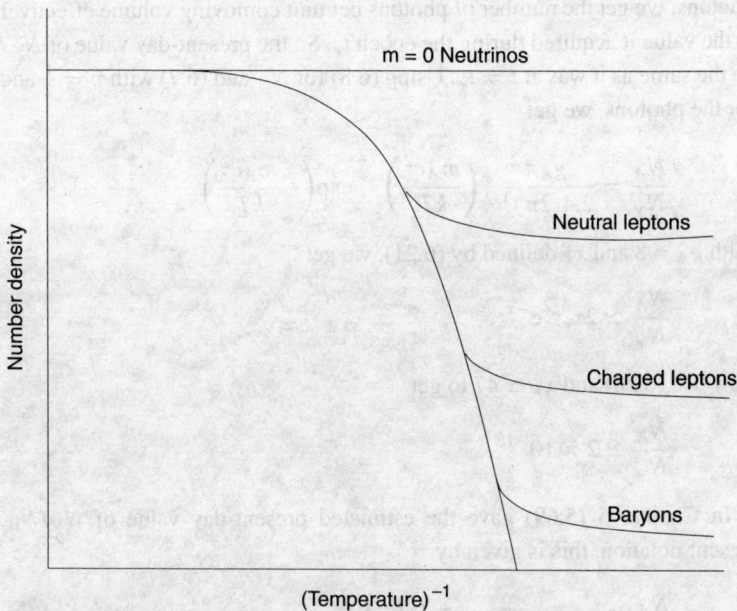

Figure 6.1 A schematic description of how the surviving number density of a particle species depends on its mass and on how strongly it interacts. This number is highest for neutrinos (with zero rest mass and weak interaction) and lowest for baryons (which are massive and strongly interacting).

Using (6.12), (6.24) and (6.23), we get

$$\Lambda = \frac{2.4 \times 10^{-15}}{(2\pi)^{3/2}} (\hbar c)^{-3} \cong 2 \times 10^{16}. \tag{6.26}$$

(Caution: here we are back to using the units seconds, centimetres and mega-electron-volts. Thus $\hbar c$ must be expressed in these units.)

Let us now apply these results to nucleons: to neutrons and protons together with their antiparticles. Then $g_A = 8$ and, with $g_\gamma = 2$ for the photons, (6.3) gives

$$g = 2 + 7 = 9. \tag{6.27}$$

The nucleon mass $m_A \simeq 940$ MeV and, from (6.12), $\bar{\beta} \cong 1$. Thus $Z \cong 313$. With $g_A = 8$ we then have from (6.23) and (6.25)

$$x_*^{-1/2} e^{x_*} \cong 5 \times 10^{19} \tag{6.28}$$

and hence $x_* \simeq 47$. Thus

$$T_* \cong 20 \text{ MeV}, \qquad t_* \cong 0.002 \text{ s.} \tag{6.29}$$

We can also use the above calculation to compute the nucleon-to-photon ratio for the present epoch. Assuming that the A–$\bar{A}$ annihilation is the main source of photons, we get the number of photons per unit comoving volume effectively frozen at the value it acquired during the epoch t_*. So the present-day value of N_A/N_γ will be the same as it was at $t = t_*$. Using (6.8) for N_A and (6.7) with $\eta = \frac{1}{2}$ and $g_A = 2$ for the photons, we get

$$\frac{N_A}{N_\gamma} = \frac{g_A \pi^2}{2.4(2\pi)^{3/2}} \left(\frac{m_A c^2}{kT} \right)^{3/2} \exp\left(-\frac{m_A c^2}{kT} \right). \tag{6.30}$$

With $g_A = 8$ and x_* defined by (6.21), we get

$$\frac{N_A}{N_\gamma} \simeq 2 x_*^{3/2} e^{-x_*}. \tag{6.31}$$

Now use (6.28) and $x_* \simeq 47$ to get

$$\frac{N_A}{N_\gamma} \simeq 2 \times 10^{-18}. \tag{6.32}$$

In Chapter 5 (5.69) gave the estimated present-day value of N_γ/N_B. In our present notation, this is given by

$$\frac{N_A}{N_\gamma} \simeq 3 \times 10^{-8} (\Omega_0 h_0^2) \left(\frac{T_0}{2.7} \right)^{-3}. \tag{6.33}$$

Since $T_0 \sim 2.7$ and $\Omega_0 h_0^2$ is not expected to be lower than $\sim 10^{-3}$ in the most extreme case, we have a large discrepancy to account for. There is one further point of criticism. If we are sure that the universe is made up predominantly of matter, then $N_A \gg N_{\bar{A}}$ and the formula (6.32) applies to $N_A (\cong N_A - N_{\bar{A}} = $ the baryon number density). However, our analysis so far is symmetric between matter and antimatter and so leads to $N_A = N_{\bar{A}}$. Clearly new inputs into the discussion given above are necessary if we are to understand why $N_A \gg N_{\bar{A}}$ and why N_A/N_γ is as high as is indicated by (6.33).

We note that the ratio N_A/N_γ given by (6.31) is a small number. In deriving it we have lost sight of the fundamental constants that went into it. It is instructive to see what (6.31) looks like in terms of $c, \hbar, G$ and m_A. Making the substitution $a = \pi^2 k^4/(15c^3\hbar^3)$ and using (6.13), (6.5), (6.20) and (6.30), we get

$$\frac{N_A}{N_\gamma} = \frac{\pi g^{1/2}}{7.2} \frac{x_*}{\zeta} \left(\frac{2Gm_A^2}{c\hbar} \right)^{1/2}. \tag{6.34}$$

We have already seen that $x_*/\zeta \sim \frac{1}{2}$ and $g^{1/2} \sim 3$, so that the coefficient in front of the expression in parentheses is of the order of unity. So the smallness of N_A/N_γ is directly related to the ratio of the strengths of the gravitational interaction and the strong interaction. Denoting this ratio by the 'gravitational fine structure constant'

$$\alpha_G = \frac{Gm_A^2}{c\hbar} \sim 6 \times 10^{-39} \tag{6.35}$$

we have

$$N_A/N_\gamma \sim \alpha_G^{1/2}. \tag{6.36}$$

The strength of the electromagnetic interaction is measured by the fine structure constant $\alpha = e^2/(\hbar c) \sim 1/137$. Notice how weak the gravitational interaction is by comparison. Had G been considerably higher than it is, we could have ended with a larger value of N_A/N_γ.

6.3   Grand unified theories and baryogenesis

Our simplified calculations of the previous section having led us into difficulties, it is evident that something more sophisticated is needed in order to understand (1) the present-day predominance of baryons over antibaryons and (2) the baryon-to-photon ratio being in the neighbourhood of 10^{-9}. Since our calculation assumed thermodynamic equilibrium and particle–antiparticle symmetry, any new input is expected to question these two assumptions. In this section we outline one of the ways in which this problem is being solved.

The solution is via the so-called grand unified theories (GUTs) – theories that seek to bring together three of the four basic interactions of physics into a single framework. The use of the plural shows that as yet there is no single theory that is universally accepted. In the present section we will follow the $SU(5)$ framework purely as an illustrative example and study its implications for the early universe. To understand what is involved, let us first have a superficial look at the three basic interactions from the group-theoretical point of view.

6.3.1   Electrodynamics

Let us begin with the simplest and best understood interaction: the electromagnetic interaction. This describes how charged leptons (the $e^{\pm}$, $\mu^{\pm}$ and $\tau^{\pm}$) interact through the exchange of photons. When an electron is shaken it emits photons. When a photon strikes an electron it accelerates. The information needed for studying this interaction requires a spinor wave function ψ for the lepton and a vector field A_i (the electromagnetic 4-potential) for the photon. The two physical effects described above are given by the following two equations (written in flat Minkowski space-time):

$$(A^{k,i} - A^{i,k})_{,k} \equiv F^{ik}_{,k} = \frac{4\pi e}{c} \bar{\psi} \gamma^i \psi, \tag{6.37}$$

$$\gamma^i \left(\nabla_i - \frac{e}{\hbar c} A_i \right) \psi - \frac{mc}{\hbar} \psi = 0. \tag{6.38}$$

γ_i are the 4×4 Dirac matrices and e is the electric charge. ∇_i denotes differentiation with respect to spacetime coordinates.

It is easy to see that these equations are invariant under the transformation

$$A_i \to A_i + \Theta_{,i}, \qquad \psi \to \psi \exp\left(-\frac{ie\Theta}{\hbar c} Q \right), \tag{6.39}$$

where Θ is any well-behaved function of spacetime coordinates and Q is the integer 1. The transformation of ψ is a unitary transformation and, since the exponent is a number (that is, a 1×1 matrix), these transformations form a *unitary group of one dimension*, denoted by $U(1)$.

6.3.2   The weak interaction

This weak interaction concerns both the charged and the uncharged leptons in pairs: (e, ν_e), (μ, ν_μ) and (τ, ν_τ). In a typical interaction the members of the pair are interchanged.[1] To describe the pair we therefore need two wave functions: for example, the combination

$$\Psi = \begin{pmatrix} \psi_e \\ \psi_\nu \end{pmatrix} \tag{6.40}$$

describes the pair (e, ν). From empirical considerations it is argued that the weak interaction is invariant under transformations of Ψ with 2×2 matrices that are

---

[1] This is the law of conservation of lepton numbers referred to in §5.2.

unitary and have determinant 1. These transformations form a group denoted by $SU(2)$. Because of parity violation and the fact that neutrinos have only one spin state – they are *left*-handed, it is customary to write a subscript L in $SU(2)_L$. A typical member of the group is denoted by

$$U = \exp(-iH),\qquad(6.41)$$

where H is a 2×2 Hermitian matrix of zero trace. The most general such matrix is

$$\begin{pmatrix} a & b+ic \\ b-ic & -a \end{pmatrix} = a\begin{pmatrix} 1 & \\ & -1 \end{pmatrix} + b\begin{pmatrix} & 1 \\ 1 & \end{pmatrix} + c\begin{pmatrix} & i \\ -i & \end{pmatrix}$$

$$(6.42)$$

Thus instead of a single number Q in (6.39) we need three real numbers a, b and c; or rather, three matrices. (These matrices are proportional to the well-known Pauli matrices.) The 'charges' in this case are the three matrices, two of which are non-diagonal. The non-diagonal matrices permit an interchange of ψ_e and ψ_v in (6.40). This means physically that e and v are interchanged. In this process a charged boson W_1 is exchanged; for example,

$$e \rightarrow W_1 + v.\qquad(6.43)$$

Corresponding to the three matrices in (6.42) there are three W particles, two with charges $\pm e$ and the third (W_3) neutral.

Although the weak interaction does not directly involve the electric charge, it still seems to demand the charged bosons W_1 and W_2. This circumstance prompted efforts to link it with the electromagnetic interaction. This link was achieved via the $SU(2)_L \times U(1)$ framework originally proposed by A. Salam and S. Weinberg (see Figure 6.2) and sometimes called *the electro-weak interaction*. The link brings the photon (which is a boson) closer to the three particles W_1, W_2, and W_3. In this unified picture it is more convenient to talk of another neutral particle Z^0 instead of W_3. Z^0 has zero mass and charge, just like the photon. However, the photon does

(a)   (b)

Figure 6.2 (a) Abdus Salam (1926–1996) (photograph reproduced by permission of Ahmed Salam) and (b) Steven Weinberg (1933–) (photograph by courtesy of S. Weinberg), who independently found the method of unifying electromagnetism with the weak interaction.

not interact with the neutrino, whereas the Z^0 does. The exchange of Z^0 does not alter the electric charge and hence such an interaction is called a *neutral current interaction*. Thus, in the electron–neutrino scattering

$$e + \nu_e \rightarrow e + \nu_e$$

we have e $\rightarrow$ e and $\nu_e \rightarrow \nu_e$ in the *neutral-current* interaction whereas e $\rightarrow \nu_e$ and $\nu_e \rightarrow$ e in the *charged-current* interaction.

The unification programme makes use of the so-called *gauge theories*. The electromagnetic theory is a gauge theory in the sense that its equations are invariant under the gauge transformation of its potential. The transformation of A_i given by (6.38) is a gauge transformation and in the Weinberg–Salam model similar gauge transformations play a pivotal role for the $SU(2)_L \times U(1)$ framework.

One reason for using the gauge theory is that it is 'renormalizable'. This is a technical term that gained currency in quantum electrodynamics (QED), which is a renormalizable gauge theory. In QED the standard calculations of probability amplitudes, average values, energy levels etc. lead to infinities because the relevant integrals diverge at high energies. Renormalization is a technique of subtracting one infinity from another so as to arrive at a finite and physically meaningful answer. Although mathematicians would baulk at such an approach, the theoretical physicist has come to accept it; its merit being that it is unambiguous to operate and the final answers after such manipulations agree very well with experiments. A discussion of this highly interesting topic will, however, take us too far from cosmology and into technical details of field theory. We simply mention that the accelerator experiments have measured the masses of the W and Z bosons and have found them to be in conformity with theoretical expectations.

6.3.3 Quantum chromodynamics

The third basic interaction of physics is the strong interaction described in the framework of quantum chromodynamics (QCD). This makes use of transformations under the $SU(3)$ group. The basic fields here are the quark fields, which are three-component vectors in an abstract space called the *colour space* with three 'dimensions': red, white and blue. Again we have a relation like (6.41) in 3×3 matrices. The matrix H now has eight independent components, so, like in (6.42), we have eight matrix charges, $T_1, \ldots, T_8$, of which two (T_3 and T_8) are diagonal. Again the general matrix character of (6.41) allows quarks to be exchanged. Corresponding to the three Ws in the $SU(2)$ framework we now have eight bosons $G_1, \ldots, G_8$ that are called the gluons. No change in colour takes place when the gluons G_3 and G_8 are exchanged.

The gluons generate an interquark force (just as the photon is responsible for the electromagnetic force between the charged particles). This force is believed to be so large that quarks are expected to be in bound states of two or three quarks. The states with two, a quark and an antiquark, form mesons (such as π^+), whereas states with three quarks are baryons. Quarks have fractional charges. The u-quark has charge $2e/3$, whereas a d-quark has charge $-e/3$. Thus a proton is made of two u-quarks and one d-quark while a neutron is made of two d-quarks and one u-quark.

6.3.4   GUTs: $SU(5)$

In a typical unification attempt we expect the participating interactions to have comparable strengths. In normal laboratory energies the strong interaction (quantum chromodynamics) is the most powerful, followed by electrodynamics and then by the weak interaction. However, as the energy is increased the gaps among the three narrow. At around 100 GeV, the last two are comparable in strength, thus making a unified 'electro-weak' theory viable. Theoretical considerations suggest that, if we extrapolate to considerably higher energies, the strong interaction decreases in strength while the electro-weak interaction gains. At around 10^{15} GeV these interactions become comparable and their unification may seem natural. Figure 6.3 illustrates the changes in strengths of the three interactions with growing energy. Figure 6.3 also shows another landmark in energy at $\sim 10^{19}$ GeV. This is the *Planck energy* given by

$$E_P = \left(\frac{c^5 \hbar}{G} \right)^{1/2} \cong 1.2 \times 10^{19} \text{ GeV}. \tag{6.44}$$

Clearly, with G and $\hbar$ in it this expression would have to do with quantum gravity. We shall consider it separately later. For the time being we exclude it from the unification attempts.

If we wished to unify the other three interactions in a grand unification scheme, we could trivially combine the three into a structure

$$SU(3) \times SU(2)_L \times U(1).$$

However, it was realized that such a structure can form part of a single larger structure denoted by $SU(5)$. Again, if we go back to (6.41) and apply it to 5×5 matrices, the matrix H has 24 arbitrary constants. Thus there are 24 bosons that now mediate among the various basic entities. Of these we already have four from the combined electro-weak interaction and eight (gluons) from chromodynamics. Thus 12 more bosons are needed to make up the list of 24. For want of any specific designation, they are referred to simply as the X-bosons.

The X-bosons are expected to link the participants of chromodynamics (that is, the quarks) with the participants of the electro-weak interaction (that is, the leptons). In the $SU(5)$ theory, therefore, it is possible to change any of the six quarks (u, d, c, s, t and b) into any of the six leptons (e, μ, τ, ν_e, ν_μ, ν_τ) or *vice versa* by the exchange of the X-bosons. This is how it becomes possible to create or destroy baryons. Figure 6.4 outlines the scenario leading to the decay of a proton.

In Figure 6.4 an $\bar{X}$ (that is, an anti-X particle) is emitted and absorbed. Assuming that the mass of this particle is m_X, the probability amplitude for the above interac-

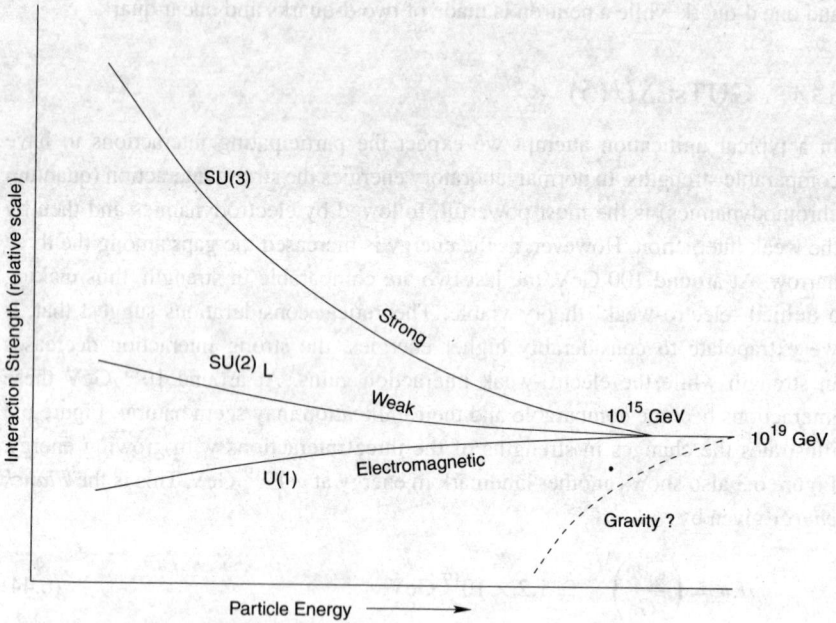

Figure 6.3 At energies of the order of $\sim 10^{15}$ GeV the strengths of the strong, the weak and the electromagnetic interactions become comparable, thus suggesting that this as a suitable energy for grand unification. Still further on we see another energy landmark, viz., 10^{19} GeV, at which level gravitation is expected to be quantized. At this stage perhaps we may have a unified theory of all four basic interactions.

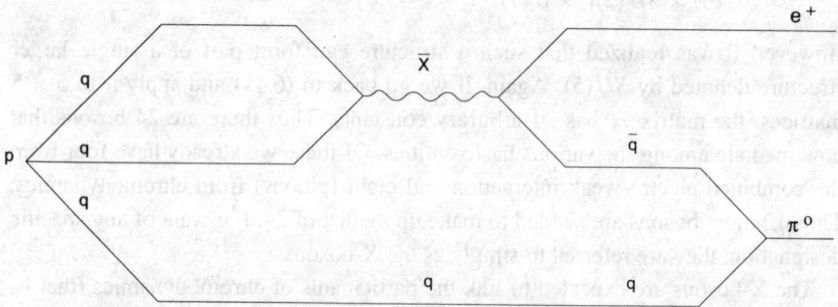

Figure 6.4 The decay of the proton takes place through the mediation of the X-bosons, which change quarks into leptons and *vice versa*. This figure illustrates how this decay could come about. Two of the three quarks (q) in the proton (p) combine to form $\bar{X}$, which decays into a positron and an antiquark. The latter combines with the third quark to form a pion.

tion will contain a factor m_X^{-2}. The proton-decay lifetime τ_p will therefore vary as the fourth power of m_X. Since we expect the lifetime to contain the constants $\hbar, c$ and the proton mass m_p, from dimensional considerations we write

$$\tau_p \sim \frac{\hbar m_X^4}{m_p^5 c^2} \tag{6.45}$$

$$\sim 2.87 \times 10^{-32} [m_X c^2 \ (\text{GeV})]^4 \ \text{years}.$$

The failure to observe the decays of protons in the laboratory experiments sets a lower limit of $\sim 10^{29}$ years on τ_p. Hence (6.45) suggests that

$$m_X c^2 \geq 10^{15} \ \text{GeV}. \tag{6.46}$$

At a value of $\tau_p \sim 10^{30}$ years, say, we have no hope of observing the decay of a particular proton. However, in a large population of protons a small fraction may decay. For example, in 1000 tons of matter about 50 protons are expected to decay every year if $\tau_p \sim 10^{30}$ years. Experiments during the 1980s failed to observe such decays in an unambiguous manner for $\tau_p \leq 10^{31}$ years. This led to the abandoning of the above simple $SU(5)$ theory in favour of more complex frameworks.

Exact dynamical theories are needed in order to quantify τ_p and m_X. However, while the proton-decay experiment is barely feasible if $\tau_p \leq 10^{33}$ years, the full testing of the predictions of GUTs is clearly beyond the scope of present-day technology. It is worth mentioning one prediction that is commonly known as the *hypothesis of asymptotic freedom*. According to this hypothesis, at extremely high energies most interactions among particles begin to lose their strength. However, even this hypothesis has still to be tested experimentally.

The other alternative, of course, is to use the hot universe for testing theoretical predictions. Even here, for a mass of 10^{15} GeV, the temperature ($= mc^2/k$) will be as high as $\sim 10^{28}$ K! A temperature of 10^{15} GeV gives, according to (6.6), an age of the universe as low as $\sim 10^{-36}$ s. We will refer to it as the GUT epoch. In the late 1970s M. Yoshimura suggested that, with GUTs, it is possible to produce a slight excess of baryons over antibaryons because the number of baryons is not conserved. However, further assumptions are needed if one is actually to produce a result in accord with observations. The following scenario is that suggested by S. Weinberg and F. Wilczek.

6.3.5 Baryogenesis in the early universe

Let us denote the mass of the X-boson (which causes non-conservation of baryons) by m_X and its coupling strength by α_X. The coupling strength may be 10^{-2} or 10^{-5}, depending on what type of particle X is. Let us denote by Γ_c the rate of collisions that do not conserve the number of baryons; that is, collisions in which the X-boson is involved. The X-boson itself does not last very long, its time scale being of the order of $\hbar/(m_X c^2)$. Denote the characteristic decay rate of the X-boson by Γ_X. We thus have three time scales to play with: Γ_X^{-1}, Γ_c^{-1} and H^{-1}. The trick lies in adjusting these time scales suitably to produce the desired answer. The argument, qualitatively, goes like this.

During the earliest epochs with temperature $\geq 10^{19}$ GeV gravity was the strongest force among the various constituents of the universe. Other interactions (including the strongest of them, QCD) were unimportant under the hypothesis of asymptotic freedom. As the universe continued to expand and its temperature dropped, there was a phase during which gravity became weaker while the other interactions still remained unimportant. Thus, for $T \leq 10^{19}$ GeV, the particles remained essentially free for some time.

During this phase it becomes necessary to examine the nature of distribution functions that are given by the formula (5.9). There we saw that, so long as $T \gg T_A$, that is, so long as we are in the relativistic regime, the distribution function preserves its equilibrium form during free expansion with $T \propto S^{-1}$. However, if $T \leq T_A$, the distribution function cannot preserve its form under free expansion. Thus it may get distorted from its equilibrium form.

Now, of the various species in the early universe, the X-bosons are probably the most massive. Thus, provided that they have a high enough value of T_X, there is a chance that the X-bosons will be the first to drop out of equilibrium. For this to happen, however, it is also necessary that they have not all decayed by then. The decay rate of the X-boson is of the order of

$$\Gamma_X \cong \alpha_X g m_X c^2 / \hbar, \tag{6.47}$$

where g is the effective number of degrees of freedom for the various particle species (g may well lie between 100 and 200; for $SU(5)$ it is $\sim$160).

The rate of expansion, on the other hand, is given by (6.1). The collision rate $\Gamma_c \approx \alpha_X \ll \Gamma_X$. A comparison of the three rates shows that $\Gamma_c < \Gamma_X < H$ soon after the Planck time t_P. Thus the universe was expanding at this stage with essentially no interaction between the species. The X-bosons began to decay when the age of the universe became comparable to Γ_X^{-1}. Using (6.1), (6.2) and (6.47), we get

$$T = \left(\frac{3 g \alpha_X^2 m_X^2 c^4}{4 \pi G a \hbar^2} \right)^{1/4}. \tag{6.48}$$

By the time the universe has cooled to the above temperature the X-bosons would have begun to decay. Were they in equilibrium until then?

As seen above, this question is decided by a comparison of T with T_X. Two cases are of interest: (1) $T \gg T_X$ and (2) $T \leq T_X$. These are illustrated by Figures 6.5 and 6.6, respectively.

In case (1) the decays occurred while the X-bosons still had their distribution functions in the equilibrium form. Under these circumstances the X-bosons could not have generated any nett excess of baryons; for thermal equilibrium implies that any decays (like that in Figure 6.4) leading to destruction of baryons would be compensated by inverse decays. In case (2), however, the distribution function of X-bosons was distorted from its equilibrium form and hence the detailed balancing between decays and inverse decays would not happen. The required new input into the early-universe scenario discussed in §6.2 is therefore provided by case (2). By departing from thermodynamic equilibrium at the right time, the X-boson distribution has a chance of producing baryon asymmetry.

The condition that T_X exceed the value of T given by (6.48) may be expressed as

$$m_X > \left(\frac{3g\alpha_X^2 k^4}{4\pi Gac^4} \right)^{1/2} = g^{1/2}\alpha_X m_p. \tag{6.49}$$

Empirical considerations of the $SU(5)$ framework suggest that, from the above inequality, $m_X c^2$ should exceed $\sim 10^{16}$ GeV. This is consistent with our earlier

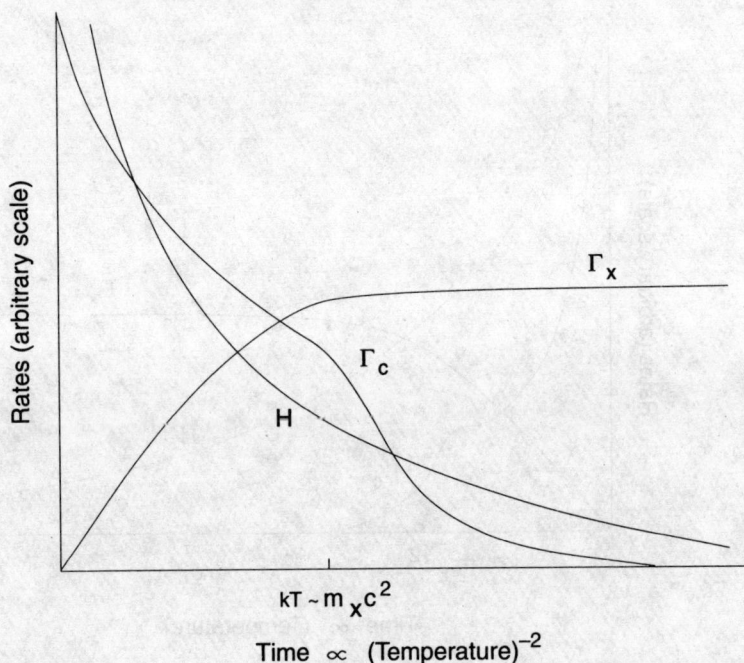

Figure 6.5 The three rates H, Γ_c, and Γ_X for the case $m_X c^2 < \alpha_X 10^{20}$ GeV. When kT drops below $m_X c^2$, Γ_X exceeds the rate of expansion H and the X-bosons decay exponentially in number while maintaining the equilibrium distribution. No nett excess of baryons is generated in this case. Adapted from D. N. Schramm and M. S. Turner, 1979, 'The origin of baryon number and related problems', in R. Balian, J. Audouze, and D. N. Schramm, eds., *Physical Cosmology*, Les Houches Lectures Sessions XXXII, p. 501 (Amsterdam: North Holland).

estimate of the mass of the X-boson from the lower limit on the lifetime of the proton.

So far we have introduced the assumption of departure from thermodynamic equilibrium. We now introduce the other assumption of baryon–antibaryon asymmetry. Suppose that the X-boson decays into two states with baryon numbers B_1 and B_2 with fractions r in state 1 and $1 - r$ in state 2. In a perfectly symmetric situation, the $\bar{X}$ boson would decay into state 1 with baryon number $-B_1$ with fraction r and state 2 with baryon number $-B_2$ with fraction $1 - r$. However, if perfect symmetry does not exist, then the fractions would be $\bar{r}$ and $1 - \bar{r}$, respectively for the $\bar{X}$-decay ($\bar{r} \neq r$). The nett number of baryons generated by these processes is therefore

$$\Delta B = (r - \bar{r})(B_1 - B_2). \tag{6.50}$$

Since the baryon-non-conserving collisions that could destroy ΔB are running at a smaller rate than H ($\Gamma_c < H$), we expect ΔB to be preserved.

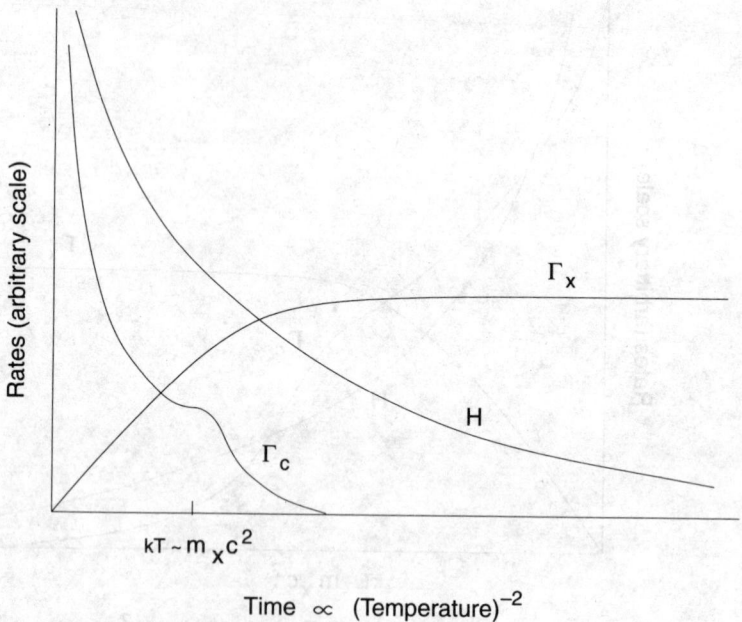

Figure 6.6 Figure 6.5 redrawn for the case $m_X c^2 > \alpha_X 10^{20}$ GeV. When kT drops below $m_X c^2$, X decays and annihilations are not effective, since both Γ_X and Γ_c are less that H. Until Γ_X exceeds H, the X-bosons do not come into equilibrium. At that stage X and $\bar{X}$ decay freely and may generate a nett excess of baryons. The excess stays since $\Gamma_c < H$. Adapted from D. N. Schramm and M. S. Turner, 1979, 'The origin of baryon number and related problems', in R. Balian, J. Audouze, and D. N. Schramm, eds., *Physical Cosmology,* Les Houches Lectures Sessions XXXII, p. 501 (Amsterdam: North Holland).

Thus, to account for the observed excess of baryons over antibaryons and to argue that the nett number density of baryons observed today is $\sim 10^9$ times the observed photon density, we have to make sure that the parameters of the GUT are such as to give appropriate quantitative expression to ΔB above. It is claimed that reasonable values of the parameters of GUTs do lead to a formula in agreement with (6.33).

Current ideas on baryogenesis carry such arguments to higher levels of sophistication. Whether or not such claims turn out to be justified, the above argument illustrates how the early universe provides an interesting arena for the application of GUTs.

6.3.6   The spontaneous breakdown of symmetry

The change from a larger group of symmetries to the subgroup $SU(3) \times SU(2)_L \times U(1)$ is spontaneous. The actual mechanism involves a set of scalar fields called the Higgs fields ϕ that change over from their initial values of zero to a set of finite values when this happens. Why and how this happens and the role the Higgs fields play in the process is a long story, which would take us into the labyrinths of gauge field theories. The explanation given below skirts the problem and provides a superficial description.

We begin with the analogy of ferromagnetism and the crucial role of the Curie temperature (770 °C for iron). Above this temperature a bar of iron exhibits no magnetism in an external field. This is because its elementary nuclear magnets are randomly aligned with no resultant magnetization. Energetically this is the lowest state for the bar and it chooses to remain in that state since it is the most stable one. Below the Curie temperature the state of lowest energy changes to that in which all the nuclei are aligned along the bar, which develops polarity at its ends. There are two states of the same lowest energy possible, depending on which (north or south) of the two poles falls at a given end. The ultimate choice of one state apparently breaks the symmetry, although theoretically and inherently the symmetry is always there.

In the very early universe something similar happens to the ϕ-field. Above a critical temperature T_c, the vacuum state, the state of lowest energy, is none other than $\phi = 0$. Below T_c, the state of lowest energy changes. It now corresponds to a situation in which ϕ has non-zero values. We will encounter explicit examples of this in §6.5.

For the time being let us suppose that there exist alternative values ϕ_i ($i = 1, 2, \ldots$) of the ϕ-field all corresponding to states of the same lowest energy which now acquire the status of vacuum. There is basic symmetry with respect to all ϕ_i but in practice the system may spontaneously acquire only one of them. This is again an apparent breakdown of symmetry.

The consequences of this for the very early universe are that, as shown in Figure 6.7(a), it is divided into distinct domains, each with a different value of ϕ_i. In this way the universe acquires regions of discontinuities separated by the domain walls. These translate into highly significant discontinuities in the distribution of matter. The fact that we do not see such discontinuities in actuality (say in the form of large

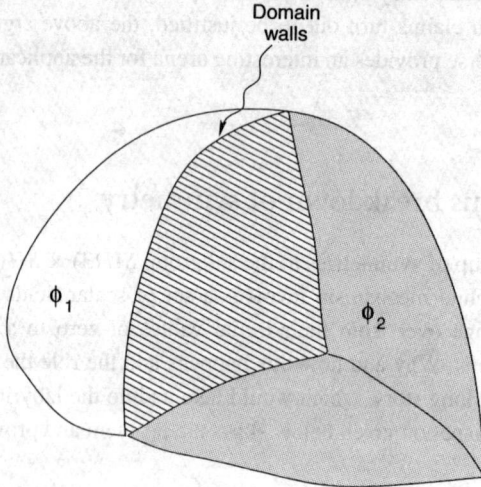

Figure 6.7 Spontaneous breakdown of symmetry can leave boundaries in space separating domains of different physical conditions. In (a) such separation is shown with the boundary in the form of a 'domain wall'. In (b) we see how two domain walls intersect in a linear structure called a 'cosmic string'.

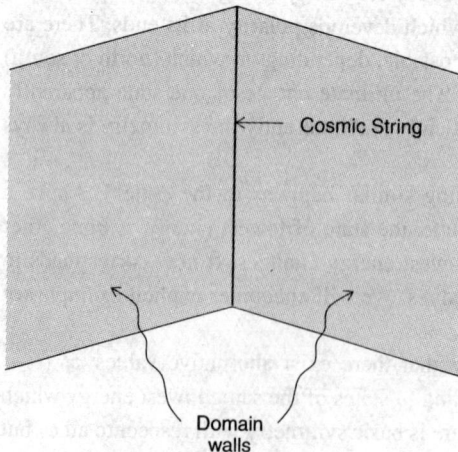

sheets of matter) is hard to explain away. This difficulty is known as the *domain-wall problem*.

The intersection of two domain walls is a linear structure known as a 'cosmic string' (see Figure 6.7(b)). Such filamentary structures have been invoked in scenarios for galaxy formation (see Chapter 7) in which large numbers of galaxies are found distributed in linear structures.

6.4 Some problems of Friedmann cosmology

It may appear from the above that, by going over to the very early universe, we have made progress in understanding some of the present-day features of the universe. In fact the situation is the exact opposite: we have acquired more problems than we managed to solve by this device. The domain-wall problem is one of them. Other, more important, problems are highlighted below.

6.4.1 The horizon problem

Let us suppose that the initial conditions for the universe were set fairly early on, during an epoch t in the radiation-dominated phase. From the considerations of Chapter 4 adapted to the scale factor $S \propto t^{1/2}$, we find that the proper radius of the particle horizon during that epoch was

$$R_P = 2ct. \tag{6.51}$$

Whatever physical processes operated during this epoch were limited in range by R_P. Hence we do not expect the homogeneity of physical quantities to extend beyond the diameter $2R_P$, unless we make the somewhat contrived assumption that the universe was *created* homogeneous. In other words, the causal limitations tell us that no region larger than $2R_P$ in size should be homogeneous.

When the initial conditions were so set, this would expand to a much larger size in the present epoch, the factor η by which it would grow is the ratio of scale factors

$$\eta = \frac{S(t_0)}{S(t)}$$

for the present and initial epochs. How do we estimate η?

The simplest method is to compare the temperatures at t and t_0 since (from considerations of Chapter 5) $S \propto T^{-1}$. Thus

$$\eta = \frac{T(t)}{T(t_0)},$$

where $T(t)$ is given by (6.6). It is convenient to express T_0 also in giga-electron-volts:

$$T_0 \text{ (GeV)} = 2.3 \times 10^{-13}\left(\frac{T_0}{2.7 \text{ K}}\right). \tag{6.52}$$

By combining (6.6) and (6.52) and writing the value of c in (6.51), we get the present limit on a homogeneous region as

$$R_{\text{Hom}}(t_0) = 2ct$$

$$= 6.2 \times 10^{17} T_{\text{GeV}}^{-1} g^{-1/2}\left(\frac{2.7 \text{ K}}{T_0}\right) \text{ cm}. \tag{6.53}$$

For $T_{\text{GeV}} \cong 10^{15}$, $g \cong 100$ and $T_0 \cong 2.7$ K we get the surprisingly small value of 62 cm! In other words we have no reason to expect homogeneity on a scale larger than, say, 1 m. The fact that the relic microwave background is homogeneous on the cosmological scale of $\sim 10^{28}$ cm tells us that there is something seriously wrong with our reasoning above. Yet, the standard model does not provide any loophole out of this so-called *horizon problem*. Notice also that the further back into the past we go (in our attempts to set the initial conditions) the larger T_{GeV} and the smaller $R_{\text{Hom}}(t_0)$ will be. Figure 6.8 illustrates the horizon problem.

6.4.2 The flatness problem

When discussing the early and the very early universe we ignored the kc^2/S^2 term in the field equations. Thus (6.1) should actually have been

$$\frac{\dot{S}^2}{S^2} + \frac{kc^2}{S^2} = \frac{8\pi G\rho}{3}. \tag{6.54}$$

Our justification for ignoring that term was that, as $S \to 0$, $\dot{S}^2 \to \infty$ and, thus, the first term far exceeds the second term on the left-hand side of (6.54). This argument is, however, *scale-dependent*. Thus, if we write $S = At^{1/2}$, then $\dot{S}^2 = A^2/(4t)$. Whether $\dot{S}^2$ exceeds c^2 for $k = \pm 1$ would depend on A. *A priori* we do not know A, unless we link it with the present size of the universe. It is more convenient to look at the density parameter Ω instead.

Writing $\rho = \Omega\rho_c$ as in (4.53) we have, for any general epoch when $S \propto t^{1/2}$,

$$\frac{kc^2}{S^2} = (\Omega - 1)\frac{\dot{S}^2}{S^2} = \frac{\Omega - 1}{4t^2}. \tag{6.55}$$

For the present epoch, on the other hand,

$$\frac{kc^2}{S_0^2} = (\Omega_0 - 1)H_0^2. \tag{6.56}$$

Dividing (6.55) by (6.56) and using $S \propto T^{-1}$, for $k = \pm 1$,

$$\Omega - 1 = 4H_0^2 t^2 \frac{T^2}{T_0^2}(\Omega_0 - 1).$$

Except for $\Omega_0 - 1$, all quantities on the right-hand side are known. Using (6.6) for t and (6.52) for T_0, we get

$$\Omega - 1 \cong 4.3 h_0^2 g^{-1} \times 10^{-21} T_{\text{GeV}}^{-2} \left(\frac{2.7 \text{ K}}{T_0} \right)^2 (\Omega_0 - 1). \qquad (6.57)$$

For $T_{\text{GeV}} = 10^{15}$ and $g \cong 100$, we get for $T_0 \cong 2.7$ K

$$\Omega - 1 \cong 4.3 h_0^2 \times 10^{-53} (\Omega_0 - 1). \qquad (6.58)$$

This expression epitomizes what has come to be known as the *flatness problem*. Suppose that the initial conditions including the density parameter Ω were set during the GUT epoch when $T \cong 10^{15}$ GeV. Then the present-day value of $\Omega_0 - 1$ is given by (6.58). Or, to invert the chain of reasoning, suppose that the present-day observational uncertainty tells us that $|\Omega_0 - 1| \sim \mathcal{O}(1)$. Then, from (6.58), during the GUT epoch Ω was differing from unity by a fraction of the order of 10^{-53}. In other words, the departure from the flat value of $\Omega = 1$ at this stage had to be

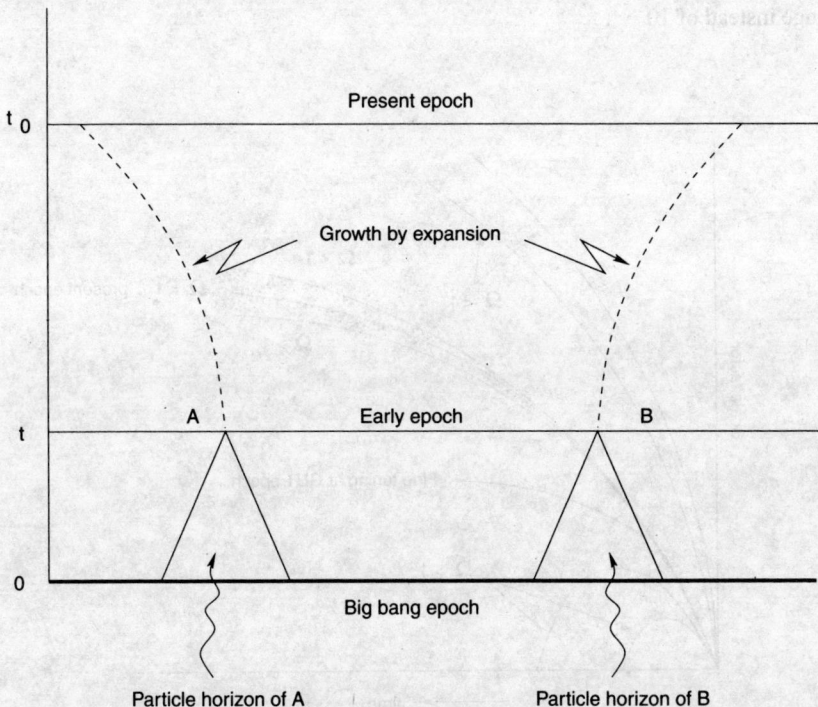

Figure 6.8 During a very early epoch t, the observers A and B have non-overlapping particle horizons. Thus there is no *a priori* reason why A and B should have the same initial conditions. Yet the universe *as seen at present* is homogeneous over distances far larger than what AB would grow to (shown by dotted segments) at present.

extremely small. Any relaxation of this fine tuning would have led to a far wider range of Ω_0 at present than is permitted by observations.

So our neglect of the curvature term kc^2/S^2 is linked with an extremely fine tuning of the universe to the flat $(k = 0)$ model. If this tuning were not there, the universe would have either collapsed $(k = 1)$ or expanded to infinity $(k = -1)$ on time scales of the order of 10^{-35} s that were characteristic of the GUT era.

Figure 6.9 illustrates this conundrum. The shaded region denotes the finely tuned set of Friedmann models that end up today within the observed range $|\Omega_0 - 1| \sim \mathcal{O}(1)$. The other curves are for the characteristic models with time scales $\sim 10^{-35}$ s which should normally have operated during the GUT stage. What made the universe get into the shaded region instead?

This problem was first highlighted by R. H. Dicke and P. J. E. Peebles in 1979, who discussed it not for the GUT epoch but at $t \sim 1$ s, when the neutrinos had decoupled and pair $(e^{\pm})$ annihilation was to begin. Thus $T \sim 10^{-3}$ GeV, $g \sim 10$ and we get $\sim 10^{-16}$ instead of 10^{-53} in (6.58). It is clear that the further back in time and closer to $t = 0$ we go the finer the tuning required. For example, if we were to initialize the problem during the Planck epoch, we would get 10^{-61} for the tuning range instead of 10^{-53}.

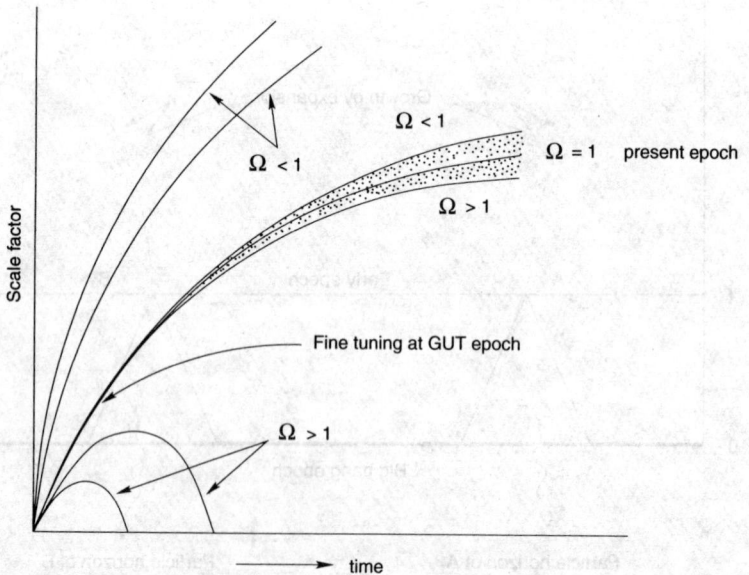

Figure 6.9 The flatness problem illustrated with the help of the expansion functions for the $k = 0, \pm 1$ models ($\Omega \geq 1$, $\Omega = 1$ and $\Omega < 1$). The observable uncertainty extends over the range of curves in the shaded region, all of which were tightly bunched together during the GUT epoch, close to the $\Omega = 1$ curve.

6.4.3   The entropy problem

This is a restatement of the flatness problem and the horizon problem in a somewhat different form. The entropy in a given comoving volume stays constant in an adiabatic expansion (see §5.2). The present photonic entropy in the observable universe of characteristic size $R \approx h_0^{-1} \times 10^{28}$ cm is given by

$$\Sigma = \frac{4\pi}{3k}aT_0^3 R^3 \approx h_0^{-3} \times 4.4 \times 10^{87}\left(\frac{T_0}{2.7}\right)^3. \tag{6.59}$$

Why such a large value? If the entropy were conserved, we would have $ST =$ constant. However, we found that, for the flatness problem, this hypothesis led to fine tuning, whereas for the horizon problem it gave an extremely small size of the homogeneity. It therefore appears that the trouble lies in $\Sigma =$ constant: it could be resolved if the assumption of adiabatic expansion were violated at some stage and Σ boosted to its present value by an enormously large factor.

6.4.4   The monopole problem

In a GUT, whenever there is a breakdown of symmetry of a larger group like $SU(5)$ to a subgroup like $SU(3) \times SU(2)_{\mathrm{L}} \times U(1)$ that contains the $U(1)$ group, there inevitably arise particles that have the characteristics of a magnetic monopole. This is a rigorous mathematical conclusion in gauge field theories. Typically the mass of the monopole (in energy units) is given by $\sim 10^{16}$ GeV. Monopoles are highly stable particles and once they have been created they are indestructible, so they would survive as relics in the present epoch.

During the GUT epoch t, the size of the horizon being $2ct$, we expect at least one monopole per horizon-sized sphere, i.e., a monopole mass density of

$$\frac{10^{16}\ \mathrm{GeV}/c^2}{(4\pi/3)(2ct)^3}.$$

At present this is diluted by the factor $(T_0/T)^3$. For T_0 in giga-electron-volt, given by (6.52) and $T = 10^{15}$ GeV, we get the present-day density of monopoles as

$$\rho_{\mathrm{M}} \cong 1.5 \times 10^{-13}\left(\frac{T_0}{2.7\ \mathrm{K}}\right)^3 g\ \mathrm{cm}^{-3}. \tag{6.60}$$

This is far in excess of the closure density $\sim 10^{-29} g\ \mathrm{cm}^{-3}$, thus making it a very awkward problem for the standard model to solve. Again, as in the earlier cases, the discrepancy grows if, instead of the GUT epoch, we use an even earlier epoch.

6.5 The inflationary universe

These difficulties of the standard big-bang model began to surface when physical cosmology was extended to the very early epochs. Their resolution seemed to require a new input during or around the GUT epoch, an input that would change the dynamics of the universe at least temporarily. Such inputs were suggested independently by D. Kazanas in 1980, by Alan Guth in 1981 and by K. Sato, also in 1981. These approaches were essentially similar; we describe below Guth's approach, which became the best known of the three.

Alan Guth proposed the so-called *inflationary phase* as the solution to these problems. The word 'inflation' is supposed to indicate a rapid expansion. Thus we envisage the following sequence:

$$t < t_1 : \text{ scale factor } S(t) \propto t^{1/2},$$

$$t_1 < t < t_2 : \text{ scale factor } S(t) \propto \exp(t/\tau), \qquad \tau = \text{constant},$$

$$t_2 < t : \text{ scale factor } S(t) \propto t^{1/2}. \tag{6.61}$$

Briefly, we have inserted a phase of rapid exponential expansion during $[t_1, t_2]$. What is this time range? How do we set the value of the time constant τ? To answer these questions let us first look at Guth's method.

6.5.1 Guth's inflationary model

As we saw earlier, the breakdown of GUT symmetry to $SU(3) \times SU(2)_L \times U(1)$ leads to a phase transition in which the vacuum state (i.e., the state of lowest energy) of the Higgs field ϕ changes. The original vacuum with $\phi = 0$ is no longer the true vacuum. The inflationary stage arises, however, if the true vacuum is not immediately attained.

An analogy to illustrate the scenario will be in order. Suppose that steam is being cooled through the phase-transition temperature of $100\,°\text{C}$. Normally we expect the steam to condense to water at this temperature. However, it is possible to supercool the steam to temperatures below $100\,°\text{C}$, although it is then in an unstable state. The instability sets in when certain parts of the steam condense to droplets of water, which then coalesce and eventually the condensation is complete. In the supercooled state the steam still retains its latent heat, which is released as the droplets form.

Suppose that similar supercooling takes place past the GUT phase-transition temperature. What happens then can be described by the steam–water analogy. Its details depend on the potential-energy function $V(\phi, T)$ which we consider next.

Consider the action principle defining the dynamics of the ϕ-field by

$$\mathcal{A}(\phi) = \int \left(\frac{1}{2} \phi_i \phi^i - V(\phi) \right) a^4 x, \tag{6.62}$$

where $\phi_i = \partial\phi/\partial x^i$ while the potential $V(\phi)$ is given as follows. First, ϕ is a scalar gauge field but it has internal degrees of freedom decided by the number of generators of the gauge group. Let the generating matrices be τ_A $(A = 1, 2, \ldots, N)$. Then we write

$$\phi = \sum_A \phi^A \tau_A \tag{6.63}$$

and consider the following *quartic* form for V:

$$V = -\tfrac{1}{2}\mu^2 \operatorname{Tr}\phi^2 + \tfrac{1}{4}a(\operatorname{Tr}\phi^2)^2 + \tfrac{1}{2}b \operatorname{Tr}\phi^4 + \tfrac{1}{3}c \operatorname{Tr}\phi^3. \tag{6.64}$$

μ, a, b and c are coupling constants.

In a typical symmetry breaking of the kind

$$SU(5) \rightarrow SU(3) \times SU(2)_L \times U(1)$$

we have

$$\langle\phi\rangle = \Phi \operatorname{diag}(1, 1, 1, -\tfrac{3}{2}, -\tfrac{3}{2}), \tag{6.65}$$

where Φ is an ordinary scalar. Alternatively, if

$$SU(5) \rightarrow SU(4) \times U(1), \tag{6.66}$$

then

$$\langle\phi\rangle = \sigma \operatorname{diag}(1, 1, 1, 1, -4), \qquad \sigma \text{ a scalar.} \tag{6.67}$$

In each case $\operatorname{Tr}\phi = 0$. If the basic GUT symmetry group is different, we will of course have different representations of ϕ. For our cosmological purpose we need to know how $V(\phi)$ will affect the geometry of spacetime via the Einstein equations. For this effect we need to average over the quantum fluctuations of the ϕ-field, which gives us an 'effective' average potential

$$V_{\text{eff}}(\phi) = \alpha\phi^2 - \beta\phi^4 + \gamma\phi^4 \ln(\phi/\sigma^2), \tag{6.68}$$

where α, β, γ and σ are parameters from particle physics.

Since this analysis is to be carried out for the hot early universe, there will be thermal fluctuations also. Their inclusion leads to the addition to $V_{\text{eff}}(\phi)$ of a thermal component to give a total potential

$$V(\phi, T) = V_{\text{eff}}(\phi) + \frac{18T^4}{\pi^2} \int_0^\infty x^2 \ln\left\{1 - \exp\left[-\left(x^2 + a\frac{\phi^2}{T^2}\right)^{1/2}\right]\right\} \mathrm{d}x. \tag{6.69}$$

Here a is a constant. Figure 6.10 plots $V(\phi, T)$ as a function of ϕ for a range of values of T.

Notice that, for a critical value $T = T_c$, the $V(\phi, T)$ curve touches the ϕ-axis at two points, $\phi = 0$ and $\phi = \phi_0$, both of which points are local minima for $V(\phi, T)$. For $T \gg T_c$, there is only one minimum, at $\phi = 0$. As T is lowered, a second minimum appears at a higher level, but, when T goes below T_c the minimum sinks to a *lower* level. In other words, for $T < T_c$ the state of lowest energy of the ϕ-field resides not at $\phi = 0$ but at a value of $\phi > 0$. This is where we have the supercooled-steam situation.

Imagine the universe being cooled through the critical value T_c. As T drops below T_c the state of lowest energy shifts in a discrete fashion signalling a phase transition. However, if the universe is supercooled, it stays in the 'false' vacuum at $\phi = 0$ until at some stage the ϕ-field tunnels across the $V(\phi) > 0$ barrier and falls down the $V(\phi)$ slope to its 'true' vacuum. Let us denote by ϵ_0 the difference between the energies of the two vacua. Until the tunnelling has taken place the universe has an extra energy density ϵ_0 at its disposal, which must have dynamical effects via the Einstein equation:

$$\frac{\dot{S}^2 + kc^2}{S^2} = \frac{8\pi G}{3c^2}(\epsilon_0 + \epsilon_r). \tag{6.70}$$

Here $\epsilon_r \propto 1/S^4$ is the energy density of radiation and relativistic particles. Since ϵ_r falls as the universe expands while ϵ_0 stays constant, the latter clearly dominates. Hence we ignore ϵ_r and solve (6.70). For $k = +1$ we get, for example,

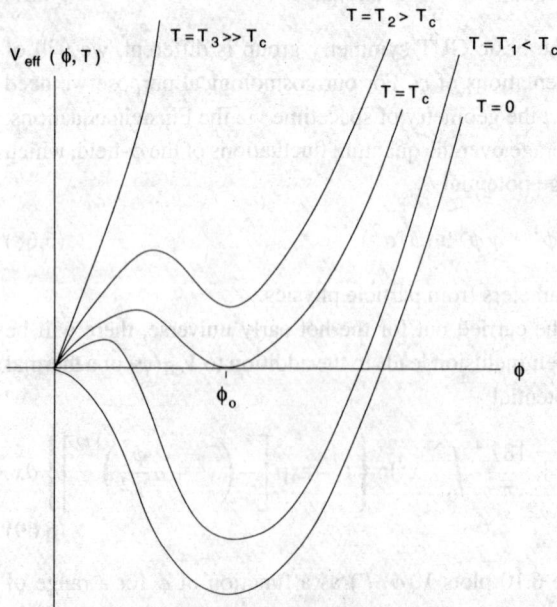

Figure 6.10 $V_{eff}(\phi, T)$, the effective potential of the Higgs field, at various temperatures. For $T < T_c$ the true vacuum, i.e., the state of lowest energy, is no longer at $\phi = 0$.

$$S = \left(\frac{3c^4}{8\pi G\epsilon_0}\right)^{1/2} \cosh\left[\left(\frac{8\pi G\epsilon_0}{3c^2}\right)^{1/2} t\right]. \tag{6.71}$$

For $k = -1$ we get a similar expression with 'cosh' replaced by 'sinh'. The main point to note is that for

$$t \gg \left(\frac{3c^2}{8\pi G\epsilon_0}\right)^{1/2}, \tag{6.72}$$

either solution approaches closely the $k = 0$ (flat) solution

$$S \propto \exp(at), \qquad a = \left(\frac{8\pi G\epsilon_0}{3c^2}\right)^{1/2}. \tag{6.73}$$

This exponential expansion is reminiscent of the de Sitter model. Indeed, the energy tensor of the false vacuum simulates the λg_{ik} term of the Einstein equations.

This rapid expansion in an exponential fashion continues until the tunnelling takes place and ϕ attains its true vacuum value. The average time τ for the tunnelling to occur can be computed quantum mechanically. It tells us the factor Z by which the scale factor S increased while inflation lasted. One finds that

$$a\tau \approx 67, \qquad Z = \exp(a\tau) \approx 10^{29}. \tag{6.74}$$

In other words, the exponential expansion or *inflation* lasts long enough for the scale factor to blow up by a large multiple, $\sim 10^{29}$. Thus, if we started with a curvature term (kc^2/S^2) comparable to the expansion term $(\dot{S}^2/S^2)$ prior to inflation, we would end up by having the former reduced by $Z^2 \sim 10^{58}$ while the latter remained constant. This large factor Z not only takes care of the fine tuning in the flatness problem but also resolves the horizon problem (by blowing up the homogeneous region by a factor Z in linear dimensions) and the monopole problem (by reducing the density of monopoles by the factor Z^3). Similarly the domain walls are blown apart so that the chance of one crossing the observable universe is negligible.

There was one serious drawback, however, which rendered the Guth model unworkable. This comes from the entropy. The entropy is also blown up by the factor $Z^3 = 10^{87}$, thus apparently explaining why the present-day universe has such a large value for Σ. However, how was this excess entropy to be dumped in the universe?

The expectation was that, as the phase transition in a bounded region is completed, it switches over to the Friedmann radiation-dominated expansion phase since it no longer has the energy ϵ_0 to draw on. The inflating region therefore breaks up into Friedmann bubbles that expand. Most of the excess energy resides on the surfaces of these bubbles so that, when two bubbles collide, the energy is thermalized. This is how wider and wider regions undergo the phase transition and acquire thermalized energy and entropy.

The expectation was nullified by the fact that, as the universe outside the bubbles expands exponentially, bubbles nucleated in different parts move away from one another so fast that they cannot collide. The thermalizing mechanism discussed above therefore does not work.

6.5.2 The new inflationary universe

A revised version of the inflationary scenario was not long in coming. A. D. Linde in the USSR and A. Albrecht and P. J. Steinhardt in the USA independently proposed what came to be known as the *new inflationary universe*. The crucial difference between the new theory and the original version of Guth was in the choice of $V_{\text{eff}}(\phi, T)$. In the new model $V_{\text{eff}}(\phi, T)$ was taken from the work of S. Coleman and E. Weinberg:

$$V_{\text{eff}}(\phi, T) = \frac{25}{16}\alpha^2 \left[\phi^4 \ln\left(\frac{\phi^2}{\sigma^2}\right) + \frac{1}{2}(\sigma^4 - \phi^4)\right]$$

$$+ \frac{18}{\pi^2}T^4 \int_0^\infty \ln\left\{1 - \exp\left[-\left(x^2 + \frac{5}{12}\phi^2\frac{g^2}{T^2}\right)^{1/2}\right]\right\}\,\mathrm{d}x$$

$$(6.75)$$

where α, σ and g are constants. Figure 6.11 plots $V_{\text{eff}}(\phi, T)$ for a characteristic value of T.

We have a false vacuum at A ($\phi = 0$) and the true vacuum at B ($\phi = \sigma$). There is a temperature-dependent bump C beyond A ($\phi > 0$), followed by a plateau portion DE that slopes down very gently before dropping down steeply from E to B. For inflation to take place for a sufficient time we need the system to remain in the upper part of the figure, i.e., during the time when it tunnels across the bump C and then slowly rolls over from D to E. Thereafter the system drops to B but, instead of staying there, it executes damped oscillations, during which energy is thermalized and entropy increased. The actual dynamics of the universe, as given by Einstein's equations, can be numerically solved. A 'satisfactory' solution is obtained by adjusting the parameters. Thus the following is a satisfactory solution:

$\phi = \phi_i \approx 0$ roll-over down the plateau begins,

$t \leq 190H^{-1}$ the roll-over time which is also the duration of inflation with $S \propto \exp(Ht)$,

$H \approx 2 \times 10^{10}$ GeV   the Hubble constant for inflation,

$Z \approx \exp(190) \approx 10^{50}$ the boost in linear size caused by inflation,

$\tau_{\text{osc}} \approx \exp(4.8) \times 10^{-4}H^{-1}$ the time of oscillation before settling down at $\phi = \sigma \approx 2 \times 10^{15}$ GeV.

Notice that the model overcorrects for the shortcomings of the standard model. For example the size of the horizon is boosted by 10^{50} so that the present size of a homogeneous region is greater than the observable universe by a factor of $\sim 10^{23}$. Notice also that, because a single region is large enough (more than enough!) to encompass the observable universe, there is no need for bubbles to collide and coalesce. In the Guth version the bubbles were small so that their collision and coalescence was necessary in order to generate a large enough region – besides the requirement of reheating.

Since $\tau_{\text{osc}} \ll H^{-1}$ the oscillations are quickly damped by the decay of ϕ into relativistic particles and radiation. With $\tau \approx 10^{13}$ GeV, the decay time Γ^{-1} is arranged to be $\ll H^{-1}$ to allow 'reheating' to take place. The temperature of the universe will rise again to $\approx 2 \times 10^{14}$ GeV.

The drawback of the new inflationary model is that it requires a fine tuning of ϕ_i/σ, where ϕ_i is the initial value from which the slow roll-over starts. We need $\phi_i/\sigma < 10^{-5}$ and the Higgs boson mass $m < 10^{-5}\sigma$. If $m \geq 10^9$ GeV then the model does not work. Since the whole concept of inflation was brought in to avoid the need for fine tuning, this requirement is like breaking the ground rules.

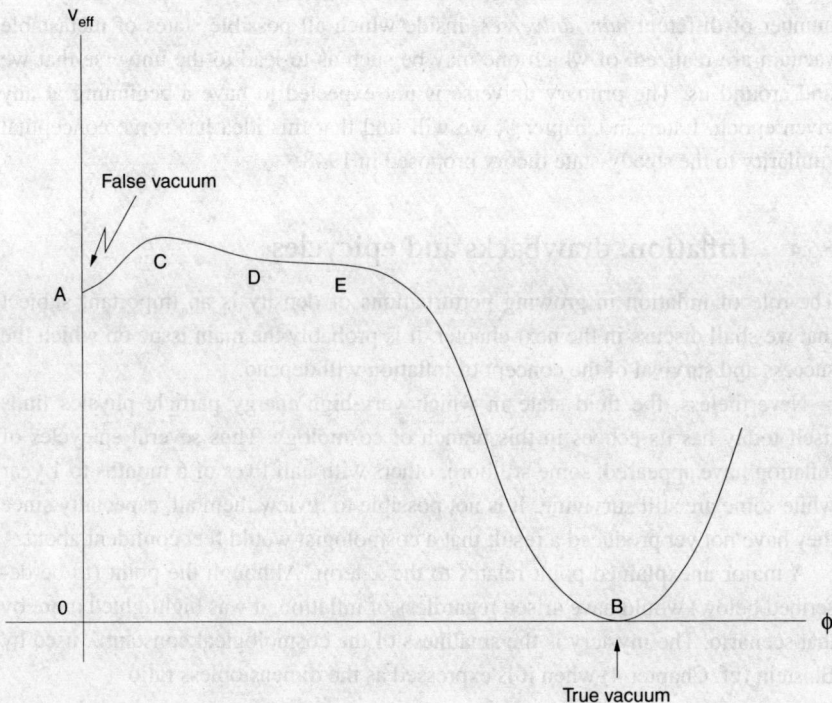

Figure 6.11 $V_{\text{eff}}(\phi, T)$ used in the new inflationary model.

6.5.3   Chaotic inflation

The original Guth model invoked a strongly first-order phase transition whereas the second model may be considered as requiring a weakly first-order or even a second-order phase transition. Can we construct an inflationary model which has *no* phase transition involved?

Yes! Such a model was proposed by Linde under the title 'chaotic inflation'. The $V(\phi)$ function here has a simple form:

$$V(\phi) = \lambda \phi^4. \tag{6.76}$$

Inflation results because of a rather slow motion of ϕ from some initial ϕ_0 towards the minimum. This initial value is believed to be due to chaotic initial conditions. Although one can produce sufficient inflation in this way, it is necessary to ensure that the initial kinetic energy of the ϕ-field is small relative to the potential energy. Detailed calculations show that this requires the field to be uniform over sizes bigger than the Hubble radius! Also the value of λ has to be fine tuned near 4×10^{-14} to get the correct density perturbations.

Going one step further, in 1887 Linde suggested the universe to be eternally existing in a chaotic inflationary stage. The universe consists of an exponentially large number of different *mini-universes*, inside which all possible states of metastable vacuum are realized, of which one may be such as to lead to the universe that we find around us. The primary universe is not expected to have a beginning at any given epoch. Later, in Chapter 9, we will find that this idea has some conceptual similarity to the steady-state theory proposed in 1948.

6.5.4   Inflation: drawbacks and epicycles

The role of inflation in growing perturbations of density is an important subject that we shall discuss in the next chapter. It is probably the main issue on which the success and survival of the concept of inflation will depend.

Nevertheless, the fluid state in which very-high-energy particle physics finds itself today has its echoes in this branch of cosmology. Thus several epicycles of inflation have appeared, some stillborn, others with half lives of 6 months to 1 year while some are still surviving. It is not possible to review them all, especially since they have not yet produced a result that a cosmologist would feel confident about.

A major unexplained point relates to the λ-term. Although the point (to be described below) would have arisen regardless of inflation, it was highlighted more by that scenario. The mystery is the smallness of the cosmological constant λ used by Einstein (cf. Chapter 4) when it is expressed as the dimensionless ratio

$$\alpha_\lambda = \frac{G\hbar\lambda_0}{c^3} \approx 10^{-126}. \tag{6.77}$$

The suffix zero on λ indicates that we are interested in its present-day value, which is of the order of H_0^2/c^2. We saw, however, that, during inflation, the energy momentum tensor of the false vacuum is that corresponding to a cosmological constant $\equiv \lambda_{GUT}$, where

$$\frac{\lambda_0}{\lambda_{GUT}} \approx 10^{-108}. \tag{6.78}$$

Thus, to start with, the cosmological constant may be as high as $c^3/(G\hbar)$, later changing to λ_{GUT} and finally to λ. (In between there is another phase transition at the breakdown of the electro-weak symmetry, wherein $\lambda/\lambda_{EW} \approx 10^{-57}$.) How does λ manage to change from such a large initial magnitude to $\sim 10^{-126}$ of its initial value? What kind of fine tuning is this? This is known as the *graceful-exit problem* for the cosmological constant.

With all its epicycles the inflationary model makes one clear-cut prediction about the present state of the universe, viz., $\Omega_0 = 1$. This is because the closeness of Ω to unity is such as to lead to $\Omega_0 = 1$ or very close to unity (for $k = \pm 1$). This prediction automatically implies that there is non-baryonic matter present, since, from Chapter 5, the baryonic density fraction $\Omega_B \ll 1$.

6.6   Primordial black holes

We now depart from the discussion of interactions among particles and GUTs to study a peculiar consequence of gravity. The study relates to black holes, which were briefly described in Chapter 2.

As the name 'black hole' implies, we do not expect any radiation to come out of such an object. For a spherical object of mass M, the black-hole condition is reached when its surface area equals $4\pi R_s^2$, where R_s, the Schwarzschild radius, is given by

$$R_s = \frac{2GM}{c^2}. \tag{6.79}$$

No particle of material or light signal emitted from $R \leq R_s$ can go into the region $R > R_s$ – at least, this is what classical general relativity tells us.

Nevertheless, in 1974 Stephen Hawking made the remarkable suggestion that a black hole can radiate. Hawking's calculation went beyond classical physics: it considered what happens when any field (for example, the electromagnetic field) is *quantized* in the spacetime exterior to a black hole. As we have already seen, the quantum mechanical description of vacuum is much more involved than the classical description, which simply states that a vacuum is empty. According to quantum field theory, the vacuum is seething with virtual particles and antiparticles whose presence cannot be detected directly. Their interference with physical processes in spacetime can, however, lead to detectable results. Hawking found that one such result in the spacetime outside a black hole is that an observer at infinity sees a flux of particles coming out from the vicinity of a black hole. We will not go through the calculations leading to this result; we will simply study the consequences of such a process in the early universe. Figure 6.12 provides a qualitative description of how the Hawking process operates. Not all aspects of the Hawking process have been worked out yet. An important issue that is still unresolved, for example, is that of back reaction: how the emission of particles by the black hole affects and alters the geometry of spacetime outside and what effect this change has on the process of radiation by the black hole.

The idea we shall use here is that a spherical black hole of mass M ejects particles in a thermal spectrum of temperature T given by

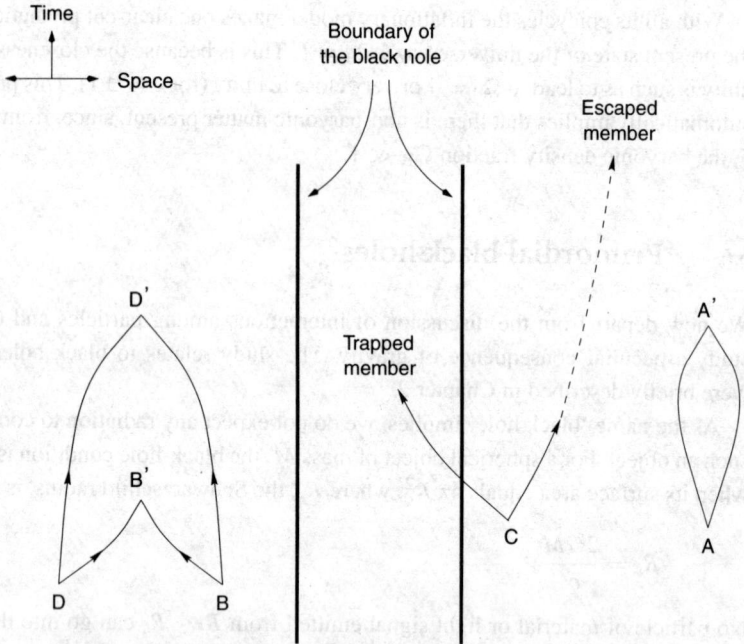

Figure 6.12 The thick lines in this spacetime diagram indicate the boundary of a black hole. The two arrows emerging from points A, B, C, . . . indicate pair creation in vacuum fluctuations. Had there been no black hole, pairs would simply annihilate and disappear as at A′, B′, etc. The black hole, however, may influence this process. For example, it may selectively attract only one member of a pair as at C and let the other member escape. A remote observer seeing this particle would conclude that the black hole has created this particle. Since the virtual particles created may have negative energies, by absorbing a negative-energy member of the pair the black hole loses part of its mass. The escaped particle carrying positive energy therefore describes the emission of energy by the black hole. This is the essence of the *Hawking process*.

$$kT = \frac{\hbar c^3}{8\pi GM} \sim 10^{26} M_{\mathrm{g}}^{-1} \qquad (6.80)$$

where $M_{\mathrm{g}} = M$ expressed in grammes. The emission of particles by the black hole leads to a rate of loss of mass given by

$$\frac{\mathrm{d}M_{\mathrm{g}}}{\mathrm{d}t} \sim -10^{26} M_{\mathrm{g}}^{-2} \text{ s}^{-1}. \qquad (6.81)$$

The $\sim$ implies that a numerical constant of the order of unity appears on the right-hand side to take account of the number of particle species emitted. If we integrate (6.81) we find that the entire mass of the black hole is radiated away in a time τ given by

$$\tau \sim 3 \times 10^{-27} M_{\mathrm{g}}^3 \text{ s}. \qquad (6.82)$$

Thus a black hole created soon after the big bang with a mass $\sim 5 \times 10^{14}$ g would last until the present day.

The process described above is slow to start with, when a black hole is massive and cold. However, as M decreases T rises and the rate of loss of mass increases until finally it reaches a catastrophically high level. This final stage is often called evaporation or explosion of a black hole. As seen above, a black hole of stellar mass ($M_{\mathrm{g}} \geq 10^{33}$) is hardly likely to explode in the lifetime of the universe! Since black holes considered in various astrophysical scenarios are at least as massive as $2M_{\odot}$, for them the Hawking process is only of academic interest.

However, it is claimed that there are scenarios in the very early universe that could lead to primordial black holes (PBHs) of masses much lower than $M_{\odot}$. Bernard Carr in 1975 was the first to explore their consequences at length. Carr investigated the formation and evaporation of PBHs in order to see whether the density of nucleons observed at present as well as the microwave background can be explained in terms of emission of baryons, leptons, photons and so on by low-mass black holes. These concepts are highly speculative and have not been integrated suitably with the other (equally speculative!) scenarios of the very early universe.

The interesting aspect of this approach is that PBHs act as sources of various particles that have somehow got to be created in the universe. The suggestion that PBHs evaporating today might account for the observed γ-ray bursts, however, does not seem to be correct, since the spectrum of γ-rays emitted in such a process is not like the spectrum observed for burst events.

There are several loose ends still to be sorted out in the PBH scenario. At the deepest level one has to understand how they can form in the first place, since the usual process of gravitational collapse that is supposed to lead to stellar or more massive black holes cannot apply here. Next one needs to express the concepts of thermodynamics and statistical mechanics in highly curved spacetime in order to give precise meaning to the notions of temperature and the blackbody spectrum:

the formulae (6.80) and (6.81) merely use a naive extrapolation of flat spacetime thermodynamics. (See the concluding section of this chapter in which this issue is elucidated further.) Furthermore, the problem of back reaction still remains unresolved. Finally, on the observational front, this bizarre concept still awaits a befitting application in the real universe.

6.7 Quantum cosmology

In this chapter we finally push our investigations into the past history of the universe back to the era $t \sim 10^{-43}$ s. Is it justified to put our faith in the standard big-bang model when the universe was so young? One way to answer this question is to look for the limit where classical theory breaks down and quantum mechanics takes over. Beyond this limit we cannot trust the classical theory of gravity – that is, the general theory of relativity.

A look at the action principle (2.103) shows that the limit sought above can be obtained by equating the gravitational action

$$A_g = \frac{c^3}{16\pi G} \int_V R\sqrt{-g}\, d^4x \tag{6.83}$$

to Planck's constant. For $A_g \gg \hbar$ we can trust our classical description of spacetime geometry, whereas for $A_g \ll \hbar$ a quantum description of cosmology is indispensable. However, to evaluate A_g we need V, the 4-volume of the spacetime manifold.

In the big-bang model we take V as the spatial volume enclosed by the particle horizon and bounded by the time span of the universe. Thus, during any epoch t, for $k = 0$, $S \propto t^{1/2}$, the particle horizon is defined by

$$rS = 2ct.$$

For $S \propto t^{1/2}$, $R = 0$ and so $A_g = 0$. However, this happens because the trace of T_k^i is zero in the early universe. As an order-of-magnitude estimate we may take R_0^0 instead of R in the computation of A_g; R_0^0 gives us an idea of how the geometrical part of the action changes with time. For $S \propto t^{1/2}$, $R_0^0 = 3/(4c^2t^2)$. Thus, up to the epoch t,

$$A_g \sim \frac{c^4}{16\pi G} \int_0^t \frac{3}{4c^2t_1^2} \frac{4\pi}{3} (2ct_1)^3 \, dt_1 = \frac{c^3}{4G} t^2.$$

By equating A_g to $\hbar$ we get

$$t = 2t_P = 2\sqrt{\frac{G\hbar}{c^5}} \cong 10^{-43} \text{ s.} \tag{6.84}$$

This time span is called the *Planck time. No classical discussion of standard big-bang cosmology can be pushed to epochs with $t < t_P$.* We already encountered this epoch in (6.44), where the temperature was $E \sim 10^{19}$ GeV. This energy, as can be seen from (6.44), is simply $\sim \hbar / t_P$.

Thus the present discussions of GUTs and cosmology already take us right up to the Planck epoch. Whether the universe did indeed have a spacetime singularity at $t = 0$ should be determined not by classical general relativity but by an appropriate theory of quantum gravity. At present the goal of having a working theory of quantum gravity seems far away. The various approaches that have been tried in order to quantize gravity do not agree on the answer to the question of whether the universe had a singular epoch. A simple approach suggests that, if we include quantum fluctuations of homogeneous and isotropic universes, then the spacetime singularity would 'most probably' be averted. The probability here is in the sense of quantum mechanics. The result can in fact be stated in a more general form proved by this author, namely that, if one considers most general quantum conformal fluctuations of a classical singular cosmological solution, then, most probably, singularity is not present in these fluctuations.

6.8   Conclusion

Before we conclude our discussion of the early universe, a discussion in which we have pushed back our incomplete knowledge as far as we possibly could, it is necessary to alert the reader to the liberties we are taking in extrapolating conventional statistical mechanics to such extreme conditions. T. Padmanabhan and M. M. Vasanthi first pointed out this aspect of the very early universe in 1982. We shall refer to it as the *small-number problem*.

6.8.1   The small-number problem

In the discussions of the early and very early universe, it is customary to ascribe a temperature corresponding to the average energy of particles during any given epoch in the universe. That is how we expressed temperature in mega- or giga-electron-volts in equation (6.6). Now this has a meaning only if we are certain that the particles are in thermal equilibrium. To settle this issue, well-known results from the statistical mechanics of flat spacetime are used. The justification for the usage of flat-spacetime formulae in a highly curved spacetime is via the strong principle of equivalence. We had discussed this principle in Chapter 2. To be able to apply it to a given volume $\mathcal{V}$, say, of spacetime, its characteristic linear dimension must be small relative to the radius of curvature of spacetime. (An analogy is with the *flat-Earth approximation* which is valid over sizes of regions on the Earth that are small in comparison with its radius.) The smaller the region the more closely the flat-spacetime approximations apply. However, we cannot make $\mathcal{V}$ arbitrarily small, since we have to have a large number $\mathcal{N}$ of particles in it so as to be able to apply the formulae of statistical mechanics. These are two conflicting requirements, so let us see whether they are satisfied in the early epochs of the big-bang universe.

During an epoch t, the typical component of the curvature tensor is $R \sim (1/c^2 t^2)$ and the corresponding characteristic linear dimension measuring curvature of space-time is ct. Let L be the characteristic linear size of our region $\mathcal{V}$. Then we should have

$$L = \alpha ct, \qquad \alpha \ll 1. \tag{6.85}$$

Now let us calculate $\mathcal{N}$ in a region of size L. Using the formulae derived in Chapter 5, e.g. equation (5.24), *based on the flat-spacetime statistical mechanics at high temperatures*, we get the number of particles of various species with an effective number of spin states g as

$$\mathcal{N} = 2.4 \times \frac{g}{\pi^2}\left(\frac{kT}{c\hbar}\right)^3 \times \frac{4\pi}{3}L^3$$

$$\sim g\left(\frac{kT}{c\hbar}\right)^3 L^3. \tag{6.86}$$

The temperature of the early universe T is related to t by an equation like (6.6), which can be written in terms of the Planck time t_P defined in equation (6.84) and the Planck temperature defined by $T_P = \hbar/(kt_P)$, as follows:

$$\mathcal{N} \sim g\alpha^3\left(\frac{45}{16\pi^3 g}\right)^3$$

$$\sim \frac{\alpha^3}{30\sqrt{g}}\left(\frac{T_P}{T}\right)^3. \tag{6.87}$$

Notice that, unless T is sufficiently small, we will not get $\mathcal{N} \gg 1$, which is necessary for application of statistical mechanics. Let us estimate how low T has to be.

In the standard Solar-System tests we detect effects of curvature with α as small as $\sim 10^{-6}$. If we use the same standard, we get for the GUT epoch, with $T = 10^{15}$ GeV and $g = 100$, say, the value

$$\mathcal{N} = 3 \times 10^{-9}, \tag{6.88}$$

which is hardly a large number! Even if we relax our flat-spacetime criterion to $\alpha = 10^{-3}$, we get $\mathcal{N} = 3$. Surely, no laws of statistical mechanics can be applied even with this relaxed value! If, for this value of α, we require $\mathcal{N}$ to be appreciably high, we need to lower the temperature T to well below 10^{15} GeV. Thus, unless we apply a rigorous curved-spacetime statistical mechanics (which has not been developed to be used in cosmology so far), we are not justified in carrying out the kind of statistical studies of high-energy interactions among particles described in this chapter. In contrast, we can satisfy ourselves that the calculations of the previous chapter were at sufficiently low temperatures to allow the use of flat-spacetime statistical mechanics.

It may happen that a rigorous curved-spacetime statistical mechanics will eventually ratify the steps described in this chapter. Until such a derivation becomes available we have to take the concept of temperature and thermal equilibrium in a universe less than 10^{-36} s old as additional assumptions that have still to be justified.

Figure 6.13 provides a schematic view of the events in the early universe that is built out of such speculations. Our next investigation will relate to the formation of discrete structures in the universe, namely to the problem of evolving a successful

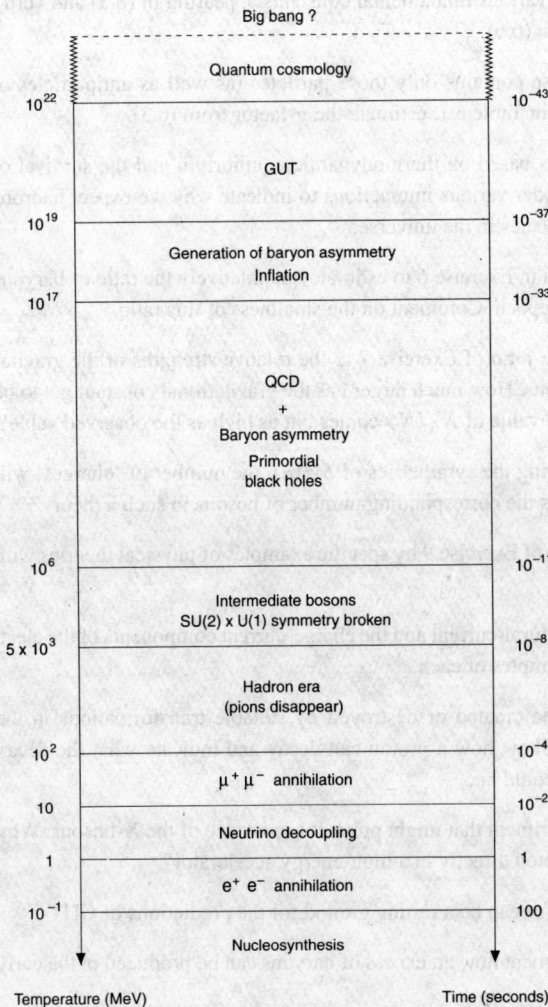

Figure 6.13 The time axis in the above diagram is assumed to start from the big-bang epoch $t = 0$. The interactions and events that govern the state of the universe are shown in the relevant temperature sections. We have to remember that rising temperatures correspond to decreasing time according to the formula (6.6). However, during the hadron era the equation of state of matter is complicated so that (6.6) does not strictly apply.

theory that starts with 'seeds' of local fluctuations implanted in the very early universe and grows them into galaxies, clusters, superclusters etc.

Exercises

1 Explain what is meant by the remark 'The early universe is the poor man's high-energy accelerator'.

2 Look up the mass and surface temperature of the Sun in an astronomy textbook and express both in mega-electron-volts.

3 Explain why the giga-electron-volt is a good unit to describe the masses and temperatures in the early universe.

4 Look up the values of the various fundamental constants appearing in (6.5) and verify the numerical coefficient in (6.6).

5 Assuming that the universe contains only those particles (as well as antiparticles of quarks and leptons) listed in Table 6.1, estimate the g-factor from (6.3).

6 Give qualitative arguments based on thermodynamic equilibrium and the survival of various particle species under various interactions to indicate why we expect hadrons to be the least abundant species in the universe.

7 Apply the arguments given in Exercise 6 to estimate quantitatively the ratio of baryons to photons for the present epoch. Comment on the smallness of this ratio.

8 Relate the smallness of the ratio of Exercise 7 to the relative strengths of the gravitational and strong interactions. How much larger has the gravitational constant got to be in order that the calculated value of N_A/N_γ comes out as high as the observed value?

9 Show that, in a theory having the symmetries of $SU(n)$, the number of 'charges' will be equal to $n^2 - 1$. What is the corresponding number of bosons in such a theory?

10 Illustrate the general result of Exercise 9 by specific examples of physical theories with $n = 2, 3$ and 5.

11 Distinguish between the natural-current and the charge-current components of the electroweak interaction. Give examples of each.

12 Show how baryons may be created or destroyed by suitable transformations in the $SU(5)$ version of GUTs. Show how a proton can decay and indicate what the decay products in such an event could be.

13 Describe a laboratory experiment that might prove the existence of the X-bosons. Why can't these bosons be detected directly in a high-energy accelerator?

14 Explain how the hot universe can be a testing ground for the predictions of GUTs.

15 Show by a qualitative argument how an excess of baryons can be produced in the early universe.

16 For the X-bosons estimate the time scale

$$\tau \sim \frac{\hbar}{m_X c^2}.$$

17 Explain why the horizon and flatness problems become more severe as we seek to set the initial conditions for earlier and earlier epochs.

18 Think of *astrophysical* reasons why we should not have free magnetic monopoles in significant quantities today.

19 In the Guth model of inflation let $\lambda(t_0)$ denote the rate at which bubbles form in a given proper volume and suppose that $p(t)$ denotes the probability that there are no bubbles engulfing a given point in space. Show that

$$p(t) = \exp\left(-\int_0^t \lambda(t_1) S^3(t_1) V(t, t_1)\, dt_1\right),$$

where

$$V(t, t_1) = \frac{4\pi}{3}\left(\int_{t_1}^t \frac{c\, dt_2}{S(t_2)}\right)^3.$$

20 If the nucleation rate may be approximated by a constant λ_0 in Exercise 19, show that

$$p(t) = \text{constant} \times \exp\left(-\frac{t}{\tau}\right),$$

where $\tau = 3a^3/(4\pi\lambda_0)$.

21 Show that, in the Coleman–Weinberg case, the evolution of ϕ during the slow roll-over is given approximately by

$$\ddot{\phi} + 3a\dot{\phi} + V'(\varphi) = 0,$$

where $\ddot{\phi}$ may be ignored and $a = 25\pi\alpha^2 G\sigma^4/(24c^2)$.

22 Show that, even if we are able to explain why the variation in Ω is $\Delta\Omega = O(1)$ for this epoch, the future epoch will again have larger and larger $\Delta\Omega$.

23 Calculate the temperature in kelvins of a spherical black hole of mass equal to the mass of (a) a proton, (b) 1 ton, (c) the Earth, (d) the Sun and (e) a star of mass $10M_\odot$.

24 Substitute the values of $\hbar$, c, G and k to verify the results (6.80) and (6.81).

25 Express in terms of the fundamental constants the time τ for which a black hole of mass M survives under its own radiation.

26 Taking $k = 0$ and $S \propto t^{1/2}$ in the early universe, calculate the size of the particle horizon. By equating this to the Schwarzschild radius of a black hole, calculate the mass M of the black hole. Show that this mass is given by

$$M = c^3 t/G.$$

27 Discuss how the primordial black holes might act as sources of particles and radiation that are now found in the universe.

28 Compute $\mathcal{A}_g(t)$ for the closed Friedmann model with given values of q_0 and h_0, taking the time interval as $(0, t)$ and the spatial extent covering the whole (spherical) space. Estimate the epoch for which $\mathcal{A}_g = \hbar$. Why do you get an answer different from t_P?

29 Relate the Planck length associated with gravity to the Compton wavelength of the proton and to the strength factor α_G defined in (6.35). Show also that, during the Planck epoch, the Schwarzschild radius of a primordial black hole filling the particle horizon is of the same order as the Compton wavelength of the black hole.

30 Comment on the limit to which classical general relativity may be pushed in discussions of the early universe.

Chapter 7

The formation of large-scale structures in the universe

7.1   A key problem in cosmology

Chapter 5 narrated the success story of the hot-big-bang model, of how the particles combined to form light nuclei as the universe cooled from 10^{10} to 10^8 K and how the relic of that hot era is today seen in the form of radiation background in microwaves. Encouraged by these achievements, the big-bang cosmologists pushed their investigations further back in time, to epochs of very-high-energy particles. These investigations, outlined in Chapter 6, brought cosmologists into contact with the very-high-energy particle theorists, leading to a variety of new but speculative inputs to classical cosmology such as inflation, non-conservation of baryons etc.

Exciting though these investigations are, we must not lose sight of the fact that cosmology is a branch of physics and as such it requires hard facts to support these speculations. One important fact is the existence of discrete structures in the universe, ranging from galaxies to superclusters. How did these structures come about? Why are they distributed in an inhomogeneous fashion when the distribution of their radiation counterpart is so smooth? This key problem of cosmology must surely have a solution buried in the early history of the universe.

In this chapter we review some attempts to come to grips with this problem. If the big-bang scenario is correct, the solution should incorporate some or all of the following epochs:

1. the Planck epoch;

2. the GUTs/inflation epoch;

3. the recombination epoch when radiation decoupled from matter;

4. the epoch when the universe switched over from being radiation-dominated to being matter-dominated;

5. the epoch of redshift ~ 5 when galaxies and QSOs may have begun to form (this epoch is fixed by the observation that the numbers of these discrete objects seem to taper off as we approach redshifts of this order);

6. the present epoch, which presents us with a hierarchy of structures in which voids also play a part, as does peculiar motion superposed on the expansion of the universe; and also

7. the role of dark matter, baryonic as well as non-baryonic.

Investigations which take into consideration all these epochs are naturally very extensive and long reviews as well as books are written on the topic of structure formation. In this introductory account we naturally cannot do justice to the topic in detail, but present here a broad brush picture which outlines the major efforts in this field.

The strategy is to consider small fluctuations of density in the very early epochs (1) or (2) and work out their growth through the successive later epochs. Since the physics of the universe changes drastically, the techniques of working out the solution also change. We will consider first the epochs from (3) onwards, which, historically, were the first to be tackled and which involve the less speculative parts of cosmology.

7.2   The Jeans mass in the expanding universe

As early as 1902, Sir James Jeans (Figure 7.1) considered the formation of galaxies in the universe as a process involving the interplay of gravitational attraction and the pressure force acting on a mass of non-relativistic fluid. Jeans' treatment used Newtonian physics and assumed a static universe. However, his ideas can be adapted to suit our problem, at any rate part of it.

7.2.1   The basic equations

Consider the universe as filled with fluid of density ρ, pressure p and velocity field v and the gravitational force field $\mathbf{F}$. We will assume Newtonian physics to hold for gravity as well as for fluid dynamics. Thus the continuity equation

$$\frac{\partial \rho}{\partial t} + \nabla \cdot (\rho \mathbf{v}) = 0 \tag{7.1}$$

and the Euler equation

$$\frac{\partial \mathbf{v}}{\partial t} + (\mathbf{v} \cdot \nabla)\mathbf{v} = -\frac{1}{\rho}\nabla p + \mathbf{F} \tag{7.2}$$

hold for fluid motion, while the Poisson equation holds for $\mathbf{F}$:

$$\nabla \times \mathbf{F} = 0, \qquad \nabla \cdot \mathbf{F} = -4\pi G\rho. \tag{7.3}$$

In the unperturbed situation of homogeneity and isotropy we get the following simple solution of the above equations, as discussed in the context of Newtonian cosmology in Chapters 3 and 4.

$$p = 0, \qquad \rho \propto \frac{1}{S^3(t)}, \qquad \mathbf{v} = \mathbf{r}\frac{\dot{S}(t)}{S(t)}, \tag{7.4}$$

$$\mathbf{F} = -\frac{4\pi G\rho}{3}\mathbf{r}. \tag{7.5}$$

The scale factor $S(t)$ satisfies the differential equation

$$\frac{\ddot{S}}{S} = -\frac{4\pi G\rho}{3}. \tag{7.6}$$

We now consider perturbations of this simple solution. Our aim in doing so is to see whether any initial clumpiness can grow in size by gravitational instability. Thus we consider small changes in the physical quantities ρ, $\mathbf{v}$, $\mathbf{F}$ and p in the above solution, denoting them by ρ_1, $\mathbf{v}_1$, $\mathbf{F}_1$ and p_1, respectively. To begin with, these

Figure 7.1 James Jeans (1877–1946).

perturbations are supposed to be small so that the above equations can be linearized. The linearized equations become

$$\dot{\rho}_1 + \frac{\dot{S}(t)}{S(t)}(\mathbf{r} \cdot \nabla)\rho_1 + 3\frac{\dot{S}(t)}{S(t)}\rho_1 + \rho \nabla \cdot \mathbf{v}_1 = 0, \tag{7.7}$$

$$\dot{\mathbf{v}}_1 + \frac{\dot{S}(t)}{S(t)}[(\mathbf{r} \cdot \nabla)\mathbf{v}_1 + \mathbf{v}_1] + \frac{1}{\rho}\nabla p_1 - \mathbf{F}_1 = 0, \tag{7.8}$$

$$\nabla \times \mathbf{F}_1 = 0, \qquad \nabla \cdot \mathbf{F}_1 = -4\pi G \rho_1. \tag{7.9}$$

We also have for *adiabatic* fluctuations

$$p_1 = c_s^2 \rho_1 \tag{7.10}$$

where c_s, is the speed of sound $(c_s = \sqrt{dp_1/d\rho_1})$.

It is not difficult to see that plane-wave solutions of the following form exist for (7.7)–(7.9):

$$\rho_1(\mathbf{r}, t) = \bar{\rho}_1(t)e^{i\chi}, \qquad \mathbf{v}_1(\mathbf{r}, t) = \bar{\mathbf{v}}_1(t)e^{i\chi}, \tag{7.11}$$

with

$$\chi = \frac{\mathbf{r} \cdot \mathbf{k}}{S(t)} \tag{7.12}$$

and

$$\dot{\bar{\rho}}_1 + \frac{3\dot{S}}{S}\bar{\rho}_1 + \frac{i\mathbf{k} \cdot \bar{\mathbf{v}}_1}{S}\rho = 0, \tag{7.13}$$

$$\dot{\bar{\mathbf{v}}}_1 + \frac{\dot{S}}{S}\bar{\mathbf{v}}_1 + \frac{ic_s^2}{S\rho}\mathbf{k}\bar{\rho}_1 - \bar{\mathbf{F}}_1 = 0, \tag{7.14}$$

$$\mathbf{k} \times \bar{\mathbf{F}}_1 = 0, \qquad i\mathbf{k} \cdot \bar{\mathbf{F}}_1 = -4\pi G \bar{\rho}_1 S. \tag{7.15}$$

It is now convenient to split $\bar{\mathbf{v}}_1$ into two parts: along and perpendicular to the wave vector $\mathbf{k}$. Thus we write

$$\bar{\mathbf{v}}_1 = \frac{\mathbf{k} \cdot \bar{\mathbf{v}}_1}{k^2}\mathbf{k} + \frac{\mathbf{k} \times (\bar{\mathbf{v}}_1 \times \mathbf{k})}{k^2} \tag{7.16}$$
$$= \mathbf{v}_\parallel + \mathbf{v}_\perp.$$

Taking the vector product of (7.14) with $\mathbf{k}$ we get

$$\left(\dot{\bar{\mathbf{v}}}_1 + \frac{\dot{S}}{S}\bar{\mathbf{v}}_1\right) \times \mathbf{k} = \mathbf{0},$$

from which our definition of $\mathbf{v}_\perp$ leads us to

$$\mathbf{v}_\perp S = \text{constant}. \tag{7.17}$$

Thus the transverse (or rotational) mode tends to decrease in the expanding universe. What about the mode parallel to $\mathbf{k}$? Taking the scalar product of (7.14) with $\mathbf{k}$ gives the following relation:

$$\dot{v}_{\parallel} + \frac{\dot{S}}{S}v_{\parallel} + \frac{i}{k}\left(\frac{c_s^2 k^2}{S} - 4\pi GS\right)\frac{\bar{\rho}_1}{\rho} = 0. \tag{7.18}$$

We now define the density-contrast parameter

$$\delta = \bar{\rho}_1/\rho. \tag{7.19}$$

Then, since $\rho \propto S^{-3}$, we get from (7.13)

$$\dot{\delta} = -\frac{ik}{S}v_{\parallel}. \tag{7.20}$$

On eliminating $v_{\parallel}$ between (7.18) and (7.20), we get

$$\ddot{\delta} + \frac{2\dot{S}}{S}\dot{\delta} + \left(\frac{c_s^2 k^2}{S^2} - 4\pi G\rho\right)\delta = 0. \tag{7.21}$$

This is the equation that tells us how or whether gravitational instability leads to the growth of condensations in the expanding universe.

We first consider (7.21) in the quasi-static approximation wherein the expansion of the universe is neglected. Thus we set $S = \text{constant}$ and $\dot{S}/S = 0$. This brings us back to the original Jeans calculation of the static universe. We define $K = k/S$ as the effective wave number for the solution (7.11) and (7.12) and call

$$K_J = \left(\frac{4\pi G\rho}{c_s^2}\right)^{1/2} \tag{7.22}$$

the *Jeans wave number*. The equation (7.21) now looks like

$$\ddot{\delta} + c_s^2(K^2 - K_J^2)\delta = 0. \tag{7.23}$$

In this approximation it is easy to see that (7.23) has sinusoidal (that is, oscillating) solutions for $K > K_J$ and exponential (growing as well as damped) solutions for $K < K_J$. If we write

$$\delta \propto e^{i\omega t} \tag{7.24}$$

then

$$\omega^2 = c_s^2(K^2 - K_J^2). \tag{7.25}$$

Notice first that, for $K < K_J$, the growth rate $|\omega|$ is maximum when $K = 0$ and is given by

$$|\omega|_{max} = c_S K_J. \tag{7.26}$$

However, the rate of expansion of the universe, which we have neglected so far, is also of this order. For, from Einstein's equations, we get (for the $k = 0$ cosmology)

$$\frac{\dot{S}^2}{S^2} = \frac{8\pi G\rho}{3};$$

that is,

$$\frac{\dot{S}}{S} = \left(\frac{2}{3}\right)^{1/2} K_J c_S. \tag{7.27}$$

Thus we cannot legitimately neglect the expansion of the universe in the present problem.

Nevertheless, we can salvage something useful from this analysis. If we set $K \gg K_J$ in (7.25) we get sinusoidal disturbances that do not grow but simply propagate like sound waves. What does this mean? To understand the meaning of $K \gg K_J$, define a mass

$$M = \frac{4\pi n m_H}{3} \left(\frac{2\pi}{K}\right)^3. \tag{7.28}$$

M is the mass of a sphere of radius $2\pi/K$ containing a number density n of hydrogen atoms, each of mass m_H. As the universe expands n decreases as S^{-3} and K decreases as S^{-1}. Thus M remains invariant. Taking $\rho \cong n m_H$ for the present, we see that the gravitational energy of this sphere is

$$\mathcal{E}_G \cong \frac{GM^2}{2\pi/K} \approx \frac{16\pi^2 G\rho^2}{9} \left(\frac{2\pi}{K}\right)^5.$$

The thermal energy of this sphere, on the other hand, is

$$\mathcal{E}_{th} = \frac{4\pi}{3} \rho c_s^2 \left(\frac{2\pi}{K}\right)^3.$$

Comparing the two expressions above, we see that

$$K \gg K_J \Rightarrow \mathcal{E}_{th} \gg \mathcal{E}_G. \tag{7.29}$$

Furthermore, $K \gg K_J$ also gives us

$$|\omega| \gg \dot{S}/S. \tag{7.30}$$

Thus, in the sound-wave approximation the gravitational forces and the expansion of the universe may be neglected.

It is convenient to express the condition $K \gg K_J$ in the form

$$M \ll \frac{4\pi n m_H}{3} \left(\frac{2\pi}{K_J}\right)^3 \equiv M_J, \tag{7.31}$$

where M_J is called the *Jeans mass*. The above result therefore means that the only disturbances that have any prospects of growth are those whose mass *exceeds* the Jeans mass M_J. For such spheres the gravitational force can dominate the thermal-pressure force and lead to compression of the sphere and hence to gravitational instability.

7.2.2 The evolution of the Jeans mass

Let us try to follow the variation of M_J as the universe goes through various phases starting with the era when e^+ and e^- annihilated (see Chapter 5). Until the electrons combined with protons to form hydrogen atoms, it is a good approximation to assume that the universe is largely made of non-relativistic ionized hydrogen in thermal equilibrium with the blackbody radiation at temperature T. In this era we may neglect the pressure and entropy of matter in comparison with that of radiation. Hence we have the density, pressure and entropy density:

$$\rho = n m_H + a T^4 / c^2, \tag{7.32}$$

$$p = \tfrac{1}{3} a T^4, \tag{7.33}$$

$$s = \tfrac{4}{3} a T^3. \tag{7.34}$$

In adiabatic changes the entropy of a comoving volume is constant, so $s/n =$ constant. Hence to evaluate c_s^2 we must calculate $dp/d\rho$ at constant s/n. A simple calculation gives

$$c_s^2 = \frac{1}{3}\left(\frac{Ts}{c^2 n m_H + Ts}\right) c^2. \tag{7.35}$$

In evaluating the Jeans mass from (7.31) we will replace ρ in (7.22) by $\rho + p/c^2$, without seriously altering any conclusions (which are order-of-magnitude results anyway!). A simple calculation gives

$$M_J = \frac{2\pi^{5/2} s^2}{9 a^{1/2} n^2 m_H^2 G^{3/2}} \left(1 + \frac{Ts}{n m_H c^2}\right)^{-3}. \tag{7.36}$$

It is more convenient to use the specific entropy

$$\sigma = \frac{s}{kn}, \qquad k = \text{Boltzmann's constant.} \tag{7.37}$$

Then

$$M_{\rm J} = \frac{2\pi^{5/2}\sigma^2 k^2}{9a^{1/2}G^{3/2}m_{\rm H}^2}\left(1 + \frac{\sigma kT}{m_{\rm H}c^2}\right)^{-3}.$$ (7.38)

As the temperature drops to ~ 3000 K the combination of electrons with ionized hydrogen is almost complete. This is the misnamed *recombination epoch* or the epoch of last scattering. If the present-day background temperature is taken as 2.7 K, this era broadly corresponds to redshifts in the range $z \sim 1000$–1500. In making any numerical estimates we will take $z_{\rm dec} = 1000$ as the redshift at last scattering when matter became decoupled from radiation.

After the decoupling, radiation pressure becomes unimportant and the gas (of H atoms) behaves as a monatomic gas is expected to behave: with $\gamma = \frac{5}{3}$ and

$$\rho = nm_{\rm H} + \frac{3}{2}\frac{nkT}{c^2},$$ (7.39)

$$p = nkT,$$ (7.40)

$$c_{\rm s}^2 = \frac{5}{3}\frac{kT}{m_{\rm H}}.$$ (7.41)

The Jeans mass then becomes

$$M_{\rm J} = 4\left(\frac{\pi}{3}\right)^{3/2}\left(\frac{5kT}{G}\right)^{3/2}n^{-1/2}m_{\rm H}^{-2}.$$ (7.42)

Just after decoupling, the temperature T of matter is the same as the radiation temperature. So we can express our answer above in terms of σ by using (7.34):

$$M_{\rm J} = \frac{2\pi^{5/2}5^{3/2}k^2\sigma^{1/2}}{9a^{1/2}G^{3/2}m_{\rm H}^2}.$$ (7.43)

How does the temperature of matter drop subsequently? In Chapter 4 we saw that random motions drop as S^{-1} so that, in this non-relativistic era, the temperature will fall as S^{-2}. Thus, starting from (7.43) at the decoupling, $M_{\rm J}$ will drop according to (7.42), that is,

$$M_{\rm J} \propto T^{3/2}n^{-1/2} \propto S^{-3/2}.$$ (7.44)

Figure 7.2 shows how $M_{\rm J}$ varies with the radiation temperature T_γ under the assumption that the present-day radiation temperature is ~ 2.7 K and is equal to that of the cosmic microwave background. The quantity that enters the expression (7.38) besides the temperature is σ, the specific entropy. In Chapter 5 we saw that σ, which is proportional to the photon-to-baryon ratio, is in the range 10^8–10^{10}. In Figure 7.2 we have taken $\sigma \cong 10^9$. It is convenient to express $M_{\rm J}$ in units of the solar mass $M_\odot$.

We see that at $T_\gamma \sim 10^9$ K, $M_{\rm J}$ was in the range 10^4–$10^5 M_\odot$. For $T \gg m_{\rm H}c^2/(\sigma k)$, $M_{\rm J}$ increases as T_γ^{-3}. The increase continues until T_γ drops to a temperature of 10^4–10^5 K ($m_{\rm H}c^2/(\sigma k) = 10^4$ for our chosen σ). The highest value reached by M_J is in the range $10^{17}-10^{19}M_\odot$. This is just before the last-scattering era, when M_J drops to a value of a few times $10^6 M_\odot$. This drop is a sharp one, but thereafter M_J drops as $S^{-3/2} \propto T_\gamma^{3/2}$. This behaviour of M_J helps us understand the difficulties of forming galaxies in the expanding universe.

Suppose that we are interested in forming a galaxy of typical mass $\sim 10^{11}M_\odot$. In terms of (7.38), M_J will be less than this value until the temperature has dropped to $\sim 10^7$ K (see also Figure 7.2). From our crude theory of the Jeans mass, we see that a fluctuation of mass $\sim 10^{11}M_\odot$ will have a chance to grow under its self-gravitation until the temperature drops to this value. The actual mode of growth must be calculated using the perturbation theory with the full general relativistic equations. This complicated problem was solved by E. Lifshitz in 1946. We will not go through the details here but simply quote the result: *in the fastest-growing normal mode $\delta\rho/\rho$ increases as t.*

In the next phase, when $M_J > 10^{11}M_\odot$, our fluctuation cannot grow. It oscillates as a sound wave until the post-recombination era, when M_J has again dropped below

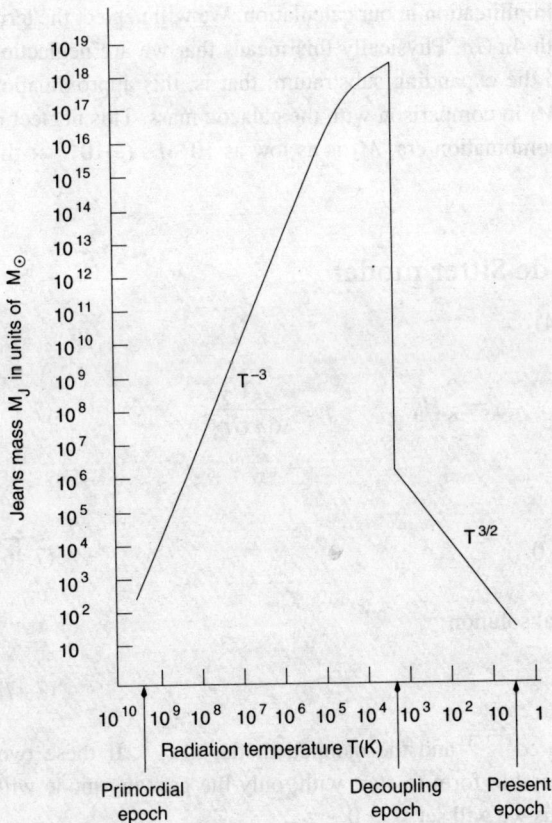

Figure 7.2 An approximate graph of the Jeans mass M_J as a function of the radiation temperature in the universe for the entropy density $\sigma \cong 10^9$.

$10^{11} M_\odot$ (see Figure 7.2). After the temperature has dropped to ~ 3000 K, growth is possible and we can use our simple Newtonian equations. We will proceed to solve this problem in the following section. It is clear, however, that, so far, within the Jeans mass theory, the number $10^{11} M_\odot$ does not seem to emerge as having any particular significance. The typical mass at recombination is of the order of the mass of a globular cluster: it is much smaller than $10^{11} M_\odot$. This was pointed out by R. H. Dicke and P. J. E. Peebles in 1968 and has been the main difficulty in trying to understand why typical discrete units of $10^{11} M_\odot$ are found in the universe.

7.3 Growth in the post-recombination era

We now try to solve (7.21) in the framework of Friedmann models. Our purpose in doing so is to try to relate any present fluctuations in temperature or number density to those in the post-recombination era, with the hope that such a calculation may give us clues as to how galaxies may have formed in that era. We will consider the problem separately for the three types $k = 0$, $k = 1$ and $k = -1$ of the Friedmann model. We will make one simplification in our calculation. We will neglect the term $c^2 k^2 / S^2$ in comparison with $4\pi G\rho$. Physically this means that we are neglecting random motions relative to the expanding substratum; that is, this approximation corresponds to neglecting M_J in comparison with the galactic mass. This neglect is valid, since, in the post-recombination era, M_J is as low as $10^6 M_\odot$ ($\cong 10^{-5} \times$ the galactic mass).

7.3.1 The Einstein–de Sitter model

In this model (see Chapter 4)

$$S(t) = \left(\frac{t}{t_0}\right)^{2/3}, \qquad t_0 = \frac{2}{3} H_0^{-1}, \qquad \rho = \frac{1}{6\pi G t^2}. \qquad (7.45)$$

Therefore (7.21) becomes

$$\ddot{\delta} + \frac{4}{3t}\dot{\delta} - \frac{2}{3t^2}\delta = 0. \qquad (7.46)$$

This equation has the general solution

$$\delta = At^{2/3} + Bt^{-1}. \qquad (7.47)$$

Thus the growing mode is $\propto t^{2/3}$ and the damped mode $\propto t^{-1}$. If these two modes are present in comparable form to start with, only the growing mode will be important eventually. Thus we will set $B = 0$.

For the epoch of decoupling t_{dec}, the redshift is z_{dec}. Taking the temperature of the epoch as ~ 3000 K, we have

$$1 + z_{\text{dec}} \cong 10^3, \tag{7.48}$$

since the radiation temperature increased in proportion to $1 + z$ in the past. Thus the density contrast δ should have grown by the factor

$$\Sigma = \frac{\delta(t_0)}{\delta(t_{\text{dec}})} = \left(\frac{t_0}{t_{\text{dec}}}\right)^{2/3} = (1 + z_{\text{dec}}) \sim 10^3. \tag{7.49}$$

7.3.2 The closed model ($k = 1$)

We use the relations (4.56)–(4.60) to write

$$ct = \tfrac{1}{2}\alpha(\Theta - \sin\Theta), \qquad S = \tfrac{1}{2}\alpha(1 - \cos\Theta),$$

$$\rho = \frac{3H_0^2}{4\pi G}\frac{q_0(1 - \cos\Theta_0)^3}{(1 - \cos\Theta)^3} = \frac{3H_0^2(2q_0 - 1)^3}{4\pi G q_0^2(1 - \cos\Theta)^3},$$

$$\alpha = \frac{2q_0}{(2q_0 - 1)^{3/2}}\left(\frac{c}{H_0}\right). \tag{7.50}$$

On changing the independent variable from t to Θ in (7.21) we get

$$(1 - \cos\Theta)\frac{d^2\delta}{d\Theta^2} + \sin\Theta\frac{d\delta}{d\Theta} - 3\delta = 0. \tag{7.51}$$

This equation has the general solution

$$\delta = A\left(\frac{5 + \cos\Theta}{1 - \cos\Theta} - \frac{3\Theta\sin\Theta}{(1 - \cos\Theta)^2}\right) + B\frac{\sin\Theta}{(1 - \cos\Theta)^2}. \tag{7.52}$$

Again, the growing mode is that multiplying the constant A. Concentrating on this mode, we first note that, for the epoch of decoupling z_{dec} given by (7.48), Θ_{dec} is small. Hence

$$1 + z_{\text{dec}} = \frac{1 - \cos\Theta_0}{1 - \cos\Theta_{\text{dec}}} \approx \frac{2}{\Theta_{\text{dec}}^2}(1 - \cos\Theta_0);$$

that is,

$$\Theta_{\text{dec}} = \left(\frac{2(1 - \cos\Theta_0)}{(1 + z_{\text{dec}})}\right)^{1/2}. \tag{7.53}$$

Thus the growth factor is given by

$$\Sigma = \frac{5(1 + z_{\text{dec}})[(5 + \cos\Theta_0)(1 - \cos\Theta_0) - 3\Theta_0\sin\Theta_0]}{(1 - \cos\Theta_0)^3}$$

$$= \frac{5(1 + z_{\text{dec}})q_0}{(2q_0 - 1)^2}\left[4q_0 + 1 - \frac{3q_0}{\sqrt{2q_0 - 1}}\sin^{-1}\left(\frac{\sqrt{2q_0 - 1}}{q_0}\right)\right], \tag{7.54}$$

where we have used (4.60) to express $\cos\Theta_0$ in terms of q_0.

7.3.3   The open model ($k = -1$)

Again using the relations (4.70)–(4.73), we write in terms of the parameter Ψ

$$ct = \tfrac{1}{2}\beta(\sinh \Psi - \Psi), \qquad S = \tfrac{1}{2}\beta(\cosh \Psi - 1)$$

$$\rho = \frac{3H_0^2(1 - 2q_0)^3}{4\pi G q_0^2 (\cosh \Psi - 1)^3},$$

$$\beta = \frac{2q_0}{(1 - 2q_0)^{3/2}}\left(\frac{c}{H_0}\right). \tag{7.55}$$

Proceeding exactly as for the closed model, we finally arrive at the growth factor

$$\Sigma = \frac{5(1 + z_{dec})q_0}{(1 - 2q_0)^2}\left[1 + 4q_0 - \frac{3q_0}{\sqrt{1 - 2q_0}}\sinh^{-1}\left(\frac{\sqrt{1 - 2q_0}}{q_0}\right)\right]. \tag{7.56}$$

Figure 7.3 plots Σ as a function of q_0 in the range $0 \le q_0 \le 5$. Notice that Σ increases up to $\sim 6 \times 10^3$. We have already seen that for $q_0 = \tfrac{1}{2}$, $\Sigma = 1 + z_{dec} \sim 10^3$. What does Σ mean in terms of galaxy formation?

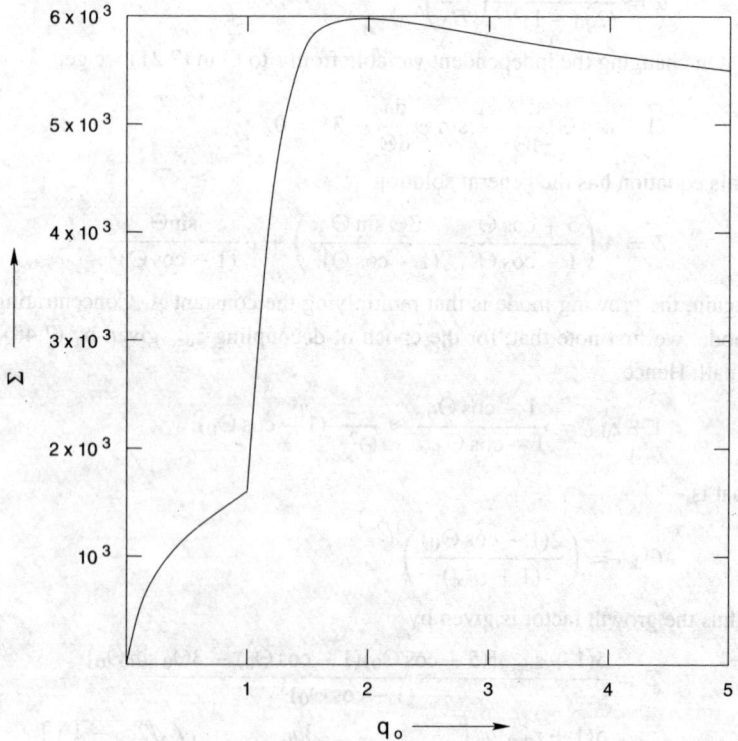

Figure 7.3 The growth function Σ plotted against q_0 for $0 \le q_0 \le 5$, for $z_{dec} \cong 1000$.

We have to admit that $\delta = \delta\rho/\rho$ representing the contrast in density between galaxies and the surrounding medium is considerably higher than unity, since the density in a galaxy is higher than the density of matter in a cluster of galaxies by at least a factor of ten. Moreover, the intracluster density is $\sim 10^{-28}$ g cm^{-3}, which is higher by another order of magnitude than the closure density of the universe ($\sim 10^{-29}$ g cm^{-3}). Thus, to apply our theory to galaxy formation, we need $\delta\rho/\rho \gg 1$ and in this region linearization of the basic equations is not valid. So we cannot use our calculations in any exact sense.

However, we can use the above analysis to demand that, in order to form galaxies, the contrast in density must at least be unity at present. Thus, if we set $\delta\rho/\rho \sim 1$ at present, we may ask for $\delta\rho_{dec}/\rho_{dec}$ to be at least Σ^{-1} at the time of recombination. Only the full non-linear theory can really tell us what $\delta\rho_{dec}/\rho_{dec}$ should have been in order to generate large contrasts in density at present. However, even the lower limit Σ^{-1} is of a magnitude that could in principle be detected by accurate measurements of the microwave background, as we shall shortly see.

7.3.4 Growth in radiation-dominated universes

Our discussion so far has centred on matter-dominated models of the universe, that is, models whose expansion is controlled by non-relativistic matter. However, as we saw in Chapter 4, there was an epoch t_{eq}, prior to which the universe was radiation-dominated. How do inhomogeneities grow in such universes?

The case of a radiation-dominated universe differs from the matter-dominated one in that the pressure p of radiation is making a significant contribution to gravitational attraction. Indeed, we have $p = \rho c^2/3$, so that the effective density is $\rho + 3p/c^2 = 2\rho$. The hydrodynamic equations set up in §7.2.1 must be modified accordingly. We will not go into details, but write the equation for δ that eventually results from the calculation:

$$\ddot{\delta} + 2\frac{\dot{S}}{S}\dot{\delta} + \left(\frac{c_s^2 k^2}{S^2} - \frac{32\pi G\rho}{3}\right)\delta = 0. \qquad (7.57)$$

Here the speed of sound $c_s = c/\sqrt{3}$. We may assume that the curvature (k) term does not produce any significant effect and may be set equal to zero. (*Caution:* refer to the flatness problem discussed in the last chapter!) The corresponding Jeans wave number is now

$$K_J = \left(\frac{32\pi G\rho}{3c_s^2}\right)^{1/2}. \qquad (7.58)$$

It is not difficult to show that the growing and damping modes can be combined in the general solution as before:

$$\delta = At + Bt^{-1}. \qquad (7.59)$$

Of course, here we are assuming that the Jeans wave number far exceeds the wave number for the inhomogeneity under consideration.

7.4   Observational constraints

The 'growth-of-fluctuations' idea outlined above encounters a problem when it is confronted with observations. We shall briefly review not only this problem but also certain other constraints that a successful theory of structure formation must satisfy. Although these constraints have posed difficulties for the standard big-bang model, we should view them in a more positive way, for they represent the remarkable progress that extragalactic observational astronomy has made in the 1980s, thanks to increasingly sophisticated observing techniques. To match these developments on the observational front, the cosmological theories have to be correspondingly more mature and less speculative. We begin with the possible impact of the fluctuations in density leading to galaxy formation on the microwave background.

7.4.1   Small-angle anisotropy

Let us estimate the effect of these fluctuations of $\delta\rho/\rho$, the contrast in density during the epoch of decoupling, on the radiation background. Assuming that the fluctuations are adiabatic, the number density of particles will vary as the cube of the radiation temperature. Therefore

$$\left(\frac{\delta T}{T}\right)_{dec} = \frac{1}{3}\left(\frac{\delta\rho}{\rho}\right)_{dec}, \qquad (7.60)$$

where the subscript (dec) denotes the epoch of decoupling.

Since the universe is optically thin after this epoch, these fluctuations will be imprinted on the radiation background and would be observed to this day. That is, if we sweep observations across the sky we should see ups and downs in the background temperature. What should the order of magnitude of this fluctuation in temperature be for the present epoch? Over what characteristic angular size should we observe these fluctuations?

Our calculations above have placed the value of $(\delta\rho/\rho)_{dec}$ in the region of Σ^{-1}. For the various cosmological models (see Figure 7.3), Σ^{-1} lies in the range $\sim 10^{-2}$ to 3×10^{-4}. Hence, from (7.60), we should have present-day fluctuations of $\Delta T/T$ in the range $\sim 3 \times 10^{-3}$ to 10^{-4}. This is, of course, true under the assumption of optical thinness mentioned earlier.

To fix the angular size of fluctuations, there are two possibilities available to us. First, we note that (7.28) relates the mass M of a typical fluctuation to the characteristic wavelength $2\pi/K$. What will the angle subtended by a length $2\pi/K$ be for the redshift z_{dec}? For this we need the formulae for angular size derived in

Chapter 4. We recall the relevant formulae (4.89), (4.92) and (4.95) and apply them in the limit of a large redshift $(1 + z_{\text{dec}} \approx 1000)$. Thus we get the angular size as

$$\Delta\theta = \frac{2\pi}{K} \frac{(1 + z_{\text{dec}})^2}{D_1},$$

where

$$D_1 \cong \frac{c}{H_0} \frac{z_{\text{dec}}}{q_0} \approx \frac{c}{H_0} \frac{(1 + z_{\text{dec}})}{q_0}.$$

Hence, from (7.28),

$$\Delta\theta = \frac{2\pi}{K} \frac{H_0 q_0}{c} (1 + z_{\text{dec}})$$

$$= \frac{H_0 q_0}{c} (1 + z_{\text{dec}}) \left(\frac{3M}{4\pi n_{\text{dec}} m_{\text{H}}} \right)^{1/3}. \tag{7.61}$$

Since $n_{\text{dec}} = n_0 (1 + z_{\text{dec}})^3$, where n_0 is the present-day number density, we get finally

$$\Delta\theta = \frac{H_0 q_0}{c} \left(\frac{3M}{4\pi n_0 m_{\text{H}}} \right)^{1/3}. \tag{7.62}$$

Using the result that $n_0 m_{\text{H}} = 3H_0^2 q_0 / (4\pi G)$, we can express the above result in the following form:

$$(\Delta\theta) \cong 23 \left(\frac{M}{10^{11} M_\odot} \right) (h_0 q_0^2)^{1/3} \text{ arcseconds.} \tag{7.63}$$

Thus galaxy formation should leave a characteristic patchiness of the angular size ~ 20 arcseconds. Certainly no such perturbations on the scale estimated above have been found.

The second possibility of discovering inhomogeneity imprinted on the microwave background is that on a larger angular scale, arising from a finite size of the particle horizon of a Friedmann universe. We shall discuss it shortly in §7.4.3 under the so-called *horizon problem*.

7.4.2   Types of perturbations

The non-discovery of temperature fluctuations $\Delta T / T$ in the range $\sim 3 \times 10^{-3}$ to 10^{-4} until the early 1990s led to considerable soul-searching on the part of standard-model cosmologists. Have the effects of matter fluctuations on the radiation background been overestimated? If so, where does the present hypothesis break down? Could some other processes be operative in structure formation, making the actual matter–radiation link weaker than that assumed here? We briefly discuss a few alternatives.

The assumption so far has been that the fluctuations are *adiabatic*, i.e., they leave the entropy invariant. Because of the large photon-to-baryon ratio noted in Chapter 5, the entropy is almost entirely carried by photons. The entropy per baryon is proportional to T^3/ρ_m, where T is the radiation temperature and ρ_m is the density of matter. The constancy of this ratio was used in equation (7.60) and this led to the relatively high expectations for $\Delta T/T$.

To reduce this expectation the other alternative is that of *isothermal* fluctuations. Here it is assumed that the radiation temperature remains invariant during pertur-bations of matter. In other words, there is *no* linkage between $\delta\rho$ and ΔT. This could be justified by arguing that the high thermal conductivity of the cosmological medium allows quick equalization of temperature. Irrespective of whether they are isothermal or adiabatic in the pre-decoupling era, the perturbations after that era will have matter fluctuations decoupled from (and hence not interacting with) radiation.

A third kind of perturbation gives rise to *isocurvature* fluctuations. These fluctua-tions preserve the curvature of spacetime by preserving the energy density. Thus, the energy densities of radiation and matter together do not change, i.e., $\delta\epsilon = -\delta\rho_m c^2$. Usually such perturbations arise when there is non-baryonic matter around. We will discuss the need for non-baryonic matter later. For the time being we may mention that, if this matter does not interact with radiation, then fluctuations in it would leave no imprint on the radiation background.

7.4.3   The horizon constraint

The second type of imprint of inhomogeneity on the microwave background is expected from the finite size of the particle horizon. Let us assume that, for any epoch t, $\bar{\rho}(t)$ denotes the smooth averaged-out density in the universe while $\rho(\mathbf{r}, t)$ denotes the actual density at any point in space with coordinate $\mathbf{r}$. To fix our ideas, as well as to simplify matters, let us illustrate the problem for the $k = 0$ model. Define the 'density constrast' $\delta(\mathbf{r}, t)$ by

$$\delta(\mathbf{r}, t) = \frac{\rho(\mathbf{r}, t) - \bar{\rho}(t)}{\bar{\rho}(t)} = \int \delta_{\mathbf{k}}(t) e^{i\mathbf{k}\cdot\mathbf{r}} \frac{d^3 k}{(2\pi)^3}. \tag{7.64}$$

If $S(t)$ is the scale factor, the proper length corresponding to $\mathbf{r}$ is $S(t)|\mathbf{r}|$. Hence the physical wave number for $\mathbf{k}$ is k/S.

The inhomogeneity denoted by $\delta(\mathbf{r}, t)$ is thus seen as a superposition of com-ponents of different wave numbers. A typical size $(2\pi/k)S(t)$ is stretched in an ex-panding universe. Besides, the amplitude for a given $\mathbf{k}$ will grow due to gravitational instability. So an inhomogeneity of characteristic size λ_0 today would correspond to a proper length

$$\lambda(t) = \lambda_0 \frac{S(t)}{S(t_0)} \tag{7.65}$$

for the epoch t. With $S(t) \propto t^n$, say (viz. $n = \frac{1}{2}$ for the radiation-dominated phase and $n = \frac{2}{3}$ for the matter-dominated one), we find that $\lambda(t) \propto t^n$. The size of the particle horizon, however, as we saw earlier, is proportional to t (cf. equations (4.87) and (6.51)). Thus with $n < 1$, for sufficiently small t, $\lambda(t)$ would exceed the size of the horizon.

Since physical processes operate under the principle of causality, it follows that any astrophysically relevant scale today demands seed fluctuations with scales not exceeding the size of the horizon during any earlier epoch. Thus there is manifestly a contradiction here. (For explicit numerical values of typical length scales see Exercises 20 and 21.) We will return to this issue in §7.5.

In general, one could argue that the size of the horizon during the epoch of last scattering t_{dec} was of the order ct_{dec}. What would the angular size of such a region be if it were observed today? The answer (see Exercise 22) turns out to be

$$\theta_H = \sqrt{\frac{8q_0}{z_{\text{dec}}}} \sim 5\sqrt{q_0} \text{ degrees.} \tag{7.66}$$

So one would expect to see some structure in the radiation background on this angular scale. However, if it is argued that the inflationary phase made everything very uniform, then this signature is not expected to be significant.

7.4.4 The scale-invariant spectrum

First we consider the two-point correlation function $\xi(r)$ for galaxies defined by the probability $\delta(r)$ of finding a galaxy in a given volume δV within a distance r from a given galaxy:

$$\delta(r) = \bar{n}[1 + \xi(r)]\delta V. \tag{7.67}$$

Here $\bar{n}$ is the mean number density of galaxies. Detailed studies of galaxy counting indicate that $\xi(r)$ has the form

$$\xi(r) = \left(\frac{r}{r_0}\right)^{-\gamma}, \tag{7.68}$$

where $r_0 = 5h_0^{-1}$ Mpc and $\gamma = 1.8$.

Now $\xi(r)$ is scale-invariant and is typical of fractals. Moreover, the galaxy, the cluster and the supercluster correlation functions have, surprisingly, the same functional form with the same γ. Thus, if we did not know what population was being described in a catalogue we would not be able to find the answer from an analysis of their correlation. Figure 7.4 illustrates this commonality. In mathematical terms all the correlation functions are adequately described by

$$\xi_L(r) = 0.3 \left(\frac{r}{L}\right)^{-1.8},$$

(7.69)

where $L = (\bar{n})^{-1/3}$. This scale-invariant spectrum has to be explained by a theory of galaxy formation.

7.4.5 The hierarchy of structures

The discrete structures range from galaxies on the scales of masses $\sim 10^{11} M_\odot$ and sizes ~ 10 kpc to superclusters on the scales of masses $\sim (10^{14}$–$10^{15}) M_\odot$ and sizes

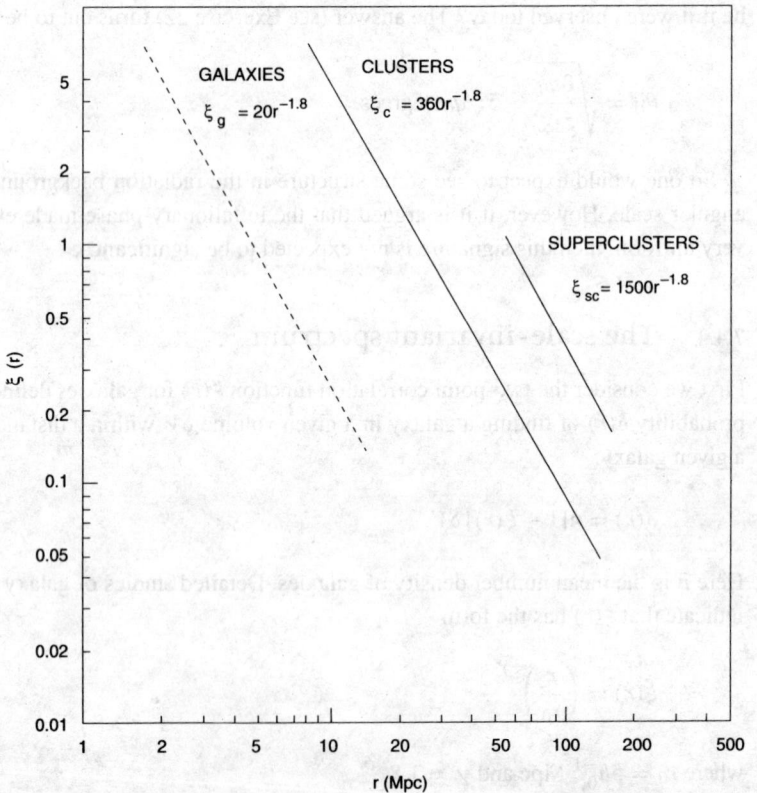

Figure 7.4 The correlation function for clusters and galaxies as well as for superclusters is seen to be the same from this observed distribution of $\xi(r)$, although their characteristic scales are different. Adapted from Figure 10 of an article by N. A. Bahcall, 1988, *Ann. Rev. Astron. Astrophys. 26*, 631.

~50–100 Mpc. We also have to understand how their large-scale filamentary structures become interspersed with giant voids ~100–200 Mpc in size. Two relatively nearby large-scale inhomogeneities are the Great Wall and the Great Attractor.

A theory of structure formation may belong to one of two types: 'top-down' and 'bottom-up'. In the top-down scenario the largest scale structures form first and later they fragment into smaller ones. The reverse is true in the bottom-up case, with smaller-scale structures forming first and accreting together in groups to form the bigger ones. One can generally relate the formation of primary structures to the type of perturbation in operation. Thus one can argue that, in the adiabatic fluctuations, the first structures to form are the massive ones on the supercluster scale and so the *top-down* scenario operates. With the isothermal fluctuations small structures on the mass scale ~$(10^5–10^6)M_\odot$ (closer to globular-cluster size than to the size of a typical galaxy), form first; this is followed by the *bottom-up* scenario.

7.4.6   The age distribution

Did all galaxies form more or less during the same epoch, or is the process of their formation a continuous and ongoing one? When did it begin? Was it related in its evolutionary sequence to the formation of QSOs?

Clues and possible checks can come from the redshift distributions of discrete objects, from the age estimates of galaxies and from their chemical evolution. Redshifts of QSOs indicate a tapering off of their numbers beyond $z = 5$. Galaxies do seem to have a variety of ages, judging by the evolutionary stages of stars therein and by the abundances of heavy elements. These clues pose important constraints on theories of structure formation.

With studies of high-redshift galaxies becoming more and more common, the colours of such galaxies give indications of their ages. Thus it is possible to relate the finding of a well-developed galaxy at high redshift to the structure-formation scenario. Figure 7.5 shows one such galaxy.

Suppose that we are working in the framework of the Einstein–de Sitter model and we see a fully formed galaxy with stars at redshift $z = 3$. Then the age of the universe at the time the galaxy was observed was

$$\tau = \frac{2}{3}h_0^{-1}(1 + z)^{-3/2} \times 10^{10} \text{ years.} \tag{7.70}$$

For $h_0 \cong 0.6$, say, this works out to be about 1.4×10^9 years. This observation therefore imposes a constraint on the theory of structure formation to produce such a galaxy within the above time limit. The constraint becomes even stricter if the galaxy is seen to have well-formed or reasonably old stars.

7.5   Density and mass fluctuations

Before proceeding further, it is useful to elaborate a little on the density-contrast formula (7.64). The Fourier coefficients of $\delta(\mathbf{r}, t)$ can be useful in relating to the sizes of inhomogeneities and their masses. In what follows we will suppress the time dependences of the density and mass fluctuations, since we shall be concerned with averages for a given epoch t.

Using (7.64) we see that the spatial average of $\delta^2(\mathbf{r})$ over a sufficiently large volume V (large enough to contain several statistically homogeneous regions so as to have the realization of $\delta(\mathbf{r})$ as a random variable) will be given by

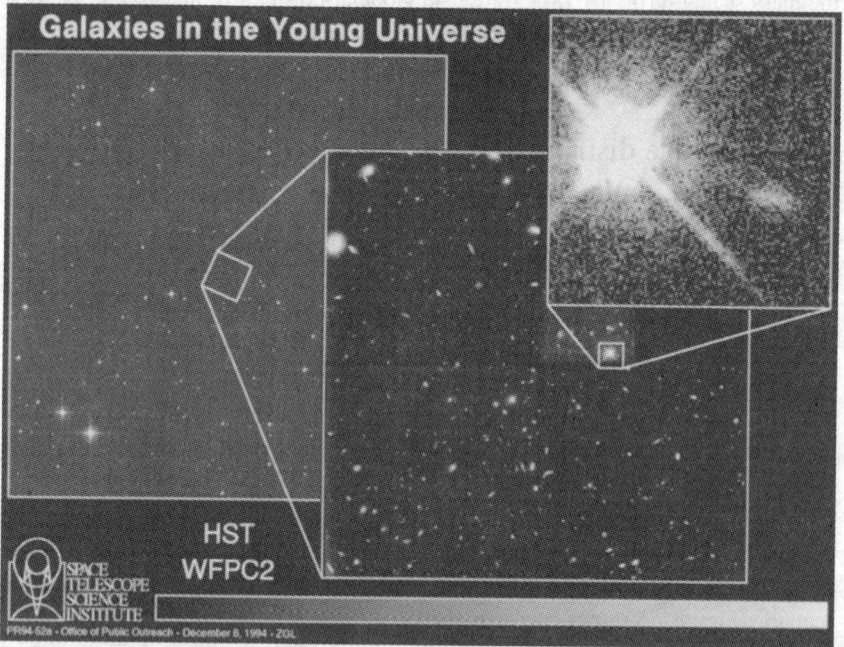

Figure 7.5 In the top right of the picture we see, enlarged, a galaxy with redshift $z = 3.330$, which is a normal galaxy in which star formation is taking place. It lies ~ 100 Mpc in front of the quasar Q 0000–263. Image by HST created with support from the Space Telescope Science Institute operated by the Association of Universities for Research in Astronomy. Reproduced with permission from AURA/STScI.

$$\langle \delta^2(\mathbf{r}) \rangle_V = \int \frac{\delta^2(\mathbf{r})}{V} d^3\mathbf{r}$$

$$= \frac{1}{V} \int |\delta_{\mathbf{k}}|^2 \frac{d^3\mathbf{k}}{(2\pi)^3}$$

$$= \frac{1}{2\pi^2 V} \int k^3 |\delta_k|^2 \frac{dk}{k}, \tag{7.71}$$

where we have assumed that the angular dependence of the fluctuations is not significant in a situation of large-scale homogeneity and isotropy. Thus the suffix of the typical Fourier component is only the magnitude $k = |\mathbf{k}|$ of the wave number. If we consider each $\delta_{\mathbf{k}}$ as a Gaussian random variable with zero mean and standard deviation σ_k, then we can call σ_k^2 the *power spectrum* of the fluctuations. Provided that V is large enough, we may take the measured values of $|\delta_k|^2$ as the power-spectrum parameters σ_k^2.

We can now relate these ideas to the two-point correlation function $\xi(r)$ introduced earlier in §7.4.4. At any point $\mathbf{r}$ it is given by

$$\xi(\mathbf{r}) = \frac{1}{V} \int |\delta_{\mathbf{k}}|^2 \frac{d^3\mathbf{k}}{(2\pi)^3}$$

$$= \int \delta(\mathbf{r} + \mathbf{r}')\delta(\mathbf{r}') \frac{d^3\mathbf{r}'}{V}. \tag{7.72}$$

Using this formalism we can show that, if $\rho(\mathbf{r})$ denotes the smoothed out density of a class of massive objects like galaxies, then the probability P_{12} of finding two galaxies at $\mathbf{r}_1$ and $\mathbf{r}_2$ with a separation of $\mathbf{r}_{12} = \mathbf{r}_1 - \mathbf{r}_2$ will be given by

$$P_{12} \propto \int \rho(\mathbf{r} + \mathbf{r}_1)\rho(\mathbf{r} + \mathbf{r}_2) d^3\mathbf{r}$$

$$\propto \bar{\rho}^2[1 + \xi(\mathbf{r}_{12})]. \tag{7.73}$$

If the existences of these galaxies were independent events, then the probability of their occurrence at the given locations would have been simply proportional to $\bar{\rho}^2$. Denoting this probability by $\bar{P}$, we get the probability in the above case as

$$P_{12} = \bar{P}[1 + \xi(\mathbf{r}_{12})]. \tag{7.74}$$

Finally we consider the structure in question as 'excess mass' in a given volume, say a sphere of radius R. To consider a bounded volume like this we need a 'window function' to cut off any contributions to the mass beyond the volume. Although the Heaviside function does exactly this, it is sometimes awkward to handle and it is more convenient to approximate it with a Gaussian. Thus we write the excess mass contained in a sphere of radius R centred on $\mathbf{r}$ as

$$\delta M_R(\mathbf{r}) = \bar{\rho} \int \delta(\mathbf{r} + \mathbf{r}_1)W(\mathbf{r}_1) d^3\mathbf{r}_1, \tag{7.75}$$

where the window function is

$$W(\mathbf{r}_1) = \text{constant} \times \exp\left(-\frac{r_1^2}{2R^2}\right). \tag{7.76}$$

The Fourier transform of $\delta M_R(\mathbf{r})$ is then given by

$$\delta M_R(\mathbf{k}) = (2\pi)^{3/2} R^3 \delta_{\mathbf{k}} \exp\left(-\frac{k^2 R^2}{2}\right), \tag{7.77}$$

showing that the contributions drop rapidly for wave numbers exceeding R^{-1}, i.e., fluctuations on scales smaller than R are rendered insignificant by the averaging. The dispersion of the mass excess $\delta M_R / M_R$ can be computed similarly to that of the density fluctuations. If we use the Gaussian window $W(\mathbf{r}_1)$, we can define a Gaussian effective volume V_G by the relation

$$M_R \equiv \int \rho(\mathbf{r}_1) W(\mathbf{r}_1) \, d^3\mathbf{r}_1 = (2\pi)^{3/2} R^3 \bar{\rho} \equiv V_G \bar{\rho}. \tag{7.78}$$

Therefore we get

$$\langle (\delta M_R / M_R)^2 \rangle = \int_0^\infty \left(\frac{W_k}{V_G}\right)^2 \frac{k^3 |\delta_k|^2}{2\pi^2 V} \frac{dk}{k}. \tag{7.79}$$

If $|\delta_k|^2 = Ak^n$, a power-law function with A a constant, then we can evaluate this integral analytically:

$$\langle (\delta M_R / M_R)^2 \rangle \equiv \sigma_M^2(R) = \Gamma\left(\frac{n+3}{2}\right) \frac{A}{2} \left(\frac{k^3 |\delta_k|^2}{2\pi^2 V}\right)_{(k=R^{-1})}. \tag{7.80}$$

Thus the fluctuation in mass is directly related to the quantity $k^3 |\delta_k|^2$ at a wavelength of the order of the size R of the region and, for a power-law behaviour of this quantity, $\sigma_M^2(R)$ behaves as $R^{-(n+3)}$.

We shall use these results later in this chapter.

7.6   Inputs from the inflationary phase

One of the attractive features claimed by the inflationary models is that they hold out the possibility of generating seed fluctuations that can grow to form the large-scale structures with a scale-invariant spectrum. To investigate this, we first discuss a scenario that produces the observed structures from seed perturbations and determine the form of the perturbations needed. Then we compute explicitly the nature of perturbations produced by inflation and see how these compare with the required ones.

7.6.1   Causal connections within the initial fluctuations

In the preceding section we saw how the physical wavelengths (of the present-day large-scale inhomogeneities) were larger than the radius of the horizon sufficiently early on in a Friedmann model, which implied that they could not be linked by causal interactions. This conclusion is altered if an inflationary phase is present. Let us see how this comes about, with an illustrative example.

Consider a wavelength λ_0 associated with a galactic mass M during the present epoch. With the mean density given by

$$\bar{\rho}_0 = \frac{3H_0^2}{8\pi G}\Omega_0$$

we have

$$M = \frac{4\pi}{3}\bar{\rho}_0\lambda_0^3,$$

i.e.,

$$\lambda_0 = \left(\frac{2GM}{H_0^2\Omega_0}\right)^{1/3} \tag{7.81}$$

We now trace this length scale back to the epoch t_f when inflation had just ended. Since the scale factor varies as the reciprocal of the radiation temperature, the length scale at t_f was

$$\lambda_f \equiv \lambda(t_f) = \lambda_0 \frac{S(t_f)}{S(t_0)} = \left(\frac{2GM}{H_0^2\Omega_0}\right)^{1/3}\frac{T_0}{T_f}. \tag{7.82}$$

Since the scale factor increased exponentially during the inflationary phase $t_1 \leq t \leq t_f$, the scale at t_i was

$$\lambda_i = \lambda_f \exp[a(t_i - t_f)] = \lambda_f Z^{-1}, \tag{7.83}$$

Z being the factor by which the universe inflated (see equation (6.74)). How does it now compare with the size of the horizon?

Assuming that the universe was in the de Sitter expansion mode during $t_i \leq t \leq t_f$, the nature of the horizon changes. The de Sitter spacetime over its full time span $-\infty < t < \infty$ does *not* have a particle horizon. It does have an event horizon c/a in radius if the expansion factor is $\exp(at)$. However, here we are dealing with a finite interval of the de Sitter expansion, so the issue is somewhat vague. For causal connections which have developed through the past light cone one should strictly talk of the particle horizon. In the absence of a clear-cut particle horizon, we may take c/a, which is also the so-called 'Hubble radius' for exponential expansion, as a length scale up to which causal connections might be established.

Because of the largeness of the factor Z we expect that, for most astrophysically relevant scales,

$$\lambda_i < c/a. \tag{7.84}$$

In other words, the Hubble radius exceeds the length scale. Thus the original causality problem of standard cosmology is circumvented.

Figure 7.6 illustrates the revised situation. It shows the Hubble radius over the inflationary epoch followed by the radius of the particle horizon in the Friedmann radiation-dominated expansion phase. The Hubble radius is constant at c/a for $t_i \leq t \leq t_f$. For $t > t_f$ the particle horizon grows as t. Relative to these scales the typical length scale for a primordial fluctuation corresponding to the wave number k grows always in proportion to the scale factor of the universe, i.e., it grows as $(2\pi/k)\exp(at)$ during $t_i < t < t_f$. Thus it exceeds and crosses out of the Hubble radius at some time t_{exit} given by

$$\frac{2\pi}{k}\exp(at_{exit}) = \frac{c}{a}. \tag{7.85}$$

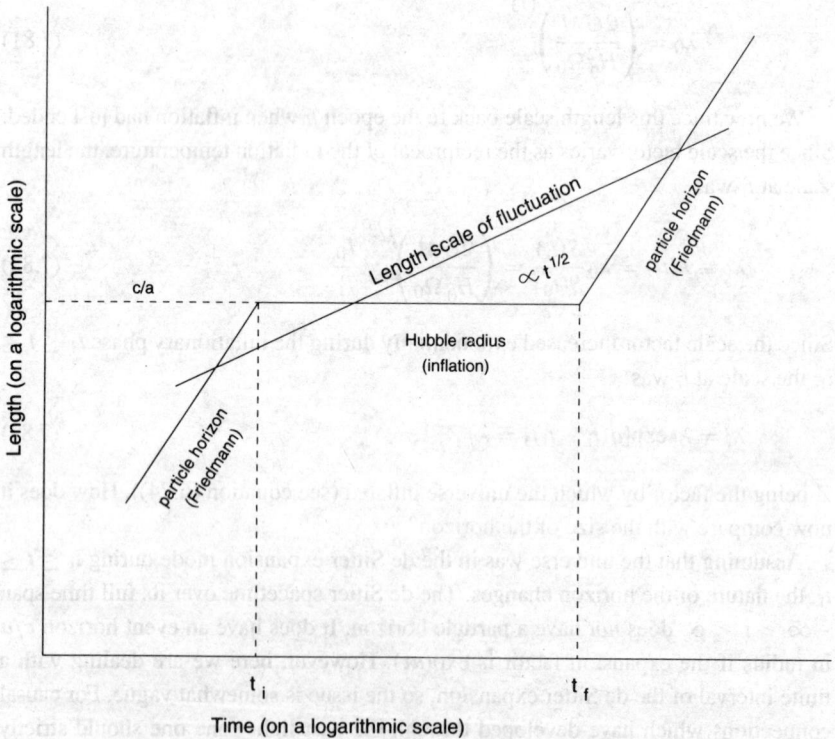

Figure 7.6 The scale size of a fluctuation grows as $S(t)$. It exceeds the Hubble radius of the inflationary model and later re-enters the particle horizon in the Friedmann phase.

For $t_{\text{exit}} < t < t_{\text{enter}}$ the scale in question exceeds both c/a during the inflationary stage and the size of the horizon during the subsequent Friedmann stage. The instant t_{enter} is given by the epoch when the proper length of the fluctuation (which grows during the Friedmann regime in proportion to $t^{1/2}$) becomes equal to the size of the horizon (which grows as t). The suffix 'enter' indicates that for $t > t_{\text{enter}}$, the length of the fluctuation will be *less* than the size of the horizon. Inflation therefore allows one to start with seed fluctuations during a very early epoch and allows them to grow to scales observed at present. It is during the interval $t_{\text{exit}} < t < t_{\text{enter}}$ that causal connections are lost and the correlations developed prior to t_{exit} are maintained intact. Notice also that both t_{exit} and t_{enter} depend on k and this circumstance plays a key role in determining the spectrum of fluctuations. We shall look at this situation next.

7.6.2 The scale-invariant spectrum

Going by the above argument, we need to know the amplitude of a typical density perturbation at the time $t_{\text{enter}}(k)$ when it enters the horizon. We use the Fourier resolution of the density perturbation given by formula (7.64). For $t > t_{\text{enter}}(k)$ we can then study its growth first by linearization techniques until its magnitude $|\delta_{\mathbf{k}}(t)|^2$ becomes comparable to unity and later by other methods. (The fact that $\Delta T/T$ in the microwave background radiation is $< 10^{-5}$ implies that $|\delta_{\mathbf{k}}(t)|^2$ was $\ll 1$ in the radiation-dominated phase of expansion and hence linearization is justified.) Thus we need to know the function

$$F(\mathbf{k}) \equiv |\delta_{\mathbf{k}}(t)|^2_{t=t_{\text{enter}}(k)}. \tag{7.86}$$

Harrison in 1970 and Zel'dovich in 1972 had independently argued, from theoretical considerations, that, at the time of entering the horizon, the perturbations should have the form $F(\mathbf{k}) \propto k^{-3}$. We will relate this argument to the present discussion next. We first recall the relationships between fluctuations of density and mass in a region of size R, which we derived in the previous section.

There we saw that the root-mean-square (RMS) fluctuation of mass M as a fraction of the average mass contained in a region of size R is proportional to $k^3|\delta_{\mathbf{k}}|^2$ at $k = R^{-1}$. Therefore, for the above $F(\mathbf{k}) \propto k^{-3}$, the (RMS) value $\langle(\delta M/M)^2\rangle$ will be independent of the scale R at $t = t_{\text{enter}}(\mathbf{k})$, thus giving equal power at all scales at the time they enter the horizon. As we saw in §7.4.4, a scale-invariant spectrum is indicated by the distribution of discrete large-scale structures.

The inflationary models discussed in Chapter 6 seem capable of producing this kind of spectrum, through fluctuations in the scalar field $\phi(\mathbf{r}, t)$ whose phase transition generates inflation. We write the fluctuations as $f(\mathbf{r}, t)$ over a smoothed average value $\phi_0(t)$. Thus

$$\phi(\mathbf{r}, t) = \phi_0(t) + f(\mathbf{r}, t). \tag{7.87}$$

These fluctuations result in fluctuations of the energy density.

Since the energy density of a scalar field is $\rho c^2 \cong \frac{1}{2}\dot{\phi}^2$, we get

$$\delta\rho(\mathbf{r}, t) \cong \dot{\phi}_0(t)\dot{f}(\mathbf{r}, t)/c^2 \tag{7.88}$$

for $|f| \ll |\phi_0|$. Writing

$$f(\mathbf{r}, t) = \int Q_{\mathbf{k}}(t)e^{i\mathbf{k}\cdot\mathbf{r}}\frac{d^3\mathbf{k}}{(2\pi)^3}, \tag{7.89}$$

we have

$$\delta\rho(\mathbf{k}, t)\, c^2 \cong \dot{\phi}_0(t)\dot{Q}_{\mathbf{k}}(t). \tag{7.90}$$

The average energy density during inflation being dominated by the constant term V_0 (say) of the Coleman–Weinberg potential, we have a contrast in density of

$$\delta_{\mathbf{k}}(t) = \frac{\delta\rho\, c^2}{V_0} = \frac{\dot{\phi}_0 \dot{Q}_{\mathbf{k}}(t)}{V_0}. \tag{7.91}$$

For $\dot{\phi}_0$ we use the mean evolution of ϕ during the slow roll-over phase: but what is f? Now, in actuality, the fluctuations in ϕ are of quantum origin but here, in a classical approximation, we are using $f(\mathbf{r}, t)$ to mimic them classically. In quantum field theory the field would be an operator $\hat{\phi}(\mathbf{r}, t)$ whose Fourier coefficients $\hat{q}_{\mathbf{k}}(t)$ are also operators. In a quantum state specified by the wave function $\psi_{\mathbf{k}}$, the fluctuations of $\hat{q}_{\mathbf{k}}$ are given by the dispersion relation

$$\sigma_{\mathbf{k}}^2(t) = \langle\psi_{\mathbf{k}}|q_{\mathbf{k}}^2(t)|\psi_{\mathbf{k}}\rangle, \tag{7.92}$$

the mean value (in $\mathbf{k} \neq \mathbf{0}$ mode) of $\psi_{\mathbf{k}}$ being zero. This is because ϕ_0, the average of ϕ, is homogeneous. Since $\sigma_{\mathbf{k}}^2(t)$ appears to be a good measure of quantum fluctuations, we may identify $Q_{\mathbf{k}}(t)$ with $\sigma_{\mathbf{k}}(t)$ and write

$$\delta_{\mathbf{k}}(t) = \frac{\dot{\phi}_0(t)}{V_0}\dot{\sigma}_{\mathbf{k}}(t). \tag{7.93}$$

Thus we have taken a semi-classical approximation to estimate the fluctuations in the energy density of the ϕ-field which act as the seed fluctuations of density during the inflationary phase $t_i < t < t_f$. For a comparison with observations we need the value of $\delta_{\mathbf{k}}(t)$ at $t = t_{\text{enter}}$. Several workers in inflation theory have found a way of relating $\delta_{\mathbf{k}}(t_{\text{enter}})$ to $\delta_{\mathbf{k}}(t_{\text{exit}})$ through an approximate conservation law,

$$\frac{\delta_{\mathbf{k}}(t_{\text{enter}})}{1 + W(t_{\text{enter}})} = \frac{\delta_{\mathbf{k}}(t_{\text{exit}})}{1 + W(t_{\text{exit}})}, \tag{7.94}$$

where $W(t)$ is the ratio of pressure $p(t)$ to density $\rho(t)$ of the average background field.

During the inflationary phase, with $\dot{\phi}_0^2 \ll V_0$,

$$p(t) = \frac{1}{2}\dot{\phi}_0^2 - V_0, \qquad \rho(t) = \frac{1}{2}\dot{\phi}_0^2 + V_0, \qquad 1 + W(t) \simeq \frac{\dot{\phi}_0^2}{V_0}. \quad (7.95)$$

In the radiation-dominated phase $1 + W = \frac{4}{3}$. Therefore,

$$\delta_{\mathbf{k}}(t_{\text{enter}}) = \frac{4}{3}\frac{V_0}{\dot{\phi}_0^2}\delta_{\mathbf{k}}(t_{\text{exit}}) \quad (7.96)$$

$$= \frac{4}{3}\frac{\dot{\sigma}_{\mathbf{k}}}{\dot{\phi}_0}\bigg|_{t=t_{\text{exit}}}. \quad (7.97)$$

For the Coleman–Weinberg potential, detailed calculations give $\sigma_{\mathbf{k}}(t)$ and $\phi_0(t)$. The final result is

$$\delta_{\mathbf{k}}(t_{\text{enter}}) \approx 10^2 k^{-3/2}. \quad (7.98)$$

In other words, the condition

$$k^3 |\delta_{\mathbf{k}}(t_{\text{enter}})|^2 = \text{constant} \quad (7.99)$$

required for a scale invariant spectrum is satisfied. Although this is undoubtedly a success for the inflationary model, the outcome is hardly satisfactory. For, after putting in numbers, we find that we have too high an amplitude for fluctuations! Instead of yielding values of the order $<10^{-4}$, the equation (7.98) leads to an amplitude of $\sim 10^2$. Some unrealistic fine tuning of the parameters of the theory is needed in order to bring the amplitude down to the required level.

7.7   The role of dark matter

The presence of dark matter also plays a significant role in the formation of structures. We have seen how the smoothness of the microwave background limits the magnitude of $\delta\rho/\rho$ during the epoch of recombination. The argument that was used in equating the $\delta\rho/\rho$ for matter and radiation depends on the matter being baryonic. Baryonic matter does interact with radiation and so we cannot have large fluctuations $\delta\rho/\rho$ of such matter coexisting with much smaller fluctuations in the radiation background.

The argument, however, breaks down if the bulk of the matter is non-baryonic and hence (possibly) not interacting with radiation. This would allow large $\delta\rho/\rho$ of non-baryonic matter during the epoch of recombination. We may then arrange for the baryonic fluctuations (which were small during that epoch) to catch up with the larger fluctuations of the non-baryonic matter during later epochs. For the two kinds of matter interact gravitationally. Because non-baryonic matter does not interact with radiation it is 'dark' for all astronomical purposes.

7.7.1 Types of non-baryonic dark matter

Non-baryonic matter can be broadly of two kinds, 'hot' and 'cold'. These adjectives indicate how fast a particle of dark matter was moving when it decoupled from the rest of the (baryonic) matter in the universe. In Chapter 5 we saw that neutrinos decoupled from the rest of the matter at temperatures $\sim 10^{10}$ K. At temperature of this order an electron (with a rest mass of ~ 0.5 MeV/c^2) would move relativistically. So, even if a neutrino has a rest mass of around 20–40 eV/c^2, it would move relativistically at decoupling.

Neutrinos are therefore an example of particles of 'hot dark matter' (HDM). At the time of decoupling they were moving with relativistic speeds. In contrast, particles whose velocities had dropped to values $\ll c$ when they decoupled are called particles of 'cold dark matter' (CDM).

Table 7.1 gives a list of candidates for dark matter, baryonic as well as non-baryonic. In the latter class the only familiar species are neutrinos, which are HDM. All other particles are conjectured by the grand unified or supersymmetric (SUSY) particle theories. Table 7.1 also lists particles of intermediate mass such as the gravitino, which have masses ~ 1 KeV and may be considered 'warm' (WDM). None has been detected in accelerator experiments. We therefore begin with a discussion of massive neutrinos. Of all those listed above, the only tangible particle so far is the neutrino; even though it is still uncertain whether the neutrino has a rest mass, it is worthwhile examining a few consequences of such a possibility.

7.7.2 Massive neutrinos

Experiments by F. Reines, H. W. Sobel and E. Pasierb as well as by V. A. Lyubimov *et al.* in 1980 suggested that neutrinos may indeed have a small rest mass. Results of subsequent experiments performed by different groups have been rather equivocal on this issue. In the early 1990s, Dennis Sciama had shown that there are several likely consequences for extragalactic astronomy if there were a neutrino of 17 keV rest mass. In 1998, the Japanese Super-Kamiokande Collaboration including scientists from 23 institutions in Japan and the USA announced indirect evidence that the muon neutrino has a small mass.

The new evidence is based upon studies of neutrinos created when cosmic rays bombard the Earth's upper atmosphere, producing cascades of secondary particles that rain down upon the Earth. Most of these neutrinos pass through the entire Earth unscathed. The Super-Kamiokande group used a large, 50 000-ton tank of highly purified water, located about 1000 m underground in the Kamioka Mining and Smelting Company's Mozumi Mine. Faint flashes of light given off by the neutrino interactions in the tank are detected by more than 13 000 photomultiplier tubes. By classifying the interactions of neutrinos according to the type of neutrino involved (electron-neutrino or muon-neutrino) and counting their relative numbers as a function of the distance from their point of creation, the scientists concluded that the muon-neutrinos are 'oscillating' between, say, the *muon*-neutrino and the τ-neutrino, or some other type of neutrino that as yet remains unknown. The experiment does not determine directly the masses of the neutrinos leading to this effect, but the rate of disappearance suggests that the difference in mass between the oscillating types is very small. At the time of writing this account there is no universally accepted value for the mass of a neutrino, so the issue remains unresolved.

However, cosmologists are interested in this possibility since it opens up a number of interesting astrophysical possibilities, including the formation of large-scale structure. As early as 1972 R. Cowsik and J. McClelland had conjectured that the 'missing mass' in the universe (that is, the dark matter) may be accounted for by relic neutrinos. What can we say today about such a possibility? Here we describe the cosmological implications of neutrinos having finite (i.e., non-zero) rest masses.

Table 7.1 Some candidates for dark matter

Candidate/particle	Approximate mass	Predicted by	Type
Axion, majoron, Goldstone boson	10^{-5} eV	QCD; symmetry breaking	CDM
Ordinary neutrino	10–100 eV	GUTs	HDM
Light higgsino, photino, gravitino, axino, sneutrino	10–100 eV	SUSY/SUGR[a]	HDM
Para-photon	20–400 eV	Modified QED	HDM or WDM
Right-handed neutrino	500 eV	Superweak interaction	WDM
Gravitino, etc.	500 eV	SUSY/SUGR	WDM
Photino, gravitino, axino, mirror particle, Simpson neutrino	keV	SUSY/SUGR	WDM or CDM
Photino, sneutrino, higgsino, gluino, heavy neutrino	MeV	SUSY/SUGR	CDM
Shadow matter	MeV	SUSY/SUGR	HDM or CDM
Preon	20–200 TeV	Composite models	CDM
Monopoles	10^{16} GeV	GUTs	CDM
Pyrgon, maximon, Perry pole, newtorities, Schwarzschild	10^{19} GeV	Higher-dimension theories	CDM
Supersymmetric strings	10^{19} GeV	SUSY/SUGR	CDM
Quark nuggets, nuclearities	10^{15} g	QCD, GUTs	CDM

[a] SUGR $\equiv$ supergravity.

Let us do the calculations taking $g_\nu = 1$ even for massive neutrinos. If the rest mass of the neutrino is larger than $\sim 2 \times 10^{-4}$ eV, neutrinos will have very small random velocities today. For convenience, we will use electron-volts to express the rest mass of the neutrino and write

$$m_\nu = M_\nu \text{ eV}.$$

From Table 5.1 we know that the number density of neutrinos is three-eighths of the number density of photons of the same temperature. We also know that the number density of photons goes as the cube of the photon temperature. Since in the post-$e^+ - e^-$-annihilation phase

$$\left(\frac{T_\nu}{T_\gamma}\right)^3 = \frac{4}{11},$$

we get the present-day ratio of the number densities of neutrinos and photons as

$$\left(\frac{N_\nu}{N_\gamma}\right)_0 = \frac{3}{22}. \tag{7.100}$$

Putting everything together, the mass density of neutrinos at present may be expressed as

$$\rho_\nu = \sum \Omega_\nu \rho_c, \tag{7.101}$$

where $\sum$ denotes a sum over all types of neutrinos and

$$\Omega_\nu \simeq \frac{M_\nu}{150} \left(\frac{T_0}{3}\right)^3 h_0^{-2}. \tag{7.102}$$

A similar contribution to the density will come from antineutrinos. If we consider all species of neutrinos (and their antineutrinos) together, we discover that their contribution to the density becomes comparable to that of baryonic matter, provided that

$$\sum_{\text{all species}} m_\nu \geq 1.5 \text{ eV}. \tag{7.103}$$

If neutrinos collapsed with the nucleons to form clusters, then we get a lower bound on the ratio of non-luminous to luminous (baryonic) matter. This lower bound is

$$\frac{\Sigma \Omega_\nu}{\Omega_B} \geq \frac{2}{3} \sum M_\nu. \tag{7.104}$$

From cluster emission of X-rays it is estimated that the mass of hot gas is related to the total mass of the cluster by the formula

$$M_{\text{HG}} \approx 0.1(2h_0)^{-3/2} M_{\text{Total}}. \tag{7.105}$$

We may take $M_{\text{Total}}/M_{\text{HG}}$ as an upper limit in (7.104). This gives

$$\sum M_\nu \leq 40 h_0^{3/2}. \tag{7.106}$$

Thus, for h_0 in the range $\frac{1}{2}$ to 1, the upper limit on $\sum m_\nu$ lies in the range ~15–40 eV.

In 1979 S. Tremaine and J. Gunn pointed out another handle on neutrino masses. A massive neutrino will have a distribution function in the momentum space of

$$dn_\nu = \frac{g}{(2\pi\hbar)^3} \left[\exp\left(\frac{p_\nu c}{kT_\nu}\right) + 1 \right]^{-1} d^3 p_\nu \tag{7.107}$$

at the time of decoupling. As they cool down $p_\nu \propto T_\nu$ and the neutrinos eventually become non-relativistic. Slow-moving neutrinos would be susceptible to being trapped by the gravitational potential wells of massive systems that eventually form clusters or single galaxies. Trapping and collapse of neutrinos changes their distribution function from (7.107) to a Maxwellian distribution of an isothermal gas. This final distribution is given by

$$dn_\nu = \frac{\rho_\nu}{m_\nu^4} \frac{1}{(2\pi\sigma^2)^{3/2}} \exp\left(-\frac{v^2}{2\sigma^2}\right). \tag{7.108}$$

In order that (7.108) represents a gas trapped by the gravitational field of a mass M at a distance R, we need

$$\langle v^2 \rangle \equiv 3\sigma^2 \simeq \frac{GM}{R};$$

that is,

$$\langle \rho \rangle \equiv \frac{3M}{4\pi R^3} = \frac{9\sigma^2}{4\pi G R^2}. \tag{7.109}$$

Expressing M in terms of $M_\odot$, R in megaparsecs and σ in units of 100 km s^{-1} $\equiv \sigma_{100}$, we get from the above

$$\frac{M}{M_\odot} \simeq 7 \times 10^{12} \sigma_{100} R_{\text{Mpc}}, \qquad \langle \rho \rangle \simeq 10^{-28} \left(\frac{\sigma_{100}}{R_{\text{Mpc}}}\right)^2 \text{ g cm}^{-3}. \tag{7.110}$$

Now, one feature of a collapse accompanied by violent relaxation is that the maximum of the phase-space density decreases. (This happens because, as the gas particles move, a mixing of states occurs, in which the maximum of the original distribution function gets mixed up with lower-density parts of the distribution function.) On comparing the maxima of (7.107) and (7.108) we therefore get

$$\frac{g_\nu}{(2\pi\hbar)^3} > \frac{\rho_\nu}{m_\nu^4(2\pi\sigma^2)^{3/2}};$$

that is,

$$m_\nu > \left(\frac{\rho_\nu(2\pi)^{3/2}\hbar^3}{g_\nu\sigma^3}\right)^{1/4} \tag{7.111}$$

Expressing this inequality in terms of (7.110) we get for $g_\nu = 1$

$$M_\nu \geq 4.5\sigma_{100}^{-1/4} R_{\text{Mpc}}^{-1/2}. \tag{7.112}$$

Relic neutrinos that are sufficiently heavy may therefore collapse and dominate the mass on the various scales given by (7.112). Tremaine and Gunn pointed out a curious aspect of this result.[1] The larger the value of M_ν the larger the ratio in (7.104), that is, the unseen mass is larger relative to the luminous mass. Yet the ratio is known to be largest for clusters of galaxies and lowest for single galaxies. Thus it would appear that relic neutrinos don't seem to solve the problem of the missing mass. To resolve this contradiction Schramm and Steigman suggested that m_ν may lie in the range 4–20 eV. Thus these neutrinos would not be massive enough to dominate gravitational clumping on the scale of a single galaxy, but may well be effective on the scale of clusters.

Very massive neutrinos will prove embarrassing for big-bang cosmology. If all neutrinos have on average a mass of ~25 eV, then $\Sigma\Omega_\nu$ is close to unity. Larger masses than this value and/or an increase in the number of relic neutrino species would increase $\Sigma\Omega_\nu$ and the overall Ω beyond the closure value $\Omega = 1$. As seen in Chapter 4, closed universes have shorter ages and an overall age $\lesssim 6 \times 10^9$ years may be embarrassingly small. It has been suggested that, under such circumstances, λ cosmologies might have to be invoked. However, at the time of writing this account, the experimental upper limits on neutrino masses are well below the above-mentioned critical value.

These calculations illustrate how astrophysics may provide valuable constraints on properties of elementary particles and *vice versa*.

7.7.3 Dark matter and structure size

An interesting relation between the mass of a non-baryonic HDM particle and the mass of the large-scale structure associated with it emerges. The ideas is as follows. Suppose that m_X is the mass of a particle X that moves in a collisionless fashion (i.e.,

---

[1] Somewhat similar arguments were used by Cowsik and McClelland in 1973 to place lower limits on neutrino masses.

it is non-interacting) with relativistic speed. Such a motion is called 'free streaming'. A population of such particles tends to wipe out any inhomogeneity. The limit on the size of the inhomogeneity is then placed by the size of the particle horizon. We estimate the effect as follows.

The particle will be relativistic until the ambient temperature drops to

$$T_X = \frac{m_X c^2}{k}. \tag{7.113}$$

The time–temperature relationship in the early universe will give the epoch as (cf. (6.5))

$$t_X = \left(\frac{3c^2}{16\pi Ga}\right)^{1/2} g^{-1/2} T_X^{-2}. \tag{7.114}$$

For this epoch the size of the horizon is

$$R_X = 2ct_X. \tag{7.115}$$

The energy density of the particles in thermal equilibrium is given by

$$\epsilon = \frac{\pi^2 (kT_X)^4}{15\hbar^3 c^3} \frac{g}{2}. \tag{7.116}$$

Therefore the total mass contained within the horizon sphere is given by putting together (7.113)–(7.116). After some manipulation we get its magnitude as

$$\mathcal{M} = \frac{4\pi}{3} R_X^3 (\epsilon/c^2)$$

$$= \frac{\sqrt{3}\pi^{3/2} g^{-1/2}}{60 G^{3/2} a^{3/2}} \frac{k^6}{\hbar^3 c^3 a^{3/2}} m_X^{-2}. \tag{7.117}$$

Writing the radiation constant and Planck mass as

$$a = \frac{\pi^2}{15} \frac{k^4}{\hbar^3 c^3}, \qquad m_P = \sqrt{\frac{c\hbar}{G}}, \tag{7.118}$$

the above expression becomes

$$\mathcal{M} = \frac{3\sqrt{5}}{4\pi\sqrt{\pi} g^{1/2}} \frac{m_P^3}{m_X^2} = \alpha \frac{m_P^3}{m_X^2}. \tag{7.119}$$

The constant α is of the order of unity. (We may take g between 10 and 100.)

Expressing the neutrino mass in units of electron-volts and $\mathcal{M}$ in units of the solar mass $M_\odot$, the above relation is

$$\mathcal{M} \approx 1.5\alpha \times 10^{15} \left(\frac{30\ \text{eV}}{m}\right)^3 M_\odot. \tag{7.120}$$

Thus, with massive electron-neutrinos as HDM, we get the characteristic scale of supercluster-like inhomogeneities. If we assume that this type of HDM exists, therefore, we have the top-down scenario to think about.

For CDM, on the other hand, the particles hardly move after they have decoupled and their masses are large. The resulting structures are therefore much smaller than those for HDM and we are in the *bottom-up* scenario. Exercise 29 at the end of the chapter shows how one may estimate the scale of a structure dominated by CDM.

7.8   The non-linear regime

The sequence of events leading to galaxy formation may be summarized as follows.

- *Stage 1* Quantum fluctuations in the primordial era were created, say, during the inflationary phase.
- *Stage 2* Fluctuations enter the horizon of the radiation-dominated universe and grow linearly until the epoch of recombination.
- *Stage 3* In the post-recombination era the growth is strongly affected by the presence and nature of dark matter.
- *Stage 4* The fluctuations grow large enough that non-linear processes become important. The end result of this stage is the large-scale structure we should be able to observe with telescopes.

We have discussed stages 1–3 and will now consider the final stage.

7.8.1   The Zel'dovich approximation

In 1970 Zel'dovich gave a simplified picture of how the growing modes of density fluctuations would lead to a non-linear regime. We briefly describe this approach.

Consider the cosmic material as made of fluid elements with trajectories given by

$$\mathbf{r} = S(t)[\mathbf{q} - b(t)\,\nabla_{\mathbf{q}}\psi(\mathbf{q})]. \tag{7.121}$$

Here $S(t)$ is the expansion factor, $\mathbf{q}$ is the comoving coordinate of the fluid element, $b(t)$ describes the growth of fluctuations and ψ is the perturbation potential.

If ρ_0 is the density in comoving coordinates and $\rho(\mathbf{r}, t)$ the proper density, then a simple mass-conservation relation gives

$$\rho(\mathbf{r}, t) = \frac{\rho_0}{S^3} \det\left\|\frac{\partial \mathbf{r}}{\partial \mathbf{q}}\right\|^{-1}. \tag{7.122}$$

The determinant is the Jacobian of transformation, the matrix of which will have eigenvalues λ_1, λ_2 and λ_3 that are continuous random functions of coordinates $\mathbf{q}$. Thus the density becomes

$$\rho(\mathbf{r}, t) = \frac{\rho_0}{S^3}(1 - b\lambda_1)^{-1}(1 - b\lambda_2)^{-1}(1 - b\lambda_3)^{-1}. \tag{7.123}$$

Without loss of generality we assume that $\lambda_1 \geq \lambda_2 \geq \lambda_3$. Then, as $b(t)$ grows, the density becomes infinite as $b\lambda_1 \to 1$. The original volume element had a cubical shape that now flattens to a two-dimensional surface, which Zel'dovich called a 'pancake'.

So far no gravity has been included. To make the picture self-consistent we need to satisfy the Poisson equation:

$$\nabla \cdot \ddot{\mathbf{r}} = -4\pi G \rho. \tag{7.124}$$

To solve the equation, write the three invariants of the transformation matrix

$$I_1 = \lambda_1 + \lambda_2 + \lambda_3, \qquad I_2 = \lambda_1\lambda_2 + \lambda_2\lambda_3 + \lambda_3\lambda_1, \qquad I_3 = \lambda_1\lambda_2\lambda_3. \tag{7.125}$$

Then the Poisson equation becomes

$$\left(3\frac{\ddot{S}}{S} + 4\pi G \frac{\rho_0}{S^3}\right) - (I_1 - 2bI_2 + 3b^2 I_3)\left(\ddot{b} + 2\frac{\dot{S}}{S}\dot{b} + 3\frac{\ddot{S}}{S}b\right) \tag{7.126}$$

$$+ 3\frac{\ddot{S}}{S}(2b^3 I_3 - b^2 I_2) = 0.$$

The first term is zero by virtue of the cosmological expansion law. The second term is zero if

$$\ddot{b} + 2\frac{\dot{S}}{S}\dot{b} + 3\frac{\ddot{S}}{S}b = 0. \tag{7.127}$$

This is the growth equation for linear fluctuations. The last term can be related to a fractional error in density given by

$$\frac{\Delta\rho}{\rho} = \frac{2b^3 I_3 - b^2 I_2}{1 - bI_1 + b^2 I_2 - b^3 I_3}. \tag{7.128}$$

For a planar collapse $\lambda_2 = \lambda_3 = 0$ and hence $\Delta\rho = 0$. This means that the Zel'dovich approximation is exact. If $\lambda_1 > 0$ then the collapse occurs when $b = \lambda_1^{-1}$. For $b\lambda_1 \ll 1$ we are in the linear regime and (7.123) approximates to

$$\rho(\mathbf{r}, t) \approx \frac{\rho_0}{S^3}[1 + b(\lambda_1 + \lambda_2 + \lambda_3)],$$

i.e., the linearized over-density is

$$\delta(\mathbf{r}, t) \simeq bI. \tag{7.129}$$

We thus have a simple picture of how transition from the linear to the non-linear regime occurs. The approximation serves as a starting point for the more exact N-body simulations on a computer.

7.8.2 N-body simulations

A general scheme for numerical simulations may be as follows. We have N particles of (generally equal) masses m_i $(i = 1, \ldots , N)$. The force on particle i located at $\mathbf{r}_i$ is calculated as a modified inverse-square law:

$$\mathbf{F}_i = Gm_i \sum_{j \neq i} \frac{m_j(\mathbf{r}_j - \mathbf{r}_i)}{(|\mathbf{r}_j - \mathbf{r}_i|^2 + \epsilon^2)^{3/2}} \qquad (7.130)$$

The small number ϵ is used in order to avoid having very large forces for close encounters. This force determines the acceleration of the ith particle. Given its position and velocity at one instant, they can then be calculated for a slightly later instant.

This method is direct but very time-consuming for large values of N. Faster approximate methods are therefore devised in order to make progress. However, the computer speeds still fall far short of giving a realistic simulation of the actual problem. Statistical techniques are, nevertheless, useful as indicators of what is going on.

In a typical project $N \geq 10^5$ and the calculations begin at $\delta\rho/\rho \approx 0.2$. The Zel'dovich approximation is used to work out the initial positions and velocities in the growing mode perturbation. The free parameters of the calculation are H_0 and Ω_0 as well as the initial amplitude of the fluctuations, given the shape of the spectrum. The spectrum is calculated by solving the linear fluctuation-growth equations for each $\mathbf{k}$.

The end product for a typical CDM scenario is illustrated in Figure 7.7. Similar pictures can be obtained for HDM also. The idea is to compare these diagrams with the actual redshift surveys that give a three-dimensional mapping of the universe. The large-scale motions are also compared with data.

Depending on the scenario used (HDM, CDM, etc.), there are physical processes that change the shape of the original power spectrum. We may denote by t_i the (primordial) time at which the fluctuations were formally specified and by t_f the time at which the influence of the processes was over. Then, writing the perturbation power spectrum at time t as $P(k; t)$, we get

$$\frac{P(k; t_f)}{[\Sigma(t_f)]^2} = T^2(k; t_f, t_i)\frac{P(k, t_i)}{[\Sigma(t_i)]^2}. \qquad (7.131)$$

The function $\Sigma(t)$ is the linear growth law for perturbations above the Jeans scale discussed in §7.3. Thus, in the absence of any other effects, $T = 1$. However, the other effects show up in the growth formula through the function $T(k; t_f, t_i)$ which is called the *transfer function*. Usually T is obtained by a numerical integration of the relevant effects.

It is fair to say that, although such exercises have given us considerable insight into how non-linear growth processes operate and how the various types of dark matter influence them differently, no successful simulation of the actual universe has yet been possible. An important question is this: do galaxies trace the mass distribution? The CDM scenario works if the galaxies do not trace the mass but form in some 'biased' regions. The biasing is introduced as follows. Consider a Gaussian distribution of fluctuations $\delta\rho$ of CDM, around an average ρ. If there were no biasing, light would trace mass, i.e., galaxies would mimic $\delta\rho/\rho$. However, if galaxies form preferentially at high values of $\delta\rho$ then there is biasing. The biasing parameter b is unity if there is no biasing and $b > 1$ for biasing. By restricting this parameter to $b > 1.5$, say, one is truncating the Gaussian to the tail where $\delta\rho$ is suitably high. The practical advantage of biasing is that it reduces $\Delta T/T$. For example, J. Silk and N. Vittorio have shown that at $4'.5$ and $1'.5$ angular scale

$$\frac{\Delta T}{T} \cong \frac{6 \times 10^{-6}}{\Omega_0 h_0 b}. \tag{7.132}$$

Thus, for $\Omega_0 = 1$, $h_0 = 1$ and $b = 1.5$ we expect $\Delta T/T$ to be around 4×10^{-6}. The present limits on $\Delta T/T$ permit $b > 0.4$.

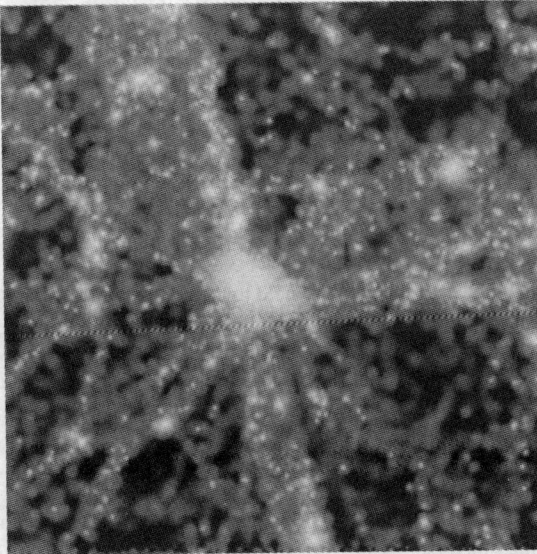

Figure 7.7 Typical results of structures obtained by N-body simulations by the Virgo collaboration between scientists from Israel and the Max Planck Institute for Astrophysics, Germany. Typically 256^3 particles were used in a simulation carried out on two Cray T3D parallel supercomputers. The above mosaic results from a CDM model with total density parameter $\Omega_0 = 0.3$ and a cosmological constant given by $\Omega_\Lambda = 0.7$. The Hubble parameter is $h_0 = 0.7$. The assumed mass per particle is $8.8 \times 10^{10} M_\odot h_0^{-1}$. Reproduced with permission from Joerg M. Colberg, Max Planck Institute of Astrophysics, Garching.

Although the CDM models survive on this count there are problems in explaining the very large-scale streaming motions and the large structures like the Great Wall and the Great Attractor. The HDM models can explain large-scale structures but find it difficult to explain galaxy-sized structures and the low values of $\Delta T/T$.

The COBE results in the early 1990s demonstrated how statistical studies of the angular inhomogeneities of the cosmic microwave background can place constraints on the scenarios for structure formation. We will discuss these findings when we consider the overall observational constraints on the various theories, in Chapters 10 and 11. There we will also mention further studies of the radiation background by more sophisticated spaceborne projects like MAP and PLANCK.

There are other scenarios besides the above CDM and HDM theories. In the cosmic-strings hypothesis the linear discontinuities at the GUT phase transition (see Chapter 6) act as seeds for growth of fluctuations. The strings untangle as the universe expands, leaving a few long stretches and closed loops within the present-day Hubble radius. In the explosions model non-gravitational processes such as shock waves generated by the explosions of supernovae are called upon to trigger the process of structure formation. Neither approach can claim full success with observations. Perhaps more daring ideas are needed! So we leave this chapter with the problem of §7.1 still unsolved.

Exercises

1 Explain what is meant by the transverse or rotational modes of the velocity field in the first-order perturbation analysis in the expanding universe. Show that these modes decrease as the universe expands.

2 Derive (7.21) satisfied by the contrast in density of the expanding universe.

3 Relate the longitudinal modes of the velocity field to the contrast in density. Comment on the fact that the contrast in density does not depend on the transverse modes.

4 What is the physical significance of the Jeans wave number? How is it related to the Jeans mass?

5 Show why we cannot neglect the expansion of the universe in a Jeans-type calculation.

6 Estimate the gravitational energy and the thermal energy of a typical spherical perturbation in the expanding universe. Relate the ratio of these energies to the ratio of the mass of the perturbation and the Jeans mass.

7 Explain the significance of the Jeans mass in relation to the perturbations that can or cannot grow in the expanding universe.

8 Using data on the Earth's atmosphere, estimate (a) the Jeans length and (b) the Jeans mass for air at normal temperature and pressure.

9 Show that, with ρ and p given by

$$\rho = nm_H + \frac{aT^4}{c^2}, \qquad p = \frac{1}{3}aT^4,$$

the speed of sound is given by

$$c_s^2 = \frac{1}{3}c^2\left(1 + \frac{m_H c^2}{\sigma kT}\right)^{-1},$$

where $k\sigma$ is the entropy per particle.

10 Show that, after the era of decoupling, the Jeans mass is given by

$$M_J \simeq \frac{4\pi}{3}\left(\frac{5\pi kT}{3G}\right)^{3/2} n^{-1/2}m_H^{-2},$$

where T is the temperature of matter.

11 Assuming that the temperature of matter in Exercise 10 equalled the radiation temperature during the epoch of decoupling, show that M_J for that epoch was given by

$$M_J \simeq \frac{4\pi}{3}\left(\frac{5\pi kT_0}{3G}\right)^{3/2} n_0^{-1/2}m_H^{-2},$$

where T_0 is the present-day temperature of the microwave background.

12 Evaluate M_J of Exercise 11 in a Friedmann universe of given (h_0, Ω_0) with $T_0 = 3$ K. Show that

$$M_J \simeq 2.54 \times 10^{39}(\Omega_0 h_0^2)^{-1/2} \text{ g}$$
$$\simeq 1.27 \times 10^6 (\Omega_0 h_0^2)^{-1/2} M_\odot.$$

13 Show that (7.43) gives, for the epoch of recombination,

$$M_J \simeq 100 M_\odot \sigma^{1/2}.$$

14 Follow the evolution of the Jeans mass in the expanding universe and discuss qualitatively how a galaxy-sized fluctuation is likely to behave in the pre- and post-decoupling eras.

15 Show, that in discussing the growth of a mass very much in excess of the Jeans mass in the post-decoupling era, the effect of pressure may be neglected. Is this a good assumption for studying the behaviour of galaxy-sized perturbations?

16 Discuss quantitatively the growth of fluctuations in the Friedmann models in the post-decoupling era.

17 Solve from first principles the differential equation

$$(1 - \cos\Theta)\frac{d^2\delta}{d\Theta^2} + \sin\Theta\frac{d\delta}{d\Theta} - 3\delta = 0$$

and relate its solutions to the behaviour of fluctuations in the post-decoupling era of the closed Friedmann universe.

18 Verify by a suitable limiting process that, as $q_0 \to \frac{1}{2}$, both (7.56) and (7.54) tend to (7.49). Plot Σ as a function of q_0 for $1 + z_{dec} = 10^3$.

19 Review some of the attempts to understand the formation of galaxies.

20 Show that the mass associated with wavelength λ measured in megaparsecs is

$$M(\lambda) = 1.5 \times 10^{11}\Omega_0 h_0^2 \lambda^3 M_\odot.$$

21 In the previous exercise a mass of the order $10^{12} M_\odot$ corresponds to $\lambda \approx 1.88$ Mpc. Show that this wavelength was bigger than the horizon at all redshifts exceeding

$$z = 1.41 \times 10^5 (\Omega_0 h_0^2)^{1/3}.$$

22 Calculate the angle subtended by the particle horizon during the epoch of decoupling at the observer today, by following these steps.

   (i) Calculate the Hubble constant and the deceleration parameter at the redshift of decoupling z_{dec} and hence the diameter d_H of the horizon. Work with the $k = 1$ model.

  (ii) Take the angular coordinate $\theta = 0$ towards the centre C of the horizon sphere and assume that the paths of rays of light from the observer O tangential to the sphere make a cone with semi-vertical angle $\theta = \theta_H/2$.

 (iii) An azimuthal plane $\phi = $ constant intersects this cone in a triangle with base ACB, say, and its vertex at O, the base being a spacelike geodesic with the equation

$$r^2 + \frac{1}{1 - kr^2}\left(\frac{dr}{d\theta}\right)^2 = \frac{r^4}{r_0^4}.$$

Here r_0 is the radial Robertson–Walker coordinate of C. Let r_1 be the radial coordinate of ends A and B.

 (iv) Relate the quantities d_H and θ_H to r_0 and r_1.

  (v) From these relations work out the relation between d_H and θ_H.

23 Show that, if the universe were dominated by three types of relic massive neutrinos during the present epoch, the average neutrino mass needed to close the universe would be

$$25\left(\frac{T_0}{3}\right)^{-3} h_0^2 \text{ eV}.$$

24 Suppose that the universe has enough baryons to make $\Omega_B = 1$ and that it has in addition three species of neutrinos of average mass 25 eV. For $T_0 = 3$ and $h_0 = 1$, calculate the age of the universe.

25 Discuss how the observation of neutrino mass affects the age of the universe. With the example given in Exercise 24, if the age of the universe comes out very low, can you think of a way out of the difficulty by using the λ cosmologies of Chapter 4?

26 A primordial neutrino has rest mass 1 eV. Estimate its random velocity relative to the cosmological rest frame during the present epoch.

27 Describe how massive neutrinos might influence the condensation of matter into galaxies or larger structures. Is it possible to think of a consistent mass range of m_ν that may account for the missing mass in galaxies and clusters of galaxies?

28 In equation (7.114) replace G by the Planck mass and arrive at $\mathcal{M}$.

29 For CDM repeat the free-streaming-motion argument given in the text for structure sizes formed by HDM, with the following changes. Assume that the CDM particle travels with speed c until it becomes non-relativistic and that this stage is reached when its temperature has dropped to $m_X c^2 / k$. Now use the ratio of the present-day number density of CDM particles n_X and n_γ to determine the ratio T_X / T and show that the free-streaming scale is

$$\lambda_{FS} = \text{constant} \times (\Omega_X h^2)^{1/3} m_X^{-4/3}.$$

By putting in the values of the constant, estimate the mass of the CDM-induced structure and show that, for $m_X \sim 1$ keV, the mass $\mathcal{M}$ is $6 \times 10^9 M_\odot$.

30 Outline the observational constraints that must be satisfied by theories of galaxy formation.

Chapter 8

Alternative cosmologies

8.1 Alternatives to Friedmann cosmologies

In 1922–24 when Friedmann produced the expanding-universe solutions of Einstein's equations, his work went largely unnoticed. Subsequent to Hubble's discovery of the nebular redshift, however, cosmologists came to regard these models as the simplest starting point for discussing their subject. The physicists, on the other hand, considered these attempts naive and speculative and so they did not pay as much attention to George Gamow's very seminal work on the early universe. The turning point for cosmology came, however, in 1965 with the discovery of the microwave background radiation (MBR). The MBR seemed to confirm the early-universe scenario and, taken together with the extended validity of Hubble's law obtained by bigger and better telescopes, laid a solid foundation for cosmology as a branch of physics. By the mid-1970s a considerable body of physicists had begun to take the Friedmann cosmology seriously, all the more so after they had realized that the big-bang cosmology provides a setting, the only setting known so far, for testing their very-high-energy physics and the grand-unification programme. Cosmologists also looked to particle physics for understanding of the primary origin of matter. Indeed, the subject of astroparticle physics has grown out of joint speculations of big-bang cosmologists and high-energy particle physicists.

Chapters 6 and 7 have given a glimpse of how the big-bang cosmology has progressed with these inputs from particle physics. The question which we will properly address in the last chapter is the following. To what extent is Friedmann cosmology a correct theory of the origin and the large-scale structure of the universe? Although the majority of today's cosmologists would put their money on the Friedmann models, there have been a few 'agnostics' from time to time, who were not satisfied with them. From their efforts there have emerged alternative theories of cosmology.

These theories have not been worked through to the depth that Friedmann cosmology can boast of. This is hardly surprising, considering the very limited number of people who worked on them. Nevertheless, they contain different perspectives and are worth taking a look at, if only because they might offer a resolution of some of the outstanding problems that the Friedmann cosmology has been unable to solve. In this chapter and the next, we describe a few such theories, in particular those based on the following concepts:

1. Mach's principle,

2. the large-numbers hypothesis and

3. the creation of matter.

Of these the last one will be described in the following chapter. Here we begin with Mach's principle, an intriguing idea that has excited considerable discussion amongst cosmologists and physicists over several generations.

8.2   Mach's principle

There are two ways of measuring the Earth's spin about its polar axis. By observing the rising and setting of stars the astronomer can determine the period of one revolution of the Earth around its axis: the period of $23^h\ 56^m\ 4^s.1$. The second method employs a Foucault pendulum whose plane gradually rotates around a vertical axis as the pendulum swings (see Figure 8.1). Knowing the latitude of the place of the pendulum, it is possible to calculate the period of the Earth's spin. The two methods give the same answer.

At first sight this does not seem surprising. If we are measuring the same quantity, we should get the same answer regardless of the method used. Closer examination, however, reveals why the issue is non-trivial. The two methods are based on different assumptions. The first method measures the period of the Earth's spin against a background of distant stars, whereas the second employs the standard Newtonian mechanics in a spinning frame of reference. In the latter case, we take note of how Newton's laws of motion have to be modified when their consequences are measured in a frame of reference spinning relative to the 'absolute space' in which these laws were first stated by Newton.

Thus, implicit in the assumption that equates the two methods is the coincidence of absolute space with the background of distant stars. It was Ernst Mach (see Figure 8.2) who pointed out in the last century that this coincidence is non-trivial. He read something deeper into it, arguing that the postulate of absolute space that allows one to write down the laws of motion and arrive at the concept of inertia is somehow intimately related to the background of distant parts of the universe. This argument is known as 'Mach's principle' and we will analyse it further.

When it is expressed in the framework of the absolute space, Newton's second law of motion takes the familiar form

$$\mathbf{P} = m\mathbf{f}. \tag{8.1}$$

Figure 8.1 The Foucault pendulum at the Inter-University Centre for Astronomy and Astrophysics, Pune. The bob of the pendulum oscillates in a vertical plane, which slowly rotates around the vertical axis in a clockwise fashion. The plane of oscillation makes a complete round in $(\sin l)^{-1}$ days, where l is the latitude of the place of the pendulum. In Pune the period is approximately 75 h.

Figure 8.2 Ernst Mach (1838–1916).

This law states that a body of mass m subjected to an external force $\mathbf{P}$ experiences an acceleration $\mathbf{f}$. Let us denote by Σ the coordinate system in which $\mathbf{P}$ and $\mathbf{f}$ are measured.

Newton was well aware that his second law has the simple form (8.1) only with respect to Σ and those frames that are in uniform motion relative to Σ. If we choose another frame Σ' that has an acceleration $\mathbf{a}$ relative to Σ, the law of motion measured in Σ' becomes

$$\mathbf{P}' \equiv \mathbf{P} - m\mathbf{a} = m\mathbf{f}'. \tag{8.2}$$

Although (8.2) outwardly looks the same as (8.1), with $\mathbf{f}'$ the acceleration of the body in Σ', something new has entered into the force term. This is the term $m\mathbf{a}$, which has nothing to do with the external force but depends solely on the mass m of the body and the acceleration $\mathbf{a}$ of the reference frame relative to the absolute space. Realizing this aspect of the additional force in (8.2), Newton termed it 'inertial force'. As this name implies, the additional force is proportional to the inertial mass of the body. Newton discusses this force at length in his *Principia*, citing the example of a rotating water-filled bucket (see Figure 8.3).

According to Mach, the Newtonian discussion was incomplete in the sense that the existence of the absolute space was postulated arbitrarily and in an abstract manner. Why does Σ have a special status in that it does not require the inertial force? How can one physically identify Σ without recourse to the second law of motion, which is based on it?

To Mach the answers to these questions were contained in the observation of the distant parts of the universe. It is the universe that provides a background reference frame that can be identified with Newton's frame Σ. Instead of saying that it is an accident that the Earth's velocity of rotation relative to Σ agrees with that relative to the distant parts of the universe, Mach took it as proof that the distant parts of the universe somehow enter into the formulation of local laws of mechanics.

One way this could happen is through a direct connection between the property of inertia and the existence of the universal background. To see this point of view, imagine a single body in an otherwise empty universe. In the absence of any forces (8.1) becomes

$$m\mathbf{f} = \mathbf{0}. \tag{8.3}$$

What does this equation imply? Following Newton we would conclude from (8.3) that $\mathbf{f} = \mathbf{0}$, that is, the body moves with uniform velocity. However, we now no longer have a background against which to measure velocities. Thus $\mathbf{f} = \mathbf{0}$ has no operational significance. Rather, the lack of any tangible background for measuring motion suggests that $\mathbf{f}$ should be completely indeterminate. It is not difficult to see that such a conclusion follows naturally, provided that we come to the remarkable conclusion, also possible from (8.3), that

$$m = 0. \tag{8.4}$$

In other words, the measure of inertia depends on the existence of the background in such a way that, in the absence of the background, the measure vanishes! This aspect introduces into mechanics a new feature not considered by Newton. The Newtonian view that inertia is the property of matter has to be augmented to give the statement that inertia is the property of matter as well as of the background provided by the rest of the universe. This general idea is known as *Mach's principle*.

Such a Machian viewpoint not only modifies local mechanics but also introduces new elements into cosmology. For, except in the universe following the perfect cosmological principle, there is no basis now for assuming that masses of particle would necessarily stay fixed in an evolving universe. This is the reason for considering

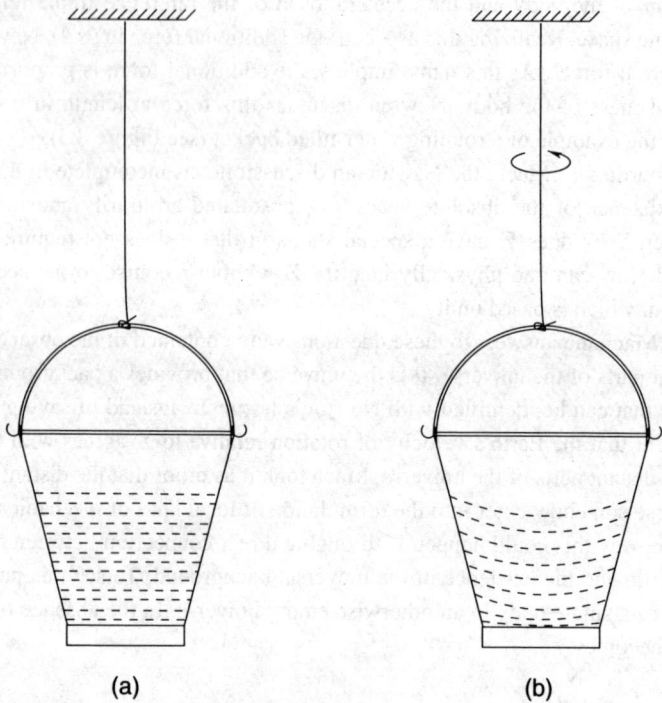

(a)   (b)

Figure 8.3 (a) A bucket full of water hanging by a rope tied to the ceiling. (b) The same bucket turning around and around as a result of the rope unwinding itself from a previously given twist. The water surface in (a) is flat and horizontal, whereas that in (b) is curved inwards. This curvature of the water surface is due to the centrifugal force that acts on the rotating mass of water. This example was discussed by Newton in his *Principia*. Newton argued that in (a) the bucket is at rest relative to the absolute space, whereas in (b) it is rotating relative to the absolute space and hence extra forces have to be postulated to explain the curvature of the water surface. The centrifugal force is the extra force in this example.

cosmological models anew from the Machian viewpoint. Presented here are some instances of how various physicists have given quantitative expression to Mach's principle and arrived at new cosmological models.

8.3   The Brans–Dicke theory of gravity

In 1961 C. Brans and R. H. Dicke (see Figure 8.4) provided an interesting alternative to general relativity that was based on Mach's principle. To understand the reasons leading to their field equations, we first note that the concept of a variable inertial mass arrived at in §8.1 itself leads to a problem of interpretation. For how do we compare masses at two different points in spacetime? Masses are measured in certain units, such as masses of elementary particles, which are themselves subject to change! We need an independent unit of mass against which an increase or decrease in mass of a particle can be measured. Such a unit is provided by gravity, by the so-called Planck mass encountered earlier:

$$\left(\frac{\hbar c}{G}\right)^{1/2} \cong 2.16 \times 10^{-5} \text{ g.} \tag{8.5}$$

Thus the dimensionless quantity

$$\chi = m \left(\frac{G}{\hbar c}\right)^{1/2} \tag{8.6}$$

measured at different points in spacetime can tell us whether masses m are changing. Or alternatively, if we insist on using mass units that are the same everywhere, a change of χ would tell us that the gravitational constant G is changing.[1] This is the conclusion Brans and Dicke drew from their approach to Mach's principle. They looked for a framework in which the gravitational constant G arises from the structure of the universe, so that a changing G could be looked upon as the Machian consequence of a changing universe.

In 1953 D. W. Sciama (Figure 8.5) had given general arguments leading to a relationship between G and the large-scale structure of the universe. We have already come across one example of such a relation in the Friedmann cosmologies:

$$\rho_0 = \frac{3H_0^2}{4\pi G} q_0. \tag{8.7}$$

If we write $R_0 = c/H_0$ as a characteristic length of the universe and $M_0 = 4\pi\rho_0 R_0^3/3$ as the characteristic mass of the universe, then the above relation becomes

---

[1] We could of course assume that $\hbar$ and c also change. However, by keeping $\hbar$ and c constant we follow the principle of least modification of existing theories. Thus special relativity and quantum theory are unaffected if we keep $\hbar$ and c fixed.

$$\frac{1}{G} = \frac{M_0}{R_0 c^2} q_0^{-1} \sim \frac{M_0}{R_0 c^2} \sim \sum \frac{m}{rc^2}. \tag{8.8}$$

Given a dynamical coupling between the inertia and gravity, a relation of the above type is expected to hold. Brans and Dicke took this relation as one that

(a)

(b)

Figure 8.4 Carl Brans (1935–) (a) (photograph by Harold Baquet) and Robert H. Dicke (1916–1997) (b) (photograph by courtesy of P. J. E. Peebles, Princeton University).

Figure 8.5 Dennis W. Sciama (1926–2000).

determines G^{-1} from a linear superposition of inertial contributions $m/(rc^2)$, the typical one being from a mass m at a distance r from the point where G is measured. Since m/r is a solution of a scalar wave equation with a point source of strength m, Brans and Dicke postulated that G behaves as the reciprocal of a scalar field ϕ:

$$G \sim \phi^{-1}, \tag{8.9}$$

where ϕ is expected to satisfy a scalar wave equation whose source is all the matter in the universe.

8.3.1 The action principle

The intuitive concepts are contained in the Brans–Dicke action principle, which may be written in the form

$$\mathcal{A} = \frac{c^3}{16\pi} \int_{\mathcal{V}} (\phi R + \omega \phi^{-1} \phi^k \phi_k) \sqrt{-g} \, d^4x + \Lambda. \tag{8.10}$$

Notice first that the coefficient of R is $c^3 \phi/(16\pi)$ instead of $c^3/(16\pi G)$ as in the Einstein–Hilbert action. The reason for this lies in the expected behaviour of G given in (8.9). The second term, with $\phi_k \equiv \partial \phi / \partial x^k$, ensures that ϕ will satisfy a wave equation, while the third term includes, through a Lagrangian density L, all the matter and energy present in the spacetime region $\mathcal{V}$. The energy momentum tensor T^{ik} of matter is related to Λ through the relation (2.97). ω is a coupling constant.

The variation of $\mathcal{A}$ for small changes of g^{ik} leads to the field equations

$$R_{ik} - \frac{1}{2} g_{ik} R = -\frac{8\pi}{c^4 \phi} T_{ik} - \frac{\omega}{\phi^2} \left(\phi_i \phi_k - \frac{1}{2} g_{ik} \phi^l \phi_l \right)$$
$$- \frac{1}{\phi} (\phi_{ik} - g_{ik} \Box \phi). \tag{8.11}$$

Similarly, the variation of ϕ leads to the following equation for ϕ:

$$2\phi \Box \phi - \phi_k \phi^k = \frac{R}{\omega} \phi^2.$$

The latter equation can be simplified by substituting for R from the contracted form of (8.11). We finally get

$$\Box \phi = \frac{8\pi}{(2\omega + 3)c^4} T, \tag{8.12}$$

where T is the trace of T_k^i. Thus (8.12) leads to the expected scalar wave equation for ϕ with sources in matter, $\Box$ being the wave operator. Because it contains a scalar field ϕ in addition to the metric tensor g_{ik}, the Brans–Dicke theory is often referred to as the *scalar-tensor theory of gravitation*.

8.3.2   Solar-System measurements of ω

It is clear from these field equations that, as $\omega \to \infty$, the Brans–Dicke theory tends to general relativity (see Exercise 8). For $\omega = O(1)$ the theory makes significantly different predictions from general relativity in a number of Solar-System tests. These tests were briefly reviewed in §2.10 in the context of general relativity.

The computation of the precession of the perihelion of the planet Mercury gives the prediction of this theory as $(3\omega + 4)/(3\omega + 6)$ times the value given by general relativity. Dicke and his colleagues suggested during the 1970s that, if the Sun is oblate, with a quadrupole-moment parameter of $\sim 2.5 \times 10^{-5}$, then the resulting change in its gravitational field would lead to a perihelion precession of about 7% of the observed (unexplained) value of ~ 43 arcseconds per century (see Exercise 9). Had this been the case the relativistic value of ~ 43 arcseconds would have been too high, whereas a Brans–Dicke value for $\omega \simeq 6$ would have correctly accounted for the residual of ~ 40 arcseconds per century. However, external studies concerning the Sun's surface do not conform with oblateness even of this order. Hence this test does not give any evidence for ω as small as 6. In fact, to the accuracy with which the effects of solar oblateness are now estimated and the measurements of perihelion shift are carried out, the parameter ω has to be at least ~ 300.

The bending angle of a ray of light grazing a massive spherical object in the Brans–Dicke theory is $(2\omega + 3)/(2\omega + 4)$ of the relativistic value. Since the accuracy of the radio and microwave measurements of the bending angle has been steadily increasing and the angle agrees with the relativistic value within the progressively decreasing error bars, the permitted value of the parameter ω had to be steadily increased over the years and is now as high as ~ 3300.

The lunar laser-ranging experiments as well as radar ranging to probe landers on Mars lead to a similarly high lower bound on ω. Here again the general relativistic value of the Earth–Moon distance is in excellent agreement with observations and any departures from it, if they are to be tolerated by the observations, have to be small enough to demand a large value of ω. It therefore follows that, at the Solar-System level, the Brans–Dicke theory has to have a large value of ω in order to survive, thus making it practically indistinguishable from general relativity. However, even for a large ω this theory can produce interesting departures from general relativity at the cosmological level. The following section outlines these differences.

8.4   Cosmological solutions of the Brans–Dicke equations

We will consider only the homogeneous and isotropic cosmological models in the Brans–Dicke theory. Accordingly we start with the Robertson–Walker line element and the energy tensor for a perfect fluid, as we did in Chapter 4. The scalar field ϕ is now a function of the cosmic time only. Thus the field equations become

$$\frac{2\ddot{S}}{S} + \frac{\dot{S}^2 + kc^2}{S^2} = -\frac{8\pi p}{\phi c^2} - \frac{2\dot{\phi}\dot{S}}{\phi S} - \frac{\omega \dot{\phi}^2}{2\phi^2} - \frac{\ddot{\phi}}{\phi}, \tag{8.13}$$

$$\frac{\dot{S}^2 + kc^2}{S^2} = \frac{8\pi\epsilon}{3\phi c^2} - \frac{\dot{\phi}\dot{S}}{\phi S} + \frac{\omega \dot{\phi}^2}{6\phi^2}. \tag{8.14}$$

Compare these equations with the corresponding ones (4.20) and (4.21) of the Friedmann cosmologies. The conservation equation corresponding to (4.24) is the same:

$$\frac{d}{dS}(\epsilon S^3) + 3pS^2 = 0. \tag{8.15}$$

In addition, we have the field equation for ϕ:

$$\frac{1}{S^3}\frac{d}{dt}(\dot{\phi}S^3) = \frac{8\pi}{(2\omega + 3)c^2}(\epsilon - 3p). \tag{8.16}$$

We expect that big-bang solutions will emerge from these equations and set the big-bang epoch at $t = 0$. Then the integral of (8.16) is

$$\dot{\phi}S^3 = \frac{8\pi}{(2\omega + 3)c^2}\int_0^t (\epsilon - 3p)S^3\,dt + C, \tag{8.17}$$

where C is a constant. Two types of solutions are obtained, depending on whether $C = 0$ or $C \neq 0$.

8.4.1 $C = 0$

We will consider a simple example of this type, with $k = 0$, $p = 0$ and $\epsilon = \rho c^2$. This solution is therefore analogous to the Einstein–de Sitter model of general relativity. Write

$$S = S_0\left(\frac{t}{t_0}\right)^A, \qquad \phi = \phi_0\left(\frac{t}{t_0}\right)^B \tag{8.18}$$

so that $\rho \propto t^{-3A}$ and the field equations give

$$A = \frac{2\omega + 2}{3\omega + 4}, \qquad B = \frac{2}{3\omega + 4} \tag{8.19}$$

and

$$\rho_0 = \frac{(2\omega + 3)B\phi_0}{8\pi t_0^2}. \tag{8.20}$$

The temporal behaviours of S and G ($\propto \phi^{-1}$) are illustrated in Figure 8.6. It can be verified that, as $\omega \to \infty$, this solution tends to the Einstein–de Sitter model.

An analogue of the radiation model can be obtained in this theory (see Exercise 12). H. Nariai obtained solutions for $p = n\epsilon$ with n in the range $0 \leq n \leq \frac{1}{3}$.

8.4.2 $C \neq 0$

In this case the ϕ-terms dominate the dynamics of the universe in the early stages. Thus, for small enough t, we have

$$\frac{8\pi}{(2\omega + 3)c^2} \int_0^t (\epsilon - 3p)S^3 \, dt \ll |C|, \tag{8.21}$$

for the cases both of dust and of radiation. For our power-law solutions for the case $p = 0$, we have at small enough t

$$3A + B = 1, \qquad t_0 = S_0^3 \phi_0 B / C. \tag{8.22}$$

In the case of a radiation-dominated universe, $p = \epsilon/3$ and we can again try a solution of the form (8.18) to get as $t \to 0$

$$A^2 = -AB + \omega B^2/6. \tag{8.23}$$

Taking into account (8.22), we can solve (8.23) to get

$$A = \frac{\omega + 1 \pm \sqrt{(2\omega/3) + 1}}{3\omega + 4}, \qquad B = \frac{1 \pm 3\sqrt{(2\omega/3) + 1}}{3\omega + 4}. \tag{8.24}$$

The upper sign holds when $C > 0$ and the lower sign when $C < 0$. For $C > 0$, $\phi \to 0$ when $S \to 0$, whereas for $C < 0$, $\phi \to \infty$ for $S \to 0$. These conclusions hold irrespective of the values of k and of the equation of state, since, for small values of S, the dynamics of the universe are controlled by the ϕ-term.

8.4.3   The production of light nuclei

Dicke and G. S. Greenstein independently investigated the problem of nucleosynthesis in the early Brans–Dicke universe. Greenstein followed the same physical

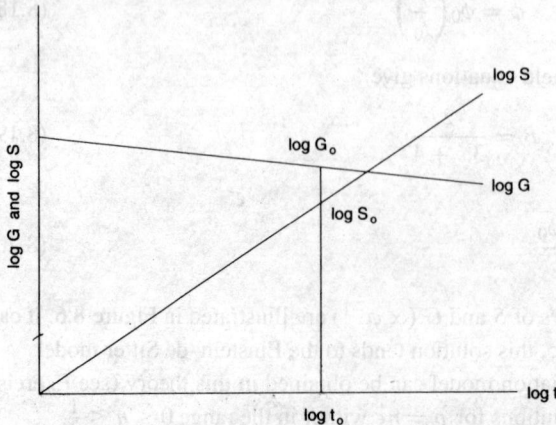

Figure 8.6 The temporal behaviours of S and G. Both are plotted on a log–log plot for $\omega = 6$. The scales are arbitrary.

approach as was outlined in Chapter 5, for the case $C = 0$. The results obtained by him for $h_0 = 1$ are given in Table 8.1.

For each of three values of the present density of matter ρ_0, Table 8.1 gives three sets of values for the abundances of deuterium and helium, corresponding to $\omega = 5$, $\omega = 10$ and $\omega = \infty$. The last case is of course that of general relativity. The differences between the Brans–Dicke theory and general relativity are noticeable for $\omega = 5$ at high values of ρ_0, for which more ^2H and ^4He are formed in the former theory. For $\omega \geq 30$, the present-day observed abundances set an upper limit of $\rho_0 \leq 5 \times 10^{-30}$ g cm^{-3} on the Brans–Dicke cosmology.

In the ϕ-dominated models the constant C can be adjusted to produce any desirable abundances, high or low. For cosmic abundances lower than the above value one has to choose a suitably low value of $|C|$. There is, however, another observational handle on C, which is described briefly below.

8.4.4   The variation of G

Since $G \propto \phi^{-1}$, a time-dependent ϕ will mean a time-dependent gravitational constant. As can be seen from (8.19), we have for $C = 0$

$$\frac{\dot{G}}{G} = -\frac{2}{3\omega + 4}\frac{1}{t} = -\frac{H}{\omega + 1}. \tag{8.25}$$

Thus $|\dot{G}|$ is of the order of Hubble's constant unless ω is large and its sign indicates that the gravitational constant should decrease with time (see Figure 8.6).

However, for a large enough $|C|$, the ϕ-dominated solutions differ significantly from the matter-dominated ones even for the present epoch. In this case, for C large and negative we can have G increasing with time even in relatively recent epochs. We will review the evidence for and against variation of G in Chapter 11.

Table 8.1 Mass fractions of ^2H and ^4He in Brans–Dicke cosmology for matter-dominated models[a]

ω	ρ_0 (g cm^{-3})		
	10^{-31}	10^{-30}	10^{-29}
5	7.6×10^{-4}	2.6×10^{-5}	3.4×10^{-8}
	0.26	0.33	0.40
10	7.6×10^{-4}	2.1×10^{-5}	$\sim 10^{-9}$
	0.26	0.30	0.35
∞	6.6×10^{-4}	1.3×10^{-5}	$\sim 10^{-11}$
	0.25	0.27	0.29

[a] The deuterium fraction is given above the helium fraction.

8.4.5   Inflation in Brans–Dicke cosmologies

Because of its relative simplicity of formulation and interpretation of observable results, the Brans–Dicke cosmology has been studied in the 'very-early-universe' phase also. C. Mathiazhagan and V. B. Johri were the first to consider the inflationary phase in this cosmology. The problem of bubble nucleation and coalescence that was faced by Guth's inflationary model represented the difficulty of what has been known as the *graceful exit* from the inflationary phase into the Friedmann radiation-dominated phase. La and Steinhardt had considered the Brans–Dicke framework to generate an *'extended inflation'*. The 'extended' phase arises because the expansion is not exponential but of a power-law type. The idea seemed to solve the problem of a graceful exit but ran into trouble because of the distortions it produced in the cosmic microwave background, distortions that were unacceptably high. Undeterred by these setbacks, the enthusiasts for inflation explored a variation on the Brans–Dicke theme by adding higher-order couplings of the scalar field with gravity. This led to the notion of *'hyper-extended inflation'*. However, none of these ideas seems to have received much following in later years.

To sum up, the Brans–Dicke theory had generated considerable interest as an alternative theory of gravity, but, with the Solar-System tests giving values very close to the predictions of general relativity with greater and greater accuracy, the parameter ω that distinguished it from general relativity had to be larger and larger, thus making it more and more indistinguishable from general relativity, at least on the scale of the Solar System. On the cosmological front the theory has given different results from standard Friedmann cosmology, but these differences do not seem to have impressed theoreticians sufficiently for them to undertake detailed studies of the cosmogony of the universe from the very early epochs.

8.5   The Hoyle–Narlikar cosmologies

We next consider another theory of gravitation that may claim to have given the most direct quantitative expression to Mach's principle. This theory was first proposed in 1964 by Fred Hoyle and the author; we will refer to it here as the HN theory and to the cosmological models based on it as HN cosmologies. Throughout this discussion we will set $c = 1$.

Like general relativity and the Brans–Dicke theory, the HN theory is formulated in the Riemannian spacetime. There is one important difference, however, between this theory and all other cosmological theories we have discussed so far. The difference lies in the fact that general relativity, the Brans–Dicke theory and so on are pure field theories, whereas the HN theory arose from the concept of *direct interparticle action*. The difference between the two types of theories is best seen from a description of electromagnetism, to which we will frequently refer in this section and the next for comparison. Until the advent of Maxwell's field theory, it was customary to describe electric and magnetic interactions as instances of direct *action at a distance* between particles. The success of Maxwell's theory established the concept of a *field* in physics at the expense of the concept of action at a distance (see Figure 8.7).

Since Mach's principle (implying as it does a connection between the local and the distant) suggests action at a distance, even an early convert to it like Einstein later became sceptical regarding its validity. Einstein's objections were based on the belief that action at a distance was supposed to be instantaneous and hence inconsistent with relativity. By the early 1960s, however, it had become clear that action at a distance can be made consistent with relativity and also successfully describe electrodynamics, besides having interesting cosmological implications. Since Hoyle and the author had played an active role in these developments, they naturally adopted an action-at-a-distance approach to Mach's principle.

Accordingly, we use here the somewhat unfamiliar notation of action at a distance. Let us denote by $a, b, \dots$ the particles in the universe, m_a and e_a being the mass and charge of the ath particle. As implied by Mach (see §8.2), the mass m_a is not entirely an intrinsic property of particle a; it also owes its origin to the background provided by the rest of the universe. To express this idea quantitatively, write

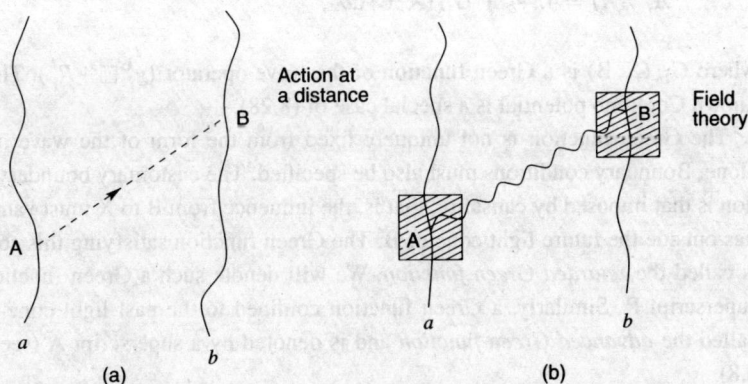

(a)   (b)

Figure 8.7 (a) In the action-at-a-distance picture the influence from the point A on the world line of particle a is transmitted directly across spacetime (along the dotted track) to the point B on the world line of particle b. (b) In field theory the field in the neighbourhood of A (shown by the shaded region) is disturbed; the disturbance propagates across spacetime as a wave in the ambient field and reaches the neighbourhood of B (also shown as a shaded region). The disturbance then exerts a force on particle b at B. This is how the influence propagates from a to b.

$$m_a(A) = \lambda_a \sum_{b \neq a} m^{(b)}(A). \tag{8.26}$$

The above expression means the following. At a typical world point A on the world line of particle a, the mass acquired by a is the nett sum of contributions from all other particles b ($\neq a$) in the universe. The contribution from b at A is given by the scalar function $m^{(b)}(A)$. The coupling constant λ_a is intrinsic to the particle a. Notice, however, that, if a were the only particle in the universe, $m_a = 0$ and we have the conclusion arrived at in (8.4).

8.5.1   A digression into electromagnetic theory

What are these functions $m^{(b)}(X)$? That they communicate the property of inertia from particles b to any particle placed at the spacetime point X is clear from the context. To arrive at a suitable form for them we take hints from action-at-a-distance electromagnetism, in which it is usual to introduce electromagnetic disturbances that arise specifically from sources, that is, from moving electric charges. Accordingly, we introduce the 4-potential $A_i^{(b)}(X)$ to denote the electromagnetic effect at X from the electric charge b. $A_i^{(b)}(X)$ satisfies the wave equation

$$\Box A_i^{(b)} + R_i^k A_k^{(b)} = 4\pi J_i^{(b)}, \tag{8.27}$$

where $J_i^{(b)}$ is the 4-current generated by the charge b. The solution of (8.27) may be written in the integral form

$$A_i^{(b)}(X) = 4\pi e_b \int G_{ik}(X, B) \, db^k, \tag{8.28}$$

where $G_{ik}(X, B)$ is a Green function of the wave operator $(g_i^k \Box + R_i^k)$. The well-known Coulomb potential is a special case of (8.28).

The Green function is not uniquely fixed from the form of the wave operator alone. Boundary conditions must also be specified. The customary boundary condition is that imposed by causality; that is, the influence from B to X must vanish if X lies outside the future light cone of B. The Green function satisfying this condition is called the *retarded Green function*. We will denote such a Green function by a superscript R. Similarly, a Green function confined to the past light cone of B is called the *advanced Green function* and is denoted by a superscript A (see Figure 8.8).

These Green functions have played a key role in action-at-a-distance theories. Because the typical Green function acts as a vector at each of its two end points, it is a *bivector*. It was originally believed that action at a distance must be instantaneous and hence inconsistent with the framework of special relativity. However, K. Schwarzschild, H. Tetrode, and A. D. Fokker demonstrated during the first three decades of the twentieth century that a relativistically consistent action-at-a-distance theory can indeed be formulated. If we consider two points in spacetime A and B with s_{AB}^2 as the invariant square of the relativistic distance between them, then $\delta(s_{AB}^2)$, where δ is the Dirac delta function, is a convenient function describing the transmission of physical influences between A and B at the speed of light, for this function acts only when A and B can be connected by a ray of light (that is, when $s_{AB}^2 = 0$). This delta function therefore necessarily occurs as the main component in any Green function in the action-at-a-distance theory. The action principle, which is the basis of the electromagnetic theory in Riemannian spacetime, is described below. We start with the action

$$\mathcal{A} = -\sum_a \sum_{<b} 4\pi e_a e_b \iint \bar{G}_{ik}\, da^i\, db^k \tag{8.29}$$

where $\bar{G}_{ik}$ is the *symmetric Green function* given by

$$\bar{G}_{ik}(A, B) \equiv \tfrac{1}{2}[G_{ik}^R(A, B) + G_{ik}^A(A, B)]. \tag{8.30}$$

Thus $\bar{G}_{ik}(A, B) = \bar{G}_{ik}(B, A)$ and each term in the action is completely symmetric between each pair of particles. The action (8.30) together with suitable cosmological boundary conditions reproduces all the electromagnetic effects of the standard Maxwell field theory.

That cosmological boundary conditions are necessary in the action-at-a-distance framework is seen from the following simple illustration. Any retarded signal emitted by particle a will get an advanced reaction back from b, as shown in Figure 8.9.

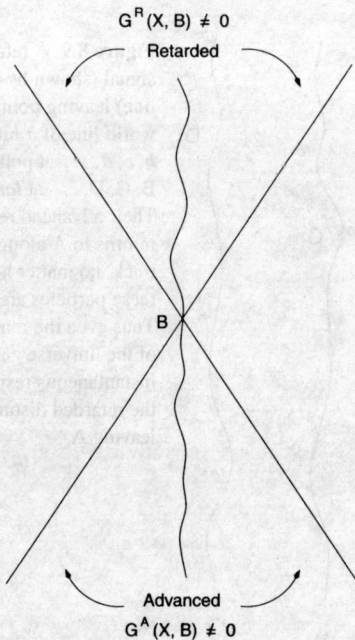

Figure 8.8 The retarded Green function of B is non-vanishing only in the future light cone of B, whereas the advanced Green function is non-vanishing only in the past light cone.

Thus the theory admits advanced signals and appears to violate causality. Moreover, in Figure 8.9 the signal from b arrives at a at the same time as that at which the original signal left a, no matter how far away b is! Thus electromagnetism ceases to be a local theory: any so-called local effect must take account of the response of the universe, which consists of reactions from all such particles b other than a. A 'correct' response can cancel out all the acausal effects. This was pointed out first by J. A. Wheeler and R. P. Feynman in 1945. Later, between 1962 and 1963, J. E. Hogarth, F. Hoyle and the author showed that this response depends on the model of the universe. In essence, to produce the correct response, the universe must be a perfect absorber in the future, i.e., it should be able to absorb all electromagnetic signals directed to the future.

What is the response of the universe? In the 1930s, it had been demonstrated by Dirac that, when an electric charge a accelerates, the force of radiative damping to which it is subjected can be calculated by evaluating half the difference between the retarded and the advanced fields of the charge *on its own world line*:

$$Q(a) = \tfrac{1}{2}[F^R(a) - F^A(a)]. \tag{8.31}$$

In the Maxwell field theory Dirac's result had remained just a curiosity without a proper understanding of why the radiative reaction must be determined by the above formula. This was linked with the more basic question that arises when we discuss electromagnetic fields of an oscillating system of electric charges. It is customary to choose the retarded solutions of the Maxwell wave equations to describe these fields,

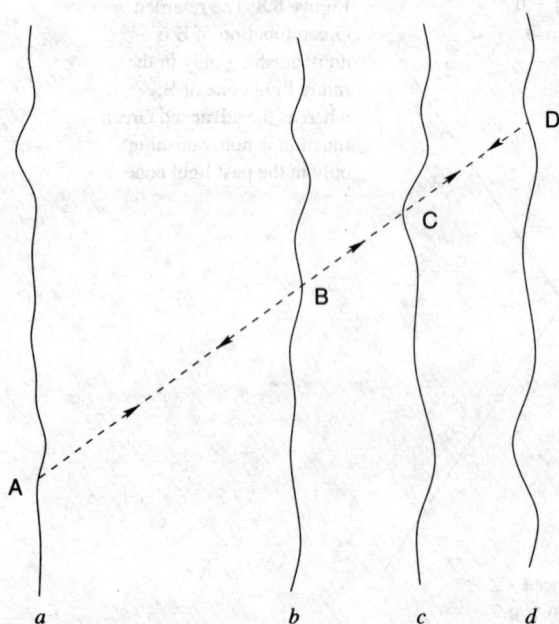

Figure 8.9 A retarded signal (shown by the broken line) leaving point A on the world line of a hits particles $b, c, d, \ldots$ at points $B, C, D, \ldots$ at *later* times. Their advanced response returns to A along the same track, no matter how far these particles are from a. Thus even the remote parts of the universe generate instantaneous responses to the retarded disturbance leaving A.

on the grounds of causality. It is because of this choice that the system radiates energy and suffers damping. So the basic question is this: why do we restrict our solutions to the retarded ones and throw away the advanced ones? Or, to put it differently, why do we have a principle of causality (that causes *precede* effects), when the basic equations of physics are time-symmetric?

The Wheeler–Feynman theory provides an answer. The theory is formulated in a time-symmetric manner with advanced solutions on an equal footing with the retarded ones. Thus a typical particle a generates a 'direct particle field' defined by

$$F(a) = \tfrac{1}{2}[F^{R}(a) + F^{A}(a)], \tag{8.32}$$

which is manifestly symmetric with regard to its advanced and retarded components. As was seen above, the universe as a whole generates a response to these individual fields of the charges and, in the Wheeler–Feynman theory, the 'correct' response from the universe to the motion of a is precisely (8.31)! It can be shown that, for the correct response, the future part of the universe (lying on the future light cone of the radiating system) must be a *perfect absorber* of all retarded, i.e., future-directed signals, and the past part of the universe an *imperfect absorber* of all advanced, i.e., past-directed signals. In such a universe, therefore, if we add this response (8.31) to the basic time-symmetric field of a, as given by (8.32), we get the nett field in the neighbourhood of a as

$$\begin{aligned} F_{\text{total}}(a) = F(a) + Q(a) &= \tfrac{1}{2}[F^{R}(a) + F^{A}(a)] + \tfrac{1}{2}[F^{R}(a) - F^{A}(a)] \\ &= F^{R}(a). \end{aligned} \tag{8.33}$$

In this way, we get the total effect in the neighbourhood of a to be a purely retarded one. A correct response therefore eliminates all advanced effects except those present in the radiation reaction. It is interesting (and significant) that the steady-state model to be discussed in the following chapter generates the correct response, whereas all Friedmann models fail to do so. Because of the crucial requirement of perfect absorption, this theory is sometimes called the 'absorber theory of radiation'. Figure 8.10 illustrates this behaviour of the future absorber.

8.5.2 Inertia and gravity

Our purpose in the above digression into electromagnetism was to show that a similar approach to inertia leads us to a Machian theory of gravity. In the case of inertia we note that the functions $m^{(b)}(X)$ are scalars and so we have to deal with *biscalar* Green functions. Thus we write

$$m^{(b)}(X) = \int \lambda_b \tilde{G}(X, B) \, ds_b \tag{8.34}$$

and the inertial action as

$$\mathcal{A} = -\sum_a \sum_{<b} \iint \lambda_a \lambda_b \tilde{G}(A, B) \, ds_a \, ds_b. \qquad (8.35)$$

What is $\tilde{G}(A, B)$? Again we proceed by analogy with electromagnetism.

From symmetry considerations, we need $\tilde{G}(A, B) = \tilde{G}(B, A)$. Furthermore, we require $\tilde{G}$ to be a Green function of a scalar wave equation. To fix $\tilde{G}$ completely we use another hitherto-undiscussed property of Maxwell's electromagnetic theory known as *conformal invariance*.

8.5.3   Conformal invariance

Let us consider the transformation

$$\bar{g}_{ik} = \Omega^2 g_{ik}, \qquad (8.36)$$

where Ω is a twice-differentiable function of coordinates x^i and lies in the range $0 < \Omega < \infty$. Such a transformation is called a *conformal transformation*. Given a spacetime manifold $\mathcal{M}$ with coordinates (x^i) and metric (g_{ik}), we have through (8.36) generated another spacetime manifold $\bar{\mathcal{M}}$ with the same coordinate system (x^i) but with a different metric $(\Omega^2 g_{ik})$. $\mathcal{M}$ and $\bar{\mathcal{M}}$ are said to be *conformal* to each other. If $\mathcal{M}$ is flat, $\bar{\mathcal{M}}$ is said to be *conformally flat*.

If we identify the corresponding points (with the same x^i) in $\mathcal{M}$ and $\bar{\mathcal{M}}$, we will find that, in general, distances between two points are stretched or compressed when

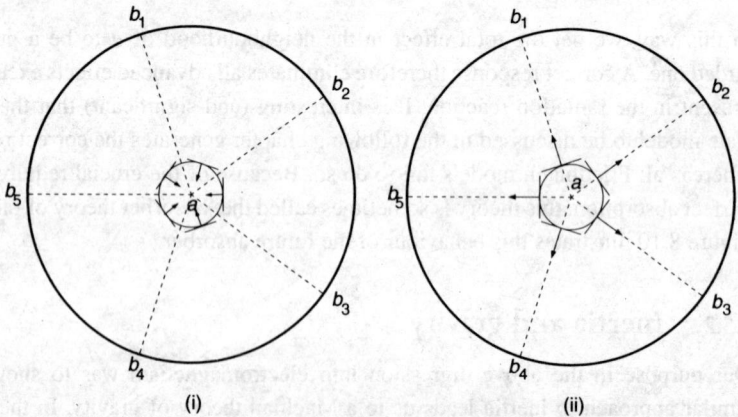

Figure 8.10 The future absorber is typically represented by a sphere centred on charge a. Advanced fields of absorber particles $b_1, b_2, \ldots$ are represented by spherical wavefronts converging respectively on them. The envelope of these wavefronts is initially (i) a spherical wave converging on a and then (ii) another diverging from it. These represent the two parts $-\frac{1}{2}F^{(A)}(a)$ and $\frac{1}{2}F^{(R)}(a)$, respectively.

we go from $\mathcal{M}$ to $\bar{\mathcal{M}}$. However, the null cones in both the manifolds are unchanged. This invariance of null cones is distinct from the invariance under coordinate transformations. The coordinate transformations preserve the null directions locally and they are important in field theories that describe physical interactions locally. The action-at-a-distance theories describe interactions globally and must take account of the global structure of null cones. Hence such theories are expected to preserve their form under conformal transformations as well.

It is easy to verify that the scalar curvature changes under the conformal transformation to

$$\bar{R} = \Omega^{-2}\left(R + 6\frac{\square\Omega}{\Omega}\right),\tag{8.37}$$

where $\square$ is evaluated with respect to the metric (g_{ik}). There are, however, certain quantities that do remain the same under a conformal transformation. These are known as *conformally invariant* quantities. It is easy to see, for example, that the action describing Maxwell's field theory is conformally invariant. Consider the changes

$$\bar{A}_i = A_i + \psi_{,i} \qquad (\psi = \text{a scalar function})$$
$$\bar{F}_{ik} = F_{ik}, \qquad \bar{J} = \Omega^{-4}J_i.$$

These changes leave the form of Maxwell's equations intact.

We now fix the form of $\tilde{G}(A, B)$ by demanding that our inertial action (8.35) be conformally invariant. Since, under the transformation (8.36),

$$d\bar{s}_a = \Omega(A)\,ds_a, \qquad d\bar{s}_b = \Omega(B)\,ds_b,\tag{8.38}$$

we must have

$$\overline{\tilde{G}(A, B)} = \Omega(A)^{-1}\Omega(B)^{-1}\tilde{G}(A, B).\tag{8.39}$$

The only scalar wave operator that permits (8.39) is then

$$\square + \tfrac{1}{6}R.$$

In other words, $\tilde{G}(X, B)$ satisfies the wave equation

$$[\square_X + \tfrac{1}{6}R(X)]\tilde{G}(X, B) = [-g(X)]^{-1/2}\delta_4(X, B).\tag{8.40}$$

$\delta_4(X, B)$ is the four-dimensional Dirac delta function, which vanishes unless $X \equiv B$. Thus we have ensured that the action-at-a-distance theory given by (8.36) does not change under conformal transformations.

8.6   Gravitational equations of HN theory

The action of HN theory is given by (8.35) and, with the help of definitions (8.26) and (8.34), we may write it as

$$A = -\sum_a \int m_a \, ds_a. \tag{8.41}$$

Written in this form, this action appears to have only the inertial term of Chapter 2 (see (2.80)). How can such an action yield any gravitational equations?

The answer to this question lies in the fact that the m_a in (8.41) are not constants but depend on spacetime coordinates *as well as on spacetime geometry*. For they are defined with the help of Green functions, which in turn are defined in terms of the geometry of spacetime. Thus, if we make a small variation

$$g_{ik} \to g_{ik} + \delta g_{ik},$$

the wave equation (8.40) will change and so will its solution. Thus we will have

$$\tilde{G}(A, B) \to \tilde{G}(A, B) + \delta \tilde{G}(A, B)$$

and hence $A \to A + \delta A$. We therefore have a non-trivial problem whose solution may be expressed in the following way. To simplify matters we will take all λ_a to be equal to unity. (Later we will relax this assumption.)

Define the following functions:

$$m(X) = \sum_a m^{(a)}(X) = \frac{1}{2}[m^R(X) + m^A(X)], \tag{8.42}$$

$$\phi(X) = m^R(X)m^A(X), \qquad m_k \equiv m_{,k}, \dots, \tag{8.43}$$

$$N(X) = \sum_a \int \delta_4(X, A)[-g(X)]^{-1/2} \, ds_a. \tag{8.44}$$

As in the electromagnetic case, we have chosen the symmetric (half R and half A) Green function. The gravitational equations then become

$$R_{ik} - \frac{1}{2}g_{ik}R = -6\phi[T_{ik} - \frac{1}{6}(g_{ik} \Box \phi - \phi_{;ik})$$
$$- \frac{1}{2}(m_i^R m_k^A + m_k^R m_i^A - g_{ik}g^{pq}m_p^R m_q^A)], \tag{8.45}$$

together with the 'source' equation for $m(X)$

$$\Box m + \frac{1}{6}Rm = N. \tag{8.46}$$

The derivation leading to the final set of equations of the theory may appear somewhat long-winded to anybody unfamiliar with the techniques of direct interparticle action. We have followed here the method used by Hoyle and the author, who arrived at this theory via their earlier work on electromagnetism. As in the electromagnetic case, the universe responds to a local event. To ensure causality and to eliminate advanced effects, the correct response should be given by

$$\sum_a m^{(a)\mathrm{A}}(\mathrm{X}) = \sum_a m^{(a)\mathrm{R}}(\mathrm{X}) = m(\mathrm{X}). \qquad (8.47)$$

Under these conditions the equations (8.45) further simplify to

$$R_{ik} - \frac{1}{2}g_{ik}R = -\frac{6}{m^2}\Bigl[T_{ik} - \frac{1}{6}(g_{ik}\,\Box\, m^2 - m^2_{;ik})$$
$$- \Bigl(m_i m_k - \frac{1}{2}g_{ik}m^l m_l\Bigr)\Bigr] \qquad (8.48)$$

Had we adopted the standard field-theoretical approach and introduced a scalar inertia field $m(\mathrm{X})$, we could have arrived at (8.46) and (8.48) from the action given by

$$\mathcal{A} = \int \Bigl(\frac{1}{12}Rm^2 - \frac{1}{2}m^i m_i\Bigr)\sqrt{-g}\,\mathrm{d}^4x - \sum_a \int m\,\mathrm{d}s_a. \qquad (8.49)$$

The action-at-a-distance approach, although it is unfamiliar to a typical theoretical physicist, is useful in that it gives direct expression to Mach's principle. The physical interpretation of the field-theoretical term (8.49) is not so easy to see. For this reason, we have discussed the former approach at some length. Notice that, in the former approach, our action (8.41) contained only the last term of (8.49), but there m was made up of non-local two-point functions. Here m is a straightforward field with sources in matter whose dynamical properties are defined through the first term in the above action.

Since the property of conformal invariance was used in the formulation of the theory, we expect the final equations (8.48) and (8.49) to exhibit conformal invariance. This expectation is borne out. If (g_{ik}, m) are a solution of these equations, then so are

$$\bar{g}_{ik} = \Omega^2 g_{ik}, \qquad \bar{m} = \Omega^{-1}m. \qquad (8.50)$$

Thus, apart from the coordinate invariance of general relativity, this theory also exhibits conformal invariance.

We saw in Exercise 33 of Chapter 2 that the coordinate invariance of the action leads to a conservation law for the energy momentum tensor. In this case the conformal invariance of the action leads to a vanishing of the trace of the field equations. It may be easily verified that the trace of (8.48) vanishes in view of (8.46). The vanishing of the trace represents the fact that the problem is underdetermined. Just as the vanishing of $T^{ik}_{;k}$ in general relativity shows that more solutions can be generated from any given solution by coordinate transformations, so we can generate more solutions through (8.50). All these solutions are physically equivalent, provided that we stick to the rule that Ω does not vanish or become infinite.

Suppose that we are allowed to choose an Ω in the above range that ensures that

$$\bar{m} = \Omega^{-1}m = \text{constant} = m_0. \tag{8.51}$$

This choice of Ω is possible provided that m does not vanish or become infinite. This conformal frame is called the Einstein frame, in which we get a simplified form for (8.48):

$$R_{ik} - \tfrac{1}{2}g_{ik}R = -\kappa T_{ik}, \tag{8.52}$$

with the constant κ given by

$$\kappa = 6/m_0^2. \tag{8.53}$$

Thus we have arrived at Einstein's equations! At first sight we don't seem to have gained anything. We have no new theory and hence no new predictions, as in the Brans–Dicke theory. Closer examination, however, reveals several ways in which this theory goes beyond relativity.

1. Our starting point was based on Mach's principle. It is only in the many-particle approximation, for which the response condition (8.47) is satisfied, that we arrive at the final Einstein-like field equations. An empty universe in relativity is given by

$$R_{ik} = 0,$$

which can have well-defined spacetimes as solutions. Test particles in such spacetimes will have well-defined trajectories. Such trajectories would not make any sense according to Mach, since we no longer have a material background against which to measure the motion of these particles. These solutions in fact correspond to the $\mathbf{f} = \mathbf{0}$ solutions of the Newtonian theory. In the HN theory an empty universe corresponds to

$$m = 0, \quad \text{indeterminate } g_{ik},$$

in accord with the Machian $m = 0$ solution of (8.3).

2. The sign of κ is fixed arbitrarily in general relativity. Neither in the heuristic derivation by Einstein nor in the Hilbert action principle is κ required to be positive. It is only when κ is determined by reference to Newtonian gravity in the weak-field approximation (see §2.9) that we conclude that $\kappa < 0$. In the HN theory (8.52) shows that κ must necessarily be positive. (This conclusion does not depend on our assumption of $\lambda_a = 1$; the result follows whatever sign the λ_a are given.)

3. In the direct interparticle approach described above, it is apparently not possible to accommodate the λ-term of cosmic repulsion. Thus Occam's razor automatically comes into play. In relativity the λ-term is still possible. However, we will see in the next chapter that, if we relax the condition that the inertial fields m satisfy linear wave equations, by permitting non-linearity, then the HN cosmology also permits a cosmological constant.

4. The transition from (8.48) to (8.52) is possible provided that $0 < \Omega < \infty$. What happens if we break this rule? Suppose that in the solution of (8.48) we had a hypersurface on which $m = 0$. If we insist on the transformation (8.51) in a region that contains such a hypersurface, we have to pay the price of $\Omega \to 0$, by admitting spacetime singularities. The work of A. K. Kembhavi in 1979 showed that the well-known cases of spacetime singularities of relativity arise because of the occurrence of zero-mass hypersurfaces in the solution of the equations (8.48). For a simple example of this conclusion let us look at the standard big-bang singularity of relativity.

Consider the Minkowski line element (with $c = 1$)

$$ds^2 = d\tau^2 - dx^2 - dy^2 - dz^2 \tag{8.54}$$

as a solution of (8.48). It is easily verified that the mass function satisfying both (8.46) and (8.48) for a uniform number density N of particles is

$$m \propto \tau^2. \tag{8.55}$$

This is the simplest possible cosmological solution in this theory.

If we now insist on going over to a frame with constant mass $\bar{m}$, then, from (8.51), we see that the appropriate Ω must be given by

$$\Omega \propto \tau^2. \tag{8.56}$$

However, Ω vanishes on the hypersurface $m = 0$. The transformation to the Einstein conformal frame is therefore 'illegal'. The price paid for insisting that $\bar{m} = $ constant is that the resulting model has a geometrical singularity at $\tau = 0$. In fact it is easily verified that the new model is none other than the singular Einstein–de Sitter model. (Make the time transformation $\tau \propto t^{1/3}$ to demonstrate this result explicitly.)

5. It is instructive to see how the phenomenon of the Hubble redshift is explained in the flat-spacetime model of (8.54) and (8.55). Clearly, a photon of light travelling in Minkowski spacetime does not undergo a redshift. Consider, however, what happens to a photon of light arriving at the observer during the present epoch τ_0 from a galaxy at a distance r. This photon originated in an atomic (or molecular) transition at time $\tau_0 - r$.

From atomic physics, the wavelength of a photon so transmitted varies inversely as the mass of the electron making the atomic transition. From (8.55) we see that, if λ is the wavelength of this photon and λ_0 the wavelength of a photon emitted in a similar transition at τ_0 at the observer, then

$$1 + z \equiv \frac{\lambda}{\lambda_0} = \frac{m(\tau_0)}{m(\tau_0 - r)} = \frac{\tau_0^2}{(\tau_0 - r)^2}. \tag{8.57}$$

Thus the redshift in the above HN cosmology arises from the variation of particle masses. Figure 8.11 illustrates this effect.

6. A variable gravitational constant arises in the HN cosmologies if we relax the assumption that λ_a are constants. If λ_a change with time it is possible to generate cosmological models in which G changes with time. We will not discuss such models in detail. The result may be stated in the form

$$\dot{G}/G = -\beta H, \tag{8.58}$$

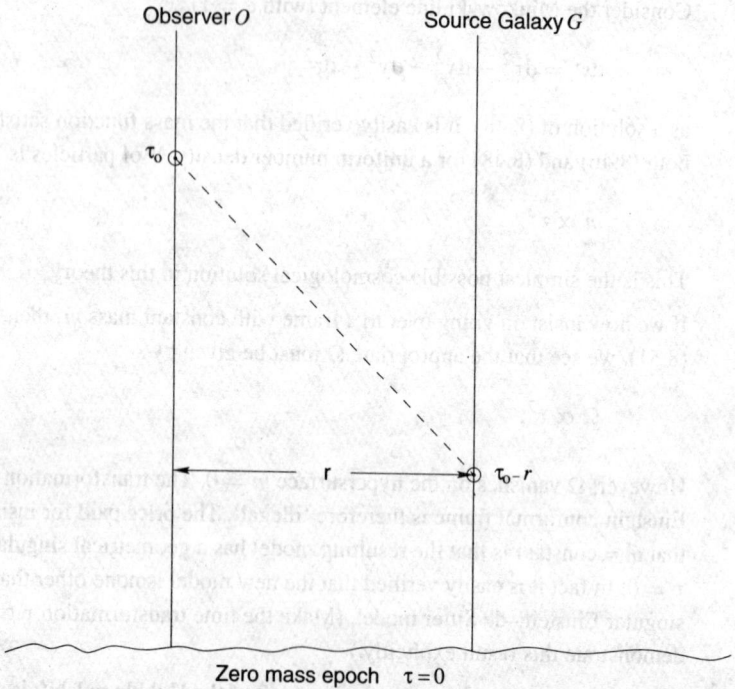

Figure 8.11 The world line of the observer (O) and that of the observed galaxy (G) are shown in Minkowski spacetime. The ray of light connecting the two is along the broken line. When the ray was emitted the masses of particles in G were smaller than those at O at the time the ray reaches O. Hence, by O's reckoning, the light from G has longer wavelengths. (The wavelength emitted is inversely proportional to the emitting mass.)

where H is the Hubble constant of the epoch of measurement and β is a constant of the order of unity.

It was shown by Hoyle and the author in 1972 that λ_a increasing with time may be interpreted as creation of new particles in the universe. They did not give a dynamical theory for the creation of matter (like the C-field theory described in Chapter 9), but instead fixed the time dependence of λ_a by an appeal to the large-numbers hypothesis. We next describe this hypothesis and its implications for cosmology.

8.7 The large-numbers hypothesis

Physics is riddled with units of various kinds and with experimentally determined quantities of various magnitudes. From this vast collection certain constants emerge as having special significance in the framing of basic physical laws; for example, the constant of gravitation G and the charge of the electron e. The numbers expressing the magnitudes of G, e and so on depend on the units used. For example

$$e = 4.803\,25 \times 10^{-10} \text{ electrostatic units},$$

$$= 1.602\,07 \times 10^{-20} \text{ electromagnetic units}.$$

Clearly these numbers by themselves cannot have absolute significance.

However, certain combinations of these physical constants have no units at all. For example, the combination of $\hbar$, c and e

$$\frac{\hbar c}{e^2} = 137.036\,02 \tag{8.59}$$

does not depend on the units used. It must therefore express some physical fact of absolute significance. Indeed, its reciprocal $e^2/(\hbar c)$, known commonly as the *fine-structure constant*, expresses the strength of the electromagnetic interaction, which we believe to be an intrinsic property of nature. In future, a more complete theory may well give a reason why this constant has this particular value.

Given e, G and the masses of the proton and the electron m_p and m_e, we can construct another dimensionless constant (that is, a constant with no units):

$$\frac{e^2}{Gm_p m_e} = 2.3 \times 10^{39} \sim 10^{40}. \tag{8.60}$$

This constant measures the relative strength of the electric and gravitational force between the electron and the proton. Like (8.59) this constant reflects an intrinsic property of nature. However, unlike (8.59), the constant in (8.60) is enormously large! Why such a large number?

Perhaps the appearance of a large dimensionless constant might be dismissed as some quirk on the part of nature. The mystery deepens, however, if we consider another dimensionless number. This is the ratio of the length scale associated with the universe, c/H_0, and the length associated with the electron, $e^2/(m_e c^2)$. This ratio is

$$\frac{m_e c^3}{e^2 H_0} = 3.7 \times 10^{40} h_0^{-1} \sim 10^{40}. \tag{8.61}$$

Not only do we have another large dimensionless number in (8.61), but it is of the same order as in (8.60).

We can generate another large number of special significance out of particle physics and cosmology. Assuming the closure density ρ_c, let us calculate the number of particles in a Euclidean sphere of radius c/H_0, the mass of each particle being m_p. The answer is

$$N = \frac{4\pi}{3m_p} \left(\frac{c}{H_0} \right)^3 \frac{3 H_0^2}{8\pi G} = \frac{c^3}{2 m_p G H_0}$$

$$= 4 \times 10^{79} h_0^{-1}$$

$$\sim 10^{80}. \tag{8.62}$$

Thus, taking N as a standard we see that the large dimensionless numbers of (8.60) and (8.61) are both of the order of $N^{1/2}$.

Reactions among physicists have varied as to the significance of all these numbers. Some dismiss it as a coincidence with the rejoinder 'So what?'. Others have read deep significance into these relationships. The latter class includes such distinguished physicists as A. S. Eddington and P. A. M. Dirac (see Figure 8.12).

Dirac pointed out in 1937 that the relationships (8.61) and (8.62) contain the Hubble constant H_0 and therefore the magnitudes computed in these formulae vary with the epoch in the standard Friedmann model. If so, the near equality of (8.60) and (8.61) has to be a coincidence of the present epoch of the universe, unless the constant (8.60) also varies in such a way as to maintain the state of near equality with (8.61) for all epochs. With this proviso, the equality of (8.60) and (8.61) is not coincidental but is characteristic of the universe *for all epochs*. The proviso also implies that at least one of the so-called constants involved in (8.60), e, m_p, m_e and G, must vary with the epoch.

This proviso was generalized by Dirac to what he called the *large-numbers hypothesis* (LNH). To understand this hypothesis we rewrite the ratio (8.61) as that between the time scale associated with the universe, $\tau_0 = H_0^{-1}$, and the time taken by light to travel a distance of the order of the classical electron radius, $t_e = e^2/(m_e c^3)$. The LNH then states that any large number that in the present epoch is expressible in the form

$$\left(\frac{\tau_0}{t_e} \right)^k$$

where k is of the order of unity, varies with the epoch t as $(t/t_e)^k$ with a constant of proportionality of the order of unity.

When it is applied to (8.60), therefore, the LNH implies that the ratio $e^2/(Gm_p m_e)$ must vary as $(t/t_e)^{-1}$. Dirac made a distinction between e, m_e and m_p on one side and G on the other in the sense that the former are atomic (microscopic quantities) whereas G has macroscopic significance. In the Machian cosmologies, G is in fact related to the large-scale structure of the universe. Dirac therefore assumed that, if we use 'atomic units' that always maintain fixed values for atomic quantities, then t_e will be constant and $G \propto t^{-1}$. That is, in terms of atomic time units, the gravitational constant must vary with the epoch t, with $|\dot{G}/G| \sim H$.

We will now explore the implications of the LNH for cosmology.

8.8 The two metrics

Clearly the variation of G predicted by the LNH goes against Einstein's theory of gravitation, which demands a constant G. As in the Brans–Dicke theory, we are forced to modify the relativistic framework to accommodate a varying G. Dirac approached this problem in the following way.

First he took note of the many Solar-System tests that are in favour of general relativity (see Chapter 2) and argued that the theory should not be abandoned altogether. Instead, Dirac proposed two scales of measurement, one holding in atomic physics and the other in macroscopic gravitation physics. If we choose the atomic

Figure 8.12 P. A. M. Dirac (1902–1984). Reproduced with permission from the website of Bob Bruen.

system, we will be able to describe atomic physics in the usual way, that is, with constant values for the atomic constants like e, $\hbar$, m_e, m_p, and so on. However, in this system G will be variable, since Dirac considered it a constant belonging to gravitation physics. If, on the other hand, we use gravitational units, then, according to Dirac, G will be constant and atomic quantities will be found to be variable; and in the latter units the gravitational phenomena can be described by the Einstein equations (2.100).

These two units can be specified in Dirac's framework by having two different spacetime metrics. We will denote these by ds_A^2 and ds_E^2, respectively, for the atomic and the gravitational systems (the subscript E in the latter case committing us to Einstein's equations of gravity). We will use these subscripts in general on any physical quantity to indicate what system of measurement is being used. Thus, according to Dirac,

$$G_E, \qquad (m_e)_A, \qquad (m_p)_A$$

are constants, whereas

$$G_A, \qquad (m_e)_E, \qquad (m_p)_E$$

are variable.

Returning to the astronomical tests of general relativity, we note that the mass of the gravitating body (for example, the Sun) occurs in the Schwarzschild solution. Clearly this mass, which is the gravitational mass, must be a constant in the gravitational units. We denote this mass by M_E. Any measurements made on the Earth, however, use atomic systems (such as spectrometers and atomic clocks), so before we interpret any experimental result we must make sure that all observable quantities are transformed to atomic units.

This argument tells us how necessary it is to know the ratio

$$\beta = \frac{ds_E}{ds_A} \tag{8.63}$$

and how the transformation of any physical quantity from one system of units to another is to be made. Here we need a quantitative theory to guide us, a theory that goes further than the above qualitative arguments have so far taken us.

We also note another outcome of our Solar-System example. If we assume that our astronomical body has N_E nucleons, each of mass m_E, then we may write

$$M_E = m_E N_E = m_E N, \tag{8.64}$$

where we have dropped the suffix E on N because it is a pure number. Whatever metric we use, we will count the same number of particles in the gravitating body. In (8.64) we have M_E = constant, but $m_E \neq$ constant, since the latter is an atomic quantity. Thus $N \neq$ constant. In other words, we are forced to conclude that the number of nucleons in the body must change with time. Again we need a quantitative theory to tell us how N changes; but creation (or destruction) of nucleons in a macroscopic object is demanded by Dirac's argument.

So far we have not used the LNH, which started us on the two-metric theory. Let us now see how it helps us in deciding how the non-conservation of the number of nucleons in the body is regulated.

8.8.1 The creation of particles

If we go back to (8.64) and apply the LNH to N, we easily find that $k = 2$, that is,

$$N(t) \sim \left(\frac{t}{t_e}\right)^2 \propto t^2. \tag{8.65}$$

In other words, the number of particles in the universe in the sense defined in §8.7 increases with t. Dirac has taken this result to imply that particles are being continually created in the universe.

The creation can occur, according to Dirac, in two possible ways. In *additive creation* the particles are created uniformly throughout space; whereas in *multiplicative creation* the new particles occur preferentially where matter already exists. Thus, in the former mode, creation occurs mostly in intergalactic space, whereas in the latter mode creation occurs mostly in the vicinity of existing astronomical objects.

Using these ideas, we return to (8.64). In additive creation the astronomical body will not acquire any significant number of new particles and thus $N = $ constant, giving

$$m_E = \text{constant} \qquad \text{(additive creation)}. \tag{8.66}$$

In multiplicative creation N must increase as t^2 and hence

$$m_E \propto t^{-2} \qquad \text{(multiplicative creation)}. \tag{8.67}$$

8.8.2 The determination of β

The connection between ds_A and ds_E can be fixed by considering the motion of a planet (such as the Earth) around a star (the Sun). The dynamical equation in the Newtonian approximation is

$$GM = v^2 r, \tag{8.68}$$

where M is the mass of the star, v is the speed of the planet and r is the radius of the orbit. The above relation is expected to hold in either of the two systems of units, since $GM/(v^2 r)$ is a dimensionless quantity. Also, with $c = 1$ the speed v is dimensionless. Thus $v = $ constant in either set of units. Next, in gravitational units $M_E = $ constant and $G_E = $ constant, hence $r_E = $ constant.

If (8.68) is used with atomic units, we have

$$G_A \sim t^{-1}.$$

(8.69)

Also, in multiplicative creation $M_A \propto t^2$ whereas for additive creation $M_A =$ constant. Hence, in these respective units,

$$r_A \sim \begin{cases} t & \text{(multiplicative creation)}, \\ t^{-1} & \text{(additive creation)}. \end{cases}$$

(8.70)

Thus we have

$$\frac{r_A}{r_E} \sim \begin{cases} t & \text{(multiplicative creation)}, \\ t^{-1} & \text{(additive creation)}. \end{cases}$$

(8.71)

In other words, measured in atomic units, the distance of the planet from the star *increases* with t if the universe has multiplicative creation of matter, whereas the distance *decreases* with t (as t^{-1}) for additive creation.

From (8.70) and (8.71) we get the behaviour of β defined in (8.63). This ratio of ds_E to ds_A behaves as t^{-1} or t, depending on whether we have multiplicative creation or additive creation in the universe.

8.9   Cosmological models based on the LNH

Using the LNH, Dirac constructed cosmological models for both the circumstances discussed above, namely for multiplicative and additive creation. As in the case of standard cosmologies, the assumptions of homogeneity and isotropy lead us to the Robertson–Walker line element in atomic units:

$$ds_A^2 = c^2\, dt^2 - S^2(t)\left(\frac{dr^2}{1 - kr^2} + r^2(d\theta^2 + \sin^2\theta\, d\phi^2) \right).$$

(8.72)

How does the LNH determine k and $S(t)$? We reproduce below the argument given by Dirac.

First we note that the metric proper distance at time t between a galaxy G at $r = 0$ and a galaxy at $r = r_1$ is given by

$$d = S(t) \int_0^{r_1} \frac{dr}{\sqrt{1 - kr^2}} \equiv S(t) f(r_1).$$

(8.73)

According to the LNH, for large t (that is, for $t \gg t_e$) the expression for $S(t)$ should be $\sim (t/t_e)^n$ or $\sim \ln(t/t_e)$. The (metric) velocity of recession corresponding to (8.73) will therefore be given by

$$\dot{d} \sim n t_e^{-n} f(r_1) t^{n-1} \quad \text{or} \quad \dot{d} \sim t^{-1} f(r_1).$$

(8.74)

The constants multiplying $(t/t_e)^n$ or $\ln(t/t_e)$ in $S(t)$ must be of the order of unity and hence the constants implied in the relation $(\sim)$ above are also of the order of unity. It is then easy to verify that, except for $n = 1$, there exists an epoch either in the past (for $n < 1$ or for $S \sim \ln t$) or in the future (for $n > 1$) when $\dot{d} = c$ for any galaxy with $r_1 > 0$. For example, for $n = \frac{1}{2}$ we find that, for a galaxy that at present has $\dot{d} \sim 10^{-3}c$, the condition $\dot{d} = c$ occurred in the past epoch given by

$$t_{\mathrm{p}} = \left(\frac{\tau_0}{t_{\mathrm{e}}}\right) \times 10^{-6} t_{\mathrm{e}} \sim 10^{34} t_{\mathrm{e}}.$$

That is, $t_{\mathrm{p}}/t_{\mathrm{e}}$ is a large number. However, by the LNH, t_{p} is a constant epoch when a significant event took place for galaxy G_1: its speed of recession became equal to c. Hence such a constant epoch should not generate a large number. Therefore only the case

$$S(t) \sim (t/t_{\mathrm{e}}) \tag{8.75}$$

is permitted by the LNH.

The arguments given above could be criticized on the following grounds. The epoch when $\dot{d} = c$ is not unique to the model as a whole; it depends on $f(r_1)$ and hence on the galaxy chosen. So it is not necessary that the LNH should apply to this epoch. Neither is it clear why $\dot{d} = c$ should be considered significant. Nothing special happens to the galaxy in question when its metric velocity of recession becomes equal to c for the observer at $r = 0$. No global property like the event horizon or the particle horizon enters the argument.

Nevertheless, if we follow the argument further, then we can write our cosmological line element as

$$\mathrm{d}s_{\mathrm{A}}^2 = c^2\,\mathrm{d}t^2 - (At)^2 \left(\frac{\mathrm{d}r^2}{1 - kr^2} + r^2(\mathrm{d}\theta^2 + \sin^2\theta\,\mathrm{d}\phi^2)\right), \tag{8.76}$$

where A is a constant. We next consider multiplicative creation. Since in this case, from §8.8,

$$\mathrm{d}s_{\mathrm{E}} \equiv t^{-1}\,\mathrm{d}s_{\mathrm{A}}, \tag{8.77}$$

it is easy to see that a transformation

$$\mathrm{d}t_{\mathrm{E}} = \mathrm{d}t/t \tag{8.78}$$

gives us

$$\mathrm{d}s_{\mathrm{E}}^2 = c^2\,\mathrm{d}t_{\mathrm{E}}^2 - A^2 \left(\frac{\mathrm{d}r^2}{1 - kr^2} + r^2(\mathrm{d}\theta^2 + \sin^2\theta\,\mathrm{d}\phi^2)\right). \tag{8.79}$$

Now we recall that the above line element must be a solution of Einstein's equations. In Chapter 3 we did obtain such a static solution for homogeneous and isotropic dust with the use of the λ-term (see §3.3), namely, the Einstein universe with $k = +1$. With a suitable scaling of the r coordinate we can express (8.79) in the form (3.18). Notice, however, that, unlike the Einstein universe, this Dirac universe does exhibit the phenomenon of a redshift of galaxies. For redshift measurements involve comparisons of the rates at which atomic clocks run in the emitting and receiving galaxies; and for such comparisons the line element (8.76) instead of (8.79) must be used.

For additive creation the situation is more complicated. In the case of multiplicative creation the gravitational mass of an astronomical object was held constant in the gravitational units in spite of creation of new particles, by letting the masses of particles decrease with time. In the case of additive creation the masses of particles remain constant even though their number increases (see (8.65)). Dirac was therefore faced with an apparent non-conservation of energy. To conserve energy Dirac proposed that, together with particles of positive mass, an equal number of particles of negative mass is also created. The negative mass distribution is homogeneous and remains undetectable by standard astronomical observations. In a completely homogeneous situation the positive and negative mass distributions compensate gravitationally to produce flat Minkowski spacetime. The formation of stars and galaxies by the accumulation of particles of positive mass in the actual universe is a result of small departures from this completely homogeneous situation.

It is worth pointing out that, when Dirac first proposed a cosmological model based on the LNH between 1937 and 1938, he assumed no creation of matter. In this model the number of particles per unit coordinate volume was constant, as in standard cosmologies. Hence the number of particles per unit proper volume went as S^{-3} and, since the proper volume of the universe goes as $(c/H)^3$, the number of particles in the universe denoted earlier by N would be

$$S^{-3}\left(\frac{c}{H}\right)^3 \propto (\dot{S})^{-3}.$$

However, by the LNH we know that

$$N \propto t^2. \tag{8.80}$$

Therefore we have

$$t^2 \dot{S}^3 = \text{constant};$$

that is,

$$S \propto t^{1/3}. \tag{8.81}$$

Thus for no creation of particles S increases much more slowly with t. (Of course, this solution is ruled out if we apply the LNH to the function S, as we did in the beginning of this section.)

8.9.1 HN cosmology revisited

Some of the ideas of Dirac are found in a version of HN cosmology proposed by its authors between 1971 and 1972. In the HN cosmology we considered the cases in which λ_a, λ_b, ..., the constants that denote the strength of the inertial interaction, are true constants. If, however, these constants vary with time, new cosmological models emerge. In these models the following properties hold: (1) there is creation of particles during all epochs in such a way that the LNH is satisfied, (2) in atomic units G varies; whereas (3) in the gravitational units G is constant and masses of particles vary. Thus this model is like the multiplicative-creation model later proposed by Dirac, although its motivation and quantitative details were different. We briefly illustrate how this model works.

Consider a homogeneous and isotropic Minkowski universe given by

$$ds_M^2 = d\tau^2 - dr^2 - r^2(d\theta^2 + \sin^2\theta \, d\phi^2), \tag{8.82}$$

where we have put $c = 1$ for convenience. Let $n(\tau)$ be the number density of particles and $\lambda(\tau)$ the time-varying inertial coupling constant of (8.34). The functions $n(\tau)$ and $\lambda(\tau)$ vary in such a way as to compensate for each other's effect; that is, to maintain

$$\lambda n = \text{constant}. \tag{8.83}$$

Thus the mass function $m(\tau)$ is the same as if we had a universe of uniform number density of particles $n = \text{constant}$ and fixed λ. As in (8.55) we then get

$$m(\tau) \propto \tau^2. \tag{8.84}$$

Since $Gm^2 = \text{constant}$, we get the gravitational constant in the Minkowski framework as

$$G_M \propto \tau^{-4}. \tag{8.85}$$

The mass of a typical particle is not, however, $m(\tau)$ but $\lambda m(\tau)$. To determine it we need to know $\lambda(\tau)$. Hoyle and the author determined $\lambda(\tau)$ from the requirement that the universe is a perfect absorber to electromagnetic radiation along the future light cone in the sense described in §8.5.1 so that the electromagnetic signals propagate via the retarded solutions in this universe.

This requirement fixes $\lambda(\tau) \propto \tau^{-1}$ and $n(\tau) \propto \tau$. It is then verified that the LNH is incorporated by the fact that the dimensionless number

$$\lambda^2(\tau^3 n)^{1/2} \equiv \text{constant} = \mathcal{O}(1). \tag{8.86}$$

A conformal transformation

$$ds_E = \Omega_E \, ds_M, \qquad \Omega_E \propto \tau^2, \tag{8.87}$$

then takes us to the gravitational framework in which $G_E = \text{constant}$. Also, the gravitational mass of an astronomical body remains constant. Thus, as in Dirac's multiplicative-creation theory, the local Solar-System tests give the same answer as in relativity.

To transform to the atomic framework we need another conformal transformation:

$$ds_A = \Omega_A \, ds_M, \qquad \Omega_A \propto \tau. \tag{8.88}$$

Writing $t \propto \tau^2$ means that the line element now becomes

$$ds_A^2 = dt^2 - 2H_0 t[dr^2 + r^2(d\theta^2 + \sin^2\theta \, d\phi^2)]. \tag{8.89}$$

In this framework the gravitational constant varies as

$$G_A \propto t^{-1}. \tag{8.90}$$

There is therefore considerable similarity between this theory and the model proposed by Dirac a few years later.

8.10 Concluding remarks

This brings us to the end of our brief excursion through some of the better-known parts of alternative cosmologies. Our survey is by no means exhaustive. We have not discussed such important models as the matter–antimatter symmetric cosmology of Alfven and Klein, the Einstein–Cartan cosmologies and Milne's kinematical relativity; nor have we discussed such unusual ideas as Segal's chronometric cosmology and McCrea's notion of cosmological uncertainty.

Our purpose here was to summarize a few non-standard cosmologies, which begin with basic philosophical ideas quite different from general relativity, viz., Mach's principle and the large-numbers hypothesis. In the next chapter we will consider other motivations related to the creation of matter that have led to new cosmological models. Thereafter we will turn to a very important part of the exercise, that of testing theories against observations.

Exercises

1 Discuss how inertial forces arise in Newtonian dynamics. A stone tied to a string is whirled around in a circle. How can the motion of the stone be understood in terms of inertial forces?

2 What observation led Mach to formulate his famous principle?

3 Why is it unsatisfactory to conclude from $mf = 0$ for a particle in an otherwise empty universe that $f = 0$? Interpret any other conclusion that could be drawn from the above equation.

4 Set up the problem corresponding to that described in Exercise 3 in general relativity, namely, the case of a test particle moving in an empty universe. Does this theory provide a satisfactory solution to the problem?

5 Construct a mass unit from the fundamental constants c, $\hbar$ and G that could be used as a standard to decide whether masses of particles change with the epoch. Under what circumstances can we assert that G is changing with the epoch?

6 Give the qualitative argument of Brans and Dicke leading to the conclusion that G^{-1} satisfies a scalar wave equation with sources in matter.

7 Derive the field equations of Brans–Dicke theory from an action principle. Why is the theory called a scalar-tensor theory?

8 Show that, in the approximation $\omega \gg 1$, the wave equation satisfied by ϕ gives a solution

$$\phi = \text{constant} + O\left(\frac{1}{\omega}\right).$$

Interpreting the constant as proportional to G^{-1}, show that the Brans–Dicke field equations take the form

$$R_{ik} - \frac{1}{2}g_{ik}R = -\frac{8\pi G}{c^4}T_{ik} + O\left(\frac{1}{\omega}\right).$$

9 In Newtonian gravity an oblate Sun will generate a gravitational potential

$$\phi = \frac{GM_\odot}{r}\left[1 - J\left(\frac{R_\odot}{r}\right)^2 P_2(\cos\theta)\right],$$

where J is the quadrupole-moment parameter and P_2 is the second Legendre polynomial. Show that the orbit of a planet precesses because of the above gravitational effect at the rate $3\pi R_\odot^2 J/l^2$, where l is the semi latus rectum of the orbit. Estimate the rate of precession for Mercury for $J = 2.5 \times 10^{-5}$. What significance does this calculation have for the Brans–Dicke theory?

10 Discuss the Solar-System tests of the Brans–Dicke theory.

11 Calculate the age of a Brans–Dicke universe for the simplest case $C = 0$, $p = 0$ and $k = 0$. Does this model have a greater or a smaller age than that in the corresponding relativistic model?

12 Show that, for a radiation universe in Brans–Dicke cosmology with $C = 0$, we have $S \propto t^{1/2}$ and $\phi = $ constant. Comment on why in Brans–Dicke cosmology this case gives exactly the same answer as relativistic cosmology.

13 Show that the inequality (8.21) is satisfied for a dust universe as well as for a radiation universe in Brans–Dicke cosmology with $C \neq 0$.

14 Derive the behaviours of S and ϕ as functions of t in the early ϕ-dominated Brans–Dicke universe.

15 Discuss primordial nucleosynthesis in the Brans–Dicke cosmology.

16 The Brans–Dicke theory can be re-expressed as a theory in which $G = $ constant but the masses of particles change with the epoch. Show that this is achieved by a conformal transformation

$$\bar{g}_{ik} = \frac{\phi}{\bar{\phi}} g_{ik}, \qquad \bar{\phi} = \text{constant.}$$

The field equations then become (in the new metric)

$$\bar{R}_{ik} - \tfrac{1}{2} \bar{g}_{ik} \bar{R} = -\kappa \bar{T}_{ik},$$

where κ is constant. Although these look like Einstein's equations, the $\bar{T}_{ik}$ contain ϕ and its derivatives. Show from the new field equations that

$$\bar{\Box} \ln \phi = \frac{8 \pi G}{(2\omega + 3)c^4} \bar{T}$$

with $G = $ constant. This form of the theory was obtained by Dicke in 1962. The masses of particles in this version vary as

$$\bar{m} = m \sqrt{\frac{\bar{\phi}}{\phi}}, \qquad m = \text{constant.}$$

17 Show that, in a ϕ-dominated Brans–Dicke cosmology, it is possible to have an increasing gravitational constant at an epoch t provided that

$$\int_0^t (\epsilon - 3p) S^3 \, dt < -c^2 \left(\frac{2\omega + 3}{8\pi} \right) C.$$

18 Illustrate the difference between a field theory qualitatively and an action-at-a-distance theory by an example from electrodynamics.

19 Verify that, in Minkowski spacetime, the electromagnetic Green function has the simple form

$$\bar{G}_{ik} = \frac{1}{4\pi} \delta(s^2) \eta_{ik},$$

where s^2 is the invariant square of the distance between the two world points at which G_{ik} is defined.

20 Use the Green function of Exercise 19 to derive the potential for a static electric charge.

21 Show how the definition of mass in the HN theory satisfies Mach's principle.

22 Show by a time transformation that Robertson–Walker spacetime with $k = 0$ is conformal to flat (Minkowski) spacetime.

23 Show, with the help of the following series of transformations, that the $k = +1$ Robertson–Walker spacetime is conformally flat:

$$r = \sin R, \qquad T = \int^t \frac{du}{S(u)}, \qquad c = 1,$$

$$\xi = \tfrac{1}{2}(T + R), \qquad \eta = \tfrac{1}{2}(T - R),$$

$$\tau = \tfrac{1}{2}(\tan \xi + \tan \eta), \qquad \rho = \tfrac{1}{2}(\tan \xi - \tan \eta).$$

What are the corresponding series of transformations to show that the $k = -1$ Robertson–Walker models are also conformally flat?

24 Show that the following tensor is conformally invariant:

$$C^h_{ijk} = R^h_{ijk} + \tfrac{1}{2}(g^h_j R_{ik} - g^h_k R_{ij} + g_{ik} R^h_j - g_{ij} R^h_k) + \tfrac{1}{6} R(g^h_k g_{ij} - g^h_j g_{ik}).$$

This tensor is known as the *Weyl conformal curvature tensor*.

25 Show that a null geodesic is invariant under conformal transformations.

26 Explain why conformal invariance should play an important role in action-at-a-distance theories.

27 Show that Maxwell's equations remain unchanged under a conformal transformation provided that the potential and the field transform as

$$\bar{A}_i = A_i + \psi_{;i}, \qquad \psi \text{ a suitable scalar,}$$

$$\bar{F}_{ik} = F_{ik}.$$

28 Verify by direct substitution that $\tilde{G}(A, B)$ defined by (8.39) does satisfy the conformal transform of the wave equation (8.40).

29 Use the conformal flatness of the Einstein–de Sitter model to calculate the explicit form of $\tilde{G}(A, B)$ in that universe.

30 Suppose that a symmetric Green function $G(A, B)$ satisfies the wave equation

$$\Box_X G(X, B) = [-g(X)]^{-1/2} \delta_4(X, B).$$

Show that a small variation of the metric tensor in a region $\mathcal{V}$ produces a small variation of $G(A, B)$ given by

$$\delta G(A, B) = \int_{\mathcal{V}} \delta(\sqrt{-g} g^{ik}) G^R(A, X)_{,i} G^A(X, B)_{,k} \, d^4 x.$$

(Note that A and B need not lie in $\mathcal{V}$.)

31 Show that the action (8.49) leads to the field equations (8.48).

32 Compare the degree of underdeterminacy of the gravitational equations of the HN theory with that of general relativity.

33 Show that any conformally invariant action leads to an energy tensor of vanishing trace.

34 Discuss the aspects in which the HN theory of gravity differs from general relativity.

35 Construct dimensionless constants from (a) e, $\hbar$ and c, (b) G, m_p, $\hbar$ and c and (c) G, m_p, c and H_0.

36 Which of the dimensionless constants of Exercise 35 are very large or very small?

37 Compute N exactly for the closed Friedmann model with $h_0 = 1$ and $q_0 = 1$. Show that N is constant for all epochs. Can this result be reconciled with the LNH?

38 Find the relation connecting the three large numbers in (8.60), (8.61) and (8.62).

39 Deduce from the LNH that the gravitational constant must decrease with the epoch at a rate such that $\dot{G}/G$ is of the order of Hubble's constant.

40 Give the arguments that led Dirac to postulate particle creation in the universe.

41 Show that, in gravitational units, multiplicative creation demands that masses of particles decrease with time t as t^{-2}.

42 In what way does the difference between additive and multiplicative creation show up in the long-term evolution of planetary orbits? How are the orbital angular speeds of the planets affected by the variation of G?

43 Give the arguments based on the LNH that lead to the conclusion that the scale factor of the expanding universe can be proportional only to the cosmic time. Comment on the plausibility of these arguments and compare them with Bondi and Gold's derivation of the steady-state line element on the basis of the perfect cosmological principle.

44 Derive the formula for the redshift in the Dirac universe with multiplicative creation. Explain how this redshift arises even though the gravitational metric is static.

45 Plot the atomic time t_A against the gravitational time t_E for the Dirac universe with multiplicative creation. Show that, although the Einstein-like universe in the gravitational metric has t_E going to $-\infty$, the atomic time goes only as far back as $t_\mathrm{A} = 0$.

46 Compare and contrast Dirac's cosmological ideas on creation of negative as well as positive mass with his ideas about vacuum as a sea of undetectable negative-energy electrons.

47 Show that, in the Dirac cosmology with no creation of particles, the gravitational constant decreases as

$$\dot{G}/G = -3H.$$

Estimate this rate in terms of the present-day estimate of the Hubble constant. How is this rate modified in the Dirac models with particle creation?

Chapter 9

Cosmologies with creation of matter

9.1   Introduction

We continue our discussion of alternative cosmologies, this time using the creation of matter as the main motivating concept. Given that we see a universe containing matter (and radiation), we should be asking the question: where did it all come from? In the standard big-bang cosmology, the answer is evaded: we are given the initial instant as a singular state of the big bang. We are not able to push our analysis beyond this epoch further into the past, since all mathematics breaks down at the singularity and no extrapolation of known physics is applicable. The best one can say is that the matter somehow came into existence at $t = 0$ and that too with infinite density and infinite energy. If at all, we may extrapolate our known physics arbitrarily close to this instant, but keeping $t > 0$ all the time.

This process has its built-in contradictions. For example, the action principle, which uses the Hilbert action (2.103), added to the action (2.80) to work out the equations of general relativity, assumes that the volume $\mathcal{V}$ is *arbitrarily chosen*. This is how the Einstein equations are derived. Nevertheless, when we solve those equations and derive the standard Friedmann models, we are led to the singular beginning. To preserve the assumptions of the variational principle, therefore, we have to *exclude* the singular epoch from any volume $\mathcal{V}$ chosen in the variational exercise. Thus we are arriving at a self-contradiction in which our conclusion disallows the assumption on which it was based.

Can we alter the basic framework of the big-bang cosmology so as to do away with the singular epoch of creation and make a beginning in our understanding of the primary phenomenon of the creation of matter and energy in the universe? We will now consider models that attempt this ambitious exercise.

317

9.2   The steady-state theory

In 1948, around the same time that George Gamow was initiating detailed studies of the physical properties of the universe close to the big-bang epoch, three
astronomers proposed an entirely new approach to cosmology. This model, now
famous (or notorious!) as the steady-state model, does not have a singular big-bang-
type epoch; indeed, it does not have either a beginning or an end on the cosmic
time axis. The cosmological scene was considerably enlivened for two decades after
the inception of the steady-state model by observers trying to shoot this rival model
down. What was the motivation that led Hermann Bondi, Thomas Gold and Fred
Hoyle (see Figure 9.1) to propose the steady-state cosmology?

First of all, in 1948 the measured value of $\tau_0 \equiv H_0^{-1}$ was only $\sim 1.8 \times 10^9$ years.
Consequently the age of a standard Friedmann model could not exceed τ_0 – a value
lower than even the geological age of the Earth! Thus there was a *prima facie* case
for doubting the conclusion that the universe began ~ 1–1.8 billion years ago.

The second reason has already been stated in the introduction above, namely
the unsatisfactory nature of the 'beginning' of the universe in a singular event.
Moreover, one may pose another philosophical question.

Have we any guarantee, when we study the past history of the universe, that the
physical laws that we use here and now have always remained the same? We could
have assumed this to be the case had the universe itself not changed considerably in
the course of time. This, however, was not the case for the Friedmann universes. A
typical standard model changes considerably in its physical content and properties
from soon after $t = 0$ to the present day (see Chapters 5–7). So the assumption
that the laws of physics have remained unchanged throughout the history of the
standard models is more an article of faith than a verifiable fact. If they have not,
then the 'guesses' and 'extrapolations' used to talk about the very early universe
remain unverifiable hypotheses.

Today, as we shall see in Chapter 10, the age problem is still with us, although
not in such a severe form as the low value of τ_0 in 1948 implied. The questions of
singularity and creation of matter still remain with the standard models: the work
discussed in Chapter 6 does not tell us what happened at $t = 0$. Although one may

Figure 9.1 T. Gold (1920–),
H. Bondi (1919–) and
F. Hoyle (1915–2001).
Courtesy Fred Hoyle.

argue that quantum gravity may ultimately resolve the issues related to the big-bang singularity, the progress to date on that front has not been very satisfactory.

Hoyle's approach to the steady-state theory was designed to attack the problem of the primary creation of matter. His colleagues Bondi and Gold, however, considered the assumption of constancy of physical laws as of paramount importance, since it pertains to the very basics of cosmology.

9.2.1 The perfect cosmological principle

Bondi and Gold argued that the cosmological principle (see Chapter 3) goes some way towards ensuring that the locally discovered laws of physics have universal validity; but it does not go far enough. This principle tells us that, at any given cosmic time t, all fundamental observers see the same large-scale features of the universe. Thus we are justified in assuming that there is no spatial variation in the basic physical laws at any given cosmic time. However, there is no justification from the cosmological principle to assume that the laws remain unchanged *with time*.

To provide such a justification Bondi and Gold strengthened and elevated the cosmological principle to what they called the *perfect cosmological principle* (PCP). The PCP states that, in addition to the symmetries implicit in the cosmological principle, the universe on the large scale is unchanging with time. Thus the geometrical and physical properties of the hypersurfaces $t =$ constant do not change with t.

It is important to emphasize the qualification 'on the large scale'. On a small enough scale the observed part of the universe *will* change. For example, stars in a galaxy will grow older, a small cluster of galaxies will evolve in shape and composition with time, and so on. However, according to the PCP the statistical properties of the large-scale features of the universe do not change.

For example, Hubble's constant should remain the same whether it is measured now or at any time past or present, since its accurate measurement involves the rate of expansion of the universe. This being a property of the large-scale structure of the universe, the constancy of H tells us immediately that

$$H = \dot{S}/S = \text{constant} = H_0, \qquad \text{i.e., } S = \exp(H_0 t). \tag{9.1}$$

Furthermore, the curvature of a $t =$ constant hypersurface is given by k/S^2. This could in principle be measured at different times and found to be changing unless $k = 0$. (See Exercise 4 for another argument leading to $k = 0$.) Thus the PCP leads us to the unique line element

$$ds^2 = c^2 \, dt^2 - e^{2H_0 t}[dr^2 + r^2(d\theta^2 + \sin^2\theta \, d\phi^2)]. \tag{9.2}$$

Notice that we have arrived at the line element of the steady-state universe without having to solve *any* field equations, as we had to do to determine $S(t)$ and k in standard cosmology. Bondi and Gold cited this result as an example of the deductive power of the PCP. Two other examples of deductions from this principle are given. Recall also, that this line element is the same as that obtained by de Sitter for his model in 1917 (Chapter 3).

Expansion of the universe

The line element (9.2) is completely characterized by H_0. It is, however, possible to have $H_0 = 0$, $H_0 < 0$, or $H_0 > 0$, all of which are consistent with the PCP. To resolve this issue, we take account of the local thermodynamic conditions to deduce that $H_0 > 0$, for our observations show that the universe in our local neighbourhood is far from being in a state of thermodynamic equilibrium. Stars radiate; regions of high and low temperatures exist within the Galaxy and outside it. If $H_0 = 0$ we would have a static, infinitely old Euclidean universe. Such a universe should have reached a thermodynamic equilibrium by now, as implied by the Olbers paradox (see Chapter 4). If $H_0 < 0$ we would have a contracting universe in which radiation from distant objects would be blueshifted. Such radiation would lead to an infinite radiation background even worse than that indicated by the calculations of Olbers. Thus our local observations preclude $H_0 \leq 0$, leaving the case $H_0 > 0$, which is consistent with the finite and low night-sky background (see Exercise 7). Hence the universe must expand: a conclusion arrived at with the help of the PCP *without looking at the spectra of any nearby galaxies!*

Creation of matter

It is easily seen that a proper 3-volume V bounded by fixed (r, θ, ϕ) coordinates increases with time as

$$V \propto \exp(3H_0 t),$$
$$\dot{V}/V = 3H_0. \tag{9.3}$$

By the steady-state hypothesis the density of the universe must remain constant at $\rho = \rho_0$. Hence the amount of matter within V must increase in mass $M \equiv V\rho_0$ as

$$\dot{M} = 3H_0 V \rho_0.$$

In other words,

$$J = 3H_0 \rho_0 \tag{9.4}$$

denotes the rate of creation of matter per unit volume. If we use cgs units we get

$$J = 2 \times 10^{-46} \left(\frac{\rho_0}{\rho_c} \right) h_0^3 \ \mathrm{g \ cm^{-3} \ s^{-1}}, \tag{9.5}$$

where ρ_c and h_0 have been defined in Chapters 3 and 4.

The small value of J shows that there is a very slow but continuous creation of matter going on, in contrast to the one-time infinite and explosive creation at $t = 0$ of the standard models.

Attractive though the above deductive approach is, it has its limitations. For example, we do not have a quantitative relation connecting H_0 to, say; the mean density ρ_0 as we have in the Friedmann cosmologies. Neither do we have any physical theory for such an important phenomenon as the continuous creation of matter. Is the sacrosanct law of conservation of matter and energy being violated in the process of creation of matter? Bondi and Gold appreciated the fact that questions like these could be answered through a dynamical theory rather than from their deductive approach. However, they felt that the PCP together with local observations already determines the large-scale properties of the universe in a form that can be tested by observations (see §9.3). Therefore they attached greater importance to the empirical approach of testing the PCP by observations than to formulating a dynamical theory that might determine H_0, ρ_0 and so on quantitatively.

9.2.2 A field theory for creation

Fred Hoyle, on the other hand, took the opposite view. He looked for a field-theoretical process that could account for the phenomenon of primary creation of matter. His 1948 paper had included a modification of the Einstein field equations of general relativity in which the right-hand side contained an explicit energy momentum tensor for a scalar 'creation field'. However, after several attempts at improving the framework, he finally adopted the formulation suggested by M. H. L. Pryce. This formulation, known as the C-field theory, was used extensively by Hoyle and the author in the early 1960s. The highlights of the C-field theory are given below.

The action principle

Like Hoyle's original approach, the C-field theory also involves adding more terms to the standard Einstein–Hilbert action (see §§2.8 and 2.9) to represent the phenomenon of creation of matter. Using Occam's razor, the additional field to be introduced is a scalar field with zero mass and zero charge. We denote this field by C and its derivative with respect to the spacetime coordinate x^i by C_i. The action is then given by

$$A = \frac{c^3}{16\pi G} \int R\sqrt{-g}\, \mathrm{d}^4x - \sum_a m_a c \int \mathrm{d}s_a$$
$$- \frac{1}{2c} f \int C_i C^i \sqrt{-g}\, \mathrm{d}^4x + \sum_a \int C_i\, \mathrm{d}a^i. \tag{9.6}$$

Instead of the electromagnetic terms (which might be present if we had charged particles), we have in (9.6) the C-field terms. To appreciate the difference between the two interactions, note that the last term of (9.6) is path-independent. If we consider the world line of particle a between the end points A_1 and A_2, we have

$$\int_{A_1}^{A_2} C_i \, da^i = C(A_2) - C(A_1). \tag{9.7}$$

Normally such path-independent terms do not contribute to any physics derivable from the action principle. So why include such a term? The answer to this question lies in the notion of 'broken' world lines. A theory that discusses creation (or annihilation) of matter *per se* must have world lines with finite beginnings or ends (or both). The C-field interaction term picks out precisely these end points of particle world lines. If we vary the world line of a and consider the change in the action $\mathcal{A}$ in a volume containing the point A_1 where the world line begins (see Figure 9.2), we get at A_1 (which is now varied)

$$m_a c \frac{da^i}{ds_a} g_{ik} - C_k = 0. \tag{9.8}$$

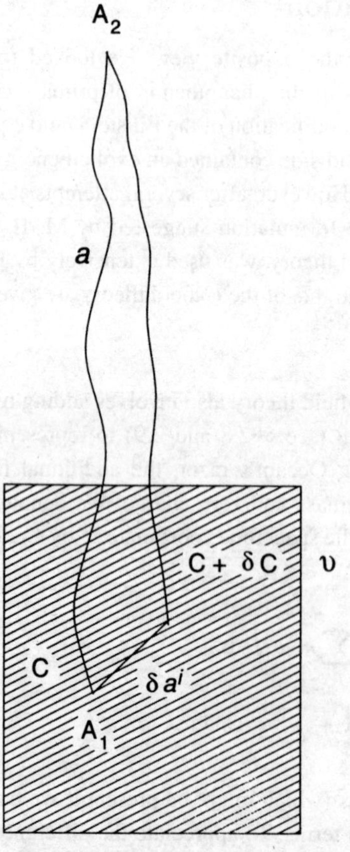

Figure 9.2 The world line of a begins at A_1 and ends at A_2. If we consider variations in the shaded region, the point A_1 shifts by δa^i. This shift produces a change in the C-field interaction term by an amount $\delta C = C_i \, \delta a^i$ The change in the inertial part of the action similarly makes a contribution at A_1 of $p_i^{(a)} \delta a^i$, where $p_i^{(a)}$ is the 4-momentum of the particle a. The result (9.8) follows by equating the nett contribution of $\delta \mathcal{A}$ at A_1 to zero.

This relation tells us that *overall energy and momentum are conserved at the point of creation*. The 4-momentum of the created particle is compensated by the 4-momentum of the C-field. Clearly, to achieve this balance the C-field must have negative energy. We will return to this point later. We also note that, since the interaction term is path-independent, the equation of motion of a is still that of a geodesic:

$$m_a\left(\frac{\mathrm{d}^2 a^i}{\mathrm{d}s_a^2} + \Gamma^i_{kl}\frac{\mathrm{d}a^k}{\mathrm{d}s_a}\frac{\mathrm{d}a^l}{\mathrm{d}s_a}\right) = 0. \tag{9.9}$$

The constant f in the action (9.6) is a coupling constant. The variation of C gives the source equation in the form

$$C^k_{;k} = cf^{-1}\bar{n}, \tag{9.10}$$

where $\bar{n}$ is the number of nett creation events per unit proper 4-volume. In calculating $\bar{n}$ we attach a $+$ sign to the points like A_1 where a world line begins and a $-$ sign to the points like A_2 where a world line ends. Again we see in (9.10) the relationship between the C-field and the creation/annihilation events.

Finally, the variation of g_{ik} leads to the modified Einstein field equations

$$R^{ik} - \frac{1}{2}g^{ik}R = -\frac{8\pi G}{c^4}\left(T^{ik}_{(m)} + T^{ik}_{(C)}\right), \tag{9.11}$$

where $T^{ik}_{(m)}$ is the matter tensor as in the earlier chapters while

$$T^{ik}_{(C)} = -f(C^i C^k - \tfrac{1}{2}g^{ik}C^l C_l). \tag{9.12}$$

Again we note that $T^{00}_{(C)} < 0$ for $f > 0$. Thus the C-field has a negative energy density that produces a repulsive gravitational effect. It is this repulsive force that drives the expansion of the universe. The above effect may resolve one difficulty usually associated with the quantum theory of negative energy fields. Because such fields have no lowest energy state, they normally do not form stable systems. A cascading into lower and lower energy states would inevitably occur if we perturb the field in a given state of negative energy. However, this conclusion is altered if we include the feedback of (9.12) on spacetime geometry through (9.11). This feedback results in the expansion of space and in the lowering of the magnitude of field energy. These two effects tend to work in opposite directions and help stabilize the system.

Cosmological equations

Using the Robertson–Walker line element and the assumption that a typical particle created by the C-field has mass m, we get the following equations out of (9.8)–(9.12):

$$\dot{C} = mc^2, \tag{9.13}$$

$$mf\left(\ddot{C} + 3\frac{\dot{S}}{S}\dot{C}\right) = \left(\dot{\rho} + \frac{\dot{S}}{S}\rho\right)c^2, \tag{9.14}$$

$$2\frac{\ddot{S}}{S} + \frac{\dot{S}^2 + kc^2}{S^2} = \frac{4\pi Gf}{c^4}\dot{C}^2, \tag{9.15}$$

$$3\frac{\dot{S}^2 + kc^2}{S^2} = 8\pi G\left(\rho - \frac{f}{2c^4}\dot{C}^2\right). \tag{9.16}$$

It is easy to verify that the steady-state solution (9.2) follows from these equations for

$$k = 0, \qquad S = e^{H_0 t}, \qquad \rho = \rho_0 = \frac{3H_0^2}{4\pi G} = fm^2. \tag{9.17}$$

Notice that both H_0 and ρ_0 are given in terms of the elementary creation process; that is, in terms of the coupling constant f and the mass of the particle created. Thus the Hoyle approach provides the quantitative information lacking in the deductive approach of the PCP.

A first order perturbation of the above equations and of the solution (9.2) also tells us that the solution is stable (see Exercise 20). Indeed, a stability analysis brings out the key role played by (9.8). This tells us that the created particles have their world lines along the normals to the surfaces $C = $ constant. Hoyle has argued that such a result gives a physical justification for the Weyl postulate: it tells us *why* the world lines of the fundamental observers are orthogonal to a special family of spacelike hypersurfaces. In the C-field cosmology these hypersurfaces are not just abstract notions but are seen to have a physical basis.

Explosive creation

Although the C-field was introduced primarily to account for the continuous creation of matter, the author showed in 1973 that it also describes explosive creation of matter such as is required in the big-bang cosmology. We illustrate below how this is achieved for the case $k = 0$.

In equations (9.13)–(9.16), we make use of the idea that all matter is created only in an explosive process at $t = 0$. Then the right-hand side of (9.14) is like a delta function $\delta(t)$, leading to the solution

$$\dot{C} = \frac{A}{S^3}, \qquad A = \text{constant}.$$

Notice that this solution is inconsistent with (9.13) except for one epoch, $t = 0$. This is hardly surprising, since we have assumed that there is no creation of matter subsequent to $t = 0$. Thus the creation condition (9.8) is not satisfied at $t > 0$.

Substituting for $\dot{C}$ in (9.16), we can integrate for S and obtain a solution

$$S(t) \propto \left(1 + \frac{(t + t_1)^2}{t_0^2}\right)^{1/3},$$ (9.18)

where t_0 and t_1 are constants related to the initial conditions at $t = 0$ (see Exercise 22).

The scale factor given by (9.18) behaves like that for the standard Einstein–de Sitter model for $t \gg t_0, t_1$. In the C-field model not only is the spacetime singularity at $t = 0$ averted but also we see the present matter as arising from a primordial explosion *that conserves energy and momentum.*

This conservation of energy and momentum must follow quite generally, for any C-field model, since the governing equations are derived from an action principle. Hence criticism based on the unexplained origin of new matter, which could validly be applied to the explosive creation of the standard cosmology or to the continuous creation in the Bondi–Gold version of the steady-state model, does not apply to the C-field cosmology.

In physical terms the creation is explained by a process of interchange of energy and momentum between the negative-energy C-field and the matter. The divergence of (9.11) gives the mathematical formula for conservation of energy:

$$T^{ik}_{(m);k} = f C^i C^k_{;k}.$$ (9.19)

It is easy to verify that the idea would not work for a positive energy field (see Exercise 23).

9.3   Observable parameters of the steady-state cosmology

Leaving aside the dynamics of the model, we now come to some of the observable features of the steady-state theory. Here we deal essentially with the line element (9.2) and the geometrical properties deducible from it. Indeed, as Bondi and Gold emphasized in their original paper, the steady-state model makes precise predictions and is therefore vulnerable to observational disproof, in contrast to the big-bang models, which can always be sustained with time-dependent parameters. (This comment will become clearer in Chapters 10 and 11 when we discuss observational cosmology.)

Since we have gone through calculations of these observable features at great length in Chapters 3 and 4, we will be brief here and simply quote the results.

The redshift

The redshift of a galaxy G_1 at (r_1, θ_1, ϕ_1) emitting light at t_1 that is received by the observer O at $r = 0$ at the present epoch t_0 is given by

$$z_1 = e^{H_0(t_0 - t_1)} - 1 = r_1 \frac{H_0}{c} e^{H t_0}.$$ (9.20)

The luminosity distance

This is given for the above galaxy by

$$D_1 = \frac{c}{H_0} z_1 (1 + z_1). \tag{9.21}$$

Equation (9.21) is the *Hubble law* for steady state cosmology. From (9.1) we also see that the deceleration parameter q_0 for this cosmology has the value -1.

The event horizon

If we look for signals beamed towards us by observers in the present epoch t_0, we will receive signals in some finite future epoch $t > t_0$, provided that their proper metric distance $r S(t_0)$ is less than c/H_0, which is therefore the radius of the event horizon of the steady-state universe. It may be verified that this model does not have a particle horizon.

Angular size

The angle $\Delta\theta$ ($\ll 1$) subtended at O by an astronomical source of projected linear size d and redshift z is given by

$$\Delta\theta = \frac{H_0}{c} d \left(\frac{1 + z}{z} \right). \tag{9.22}$$

Thus the angular size tends to a finite minimum as $z \to \infty$.

Flux density

The formula (3.57) becomes in this case

$$F_{bol} = \frac{L_{bol}}{4\pi \left(\dfrac{c}{H_0} \right)^2 z^2 (1 + z)^2}, \tag{9.23}$$

Whereas for (3.56) we get

$$\mathcal{F}(v_0) = \frac{L J(v_0 (1 + z))}{4\pi \left(\dfrac{c}{H_0} \right)^2 z^2 (1 + z)}. \tag{9.24}$$

Number count

In the notation of §3.11 the number of sources with redshift less than z is given by

$$N(z) = 4\pi n \left(\frac{c}{H_0} \right)^3 \left(\ln(1 + z) - \frac{3z^2 + 2z}{2(1 + z)^2} \right). \tag{9.25}$$

The age distribution of galaxies

New galaxies are always being formed in the steady-state universe. Since the universe expands, the galaxies, once they have formed, move away from each other. Thus the older a population of galaxies the more sparse its distribution. Since the volume bounded by galaxies increases with time as $\exp(3H_0t)$, we have the following simple result for the age–density relation of galaxies:

$$Q(\tau) \propto e^{-3H_0\tau}, \tag{9.26}$$

where $Q(\tau)\,d\tau$ is the proper number density of galaxies with ages in the range τ, $\tau +$ $d\tau$. The average age is therefore $(3H_0)^{-1}$.

9.4 Physical and astrophysical considerations

This section briefly outlines some of the ideas proposed from time to time in the context of the steady-state theory to discuss such problems as the nature of created particles, the formation of galaxies and the origin of the microwave background radiation. Some of these concepts might still be relevant even if the original steady-state cosmological picture does not survive today.

9.4.1 The hot universe

In 1958 Gold and Hoyle proposed the hypothesis that the created matter was in the form of neutrons. The creation of neutrons does not violate any standard conservation laws of particle physics except the constancy of the number of baryons. Although this was considered an objection in 1958, today the number of baryons is no longer regarded as invariant. Indeed, in Chapter 6 we saw how scenarios based on non-conservation of baryons are being proposed in the context of the very early universe to account for the observed number of baryons in the universe.

In the Gold–Hoyle picture the created neutron undergoes a β-decay:

$$n \rightarrow p + e^- + \bar{\nu}. \tag{9.27}$$

The conservation of energy and momentum results in the electron taking up most of the kinetic energy and thereby acquiring a high kinetic temperature of $\sim 10^9$ K (see Exercise 29). Gold and Hoyle argued that such a high temperature produced inhomogeneously would lead to the working of heat engines between the hot and cold regions, which provide pressure gradients that result in the formation of condensations of size ≥ 50 Mpc (see Exercise 30). As we have already seen in Chapter 7, pure gravitational forces are not able to provide a satisfactory picture of galaxy formation. The temperature gradients set up in the hot universe of Gold and Hoyle help in this process.

The resulting system, however, is not a single galaxy, but a supercluster of galaxies containing $\sim 10^3$–10^4 members. Such large-scale inhomogeneities in the distribution of galaxies were referred to in Chapter 1. Inhomogeneities on such a large scale ≥ 50 Mpc caution us against applying the cosmological principle too rigorously. For example, the formula (9.26) for the age distribution of galaxies will hold over a region considerably larger than 50 Mpc in such a model. If we are in a particular supercluster, we expect to see a preponderance of galaxies of ages similar to that of ours in our neighbourhood out to say 20 or 30 Mpc. Thus it will not be surprising if our local sample yields an average age much larger than the universal average of $(3H_0)^{-1} \approx 3 \times 10^9 h_0^{-1}$ years.

Although newly created electrons have a kinetic temperature of $\sim 10^9$ K, the temperature tends to drop because of expansion. The average temperature is three fifths of this value, that is, around 6×10^8 K. It was suggested by Hoyle in 1963 that such a hot intergalactic medium would generate the observed X-ray background. However, quantitative estimates by R. J. Gould soon showed that the expected X-ray background in the hot universe would be considerably higher than what is actually observed, thus making the hot universe untenable. The present-day background measurements do not rule out such a hot universe for $h_0 \simeq 0.5$. Astrophysicists are, however, inclined to look for other explanations for the origin of the X-ray background.

Although it is now discredited, the hot universe model was the first exercise in linking particle physics (neutron decay) to the formation of large-scale structures in the universe.

9.4.2   The bubble universe

In 1966 Hoyle and the author discussed the effect of raising the coupling constant f by $\sim 10^{20}$. As the formulae (9.17) show, we would then have a steady-state universe of very large density ($\rho_0 \simeq 10^{-8}$ g cm^{-3}) and very short time scale ($H_0^{-1} \simeq 1$ year!). If in such a dense universe creation is switched off in a local region, that is, if we locally have a phase transition from the creative to the non-creative mode,

$$C^i_{;i} = 0, \tag{9.28}$$

then this local region will expand according to (9.18). Being less dense than the surroundings, such a region will simulate an air bubble in water. The reader may look back to Chapter 6 and discover the similarity between this model and the inflationary model that came into fashion 15 years later.

According to this model, this bubble is all that we see with our surveys of galaxies, quasars and so on. Hence our observations tell us more about this unsteady perturbation than about the ambient steady-state universe. There are, however, observable effects that give indications of the high value of f. For example, these authors showed that particle creation is enhanced near already existing massive objects and that the resulting energy spectrum of the particles would simulate that of high-energy cosmic rays. The actual energy density of cosmic rays requires the high value of f chosen here.

9.4.3 The origin of elements and the microwave background

Between 1964 and 1965 the steady-state model received two near-fatal blows. Both came from considerations of issues of the early universe, (i) the abundances of light nuclei and (ii) the discovery of the microwave background. We will discuss them briefly in that order.

As we discussed in Chapter 5, Gamow's ideas on primordial nucleosynthesis had turned out to work for light nuclei only, producing mainly helium and small quantities of deuterium and a few other light nuclei. For the majority of elements, stars provided the right setting for successive nucleosynthesis. This had led most astronomers in the 1950s to believe that stellar nucleosynthesis was the key process for all elements, since even helium was seen to be produced in stars. However, it was slowly realized that the observed abundance of helium in several parts of the Galaxy is considerably higher than that generated in the stars. This in turn led astronomers back to Gamow's ideas once again (see Exercise 34). Apart from helium, the discovery of deuterium made the primordial version more credible since there was no known process for making even tiny quantities of deuterium in stars. So the steady-state theory, which had no early hot era to make such light nuclei, faced difficulties in explaining their observed abundances.

The case for the hot big bang became even stronger with the discovery of the microwave background in 1965. The steady-state model had no natural process to maintain a radiation background at ~ 3 K and it never quite recovered from these two blows. Nevertheless, the model made a comeback in a modified form in the 1990s and we describe its main features next.

9.5 The quasi-steady-state cosmology

In 1993, Fred Hoyle, Geoffrey Burbidge and the author (HBN hereafter) proposed a new cosmology, which drew considerably on the earlier steady-state cosmology, but also allowed the possibility of evolution on shorter time scales. This cosmology, known as the *quasi-steady-state cosmology*, or the QSSC, arises from the considerations of the Hoyle–Narlikar theory described in Chapter 8. We begin our discussions with a general comment.

General relativity is known to give an accurate description of gravitational phenomena in the limit of weak gravitational fields. We have seen in Chapter 2 its various successes on the observational front. However, there is no evidence, either experimental or observational, to show the precise quantitative correctness of general relativity for strong gravitational fields. As we have already seen, the big-bang cosmology is engaged in working from assumed initial conditions along with sweeping extrapolations in the domain of gravitation and high-energy physics, for which there is no clear evidence. In considering any alternative to it, therefore, it is desirable to identify any conceptual shortcomings of the standard approach and find ways of avoiding them. HBN start from what they consider to be a major defect of general relativity, namely that, unlike the rest of physics, it is not scale invariant.

We are well used to physical results being independent of the units in which quantities are expressed. This is because results are always dimensionless numbers. This usual situation is for units that stay the same at every point X in spacetime. However, should anything in physics be altered if units were changed differently at different points in spacetime? Using the Compton wavelength of some specified particle as the length unit and noting that $c = 1$ requires the time and space units to be the same, this question can be discussed by a general change of the length scale achieved by transforming from

$$ds^2 = g_{ik}\, dx^i\, dx^k \tag{9.29}$$

to

$$ds^{*2} = \Omega^2(x) g_{ik}\, dx^i dx^k. \tag{9.30}$$

We briefly discussed this transformation, known as *conformal transformation*, in Chapter 8. In such a transformation the spacetime coordinates of points X on the path of a particle stay fixed. It is the proper distance between adjacent points that changes according to the choice of the twice-differentiable scalar function $\Omega(X)$. Experiments confined to a locality over which Ω does not change appreciably will evidently be unaffected. It is possible, however, for events in one locality to be related to distant localities through the propagation of some field, for example the electromagnetic field. The possibility that physics will be unaffected by $\Omega(X)$ even for observations related to widely separated localities is suggested by the circumstance that light cones are not affected by the transformation (9.30), provided that Ω is restricted to the so-called conformal condition $\Omega \neq 0$. Indeed, Maxwell's equations are invariant with respect to (9.30), with the electromagnetic field tensor unchanged,

$$F_{ik}^* = F_{ik}, \tag{9.31}$$

and with $A_i^* = A_i$ followed by a suitable gauge transformation leaving the wave equation for the 4-potential also invariant. Such an invariance is called *conformal invariance*.

Since the coordinate positions of particles remain unchanged, the number of particles counted in a specified three-dimensional coordinate volume must be unaltered, despite the proper three-dimensional volume being changed by Ω^3. Because $|\psi|^2$ measures particle probabilities per unit proper 3-volume, the wave function ψ in quantum mechanics is therefore required to transform from ψ to ψ^* according to

$$\psi^* = \Omega^{-3/2}\psi. \tag{9.32}$$

Moreover, the spatial coordinate distance between two particles remains unchanged whereas the proper distance is altered by Ω. Since the Compton wavelength m^{-1} of some standard particle measures the latter, it is necessary for m to transform to m^* according to

$$m^* = \Omega^{-1}m. \tag{9.33}$$

The number of Compton wavelengths separating two particles then remains the same. This inference can be tested strictly by considering the behaviour of the Dirac equation

$$i\gamma^k \frac{\partial\psi}{\partial x^k} + m\psi = 0 \tag{9.34}$$

under the transformations (9.30), (9.32) and (9.33). To this end it is necessary first to generalize (9.34) to Riemannian space, writing

$$i\gamma^k \psi_{;k} + m\psi = 0 \tag{9.35}$$

in place of (9.34), with non-Euclidean terms entering into the covariant derivative of the spinor field ψ. (We have not touched upon spinor fields in this book; however, we can define covariant differentiation of spinors along the lines of vectors discussed in Chapter 2.) A demonstration of the conformal invariance of the Dirac equation, i.e., the transformation of (9.35) to

$$i\gamma^{*k} \psi^*_{;k} + m^*\psi^* = 0, \tag{9.36}$$

can then be given.

With quantum mechanics invariant with respect to (9.30) and subject to (9.33) and (9.34), and with the electromagnetic field also conformally invariant, so is quantum electrodynamics. More recent developments in physics concerned with abstract particle spaces are also considered to be scale-invariant. So why should gravitation be the only aspect of physics that is not? HBN felt that the first step towards a better understanding of cosmology is to remedy this deficiency and for this purpose they found that the formalism of the Hoyle–Narlikar theory of gravity described in the previous chapter was admirably suited. We will therefore restate the formalism with the QSSC in mind.

We begin by finding an action $\mathcal{A}$ that is unaffected in its value by a scale transformation. For a set of particles $a, b, \ldots$ of masses $m_a, m_b, \ldots$ the form of the action is

$$\mathcal{A} = -\sum_a \int m_a(\mathrm{A})\, \mathrm{d}s_a, \tag{9.37}$$

where the possibility of the masses of particles varying with the position in spacetime requires the mass $m_a(\mathrm{A})$ of particle a to vary with the point A on its path; and similarly for the other particles.

With $\mathrm{d}s_a^* = \Omega\, \mathrm{d}s_a$ and $m_a^* = \Omega^{-1} m_a$ it is clear that (9.37) is invariant with respect to a conformal (scale) transformation. We have already seen in the previous chapter that the scalar curvature transforms as

$$R^* = \Omega^{-2}(R + 6\Omega^{-1}\,\Box\,\Omega), \tag{9.38}$$

and it is equally clear that the usual Hilbert term that leads to the Einstein tensor is not conformally invariant. This is why the equations of general relativity are not conformally invariant. Since one can already remark that, in the action functional for general relativity, we have a strange combination of physical and geometrical quantities, it is surely here that a change must be made. Investigation shows that an attempt to replace the usual Hilbert term by some other geometrical quantity does not succeed, leaving (9.37) as seemingly the only possibility, a possibility that is startling in its simplicity. The gravitational equations are to be obtained in the usual way, by making a slight change of the metric tensor, $g_{ik} \to g_{ik} + \delta g_{ik}$, in a general four-dimensional volume with $\delta g_{ik} = 0$ on the boundary of the volume. The outcome can be written in the form

$$\delta\mathcal{A} = -\frac{1}{2}\int [T^{ik} + (?)]\, \delta g_{ik}\, \sqrt{-g}\, \mathrm{d}^4 x, \tag{9.39}$$

where the energy momentum tensor T^{ik} has its usual form,

$$T^{ik}(\mathrm{X}) = \sum_a \int m_a(\mathrm{A}) \frac{\delta_4(\mathrm{X, A})}{\sqrt{-g(\mathrm{A})}} \frac{\mathrm{d}a^i}{\mathrm{d}s_a} \frac{\mathrm{d}a^k}{\mathrm{d}s_a}\, \mathrm{d}s_a, \tag{9.40}$$

and (?) is a tensor whose form is determined by the properties of the masses of the particles, namely by the variations of $m_a(\mathrm{X}), m_b(\mathrm{X}), \ldots$ with respect to the position in spacetime. The gravitational equations following from the principle of stationary action, $\delta\mathcal{A} = 0$ for all δg_{ik}, are simply

$$T^{ik} + (?) = 0. \tag{9.41}$$

Now choose a 'mass field' $M(\mathrm{X})$ to satisfy

$$\Box_{\mathrm{X}} M(\mathrm{X}) + \frac{1}{6} R M(\mathrm{X}) = \sum_a \int \frac{\delta_4(\mathrm{X, A})}{\sqrt{-g(\mathrm{A})}}\, \mathrm{d}s_a. \tag{9.42}$$

Equation (9.42) has both advanced and retarded solutions. We particularize an advanced solution $M^{\mathrm{adv}}(\mathrm{X})$ and a retarded solution $M^{\mathrm{ret}}(\mathrm{X})$ in the following way.

$M^{\text{ret}}(X)$ is to be the so-called fundamental solution in the flat spacetime limit. This removes from $M^{\text{ret}}(X)$ the ambiguity that would obviously arise from the homogeneous wave equation. The corresponding ambiguity for $M^{\text{adv}}(X)$ is removed by the physical requirement that fields without sources are to be zero. Since

$$\Box[M^{\text{adv}} - M^{\text{ret}}] + \tfrac{1}{6}R[M^{\text{adv}} - M^{\text{ret}}] = 0, \tag{9.43}$$

the immediate consequence of this boundary condition is that $M^{\text{adv}} - M^{\text{ret}}$, being without sources, must be zero, so that

$$M^{\text{adv}}(X) = M^{\text{ret}}(X) = M(X), \tag{9.44}$$

say. The gravitational equations are now obtained by putting

$$m_a(A) = M(A), \qquad m_b(B) = M(B), \dots . \tag{9.45}$$

The tensor (?) can then be determined and it can also be shown that, in a conformal transformation, the mass field $M(X)$ transforms as in (9.33),

$$M^*(X) = \Omega^{-1}(X)M(X), \tag{9.46}$$

a result that follows from the form of the wave equation (9.43) (for details of the derivation, see the book by Hoyle and Narlikar (1974, p. 111) listed in the bibliography):

$$K(R_{ik} - \tfrac{1}{2}g_{ik}R) = -T_{ik} + M_i M_k - \tfrac{1}{2}g_{ik}g^{pq}M_p M_q + g_{ik}\Box K - K_{;ik}, \tag{9.47}$$

where

$$K = \tfrac{1}{6}M^2. \tag{9.48}$$

These gravitational equations are scale invariant. It may seem curious that, from a simpler beginning than the usual Einstein–Hilbert action, the outcome is more complicated; but this seems to be a characteristic of physical laws: they become simpler and more elegant in their initial global statement but more complicated in their detailed and specific consequences. However, the conformal invariance of the theory comes to our rescue. For, make the scale change

$$\Omega(X) = M(X)/\tilde{m}_0, \tag{9.49}$$

where $\tilde{m}_0$ is a constant with the dimensionality of $M(X)$. After the scale change, the particle masses simply become $\tilde{m}_0$ everywhere and, in terms of transformed masses, the derivative terms drop out of the gravitational equations. Then, defining the gravitational constant G by

$$G = \frac{3}{4\pi \tilde{m}_0^2},$$ (9.50)

the equations (9.47) take the form of general relativity:

$$R_{ik} - \tfrac{1}{2} g_{ik} R = -8\pi G T_{ik}.$$ (9.51)

It now becomes clear why the equations of general relativity are not scale-invariant. These are the special form to which the scale-invariant equations (9.47) reduce with respect to a particular scale, namely that in which masses of particles are everywhere the same.

9.6   Planck particles

It is easily seen from the wave equation (9.47) that $M(X)$ has dimensionality $(\text{length})^{-1}$ and so has $\tilde{m}_0$. Units are frequently used in particle physics for which both the speed of light c and Planck's constant $\hbar$ are unity and in these units mass has dimensionality $(\text{length})^{-1}$. If we suppose that these units apply to the above discussion, then, from (9.50),

$$\tilde{m}_0 = [3/(4\pi G)]^{1/2},$$ (9.52)

which, with $c = \hbar = 1$, is the mass of the Planck particle. This suggests that, in a gravitational theory without other physical interactions, the particles must be of mass (9.52), which in ordinary practical units is about 10^{-5} g, the empirically determined value of G being used.

Let m_P be the mass of the particles with respect to some practical unit and consider practical units also for time and length, so that neither c nor $\hbar$ is unity. Then we have

$$\tilde{m}_0 = m_P c / \hbar.$$ (9.53)

The gravitational equations (9.51) are

$$\frac{1}{6} \tilde{m}_0^2 \left(R^{ik} - \frac{1}{2} g^{ik} R \right) = -\sum_a \tilde{m}_0 \int \frac{\delta_4(X, A)}{\sqrt{-g(A)}} \frac{da^i}{ds_a} \frac{da^k}{ds_a} ds_a.$$ (9.54)

The right-hand side is the energy momentum tensor for particles $a, b, \ldots$. Using (9.53) to replace $\tilde{m}_0$ by m_P we get

$$R^{ik} - \frac{1}{2} g^{ik} R = -\frac{6\hbar}{c^3 m_P^2} \sum_a m_P c^2 \int \frac{\delta_4(X, A)}{\sqrt{-g(A)}} \frac{da^i}{ds_a} \frac{da^k}{ds_a} ds_a.$$ (9.55)

Identifying $8\pi G/c^4$ with $6\hbar/(m_{\mathrm{P}}^2 c^3)$, these are the equations of general relativity, i.e., for

$$m_{\mathrm{P}} = \left(\frac{3\hbar c}{4\pi G}\right)^{1/2}. \tag{9.56}$$

When the empirically determined value of G is used in (9.56) this is the mass of the Planck particle, the value 1.06×10^{-5} g quoted above. We will shortly see that the Planck particle plays a key role in the QSSC, as the basic particle to be created. However, before we come to that aspect of the theory, we wish to have a second look at the possibility of having a cosmological constant in this cosmology.

9.7   The cosmological constant

Recall that, in the previous chapter, we had concluded that, if the driving wave equation for inertia is linear, then the above formulation leads to Einstein's equations *without* the cosmological constant. We now show that, if the assumption of linearity is dropped, then a cosmological constant is naturally allowed by the theory. Moreover, its magnitude is determined by the large-scale structure of the universe.

Writing $M^{(a)}(\mathrm{X})$, $M^{(b)}(\mathrm{X})$, ... as the mass fields produced by the individual Planck particles $a, b, \ldots$, the total mass field

$$M(\mathrm{X}) = \sum_a M^{(a)}(\mathrm{X}) \tag{9.57}$$

satisfies the wave equation (9.42) when $M^{(a)}, M^{(b)}, \ldots$ satisfy

$$\Box M^{(a)} + \frac{1}{6} R M^{(a)} = \int \frac{\delta_4(\mathrm{X}, \mathrm{A})}{\sqrt{-g(\mathrm{A})}} \, \mathrm{d}s_a, \ldots \tag{9.58}$$

Scale invariance throughout requires all the mass fields to transform as

$$M^{*(a)} = M^{(a)} \Omega^{-1} \tag{9.59}$$

with respect to the scale change Ω, when both the left- and the right-hand side of every wave equation transform to its starred form multiplied by Ω^{-3}; i.e., the left-hand side of (9.58) goes to $(\Box M^{*(a)} + \frac{1}{6} R^* M^{*(a)})\Omega^{-3}$ and the right-hand side to

$$\Omega^{-3} \int \frac{\delta_4(\mathrm{X}, \mathrm{A})}{\sqrt{-g^*(\mathrm{A})}} \, \mathrm{d}s_a^*.$$

Then the factor Ω^{-3} cancels out to give the appropriate invariant equation. This cancellation is evidently unaffected if, instead of (9.58) for the wave equation satisfied by M^a, we have

$$\Box M^{(a)} + \frac{1}{6} R M^{(a)} + M^{(a)^3} = \int \frac{\delta_4(\mathrm{X}, \mathrm{A})}{\sqrt{-g(\mathrm{A})}} \, ds_a. \tag{9.60}$$

Since the cube term transforms to $M^{*(a)^3} \Omega^{-3}$ with respect to Ω, changing (9.58) to (9.60) preserves scale invariance in what appears to be its widest form. Since in other respects the laws of physics always seem to take on the widest ranging properties that are consistent with the relevant forms of invariance, we might think it should also be so here, in which case (9.60) rather than (9.58) is the correct wave equation for $M^{(a)}$; and similarly for $M^{(b)}, \ldots$, the mass fields of the other Planck particles.

However, this departure from linearity in the wave equations for the individual particles prevents a similar equation being obtained for $M = \sum_a M^{(a)}$. Nevertheless, the addition of the individual equations can be considered in a homogeneous universe to lead to an approximate wave equation for M of the form

$$\Box M + \frac{1}{6} R M + \Lambda M^3 = \sum_a \int \frac{\delta_4(\mathrm{X}, \mathrm{A})}{\sqrt{-g(\mathrm{A})}} \, ds_a, \tag{9.61}$$

$$\Lambda = N^{-2}, \tag{9.62}$$

where N is the effective number of particles contributing to the sum $\sum_a M^{(a)}$. The latter can be considered to be determined by an Olbers-like cut-off, contributed by the portion of the universe surrounding the point X in $M(\mathrm{X})$ to a redshift of the order of unity. In the observed universe this total mass is $\sim 10^{22} M_\odot$, sufficient for $\sim 2 \times 10^{60}$ Planck particles. With this value for N

$$\Lambda \simeq 2.5 \times 10^{-121}. \tag{9.63}$$

The next step is to notice that the wave equation (9.60) would be obtained in usual field theory from $\delta \mathcal{A} = 0$ for a scalar function $M \to M + \delta M$ when it is applied to

$$\mathcal{A} = -\frac{1}{2} \int \left(M_i M^i - \frac{1}{6} R M^2 \right) \sqrt{-g} \, d^4 x + \frac{1}{4} \Lambda \int M^4 \sqrt{-g} \, d^4 x$$

$$- \sum_a \int \frac{\delta_4(\mathrm{X}, \mathrm{A})}{\sqrt{-g(\mathrm{A})}} M(\mathrm{X}) \, ds_a. \tag{9.64}$$

In the scale in which M is m_P everywhere the derivative term in (9.64) vanishes and, since $G = 3/(4\pi m_\mathrm{P}^2)$ the term in R is the same as in the Hilbert term, as are also the line integrals; requiring the remaining term to be the same gives

$$\lambda = -3\Lambda m_\mathrm{P}^2. \tag{9.65}$$

Thus we have obtained not only a cosmological constant but also its magnitude, something that lies beyond the scope of the usual theory. With 2.5×10^{-121} for Λ as in (9.63) and with m_P the inverse of the Compton wavelength of the Planck particle, $\sim 3 \times 10^{32}$ cm^{-1}, (9.65) gives

$$\lambda \simeq -2 \times 10^{-56} \text{ cm}^{-2}, \tag{9.66}$$

agreeing closely with the magnitude that has previously been assumed for λ. In the classical big-bang cosmology there is no dynamical theory to relate the magnitude of λ to the density or other physical properties of matter. For observational consistency it is assumed that λ today is of the order of (9.66). The sign of the cosmological constant derived here is, however, negative. As we shall see shortly, this important difference is responsible for producing bounded oscillations. Although, in standard cosmology, λ is taken to be positive, we will see in later chapters that, within the framework of the cosmology discussed here, a negative λ is not inconsistent with observations.

A dynamical derivation of λ within the standard cosmology is possible if one goes into the very early inflationary epochs. However, the values of λ deduced from those calculations are embarrassingly large, being 10^{108}–10^{120} times the value given by (9.66). The problem then becomes that of how to reduce λ from such high values to the range acceptable at present. In contrast, the present derivation leads to the acceptable range of values with very few theoretical assumptions. Also, we have the satisfaction of seeing its smallness related to the large number N.

Planck particles are subject to rapid decay, when their mass couplings are replaced by those of more stable secondaries. Provided that the secondaries interact together in the same way in the cube terms as the Planck particles from which they are derived, the result (9.66) continues to hold.

9.8 The creation of matter

We now come to the basic difference of the QSSC from standard cosmology, namely the possibility of creation of new matter from time to time. As we will see, the action principle allows this possibility simply by the assumption that the particle world lines are not endless. Thus a world line may begin at a finite point in spacetime A_0 and end at another finite point in spacetime A_1. This means that a particle was created and then, after a finite lifetime, annihilated. Can the theory describe the dynamics of this 'creation/annihilation' process through the same action principle? The answer is 'yes'. As we saw earlier, the characteristic mass of a particle that is built out of the fundamental constants c, $\hbar$ and G is the Planck mass. So we will *assume* that the typical particle to be created in this primary process is the Planck particle.

A typical Planck particle a exists from A_0 to $A_0 + \delta A_0$, in the neighbourhood of which it decays into n stable secondaries, $n \simeq 6 \times 10^{18}$, denoted by $a_1, a_2, \ldots, a_n$. Each such secondary contributes a mass field $m^{(a_r)}(X)$, say, which is the fundamental solution of the wave equation

$$\Box m^{(a_r)} + \frac{1}{6} R m^{(a_r)} + n^2 m^{(a_r)^3} = \frac{1}{n} \int_{A_0+\delta A_0} \frac{\delta_4(X, A)}{\sqrt{-g(A)}} \, da, \qquad (9.67)$$

while the brief existence of a contributes $c^{(a)}(X)$, say, which satisfies

$$\Box c^{(a)} + \frac{1}{6} R c^{(a)} + c^{(a)^3} = \int_{A_0}^{A_0+\delta A_0} \frac{\delta_4(X, A)}{\sqrt{-g(A)}} \, da, \qquad (9.68)$$

Summing $c^{(a)}$ with respect to $a, b, \ldots$ gives

$$c(X) = \sum_a c^{(a)}(X), \qquad (9.69)$$

the contribution to the total mass $M(X)$ from the Planck particles during their brief existence, while

$$\sum_a \sum_{r=1}^{n} m^{(a_r)}(X) = m(X) \qquad (9.70)$$

gives the contribution of the stable particles.

Although $c(X)$ makes a contribution to the total mass function

$$M(X) = c(X) + m(X) \qquad (9.71)$$

that is generally small relative to $M(X)$, there is the difference that, whereas $m(X)$ is an essentially smooth field, $c(X)$ contains small exceedingly rapid fluctuations and so can contribute significantly to the derivatives of $c(X)$. The contribution to $c(X)$ from Planck particles a, for example, is largely contained between two light cones, one from A_0, the other from $A_0 + \delta A_0$. Along a timelike line cutting these two cones the contribution to $c(X)$ rises from zero as the line crosses the light cone from A_0, attains some maximum value and then falls back effectively to zero as the line crosses the second light cone from $A_0+\delta A_0$. The time derivative of $c^{(a)}(X)$ therefore involves the reciprocal of the difference in time between the two light cones. This reciprocal cancels out the short duration of the source term on the right-hand side of (9.68). The factor in question is of the order of the decay time τ of the Planck particles, $\sim 10^{-43}$ s. No matter how small τ may be, the reduction in the source strength of $c^{(a)}(X)$ is recovered in the derivatives of $c^{(a)}(X)$, which therefore cannot be omitted from the gravitational equations (see Figure 9.3).

The derivatives of $c^{(a)}(X), c^{(b)}(X), \ldots$ can equally well be negative as positive, so that, in averaging over many Planck particles, linear terms in the derivatives do disappear. Omitting for the moment the cosmological constant, it is therefore not hard to show that, after such an averaging, the gravitational equations become

$$R_{ik} - \frac{1}{2} g_{ik} R = \frac{6}{m^2} \left[-T_{ik} + \frac{1}{6}(g_{ik} \Box m^2 - m^2_{;ik}) + \left(m_i m_k - \frac{1}{2} g_{ik} m_l m^l \right) \right.$$
$$\left. + \frac{2}{3} \left(c_i c_k - \frac{1}{4} g_{ik} c_l c^l \right) \right]. \qquad (9.72)$$

Since the same wave equation is being used for $c(X)$ as for $m(X)$, the theory remains scale-invariant. One can therefore introduce a scale change that reduces $M(X) = m(X) + c(X)$ to a constant, or one that reduces $m(X)$ to a constant. Only that which reduces $m(X)$ to a constant, viz.

$$\Omega = m(X)/m_P, \tag{9.73}$$

has the virtue of not introducing small very rapidly varying ripples into the metric tensor. Although they are small in amplitude, such ripples produce non-negligible contributions to the derivatives of the metric tensor, causing difficulties in the evaluation of the Riemann tensor, so they are better avoided. Simplifying with (9.73) does not bring in this difficulty, which is why separating out the main smooth part of $M(X)$ in (9.72) now proves an advantage, with the gravitational equations simplifying to

$$8\pi G = \frac{6}{m_P^2}, \qquad m_P \text{ a constant}, \tag{9.74}$$

$$R_{ik} - \tfrac{1}{2}g_{ik}R = -8\pi G[T_{ik} - \tfrac{2}{3}(c_i c_k - \tfrac{1}{4}g_{ik}c_l c^l)]. \tag{9.75}$$

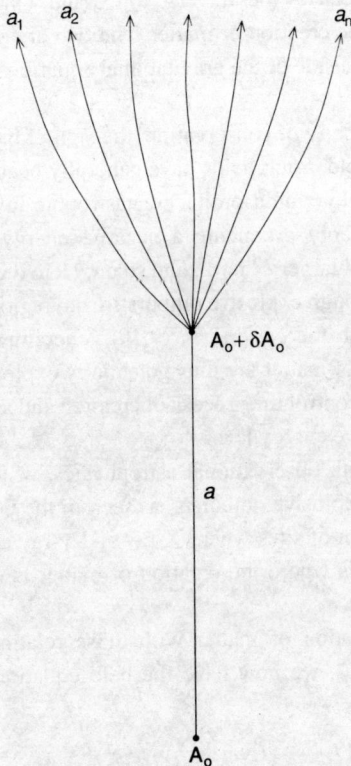

Figure 9.3 A Planck particle is created at A_0 and it shortly thereafter decays at $A_0 + \delta A_0$ into a large number of baryons, mesons, photons etc. The C-field arising from this stretch has a small magnitude but large derivatives.

Using the metric (3.45) with $k = 0$, the dynamical equations for the scale factor $S(t)$ are

$$\frac{2\ddot{S}}{S} + \frac{\dot{S}^2}{S^2} = \frac{4\pi}{3} G \bar{\dot{c}^2}, \qquad (9.76)$$

$$\frac{3\dot{S}^2}{S^2} = 8\pi G \left(\bar{\rho} - \frac{1}{2}\bar{\dot{c}^2} \right), \qquad (9.77)$$

with $\bar{\rho}$ the average particle mass density and $\bar{\dot{c}^2}$ being the average value of $\dot{c}^2$, the average value of terms linear in c and $\ddot{c}$ being zero. It is easily shown from the above that

$$\frac{\partial \bar{\rho}}{\partial t} + \frac{3\dot{S}}{S} \bar{\rho} = \frac{1}{2} \left(\frac{\partial \bar{\dot{c}^2}}{\partial t} + \frac{4\dot{S}}{S} \bar{\dot{c}^2} \right). \qquad (9.78)$$

If at a particular time there is no creation of matter then at that time the left-hand side of (9.78) is zero with $\bar{\rho} \propto S^{-3}$. With the right-hand side also zero at that time, $\bar{\dot{c}^2} \propto S^{-4}$. The sign of the $\bar{\dot{c}^2}$ term in (9.77) is that of a negative pressure, a characteristic of the fields introduced into inflationary cosmological models. The concept of Planck particles forces the appearance of a negative pressure. In effect the positive energy of created particles is compensated for by the sign of the $\bar{\dot{c}^2}$ terms, which in (9.76) increases $\ddot{S}/S$ and so causes the universe to expand. One can say that the universe expands because of the creation of matter. The two are connected because the divergence of the right-hand side of the gravitational equations (9.75) is zero.

As would be expected from this property of conservation, the sign of the $\bar{\dot{c}^2}$ term in (9.77) is that of a negative energy field. Such fields have generally been avoided in physics because in flat spacetime they would produce catastrophic instabilities – creation of matter with positive energy producing a negative-energy $\bar{\dot{c}^2}$ term producing more matter, producing a still larger $\bar{\dot{c}^2}$ term, and so on. Here the effect is to produce expansion of local space through explosive outbursts from regions where any such instability takes hold, through the $\bar{\dot{c}^2}$ term in (9.76) generating a sharp increase of $\ddot{S}$. The sites of the creation of matter are thus potentially explosive. The explosive expansion of space serves to control the process of creation and avoids the catastrophic cascading down the negative energy levels.

The requirement is in agreement with observational astrophysics, which in respect of high-energy activity is all of explosive outbursts, as seen in the QSOs, the active galactic nuclei, etc. The profusion of sites where X-ray and γ-ray activity is occurring are in the present theory sites where the creation of matter is currently taking place.

Combining the above ideas on creation of matter with those relating to the cosmological constant described in §9.7, we now have the field equations in the form

$$R_{ik} - \tfrac{1}{2} g_{ik} R + \lambda g_{ik} = -8\pi G [T_{ik} - \tfrac{2}{3}(c_i c_k - \tfrac{1}{4} g_{ik} c_l c^l)]. \qquad (9.79)$$

It has been on (9.79) that the discussion of what has been called the quasi-steady-state cosmological model (QSSC) has been based. A connection with the C-field of the earlier steady-state cosmology can also be given. Writing

$$C(X) = \tau c(X), \tag{9.80}$$

where τ is the decay lifetime of the Planck particle, the action contributed by Planck particles $a, b, \ldots$,

$$-\sum_a \int_{A_0}^{A_0+\delta A_0} c(A)\,da, \tag{9.81}$$

can be approximated as

$$-C(A_0) - C(B_0) - \cdots, \tag{9.82}$$

which form was used in the steady-state cosmology.

Thus the equations (9.79) are replaced by

$$R_{ik} - \tfrac{1}{2} g_{ik} R + \lambda g_{ik} = -8\pi G[T_{ik} - f(C_i C_k - \tfrac{1}{4} g_{ik} C_l C^l)], \tag{9.83}$$

with the earlier coupling constant f defined as

$$f = \frac{2}{3\tau^2}. \tag{9.84}$$

(We remind the reader that we have taken the speed of light as $c = 1$.)

The question of why astrophysical observation suggests that the creation of matter occurs in some places but not in others now arises. For creation to occur at the points $A_0, B_0, \ldots$ it is necessary classically that the action should not vary with respect to small changes in the positions in spacetime of these points, which was shown earlier to require

$$C_i(A_0)C^i(A_0) = C_i(B_0)C^i(B_0) = \cdots = m_P^2. \tag{9.85}$$

More precisely, the field $c(X)$ is required to be equal to m_P at $A_0, B_0, \ldots$,

$$c(A_0) = c(B_0) = \cdots = m_P. \tag{9.86}$$

(For equation (9.80) tells us that the connection between c and C is through the lifetime τ of the Planck particle.)

As already remarked above, this is in general not the case: in general the magnitude of $c(X)$ is much less than that of m_P. However, close to the event horizon of a massive compact body $C_i(A_0)C^i(A_0)$ is increased by a relativistic time dilatation factor, whereas m_P stays fixed. Figure 9.4 illustrates this. Hence, near enough to an event horizon the required conservation conditions can be satisfied, which has the consequence that creation events occur only in compact regions, agreeing closely with the condensed regions of high excitation observed so widely in astrophysics.

The theory would profit most from a quantum description of the individual process of creation. The difficulty, however, is that Planck particles are defined as those for which the Compton wavelength and the gravitational radius are essentially the same, which means that, unlike other quantum processes, flat spacetime cannot be used in the formulation of the theory. A gravitational disturbance is necessarily involved and the ideal location for triggering creation is that near a compact massive object. The C-field boson far away from a compact object might not be energetic enough to trigger the creation of a Planck particle. On falling into the strong gravitational field of a sufficiently compact object, however, the boson energy is multiplied by a dilatation factor, viz. $(1 - \mu/r)^{-1/2}$ for a local Schwarzschild metric

$$(1 - \mu/r)\,dt^2 - \frac{dt^2}{1 - \mu/r} - r^2(d\theta^2 + \sin^2\theta\,d\phi^2). \qquad (9.87)$$

Bosons then multiply up in a cascade, one makes two, two makes four, ..., as in the discharge of a laser, with production of particles multiplying similarly and with negative-pressure effects ultimately blowing the system apart. Such an explosive

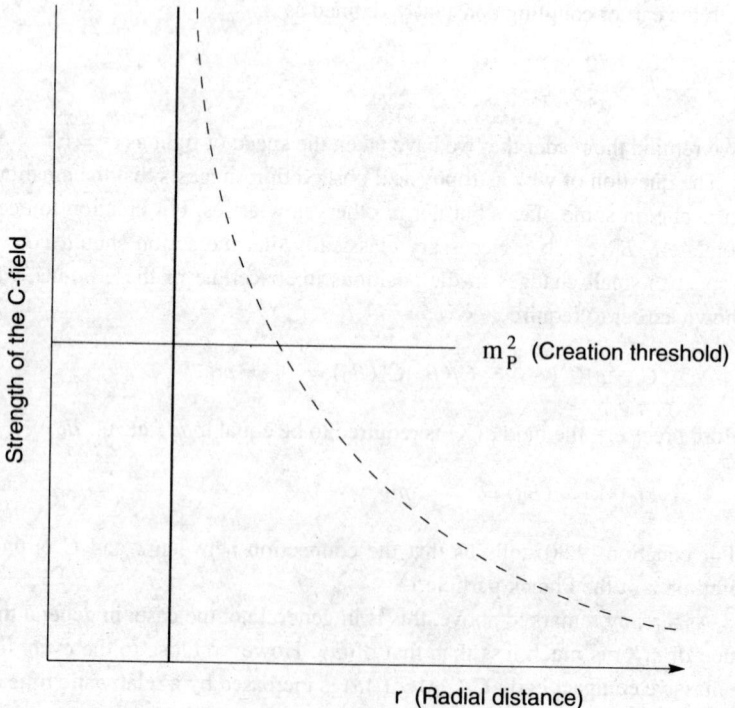

Figure 9.4 How the strength of the C-field rises near a collapsed massive object.

event may be called a *mini-bang*, or a *mini-creation event*. Unlike the big bang, however, the dynamics of this phenomenon is well defined and non-singular.

Although it is still qualitative, this view agrees well with the empirical facts of observational astrophysics, for we do see several explosive phenomena in the universe, such as jets from radio sources, bursts of γ-rays, X-ray bursters, quasars and active galactic nuclei. Generally it is assumed that a black hole plays the lead role in such an event by somehow converting its huge gravitational energy into large kinetic energies of the 'burst' kind. In actuality, one does not see infalling matter that is the signature of a black hole. Rather, one sees outgoing matter and radiation, which agrees very well with the picture presented above.

9.9 Cosmological models

We write the field equations (9.83) for the Robertson–Walker line element and for matter in the form of dust, whereupon they reduce to essentially two independent equations:

$$2\frac{\ddot{S}}{S} + \frac{\dot{S}^2 + k}{S^2} = 3\lambda + 2\pi Gf\dot{C}^2, \tag{9.88}$$

$$\frac{3(\dot{S}^2 + k)}{S^2} = 3\lambda + 8\pi G\rho - 6\pi Gf\dot{C}^2, \tag{9.89}$$

where we have set the speed of light as $c = 1$ and the density of dust is given by ρ. From these equations we get the conservation law in the form of an identity:

$$\frac{\mathrm{d}}{\mathrm{d}S}[S^3(3\lambda + 8\pi G\rho - 6\pi Gf\dot{C}^2)] = 3S^2(3\lambda + 2\pi Gf\dot{C}^2). \tag{9.90}$$

This law incorporates 'creative' as well as 'non-creative' modes. We will discuss both, in that order.

9.9.1 The creative mode

This has

$$T^{ik}_{;k} \neq 0, \tag{9.91}$$

which, in terms of our simplified model becomes

$$\frac{\mathrm{d}}{\mathrm{d}S}(S^3\rho) \neq 0. \tag{9.92}$$

For the case $k = 0$, we get a simple steady-state de Sitter-type solution with

$$\dot{C} = m, \qquad S = \exp(t/P) \tag{9.93}$$

and from (9.88) and (9.89) we get

$$\rho = fm^2, \qquad \frac{1}{P^2} = \frac{2\pi G\rho}{3} + \lambda. \tag{9.94}$$

Since $\lambda < 0$, we expect that

$$\lambda \approx -\frac{2\pi G\rho}{3}, \qquad \frac{1}{P^2} \ll |\lambda|, \tag{9.95}$$

but will defer the determination of P to after we have looked at the non-creative solutions.

For the sake of completeness we may mention the cases $k = \pm 1$, for which the scale factor is different, although the rest of the quantities remain the same. Thus we have

$$S = \frac{1}{P} \cosh\left(\frac{t}{P}\right), \qquad \text{for } k = 1;$$

$$S = \frac{1}{P} \sinh\left(\frac{t}{P}\right), \qquad \text{for } k = -1. \tag{9.96}$$

Both these are in fact variations on the de Sitter metric. In both cases timelike Killing vectors exist, corresponding to the 'steady-state postulate'. However, we shall take the $k = 0$ case further.

The rate of creation of matter is given by

$$J = 3\rho/P. \tag{9.97}$$

As will be seen in the quasi-steady-state case, this rate of creation is an overall average made of a large number of small events. Furthermore, since the creation activity has ups and downs, we expect J to denote some sort of temporal average. This will become clearer after we consider the non-creative mode and then link it to the creative one.

9.9.2 The non-creative mode

In this case $T^{ik}_{;k} = 0$ and we get a different set of solutions. The conservation of matter alone gives

$$\rho \propto 1/S^3, \tag{9.98}$$

while, for (9.98) and a constant λ, (9.90) leads to

$$\dot{C} \propto 1/S^2. \tag{9.99}$$

Therefore, equation (9.89) gives

$$\frac{\dot{S}^2 + k}{S^2} = \lambda + \frac{A}{S^3} - \frac{B}{S^4}, \tag{9.100}$$

where A and B are positive constants arising from the constants of proportionality in (9.98) and (9.99).

We will next consider the solutions of (9.89) for the cases $k = 0$ and $k = \pm 1$, taking the former first.

9.9.3 Solutions for $k = 0$

We now find that the exact solution of (9.89) in the case $k = 0$ is given by

$$S = \bar{S}[1 + \eta \cos \theta(t)], \tag{9.101}$$

where η is a parameter and the function $\theta(t)$ is given by

$$\dot{\theta}^2 = -\lambda(1 + \eta \cos \theta)^{-2}[6 + 4\eta \cos \theta + \eta^2(1 + \cos^2 \theta)]. \tag{9.102}$$

Here, $\bar{S}$ is a constant and the parameter η satisfies the condition: $|\eta| < 1$. Thus the scale factor never becomes zero and the model oscillates between finite scale limits

$$S_{min} \equiv \bar{S}(1 - \eta) \leq S \leq \bar{S}(1 + \eta) \equiv S_{max}. \tag{9.103}$$

The density of matter and the C-field energy density are given by

$$\bar{\rho} = -\frac{3\lambda}{2\pi G}(1 + \eta^2), \tag{9.104}$$

$$f\dot{C}^2 = -\frac{\lambda}{2\pi G}(1 - \eta^2)(3 + \eta^2), \tag{9.105}$$

while the period of oscillation is given by

$$Q = \frac{1}{\sqrt{-\lambda}} \int_0^{2\pi} \frac{(1 + \eta \cos \theta)\, d\theta}{[6 + 4\eta \cos \theta + \eta^2(1 + \cos^2 \theta)]^{1/2}}$$

$$\equiv \frac{1}{\sqrt{-\lambda}}\xi(\eta). \tag{9.106}$$

The oscillatory solution can be approximated by a simpler sinusoidal solution with the same period:

$$S \approx 1 + \eta \cos\left(\frac{2\pi t}{Q}\right). \tag{9.107}$$

Thus the function $\theta(t)$ is approximately proportional to t.

Figure 9.5 shows the typical exact oscillatory cycle which is approximated by the solution (9.107), adjusted to have the same period and amplitude. We find that the exact solution has flatter crests and narrower troughs than does the approximate solution, which otherwise simulates the former well.

9.9.4 Solutions for $k = \pm 1$

In this case we have (9.100) given by

$$\dot{S}^2 = \pm 1 + \lambda S^2 + \frac{A}{S} - \frac{B}{S^2}, \tag{9.108}$$

where the plus sign corresponds to $k = -1$ and the minus sign to $k = +1$.

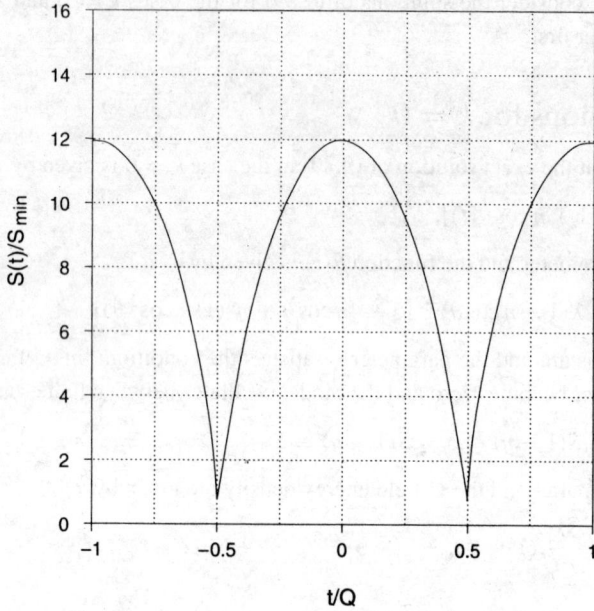

Figure 9.5 The scale factor of the exact solution for the flat case.

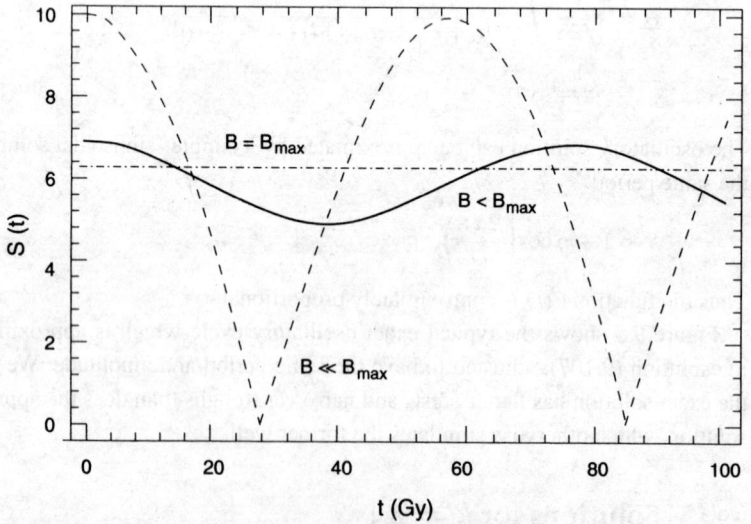

Figure 9.6 Typical scale factors for various parameters. The horizontal line denotes a static universe with $B = B_{max}$ ($- \cdot - \cdot -$). For $B \lesssim B_{max}$ we get small oscillations in the scale factor (———) and for $B \ll B_{max}$ we get large oscillations ($- - -$).

The various dynamical possibilities for the scale factor are shown in Figure 9.6 by the curves for the right-hand side of (9.108) as the parameters take different values. Notice that there is a special class of solutions in which the universe is static. Since the right-hand side of (9.108) must be non-negative, these curves tell us that the scale factor never becomes zero. Likewise, because of the negative cosmological constant, there is a finite upper bound on S. Hence the models all oscillate between finite radii. The generic solution is again of the kind given by (9.102) but with $\theta(t)$ satisfying the equation:

$$\dot{\theta}^2 = \pm \frac{1}{S^2(1 + \eta \cos \theta)^2} - \frac{\lambda[6 + 4\eta \cos \theta + \eta^2(1 + \cos^2 \theta)]}{(1 + \eta \cos \theta)^2}. \qquad (9.109)$$

In Figure 9.7 we show typical scale factors as functions of time for the cases $k = 0, \pm 1$. As expected, in the positive-curvature case the contracting phase is shorter and faster than it is for the zero- and negative-curvature cases. Also, the time period is shorter in the $k = +1$ case that it is in the $k = 0, -1$ cases.

9.10 The quasi-steady-state solution

The quasi-steady-state cosmology is described by a combination of the creative and the non-creative modes. For this the general procedure to be followed is to look for a solution of the form

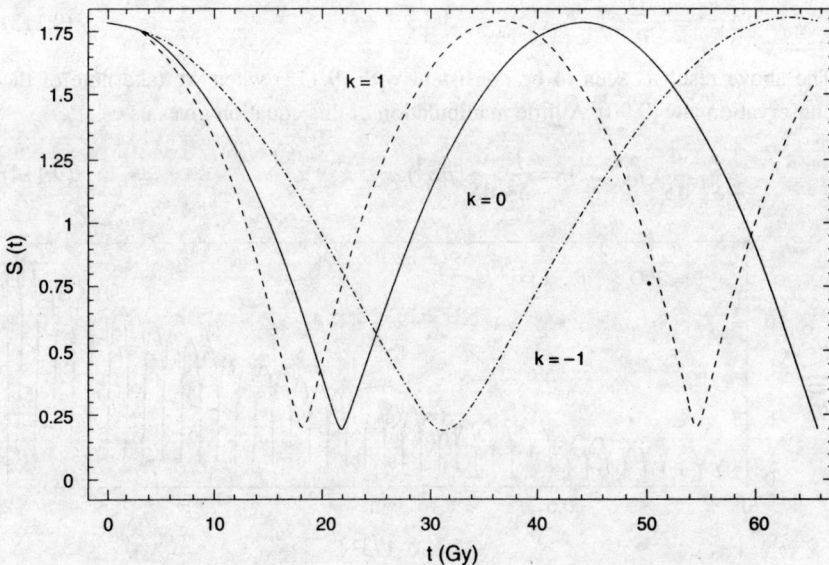

Figure 9.7 The influence of k on oscillations. The continuous line is the $k = 0$ case, the dashed line is for $k = +1$ and the dot–dashed line is for the $k = -1$ case.

$$S(t) = \exp\left(\frac{t}{P}\right)[1 + \eta \cos \theta(t)] \qquad (9.110)$$

wherein $P \gg Q$. Thus, over a period Q as given by (9.106), the universe is essentially in a non-creative mode. However, at regular instances separated by the period Q, it has injection of new matter at such a rate as to preserve an average rate of creation over period P given by J in (9.97). It is most likely that these epochs of creation are those of the minimum value of the scale factor during oscillation when the level of the C-field background is the highest (see Figure 9.8). Note that what appears as a cusp at a typical minimum in Figure 9.8 is in fact a continuous curve that turns around rapidly at $S = S_{\min}$.

Suppose that creation of matter takes place at the minimum value of $S = S_{\min}$ and that N particles are created per unit volume with mass m_0. Then the extra density added during this epoch in the creative mode is

$$\Delta \rho = m_0 N. \qquad (9.111)$$

After one cycle the volume of the space expands by a factor $\exp(3Q/P)$ and, to restore the density to its original value we should have

$$(\rho + \Delta \rho)e^{-3Q/P} = \rho, \qquad \text{i.e.,} \quad \Delta \rho / \rho \cong 3Q/P. \qquad (9.112)$$

The C-field strength likewise takes a jump at creation and declines over the following cycle by the factor $\exp(-4Q/P)$. Thus the requirement of a 'steady state' from cycle to cycle tells us that the change in the strength of $\dot{C}^2$ must be

$$\Delta \dot{C}^2 = \frac{4Q}{P} \dot{C}^2. \qquad (9.113)$$

The above result is seen to be consistent with (9.112) when we take note of the conservation law (9.90). A little manipulation of this equation gives us

$$\frac{3}{4} \frac{1}{S^4} \frac{d}{dS}(f \dot{C}^2 S^4) = \frac{1}{S^3} \frac{d}{dS}(\rho S^3). \qquad (9.114)$$

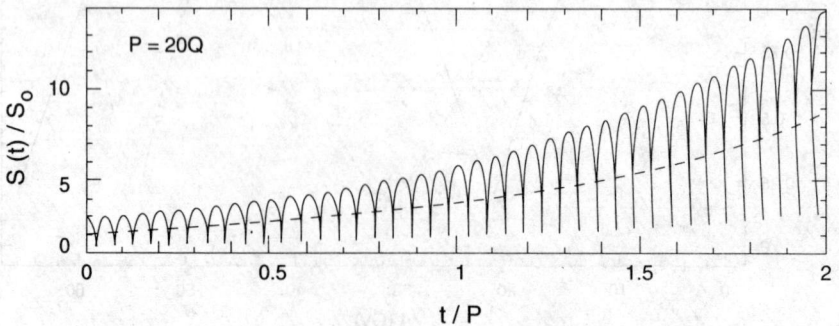

Figure 9.8 A curve showing the composite QSSC solution. The epochs of creation of matter are close to the minima of the oscillations of S.

However, the right-hand side is the rate of creation of matter per unit volume. Since from (9.113) and (9.114) we have

$$\frac{\Delta \dot{C}^2}{\dot{C}^2} = \frac{4}{3} \frac{\Delta \rho}{\rho} \tag{9.115}$$

and from (9.94) and (9.93) we have $\rho = f\dot{C}^2$, we see that (9.115) can be deduced from (9.112) and (9.113).

To summarize, we find that the composite solution properly reflects the quasi-steady-state character of the cosmology in that, although each cycle of duration Q is exactly a repeat of the preceding one, over a long time scale the universe expands with the de Sitter expansion factor $\exp(t/P)$. The two time scales P and Q of the model thus turn out to be related to the coupling constants and the parameters λ, f, G and η of the field equations. Further progress in the theoretical problem can be made after we understand the quantum theory of creation by the C-field.

These solutions contain a sufficient number of arbitrary constants to assure us that they are generic, once we make the simplification that the universe obeys the Weyl postulate and the cosmological principle. The composite solution can be seen as an illustration of how a non-creative mode can be joined with the creative mode. There may be more possibilities of combining the two within the given framework. We have, however, followed the simplicity argument (also used in the standard big-bang cosmology) by limiting our present choice to the composite solution described here. HBN have used (9.110), or its approximation

$$S(t) = \exp\left(\frac{t}{P}\right)\left[1 + \eta \cos\left(\frac{2\pi t}{Q}\right)\right], \tag{9.116}$$

to work out the observable features of the QSSC, which we shall highlight next.

9.11 The astrophysical picture

Coming down from these mathematical solutions to physical interpretation, we can visualize the above model in terms of the following values of its parameters:

$$P = 20Q, \qquad Q = 4.4 \times 10^{10} \text{years}, \qquad \eta = 0.8,$$
$$\lambda = -0.3 \times 10^{-56} \text{ cm}^{-2}, \qquad t_0 = 0.7Q. \tag{9.117}$$

These parametric values are not uniquely chosen; they are rather indicative of the magnitudes that may describe the real universe. In later chapters to follow, we will look at the various observational tests that can be employed to place constraints on these values. It is only then that we will know whether the QSSC provides a realistic and viable alternative to the big bang.

It is worth commenting, however, on a few issues of astrophysical nature. For example, the typical cycle has a lifetime long enough for most stars of masses exceeding $\sim$0.5–0.7$M_\odot$ to have burnt out. Thus stars from previous cycles would be mostly extinct as radiators of energy. Their masses will continue, however, to exert a gravitational influence on visible matter. The so-called dark matter seen in the outer reaches of galaxies and within clusters may very well be these stellar remnants.

To what extent does this interpretation tally with observations? Clearly, in the big-bang cosmology the time scales are not long enough to allow such an interpretation. Neither does that cosmology permit dark matter to be baryonic to such an extent. The constraints on baryonic dark matter in standard cosmology come from (i) the origin and abundance of deuterium and (ii) considerations of large-scale structure. In the QSSC, as we shall see in this chapter, these constraints are not relevant. We will discuss the origin of light nuclei next.

9.12   The origin of the light nuclei

We began the discussion of this cosmology with the creation of the Planck particle as the basic building block of matter. However, being unstable, the particle decays over times of the order of $\tau \sim 6 \times 10^{-44}$ s. Decays into what? In terms of mass, its decay can eventually produce $\sim 6 \times 10^{18}$ nucleons. However, such a decay cannot happen immediately. Considerations of degeneracy prevent such a dense packing, so that the resulting aggregate of secondaries is required to spread out initially as bosons. Only when a dimension of $\sim 10^{-7}$ cm has been reached can a swarm of $\sim 2 \times 10^{19}$ quarks appear. The situation in this respect is the same as that in the first moments of big-bang cosmology. Instead of the hot big bang we have a fireball around each Planck particle created.

However, whereas in big-bang cosmology the early Planck particles are contiguous to each other, here they are well separated, each producing its own cluster of expanding secondaries. A further difference is that the Planck particles here consist of matter decaying largely into matter, so that the overwhelming conversion of energy into radiation that occurs in big-bang cosmology does not happen here. It happens in big-bang cosmology because of an essentially contradictory double assumption. First the decay products of the Planck particles are said to be balanced between matter and antimatter and then the balance is said to be not quite perfect, as if to say -1 does not quite balance $+1$. The surviving remnant of matter embedded in a greatly dominant radiation field produces the low mass density in a high-temperature radiation field which is the essential feature of the relation (5.69) of Chapter 5, which, with the preferred value of the parameter η, the value that is considered necessary for the synthesis of the light elements in big-bang cosmology, gives

$$\rho \cong 10^{-32} T^3. \tag{9.118}$$

The challenge here is to produce a synthesis of the light elements with a very much larger coefficient than 10^{-32} in such a relationship. Energy lost into radiation early in the expansion of a Planck fireball goes quickly into the dynamical expansion of the secondary particles. Without knowing this loss at all precisely, it can be taken simply to produce a speed of expansion of the order of c. Thus the time scale for the expansion to a dimension of 10^{-7}–10^{-6} cm is of the order of 10^{-16} s and the density of particles when large numbers of baryons are at last produced is in the range 10^{36}–10^{40} cm^{-3}.

The discussion begins, however, at a much higher density than this, during a quark phase with the density of quarks high enough, above 10^{42} quarks cm^{-3}, for degeneracy to force the six quark flavours to be represented nearly equally. However, as expansion continues and degeneracy becomes less relevant, the high masses of the top, bottom and charmed quarks become an increasing energy burden and it is considered that transitions to the up, down and strange quarks occur within the available time scale. Because we are outside known laboratory physics, however, this step is necessarily conjectural, not in principle but in the quantitative details of the time scale for the occurrence of the relevant transitions.

The only electrons present are from pair production from γ-rays and, because electrons have negligibly short mean free paths, charge neutrality must be maintained among the quarks themselves. That is to say, an excess charge among the quarks cannot be balanced by electrons and, since any such excess must be small, again for energetic reasons, the strange quark cannot be dispensed with in the manner of the charmed, top and bottom quarks. To obtain charge neutrality, equal densities of up, down and strange quarks with charges $+\frac{2}{3}$, $-\frac{1}{3}$ and $-\frac{1}{3}$, respectively, are needed. This has the effect, when baryons are eventually formed in their lowest energy state through association of the quarks, of producing p, n, Λ, Σ^+, Σ^0, Σ^-, Ξ^- and Ξ^+ almost equally. There is a little Ω^- but this is effectively negligible. Hence, with n and p eventually combining into stable nuclei of ^{4}He and with the other six baryons not forming stable nuclei – with them eventually decaying into protons – the ultimate mass ratio of ^{4}He to H is $\sim$0.25.

These later stages in the expansion of Planck fireballs can be calculated quantitatively, as was first done by Hoyle in 1992. A radiation field becomes established through the 75 MeV decay of Σ^0 into Λ. Using this radiation field, it can be shown from detailed nuclear-reaction rates that only small fractions of the n and p remain uncombined into ^{4}He, giving a more precise mass ratio of ^{4}He and H of between 0.23 and 0.24. The relative abundances of the other light nuclei agree very well with the meteoritic values for the Solar System, as shown in Table 9.1.

Our purpose here is to raise two points about these results. First, it is remarkable that such results can be obtained under conditions relating density, temperature and time scale greatly different from those in the big-bang cosmology – if the present discussion were illusory this would hardly be expected. The second is that, with (9.118) no longer necessary, there is no reason why the present-day cosmological density, $\sim$$10^{-29}$ g^{-3}, should not be entirely baryonic. Already in 1967 the need for (9.118) to be understood in basic physical terms was strongly emphasized by

Wagoner, Fowler and Hoyle when they reworked the earlier Gamow problem of primordial nucleosynthesis. However, this understanding has not as yet been forthcoming in that theory.

The observation that ^{4}He/H is indeed about 0.23 by mass requires the observed matter to have emerged in the present theory from the decays of Planck particles and therefore to have been created at times in the past. Big-bang cosmology requires the same, with creation of particles occurring during the brief moments following the big bang. Here creation can be distributed in time, it can be an ongoing process.

9.13   The microwave background

In §9.4.3, in the context of the steady-state theory we had mentioned that simple considerations of the kind given in Exercise 34 suggest that the universe should have a radiation background with temperature close to 2.7 K. We now demonstrate this effect in the context of the QSSC.

9.13.1   The temperature of the background

Needed in this calculation is the present-day energy density of intergalactic starlight, which we regard as coming largely from dwarf stars, i.e. from old star populations, typically of spectral type K with a bolometric correction of about 0.75 mag. Thus the usual estimate of $\sim 10^{-14}$ erg cm^{-3} for the energy density in the visual spectrum becomes $\sim 2 \times 10^{-14}$ erg cm^{-3} for the total energy density of starlight.

Averaging over large volumes, write ε for the rate of production of starlight per unit volume per unit time. With starlight coming from suitably old stars, the rate of production ε can be considered approximately constant over the time scale Q, i.e., over an oscillatory cycle. We will use the approximate formula (9.116) for this computation. We will take $\eta = 0.75$ and $t_0 = 0.85Q$ to illustrate the calculation. These parametric values are slightly different from those in (9.117), but the difference does

Table 9.1 Ratios of abundances of light elements emerging from Planck fireballs

D/H	2×10^{-5}
^{3}He/H	2×10^{-5}
^{7}Li/H	10^{-9}
^{9}Be/H	3×10^{-11}
^{11}B/H	10^{-10}
^{6}Li/^{7}Li	10^{-1}
^{10}B/^{11}B	10^{-1}

not significantly affect the outcome. Then, with the time t in units of Q, the average production since the last oscillatory minimum is

$$\varepsilon \int_{0.5}^{0.85} \frac{dt}{1+z}. \tag{9.119}$$

This is per unit coordinate volume, the factor $1/(1+z)$ appearing in the integral because of the redshifting of every quantum. Setting (9.119) equal to the present-day total energy density of starlight, we determine ε via

$$\varepsilon \int_{0.5}^{0.85} \frac{1+0.75\cos(2\pi t)}{1+0.75\cos(2\pi t_0)}\, dt = 2 \times 10^{-14}, \tag{9.120}$$

whence, after evaluating the integral, we have

$$\varepsilon = 1.14 \times 10^{-13}\ \text{erg cm}^{-3}\ \text{per unit time,} \tag{9.121}$$

with unit time being Q. Thus the total production of starlight from the previous oscillatory minimum at $t = -0.5$ to that at $t = 0.5$ is

$$\varepsilon \int_{-0.5}^{0.5} \frac{dt}{1+z} = 4.56 \times 10^{-13}\ \text{erg cm}^{-3}, \tag{9.122}$$

(9.121) being used for ε.

Now, the weakening in such a cycle of the energy density, W_{min} say, at oscillatory minima of the microwave background due to the factor $\exp(t/P)$ in $S(t)$ is $4QW_{min}/P$ and, to maintain the background from cycle to cycle at a steady level, the input (9.122) arising from the thermalization of starlight must just compensate for this loss rate. Thus the steady requirement for $P/Q = 20$ gives

$$W_{min} = 1.14 \times 10^{-13} \frac{P}{Q} = 2.28 \times 10^{-12}\ \text{erg cm}^{-3}. \tag{9.123}$$

Since (9.122) is with respect to coordinate volume, there is a weakening of the background energy density by the fourth power of $S(0.5)/S(t_0)$ between the last oscillatory minimum and the present time $t_0 = 0.85$, i.e., by a factor 0.1734 to

$$W_{present} = 3.96 \times 10^{-13}\ \text{erg cm}^{-3}. \tag{9.124}$$

The cosmological scale factor being adjusted to be unity at $t = t_0$, this is also the present-day energy density of the microwave background per unit proper volume. Dividing (9.124) by the radiation-density constant and then taking the fourth root, we obtain the present-day radiation temperature as

$$T_0 = 2.68\ \text{K}, \tag{9.125}$$

which is very close to the observed value of 2.73 K. Considering the approximate nature of the estimates used for calculating the temperature, the agreement is striking indeed. This is especially so when we recall that, in the standard 'relic' interpretation of this background, its present-day temperature cannot be estimated at all. Although Alpher and Hermann had a good guess at it by proposing 5 K, in their 1948 paper, Gamow himself gave various estimates (all guesses) ranging from ~7 to ~50 K! However, the crucial aspect of this result, still to be justified, is a process of thermalization that will convert the relic starlight into a thermal spectrum.

9.13.2 Thermalization

In the big-bang cosmology thermalization of primordial radiation took place in the very early epochs and generally through the scattering of photons by electrons. In the QSSC, free electrons are not available to carry out the thermalization. Instead one has to look to intergalactic dust of a special kind for this purpose. In the 1980s, Fred Hoyle and Chandra Wickramasinghe proposed whisker-shaped particles of metallic dust as the ideal agents for this job. Where can such dust grains arise? As it turns out, supernovae provide the ideal setting for this purpose. The mini-creation events of the QSSC can also act as sources.

It is estimated that, in a typical type-II supernova of the kind SN 1987A, the total mass of iron produced is $\sim 0.1 M_\odot$. The metal is produced in an iron-rich shell and cools down from a vapour state at an initial temperature of 10^{10} K and density 10^9 g cm^{-3} to a state of temperature 10^3 K and density 10^{-15} g cm^{-3}. During this period nuclei of elements of the iron group, especially Ni^{56}, would decay to Fe^{56} too and condensation of the iron vapour begins. Details of the growth process indicate that initial condensation is spherical to a size of 0.01 μm radius. Thereafter radioactive decays of Co^{56} define a dislocation axis along which subsequent whisker growth takes place. This growth is linear and exponential with respect to time, the length l at time t being

$$l = l_0 \exp(t/t_1), \tag{9.126}$$

where the characteristic time t_1 is given by the various parameters of the condensation process. In the steady state, therefore, the condensates will be whiskers of varying lengths with a length distribution given by

$$n(l)\,dl = B\,dl/l, \qquad l_{min} \le l \le l_{max}, \qquad B = \text{constant}, \tag{9.127}$$

$n(l)\,dl$ being the number density of whiskers of length between l and $l + dl$. The upper limit on their length arises from breakages of whiskers through collisions. Apart from iron, carbon produced in and ejected from the supernova also can appear in whisker shape.

There is laboratory evidence for metallic vapours condensing into whiskers. The process of condensation results in growth of metal grains not in an isotropic form, but, because of the crystal defects, in a linear form. Thus whiskers rather than spherical balls of dust condense out.

Although some of the whiskers so produced may remain confined to the neighbourhood of the source, not all of them need do so and most may escape into the intergalactic space because of their high velocities acquired from the supernova shock wave. The estimated mean mass density of these whiskers may be as low as 10^{-35} g cm^{-3}. They nevertheless may produce several observable effects. To understand them we need to look at their extinction properties.

Typically an iron whisker may be a cylinder of cross-sectional radius 0.01 μm, and length in the range ∼5–10 mm. Figure 9.9 shows the mass extinction coefficient for a population of such grains distributed according to the law given by (9.127) above. Notice the significant peak in the millimetre–centimetre range, which is supported by the observations outlined below.

1. *The Crab pulsar.* The continuum spectrum of the Crab pulsar has a dip in the
   wavelength range 30 μm to 30 cm, which can be best accounted for by the
   presence of whiskers originating in the Crab supernova and still remaining in
   the neighbourhood. Figure 9.10 shows this effect.

2. *The Galactic Centre.* Since it is expected to be rich in supernovae, we expect
   the Galactic Centre to have a greater than normal density of whiskers. The
   C II and N II lines at wavelengths 157.7 and 205.3 μm, respectively, exhibit a
   dip in the surveys carried out in the region of the Galactic Centre, a fact that

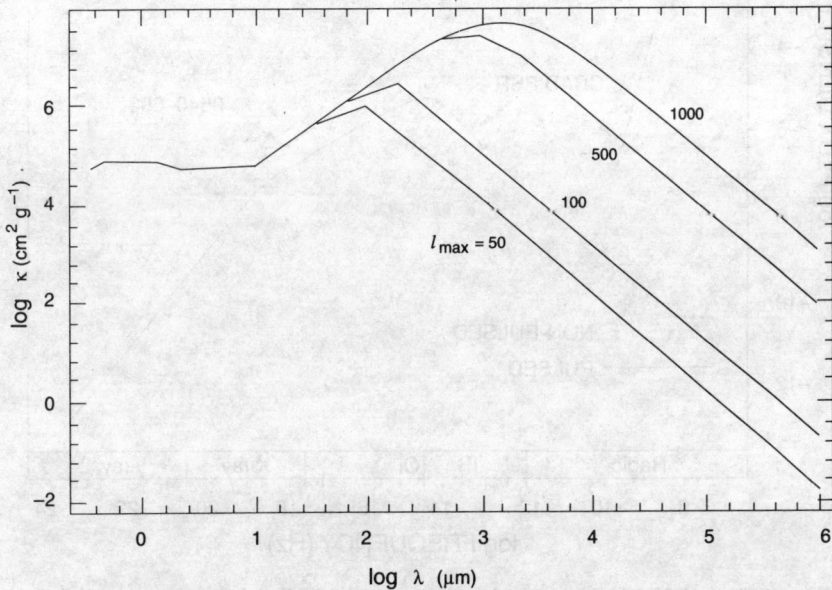

Figure 9.9 The mass extinction coefficient plotted against wavelength, for whiskers of various lengths expressed in micrometres. Notice that the predominant absorption is in the millimetre wavelengths, whereas it is negligible for radio waves.

is readily explained by invoking the presence of whiskers of lengths up to 100 μm.

3. *Dust mass in high-redshift sources.* Wickramasinghe and Ramadurai have shown that the requirement of dust mass needed to explain the observed high millimetre-wave fluxes in sources of high redshift comes down considerably (by a factor of $\sim 10^3$) if the more efficient convertors of ambient radiation into these waves, viz. metallic whiskers are assumed to be present. The recent analyses of high-redshift dusty galaxies otherwise demand a very high proportion of dust mass, if it is assumed to be in the normal spherical form. The mass estimates come down by a large factor if whiskers are assumed to constitute the bulk of the dust.

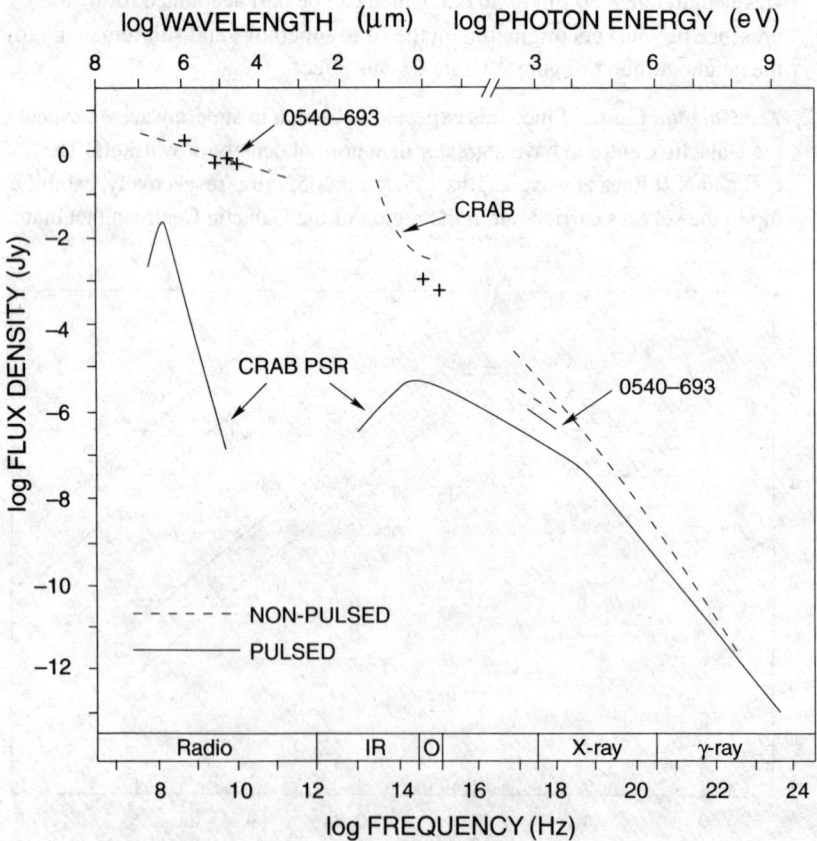

Figure 9.10 This multi-wavelength spectrum of the Crab Pulsar shows that there is a dip in the millimetre range, which is indicative of an extinguishing agent in the vicinity. Metallic whiskers offer a natural explanation of this observation when we consider their mass absorption coefficient shown in Figure 9.9.

4. *Spectra of radio quasars and active galactic nuclei.* As in the case of the Crab pulsar, the spectra of radio sources and active galactic nuclei exhibit dips in the continuum, which are indicative of absorption, in the millimetre-wavelength region. This is most easily explained as the result of extinction by whiskers.

5. *Intergalactic absorption.* As will be discussed in Chapter 12, the existence of intergalactic dust in the form of metallic whiskers appears to provide a good explanation for the observed magnitude–redshift relation of type-Ia supernovae at large redshifts.

While iron whiskers are likely to be the dominant source of opacity in the microwave region, carbon whiskers considerably dominate at shorter wavelengths, for carbon is produced in greater quantities by stars and its efficiency is greater at shorter wavelengths. In the following argument we denote by the parameter x the ratio $S(t)/S_{\min}$ during a typical cycle. Thus, if the maximum redshift observable during this cycle today is 5, then $x = 6$ at present.

The absorption coefficient for graphite whiskers is essentially constant for all wavelengths longer than ~ 1 μm, extending even to long radio wavelengths, and is 10^5 cm^2 g^{-1} for whiskers of diameter 0.01 μm and length ~ 1 mm. On the other hand, the absorption coefficient is about three times greater than this for ultraviolet radiation, $\sim 3 \times 10^5$ cm^2 g^{-1}, requiring an intergalactic density of $\sim 10^{-34}$ g cm^{-3} to produce appreciable absorption of ultraviolet light as the parameter x decreases below 6, as the minimum-scale epoch is approached in the contracting phase of each cycle. Thus an intergalactic density of $\sim 10^{-34}$ g cm^{-3} at $z_{\max} = 5$ would rise to $\sim 2 \times 10^{-32}$ g cm^{-3} at $z = 0$. Over a cosmological distance of 10^{27} cm for this epoch this would give optical depths of ~ 6 for ultraviolet light and ~ 1 for wavelengths longer than 1 μm.

The great bulk of the optical radiation in the form of starlight that is thermalized has thus been travelling for 10^{11} years in each contracting phase and for even longer in the case of microwaves. The radiation incident on a carbon whisker has mostly been in propagation through the last maximum phase of the oscillation. This includes all the microwave radiation existing before the present oscillation and all the starlight generated by galaxies during the first half of the current cycle, with the consequence that all such radiation is exceedingly uniform in its energy density. It is not necessary, however, that the carbon whiskers responsible for absorbing the starlight and re-emitting it as microwaves are uniformly distributed. The carbon whiskers can be lumpy on the scale of clusters of galaxies. This means that the conversion of starlight into microwaves will occur lumpily as the minimum of the oscillation is approached.

Nevertheless, each carbon whisker, wherever it is situated, finds itself in a radiation bath of uniform energy density, a radiation bath of which the major fraction already consists of microwaves from previous cycles and the rest is mostly starlight that has still to be converted into microwaves. Had the whole of the radiation been in microwaves, then the flatness of the absorption coefficient with respect to wavelength through the range longward of ~ 1 μm implies that the temperature attained by the particles would be simply the standard microwave temperature observed at present, which is ~ 2.73 K. However, because a fraction, say 10%, of the radiation is ultraviolet, blueshifted from the decreasing values of the parameter x, the stellar component has the higher value of the absorption coefficient discussed above. This forces up the temperature T_g of the grains, to a value slightly higher than ~ 2.73 K, namely to

$$T_g \sim [0.9 + (3^{1/4} \times 0.1)]2.73 \simeq 2.82 \text{ K}, \tag{9.128}$$

the second term in the brackets being contributed by the absorption of the starlight. As the starlight is progressively absorbed with optical depth τ, the factor 0.1 in this equation is replaced by $0.1 \exp(-\tau)$ and the factor 0.9 is replaced by $1 - 0.1 \exp(-\tau)$, so that the temperature of grains varies according to

$$T_g[1 + 0.1e^{-\tau}(3^{1/4} - 1)] \times 2.73 \text{ K}. \tag{9.129}$$

Thus, as the starlight begins to be absorbed, the temperature of whiskers goes about 0.1 K higher and then falls back to 2.73 K as the starlight is progressively absorbed.

Since the emissivity of the particles has no wavelength dependence, they simply emit a Planck distribution $1/\{\exp[h\nu/(kT)] - 1\}$ at whatever value of T_g they may happen to have, according to the above equation giving temperatures that may range up to about 2.82 K. However, in general they do not produce the Planck intensity $\nu^3/\{\exp[h\nu/(kT)] - 1\}$. When T is raised slightly to 2.82 K the intensity distribution is slightly diluted.

So what is the outcome from this first absorption of the starlight? It is a uniform energy density of microwaves with a distribution approximately that of a blackbody at 2.73 K, but with some fluctuations in the details of the intensity curve, with those details having initially a somewhat uneven distribution to the extent that the carbon whiskers are distributed unevenly.

Now, with the starlight absorbed and the temperature of the particles everywhere the same, further absorption and re-emission inevitably generates the strict Planck distribution for 2.73 K. Only a few further absorptions at the oscillatory minimum are sufficient for this last step. It can be done with carbon whiskers, as Narlikar, Edmunds and Wickramasinghe suggested as long ago as 1975. However, the addition of even a small quantity of iron whiskers, with very high opacity at the centre of the microwave distribution, would make this final step even more decisive.

The essential point is the one already emphasized, that expansion through an oscillatory maximum produces mixing of radiation coming from very large distances, $\sim 10^{29}$ cm, and it does so with the intergalactic density of particles low, permitting radiation to travel freely.

9.13.3 Anisotropy of the background

The effect of τ above, being lumpily distributed on the scale of clusters of galaxies, causes this process of a slight rise in temperature followed by a fall back to 2.73 K to be correspondingly lumpy. *But what it does not do is make the total energy density of radiation lumpy at all.* Once the energy density of radiation is uniform in total, this essential property is not changed by absorption and re-emission due to particles. Because of course each particle emits just as much energy as it absorbs – the total assembly of particles has itself only a negligible heat content – the particles do not store heat except in a very small amount. Thus, the first impression that a lumpy distribution of absorbers will produce a lumpy distribution of radiation is not correct.

However, there will always be some radiation emitted by some local source, like a galaxy, that is capable of producing a small fluctuation in the energy density. Galaxies in which the light comes mostly from main-sequence dwarfs would have an absolute magnitude of about -21, that is to say an emission of about 10^{44} erg s^{-1}. A nearby galaxy, at a characteristic distance of about 3 Mpc, would contribute an energy density of $\sim 3 \times 10^{-18}$ erg cm^{-3}, compared with a cosmological energy density of starlight of $\sim 10^{-14}$ erg cm^3, i.e., a local variation of the energy density amounting to one part in 3000. This leads to variability in the temperature of the thermalizing particles by one part in 12 000, which, however, becomes negligible when one takes into account all the thermalizing particles reaching out to the Olbers limit.

A more extreme case of fluctuation generated by local sources comes from rich clusters of galaxies. Thus, for a dust particle lying 3 Mpc away from a rich cluster of 1000 galaxies, the local modification of the cosmic field of starlight (10^{-14} erg cm^{-3}) is $1000 \times 3 \times 10^{-18}$ erg cm^{-3} = 3×10^{-15} erg cm^{-3}. This is a fluctuation in the energy density of about 30%, which is capable of changing the temperatures of particles from 2.73 K by an amount that is 30% of the change calculated above, which was from 2.73 K to about 2.83 K. The change is thus by about 3×10^{-2} K.

When the sky is examined on an angular scale such that beam widths are compared, one containing a rich cluster of galaxies and the other not, then we shall expect to find a detectable variation in the effective background temperature. However, the amount of the variation will be by no means as large as the value of 3×10^{-2} K just calculated. This is because the particles for which this variation applies must lie within about 3 Mpc of the cluster of galaxies and this purely local fluctuation has to be considered together with all the thermalizing particles along the line of sight out to the Olbers limit. The latter reduce the relative magnitude of the fluctuation by a factor of the order of 1000, to of the order of 3×10^{-5} K, or about 30 μK, in good general agreement with the observed fluctuations, as we shall see in the next chapter.

To summarize, therefore, unlike the situation in the big-bang cosmology, the fluctuations of the microwave background are nothing but minor deviations occasioned by spatial variations in the energy density of starlight due to the inhomogeneous distribution of galaxies.

We have therefore deduced the energy of the microwave background, its spectrum and also the amplitude of its fluctuations. The QSSC therefore solves the problem which the steady-state theory failed to solve. However, there are other cosmological tests too and it will be interesting to see how the cosmology fares *vis-à-vis* the standard cosmology there. In the following two chapters we will take up the observational evidence of which any cosmological model must take cognizance. We will return to the QSSC in Chapter 12.

9.14   Large-scale structure

Unlike the big-bang cosmology, in which structures have to evolve out of primordial inhomogeneities that are put in by hand, here the problem is that of how to reproduce the structure in the present cycle from what existed in the previous ones. Since the mini-creation events play a pivotal role in this cosmology, it is expected that new nuclei of creation would grow out of matter ejected from them.

Nevertheless, it is worth seeing first how the gravitational instability grows in this cosmology. In 1997, Banerjee and Narlikar adopted the following approach. The metric, the density and the C-field were perturbed and, by restricting to only first-order quantities, the changes in these perturbations were calculated in the background spacetime. Predictably, the density inhomogeneities grew during the contracting phase of an oscillation and were damped during the expanding phase. However, overall there was no significant instability in the solution. While this generates confidence in the robustness of the basic solution, it also forces one to look for non-gravitational effects to produce structure. The process of creation provides a possibility. In a recent attempt to understand how structures may grow and become distributed in space, the following numerical experiment suggested by Fred Hoyle was tried by Ali Nayeri, Sunu Engineer and the author.

A large number of points ($N \sim 10^5$–10^6) was distributed over a square area at random. Each point was made to produce a random neighbour within a specified fraction x of the average interparticle distance of the original set. The area was then scaled to twice the original size, so that the density of particles remained the same. Then, from the expanded area, a central portion corresponding to the original area was retained, the rest being thrown away. With this new square the experiment was repeated.

Very soon, i.e., after three or four iterations of the above procedure, clusters and voids began to appear in the picture and voids grew in size while the clustering became denser as the experiment was repeated. If the creation of the new neighbour

B around a typical point A was not entirely random, but linked to the previous history of creation of A, so that the direction AB was broadly aligned with the direction in which A had been ejected, then the filamentary structure grows together with voids. The latter alignment may be explained by assuming that the creation centre houses a spinning collapsed object (like a Kerr black hole); for then the created particles are preferentially ejected along the axis of spin. Pictures generated in this way are very suggestively similar to the observed large-scale structure.

The experiment has been repeated in three dimensions and slices of two dimensions examined for structures. Again these look remarkably similar to the filaments and voids found in redshift surveys. To bring the experiment closer to the dynamics of the QSSC, the initial cube is expanded by a factor $\exp(Q/P)$ in each direction and only a fraction

$$f = [1 - \exp(-3Q/P)] \tag{9.130}$$

of the original set of points is allowed to produce new neighbours. Preliminary work by A. Nayeri shows that filaments and voids begin to appear after a few iterations. See Figure 9.11, for example.

How does the clustering seen in simulations like that in Figure 9.11 compare with the observed one? An exercise to measure the two-point correlation function

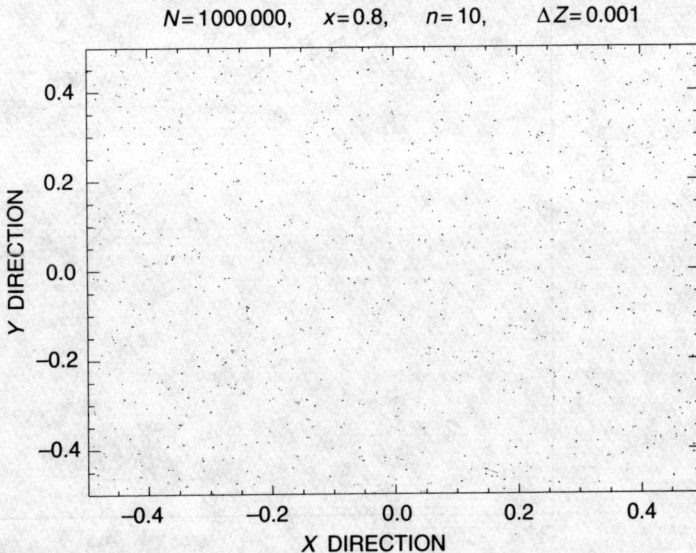

Figure 9.11 A simulation of large-scale structure in the QSSC, based on the creation of one generation of clusters in the vicinity of those of the previous generation, keeping aligned ejection from one generation to the next. Filaments and voids develop on going from one generation to next. The figure shows the result after ten iterations, for the parametric values given at the top. The points represent a distribution in a thin slice in the Z direction of a cubic distribution.

ξ, defined in Chapter 7, for observed simulations after successive iterations was undertaken. As shown in Figure 9.12, the $\xi-r$ relationship quickly approaches the observed power-law spectrum with index -1.8. This therefore suggests that the present approach to structure formation, although it is simple in nature, could be close to reality.

9.15   Concluding remarks

The above ideas reflect efforts to understand the phenomenon of creation of matter within a well-defined physical framework. The steady-state theory of Bondi and Gold deduced that the creation of matter was an inevitable consequence of the perfect cosmological principle. Hoyle's approach brought in the more familiar field

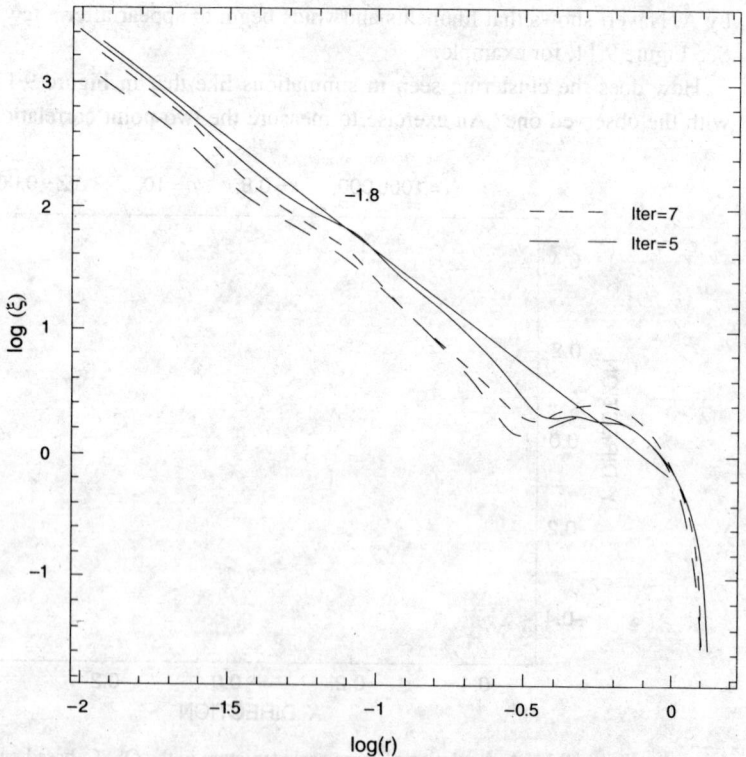

Figure 9.12 A two-point correlation function for simulations of the kind shown in Figure 9.11 has a power-law-type distribution with index approaching -1.8 after a few iterations of the process of creation. The evolution of the index towards the ideal -1.8 slope line (shown by the continuous line) is shown as the process goes on for up to ten generations.

theory to explain the phenomenon within the framework of general relativity. The quasi-steady-state cosmology, on the other hand, starts with Mach's principle and from the more general gravitational equations 'descends' to general relativity, but with a cosmological constant and a scalar field signifying creation of matter. The QSSC essentially begins where the old steady-state theory left off, and can claim to have resolved some of the outstanding issues that troubled the former version.

The immediate successes of the theory are the following.

1. The circumstance that G as determined by all the particles in the universe is necessarily positive requires gravitation to act as an attractive force. Aggregates of matter must tend to pull together. This is unlike general relativity, for which gravitation can equally well be centrifugal, with aggregates of matter always blowing apart, as follows if G in the action of general relativity is chosen to be negative.

2. In the cosmological case with homogeneity and isotropy, the pressure contributed by the C-field term in the gravitational equations is negative, explaining the expansion of the universe. Thus the expansion is not ascribed to a mysterious primordial process like the 'big bang'.

3. Also in the cosmological case, the energy contribution of the C-field is negative, which ensures that, when the conditions for creation are satisfied, creation tends to cascade, with explosive consequences.

4. The magnitude of the constant λ is shown to be of the order needed for cosmology. Unlike big-bang cosmology this is a deduction, rather than an assumption; and it results in an answer that is observationally of the right order of magnitude. The sign of λ is, however, negative, i.e., opposite to that assumed usually.

5. In the typical QSSC model the universe has a long-term de Sitter-type expansion, superposed on short-term oscillations between finite minimum and maximum scale factors. Thus the solution represents a non-singular universe.

6. With newly created particles shown to be Planck particles, their decay provides a means for generating the light elements.

7. The microwave background is related directly to the relic starlight from previous cycles of the QSSC, its thermalization resulting in a blackbody distribution with temperature ~ 2.7 K. Unlike the big-bang cosmology, we have here an astrophysical explanation of why the present background has a temperature of this magnitude.

8. The presence of dark matter can be understood in terms of remnants of stars of previous cycles whose light appears as the microwave background.

9. The success of the toy model developed for structure formation via mini-creation events encourages further steps in that direction.

This revised steady-state concept, which is being used now in a more effective way than in the old classical steady-state theory, shifts the cosmological problem to one in which the parameters on which calculations are based can be determined from astronomical observations. They are not guesses as in big-bang cosmology. In the remaining chapters we will subject the big-bang cosmology and also the QSSC to more rigorous observational scrutiny and compare how well they perform.

Exercises

1 Discuss the considerations that led to the formulation of the steady-state cosmology. Are any of these considerations valid today?

2 What is the perfect cosmological principle (PCP)? What shortcoming of the ordinary cosmological principle is it designed to remove?

3 By considering various astronomical objects, arrive at a length scale over which you would expect the PCP to apply. What are the corresponding time scales over which you would expect the universe to obey this principle?

4 Compute the scalar curvature R for the Robertson–Walker model. Use the PCP to demand that R is constant and show that this leads to S as a specific function of t for $k = 0, \pm 1$. Use the constancy of Hubble's constant to deduce that $k = 0$.

5 Show that the deceleration parameter for the steady state universe is equal to -1 for all epochs.

6 Deduce from the PCP and the local observation of a departure from thermodynamic equilibrium that the steady-state universe must expand.

7 Show that, if the steady-state universe has a proper number density n of sources each radiating with luminosity L, then the total intensity of light in a solid angle $d\Omega$ of the sky is given by $F\,d\Omega$, where

$$F = \frac{1}{16\pi} Ln \frac{c}{H_0}.$$

Estimate F by substituting characteristic values of L and n for galaxies and deduce that the night sky is quite dark.

8 Discuss the validity of the following statement: 'Of the various ways of resolving Olber's paradox, the only way open to the steady-state model is that of the expansion of the universe'.

9 Show that, according the the C-field cosmology, the rate of creation of matter needed to sustain the steady-state model is given by

$$Q = 4 \times 10^{-46} h_0^3 \, \text{g cm}^{-3} \, \text{s}^{-1}.$$

10 Using the result of Exercise 9, express the rate of creation in terms of solar masses per year per cubic megaparsec.

11 Discuss the merits and limitations of the C-field cosmology.

12 It is claimed that the steady-state theory is readily testable and therefore more prone to an observational disproof than is the standard big-bang cosmology. Give examples to justify this claim.

13 Compare and contrast the C-field and the electromagnetic field.

14 Show with the example of the C-field that path-independent terms in the action can lead to non-trivial results.

15 Show that, in the absence of any other forces, a particle created in the C-field cosmology follows a geodesic.

16 Show that, in the C-field theory, overall energy and momentum are conserved when a particle is created.

17 Derive the form of the energy momentum tensor of the C-field by considering the variation of g_{ik} and setting the first-order variations of the action equal to zero. Show that

$$T^{ik}_{(C);k} = -fC^i C^k_{;k}.$$

Evaluate this relation near the point in spacetime where a particle is created and deduce the law of conservation of the 4-momentum.

18 Discuss the physical implications of the negativity of the C-field energy.

19 Obtain the cosmological equations (9.13)–(9.16). Derive the general solution of these equations for the case $k = 0$.

20 Consider a perturbation of the steady-state line element of the following form:

$$ds^2 = g_{ik} \, dx^i dx^k, \qquad x^0 = ct,$$

$$g_{00} = (1 + h_{00}), \qquad g_{0\mu} = 0, \qquad g_{\mu\nu} = -(\delta_{\mu\nu} + h_{\mu\nu})e^{2H_0 t},$$

where the h_{ik} are general functions of spacetime coordinates. Furthermore, take the density and the flow vector as

$$\rho = \frac{3H_0^2}{4\pi G} + \rho_1, \qquad u^i = (1, 0, 0, 0) + u_1^i.$$

Treating ρ_1, u_1^i and $h_{\mu\nu}$ as small quantities of the first order, show that they decay with time as

$$\rho_1 = Ae^{-3H_0 t} + Be^{-5H_0 t}, \qquad u_1^i = \bar{u}^i e^{-5H_0 t},$$

$$h_{00} = 0, \qquad h_{\mu\nu} = \alpha_{\mu\nu} + \beta_{\mu\nu}e^{-2H_0 t} + \gamma_{\mu\nu}e^{-3H_0 t} + \epsilon_{\mu\nu}e^{-5H_0 t}.$$

The functions A, B, $\bar{u}^\mu$, $\alpha_{\mu\nu}, \dots, \epsilon_{\mu\nu}$ depend on x^μ. Prove that even the inhomogeneity corresponding to $\alpha_{\mu\nu}$ becomes less and less important as the universe expands.

21 Deduce that the flow vector of created matter has zero spin. What implication does this result have for Mach's principle?

22 Deduce the solution (9.18) for the process of explosive creation. Relate the parameters of your solution to the amount of matter created in the universe. In particular, show that

$$t_1 = t_0$$

and that the maximum density occurs at $t = 0$ and is given by fm^2.

23 Consider a reservoir of energy ξ in a volume V that expands. Show that, if $\xi > 0$, expansion as well as conversion of energy to matter will reduce ξ to zero in a finite time. Show further that, if $\xi < 0$, this conclusion is drastically altered.

24 Show that, in the steady-state cosmology, the redshift of a galaxy is proportional to its radial proper distance from us.

25 Show that steady-state cosmology does not have a particle horizon, but that it does have an event horizon of proper radius c/H_0. That is, show that a galaxy whose radial proper distance from us exceeds c/H_0 cannot ever communicate with us.

26 Estimate the difference between the apparent bolometric magnitudes of a galaxy of redshift $z = 1$ computed according to the steady-state model and the Friedmann model with $q_0 = 1$.

27 A family of radio sources with the same luminosity and with an energy spectrum given by $\sim \nu^{-1}$ as a function of frequency ν are being counted in the steady-state universe. Show that the flux density S varies with the source redshift z as

$$S \propto z^{-2}(1+z)^{-2}.$$

Calculate the slope $d\log N/d\log S$ as a function of z, where N is given by (9.25). Tabulate this function for $z = 10^{-n}$, $n = 4, 3, 2, 1, 0$. What do you conclude from this table?

28 In a universe with the line element

$$ds^2 = c^2\,dt^2 - S^2(t)[dr^2 + r^2(d\theta^2 + \sin^2\theta\,d\phi^2)],$$

$Q(t, \tau)\,d\tau$ denotes the proper number density of galaxies during epoch t with ages between τ and $\tau + d\tau$. Suppose that $\eta(t)$ denotes the rate (per unit proper volume) at which new galaxies are being injected into the universe. Show that $Q(t, \tau)$ satisfies the differential equation

$$\frac{\partial Q}{\partial t} + \frac{\partial Q}{\partial \tau} + 3\frac{\dot{S}}{S}Q = \eta(t)\delta(\tau).$$

Deduce from this equation the age distribution of galaxies in the steady-state universe.

29 Look up the rest-mass energies of the neutron and the proton in the table of constants at the end of this book. Assuming that $\sim$20% of this difference in energy is acquired by the electron in a β-decay of the neutron, estimate the velocity and the kinetic temperature of the electron.

30 For a density of hydrogen atoms of 2×10^{-5} cm^{-3} and a kinetic temperature of 10^9 K, estimate the velocity of sound. Equate this to the expansion velocity $H_0 D$ according to Hubble's law and estimate the distance scale D of the irregularity that would develop in the Gold–Hoyle hot universe in which thermal pressures are pitted against the force of universal expansion.

31 Compare the bubble universe with the inflationary models.

32 Discuss the bubble universe. Do you see any similarity between the way the hot universe generates spatial inhomogeneities and the way the bubble universe generates temporal unsteadiness? What happens to the PCP in these models?

33 An expanding bubble may be considered as a cloud of gas moving radially outwards. In a uniform spherical bubble with mass $M(r)$ within radius r the expansion is given by

$$\dot{r}^2 = 2GM(r)/r.$$

Suppose next that a supermassive object of mass μ appears at the origin when $r = r_0$. Show that the cloud now expands to a maximum radius given by

$$r_{max} = \left(1 + \frac{M}{\mu} \right) r_0.$$

In the bubble-universe theory this idea serves as a basis for forming an elliptical galaxy as a cloud of gas out to r_{max} that is gravitationally controlled by a supermassive object at the galactic nucleus.

34 Assuming that our Galaxy has been radiating at the rate of 4×10^{43} erg s^{-1} for 3×10^{17} s and that this energy is derived from conversion of hydrogen into helium, estimate how much helium is formed in this way. (6×10^{18} erg g^{-1} is released when hydrogen is converted into helium.) Comment on this answer in relation to the primordial mass fraction of helium obtained in Chapter 5.

35 Assuming that there is thermal radiation present today, with temperature 2.7 K, place a lower limit on the parameter B in equation (9.100) to guarantee a non-singular model.

36 Determine the parametric values that ensure static solutions of the equation (9.100). To what extent is the condition $\lambda < 0$ needed for such solutions?

37 Why does the abundance of deuterium not restrict the density of baryons in the QSSC?

38 Write down the expression for the angle subtended at the observer by a spherical cluster of radius R at $z = z_{max}$ in the QSSC. Relate this expression to the angular scale of anisotropy of the microwave background in the QSSC.

39 Explain why the QSSC does not have an Olbers type problem of darkness of the night sky.

40 Show that the QSSC model has an event horizon, but no particle horizon.

Chapter 10

Local observational tests of cosmological significance

10.1   **Introduction**

The discussions so far have largely been theoretical: it is high time that we related theories described so far to reality. This can be done through various observational checks on the predictions of the cosmological models. Recall that, in the very early days, the static universe of Einstein was disproved by Hubble's observations of redshifts, observations that led eventually to the acceptance of the basic hypothesis that the universe is expanding.

This example tells us about another aspect of the theory-versus-observation debate, that is sometimes forgotten. Hubble's observations *ipso facto* did not say that the universe is expanding. The factual part of the observations was that there is a nebular redshift that increases with apparent magnitude. Given this fact, we interpret the redshift as arising from the expansion of the universe, through equation (3.51), and likewise relate the apparent magnitude to distance through equations (3.55)–(3.58). Thus it is our interpretation of Hubble's observations through the expanding models that needs to be checked further by more sophisticated observations. For it may be that there are alternative interpretations of Hubble's observations, such as the one described in Chapter 8 (see equation (8.57)) and at some stage the cosmologist may wish to check which of the two interpretations is correct. Another interpretation, that we have not touched at all in this book, is that of the 'tired' photon. If the photon has a small rest mass and it loses energy through interaction with matter as it travels across the universe, it will be redshifted, with the redshift increasing with distance. How do we test this hypothesis? In short, the same basic data may find different interpretations in different theories, so it then becomes necessary to go further and think of more sophisticated tests to distinguish among the various interpretations.

We will go by Karl Popper's view that a scientific theory can be disproved but never proved by any tests. If a test gives results significantly different from the predictions of a theory, then the theory stands disproved. However, if the test gives results as predicted by the theory then all we can say is that the theory is *consistent with* the results of the test. While being proved consistent will improve the credibility rating of the theory, it continues to remain on probation.

In this chapter we will review those astronomical observations that attempt to determine the large-scale structure of the universe from relatively local surveys. These tests do not tell us directly about the large-scale geometrical structure of the universe, since they do not extend far enough. Nevertheless, we shall see how even local measurements indirectly place restrictions on what can be said about the distant parts of the universe. This may sound paradoxical, but it is a consequence of the symmetry assumptions made by most models of the universe, in particular the cosmological principle.

While relating cosmological models to such observations we will mainly discuss the standard hot-big-bang cosmology. As described in the earlier chapters, these models can be classified in the following way.

- Mark I: The dust-dominated Friedmann–Lemaître model with $k = 0$ and $\lambda = 0$. This is also called the *Einstein–de Sitter model* since Einstein and de Sitter (1932) had jointly advocated it after the Hubble expansion of the universe became established.

- Mark II: The same as Mark I plus the radiation-dominated expansion prior to t_{eq}. George Gamow had used this model for discussing primordial nucleosynthesis.

- Mark III: The same as Mark II plus inflation that leads to $\Omega_m = 1$. This became popular with the advent of inflation.

- Mark IV: The same as Mark III plus inflation that leads to $\Omega_m + \Omega_\Lambda = 1$. This was advocated when observational difficulties seemed to rule out Mark III.

- Mark V: The Friedmann–Lemaître model with $k = -1$ plus an early radiation-dominated phase but no inflation. This is sometimes called the low-density universe and was popular in the 1970s before inflation and dark matter, but has crept back into contention recently.

We also discussed several alternative cosmologies. However, we will consider only one of them, namely the quasi-steady-state cosmology (QSSC) for a comparison with the big-bang models, against the backdrop of observations.

Briefly these tests are as follows:

1. the measurement of Hubble's constant,

2. the anisotropy of large-scale velocity fields,

3. the distribution and density of matter in our neighbourhood,

4. the age of the universe,

5. the abundance of light nuclei,

6. the evidence for antimatter and

7. the microwave background.

A survey of observational cosmology today reveals a number of issues on which there are disagreements among various observers and theoreticians. Sometimes the more important points of physical significance get buried under heaps of numerical data. In some cases new data have replaced old data, so that fresh interpretation becomes necessary. The approach adopted in this text emphasizes the significant issues that the observations are supposed to reveal rather than the many controversial numerical details. While every attempt is made to present 'up-to-date' data, newer observations than those discussed here are bound to arise in the course of time. We hope that whenever this happens the readers will be able to relate the arguments to new data and see how the conclusions are affected.

10.2   The measurement of Hubble's constant

Modern cosmology began with Hubble's observations, which were referred to in Chapter 1. Hubble obtained a value of $h_0 \sim 5.3$ from his original observations, whereas present-day observations suggest that h_0 lies in the range $0.5 \leq h_0 \leq 1$. The reader may wonder not only at such a drastic change in h_0 over the last six decades but also at the fact that, even today, there is considerable uncertainty about the true value of this important parameter of modern cosmology. This section attempts to clarify the situation.

To begin with, let us recall that the Hubble constant H_0 relates the redshift z of a nearby galaxy to its distance D from us:

$$cz = H_0 D. \tag{10.1}$$

Therefore, if we measure z and D for a number of galaxies (as Hubble did), we should be able to estimate H_0. The observations measure z fairly accurately. The difficulties arise while one is estimating D. The large value obtained by Hubble was due to the fact that he grossly underestimated the distances from us of the galaxies in his survey.

Figure 10.1 shows, for example, the original relation of Hubble alongside the plot of the same extragalactic objects with modern revised distance estimates. The reader may draw his own conclusion as to whether Hubble would have got a linear relation if he had had access to the revised data.

How does an astronomer measure distances from us of galaxies? We will outline below the methods available to him, all of which follow the philosophy outlined by S. van den Bergh in 1975: 'All determinations of the extragalactic distance scale are ultimately based on the assumption that recognizable types of distant objects are similar to nearby objects of the same type'. We will see how this philosophy operates in practice.

10.2.1 The distance modulus

Before we begin with specific methods, it is useful to introduce the concept of a *distance modulus*, which is familiar to the stellar astronomer. Recall that, for an object of luminosity L at a distance D from us, the apparent and absolute magnitudes are defined by the formulae

$$m = -2.5 \log\left(\frac{L}{4\pi D^2}\right) + \text{constant}, \tag{10.2}$$

$$M = -2.5 \log L + \text{constant}. \tag{10.3}$$

The constant in (10.2) is fixed by assigning a given magnitude $m = 0$ to an object with $L/(4\pi D^2) = 2.52 \times 10^{-5}$ erg cm^{-2} s^{-1}. The constant in (10.3) is fixed by

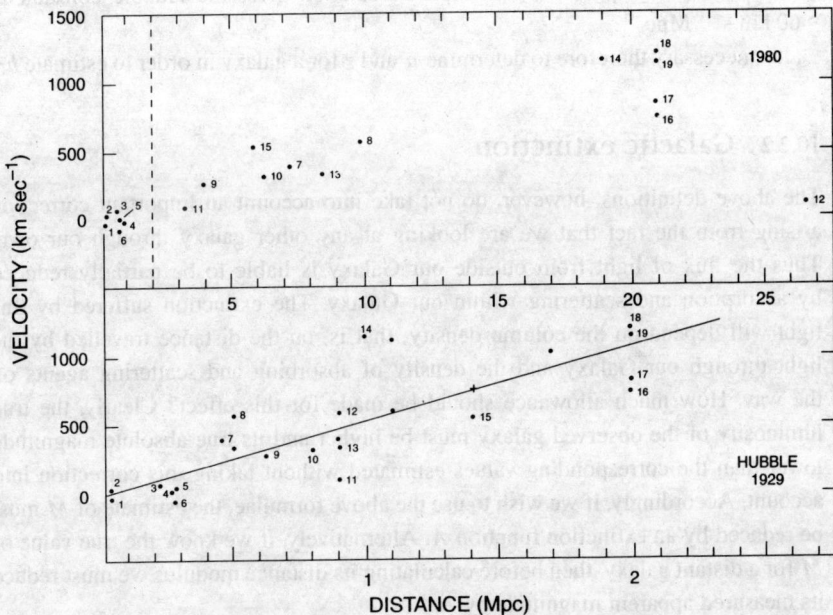

Figure 10.1 The Hubble plot obtained by Hubble side by side with the modern plot with revised distances from us of the same objects (by courtesy of A. Hewitt).

defining M as the apparent magnitude of an object if it were viewed from a distance of 10 pc. Hence, if D is measured in parsecs, (10.2) and (10.3) give

$$m - M = 5 \log D_{pc} - 5. \tag{10.4}$$

The stellar astronomer usually measures distances in parsecs. Hence the above relation is convenient to him. The cosmologist, on the other hand, measures distances in *megaparsecs*. For him the convenient form of (10.4) is therefore

$$m - M = 5 \log D_{Mpc} + 25 = \mu. \tag{10.5}$$

μ is called the *distance modulus*.

If we substitute the Hubble relation (10.1) into (10.5) with the values $H_0 = 100 h_0$ km s^{-1} Mpc^{-1} and $c = 2.997\,929 \times 10^5$ km s^{-1}, we arrive at the following relation for the Hubble law:

$$5 \log h_0 = 42.38 + (M - m) + 5 \log z$$
$$= 42.38 - \mu + 5 \log z. \tag{10.6}$$

Example

A galaxy with redshift 0.002 has a distance modulus ≈ 30. The above relation can be used to compute h_0. A simple calculation gives the Hubble constant as ≈ 60 km s^{-1} Mpc^{-1}.

It is necessary therefore to determine μ and z for a galaxy in order to estimate h_0.

10.2.2 Galactic extinction

The above definitions, however, do not take into account an important correction arising from the fact that we are looking at any other galaxy through our own. Thus the flux of light from outside our Galaxy is liable to be partially reduced by absorption and scattering within our Galaxy. The extinction suffered by this light will depend on the column density; that is, on the distance travelled by the light through our Galaxy and the density of absorbing and scattering agents on the way. How much allowance should be made for this effect? Clearly, the true luminosity of the observed galaxy must be higher and its true absolute magnitude lower than the corresponding values estimated without taking this correction into account. Accordingly, if we wish to use the above formulae, the estimate of M must be reduced by an extinction function A. Alternatively, if we know the true value of M for a distant galaxy, then before calculating its distance modulus we must reduce its measured apparent magnitude by A.

Observers are not unanimous on the value of A. A. Sandage and G. Tammann have used the following extinction law for blue magnitudes:

$$A = \begin{cases} 0 & \text{for } |b| > 50°, \\ 0.13(|\cosec b| - 1) & \text{for } |b| \le 50°, \end{cases} \qquad (10.7)$$

whereas G. de Vaucouleurs used a uniform cosecant law,

$$a = 0.20(|\cosec b| - 1) \qquad (10.8)$$

for all galactic latitudes b. Figure 10.2 illustrates the galactic models underlying these formulae. Already, it is clear that different corrections for extinction are liable to lead to different answers for h_0.

10.2.3 Measurements of extragalactic distance

The distances of planets and satellites within the Solar System are accurately measured with the help of trigonometry and Kepler's laws. The distances of stars up to ~25–50 pc away can be measured with the help of trigonometric parallax. Going still further, a more reliable method is that based on the Hyades main sequence. A comparison of the main sequence of the Hyades cluster with the main sequences of more remote clusters in our Galaxy allows us to measure the distances from us of stars in these clusters. These methods, however, do not work beyond our Galaxy. New techniques are needed for measurements of extragalactic distances. We discuss some of them below.

(a) Cepheid variables. Cepheid variables are members of a class of stars whose luminosities vary by about 10%, but with a great deal of regularity. One can associate a period P with each cycle of variation of a Cepheid variable. The first of these variable stars, the star known as δ Cephei, was discovered as early as 1784 by John Goodricke. He was a deaf mute as a result of a fever contracted in infancy, but conquered his disabilities to become an observational astronomer. His discovery of 1784 was followed by confirmation that the star varies in its intensity by as much

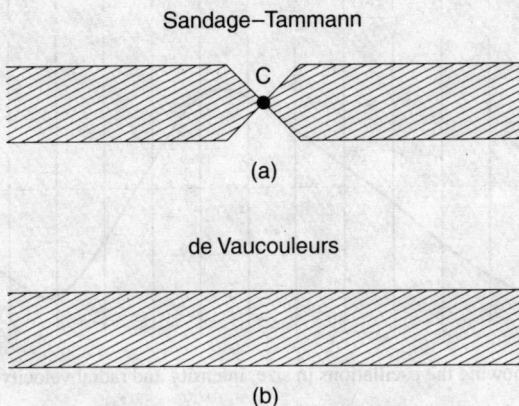

Sandage–Tammann

(a)

de Vaucouleurs

(b)

Figure 10.2 The Galaxy models assumed by (a) Sandage and Tamman and (b) de Vaucouleurs to compute the extinction of visual light from outside our Galaxy. In the former case the absence of a shaded region at high latitude describes the assumption that there is no extinction for $|b| > 50°$.

as a factor of two between maximum and minimum. Since then stars similar to this one came to be called *Cepheid variables* or simply *Cepheids*. Figure 10.3 shows how a Cepheid pulsates in size and intensity. Typically a star like δ Cephei pulsates when there is an internal perturbation in its structure upsetting the balance between its hydrostatic pressure force and gravity. We will not go into details of the process here, except to state that, with the current understanding of stellar structure, it is possible to give an explanation not only for the pulsation of the star but also for the crucial property that makes the Cepheids so useful for extragalactic distance measurement.

This property was discovered in 1912 by Henrietta Leavitt and represents a unique relationship between P and the average luminosity L of the star. Figure 10.4 illustrates this relationship, which can be stated in the form:

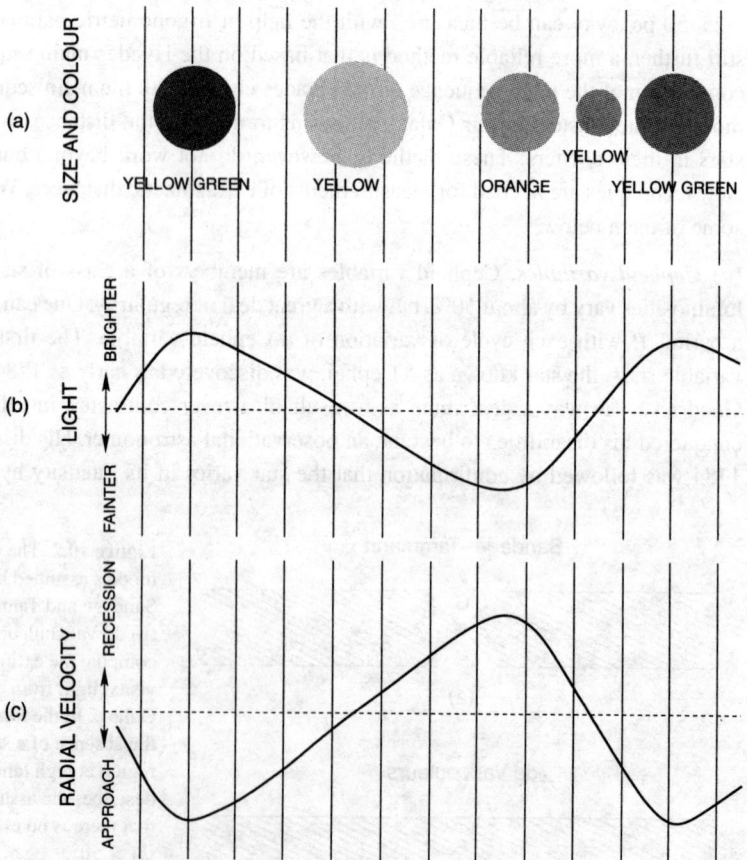

Figure 10.3 A graph showing the oscillations in size, intensity and radial velocity of a Cepheid variable star.

$$\langle M_V \rangle = a + b \log_{10} P, \tag{10.9}$$

where $\langle M_V \rangle$ is the average absolute magnitude of the Cepheid and the period of oscillation is measured in days. Again we will not go into the details of attempts to measure a and b accurately amidst several observational uncertainties. Pioneering work in this field was done by Ejnar Hertzsprung and Harlow Shapley.

Because Cepheids are bright and variable, they can be detected in nearby galaxies with relative ease. Thus, if we detect a Cepheid in a galaxy and measure its period, we can accurately estimate its luminosity L and hence its absolute magnitude M. Then (10.5) gives its distance modulus and hence the distance modulus of the galaxy in which it is located.

It was with the help of Cepheids that Hubble established the fact that galaxies exist outside our own. His early work leading to the discovery of the expansion of the universe was also based on Cepheids. This method takes us to distances of $\sim$10 Mpc; that is, to galaxies in our local neighbourhood.

However, there was a pitfall that led Hubble to under-estimate distances grossly by this method! There are *two* types of Cepheids, one belonging to *population-I* stars in the Galaxy and the other to *population-II* stars. The light curves and spectra of the two types differ and the former are about four times more luminous than the latter. This difference was noticed and pointed out by Walter Baade in 1952 and

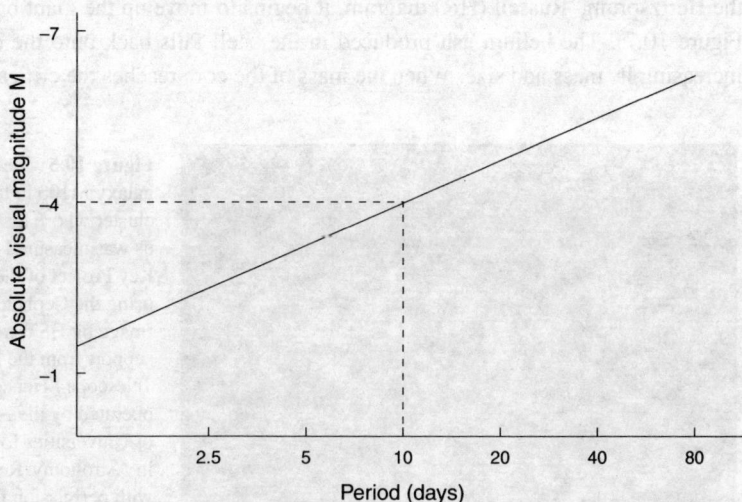

Figure 10.4 Average luminosity plotted on a logarithmic scale against period shows the idealized period–luminosity relation for Cepheid variables. The straight line illustrates the fact that their average luminosity increases with period, which allows us to calculate the luminosity of a distant Cepheid by measuring its period. Thus, a Cepheid with a period of 10 days will have absolute magnitude $M = -4$. Based on H. Arp, 1960, 'Southern hemisphere photometry VIII: Cepheids in the Small Magellanic Cloud', *A.J. 65*, p. 404.

this led to a downward revision of the value of Hubble's constant. Even today there are differences among the estimates of a and b derived by various observers. For instance, Feast and Catchpole (1997) find $a = -1.43$ and $b = -2.81$, whereas Tanvir (1997) finds $a = -1.38$ and $b = -2.77$.

On the other hand, the use of the Hubble Space Telescope (HST) made available the possibility of seeing Cepheids in relatively more distant galaxies. Indeed, one of the *key projects* of the HST was to estimate the value of H_0 to within 10% error bars. The principal investigators in this very large team were Wendy Freedman, Robert Kennicutt Jr and Richard Mould. The first step in their investigation was the use of Cepheids in the spiral galaxy M100 in the Virgo cluster (see Figures 10.5 and 10.6). In 1994 they announced a result based on 20 Cepheids found in M100. Their value of Hubble's constant was 80 ± 17 km s^{-1} Mpc^{-1}. This value has subsequently come down to the range 70–80 km s^{-1} Mpc^{-1}, using the method involving Cepheids.

(*b*) *Brightest star.* This method of measuring distances from us of distant galaxies makes use of the assumption that, in similar spiral Sc galaxies of comparable luminosities, the brightest stars also have comparable luminosities. What are these stars? It can be shown that, when a star of relatively low mass is close to finishing its core hydrogen fuel, it begins fusing the hydrogen in the shell surrounding the helium core. As the shell increases in size the star's luminosity increases and, in the Hertzsprung–Russell (HR) diagram, it begins to move up the giant branch (see Figure 10.7). The helium ash produced in the shell falls back onto the core, thus increasing its mass and size. When the mass of the core reaches the critical value of

Figure 10.5 The spiral galaxy M100 in the Virgo cluster whose distance from us was measured under the Key Project of the HST, by using the Cepheids methods. Image by HST created with support from the Space Telescope Science Institute operated by the Association of Universities for Research in Astronomy. Reproduced with permission from AURA/STScI.

$\sim 0.4 M_\odot$, helium fusion starts rapidly. This phenomenon is called the *helium flash*. The large luminosity of the star lasts, however, only for a short time, whereafter the star becomes dimmer and moves back down the giant branch. Ultimately it moves along the horizontal branch in the HR diagram until its helium supply has been consumed. It is the tip of the giant branch that denotes the phase when the star is at its brightest. Since its luminosity comes from a helium shell of the critical mass, it is constant from star to star. This is the standard candle that the cosmologist uses for estimating distances.

Since these are the brightest stars, they can be spotted in more distant galaxies, in which the Cepheids (being too faint) cannot be seen and also one need not wait for multiple exposures of images as required for Cepheids, so this method has distinct advantages. For example, Hubble found that the brightest stars in the galaxies M31 and M33 are significantly brighter than the brightest Cepheids. Indeed, this method takes us as far as the centre of the Virgo cluster of galaxies; that is, to distances of $\sim 10\text{--}15$ Mpc, with observations from ground-based telescopes alone.

(*c*) *Planetary nebulae.* Planetary nebulae (PN; see Figures 10.8–10.10) are formed

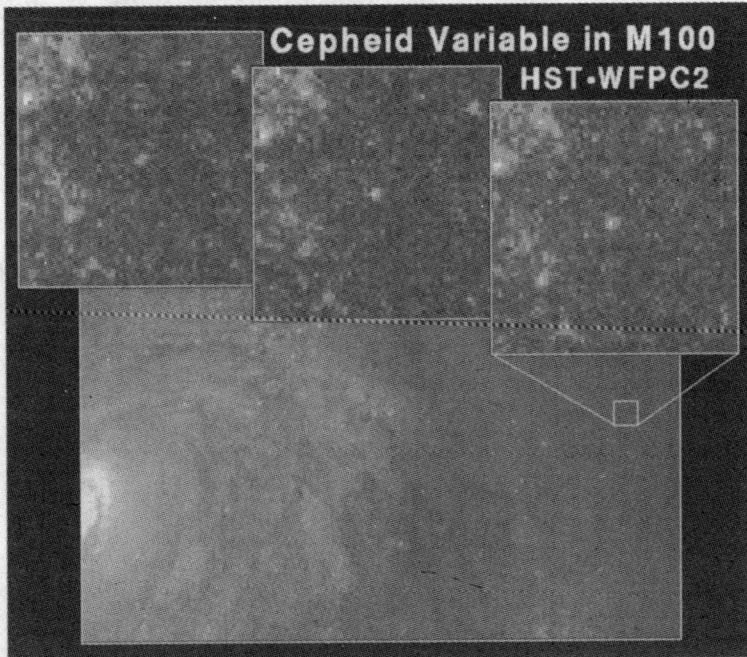

Figure 10.6 The intensity of a Cepheid variable star detected in M100 by the HST. Notice the variation of intensity in different frames put together. Image by HST created with support from the Space Telescope Science Institute operated by the Association of Universities for Research in Astronomy. Reproduced with permission from AURA/STScI.

by gas shells ejected by stars from the outer parts of their envelopes, in the closing stages of their evolution. At this stage the star is a red giant whose envelope expands and becomes diffuse while the core contracts and gets hotter. So much so that at some stage it begins to emit ultraviolet photons, which ionize the emitted shell and make it shine. The stage of a planetary nebula is relatively transient, lasting no longer than about 25 000 years. It is estimated that there may be as many as 50 000 planetary

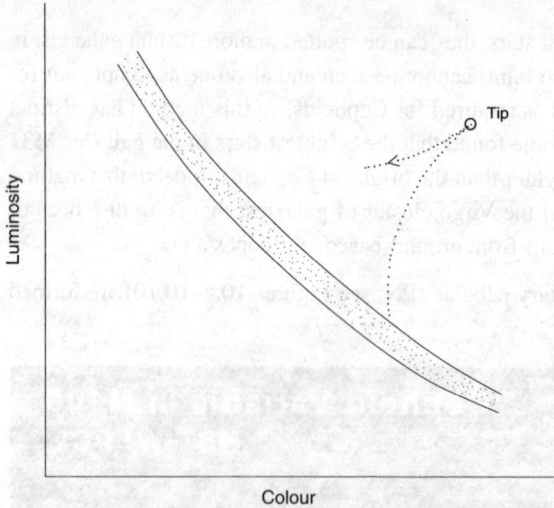

Figure 10.7 The tip of the giant branch in this HR diagram is shown, where the luminosity of the star is maximum.

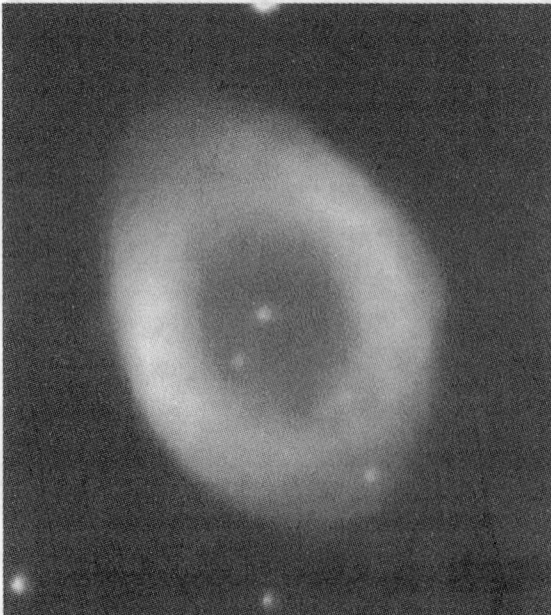

Figure 10.8 The planetary nebula M57, called the 'Ring Nebula'. CCD picture by Nelson Caldwell.

nebulae in our Galaxy. Their luminosity function, that is, the number of PN in the absolute magnitude range $(M, M + dM)$, is given by

$$N(M)\,dM = N_0 \exp(0.307M)\{1 - \exp[3(M^* - M)]\}\,dM, \qquad (10.10)$$

where N_0 is a constant and M^* is the absolute magnitude of the brightest PN.

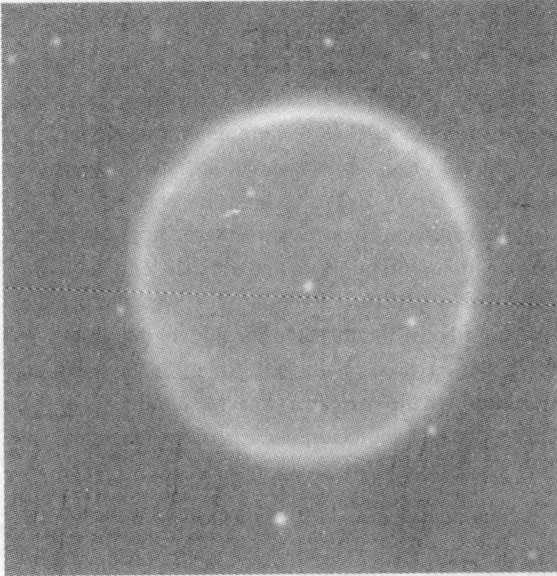

Figure 10.9 The planetary nebula Abell 39. From the website http://www.noao.edu/ image_gallery/html/ im0636.html of WIYN/NOAO/NSF.

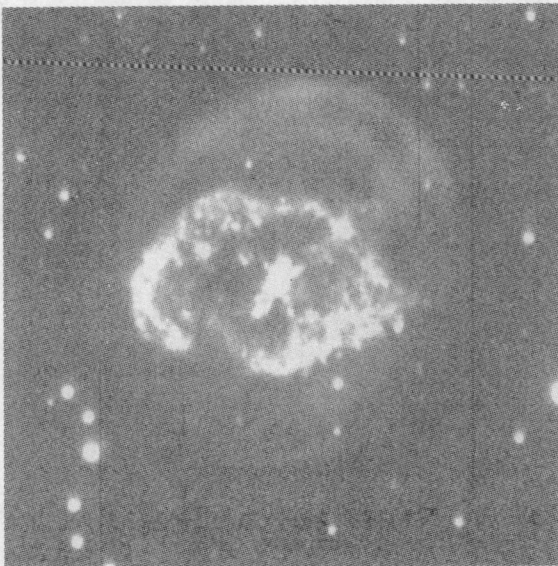

Figure 10.10 The planetary nebula Abell 78. Courtesy of George Jacoby/ WIYN/NOAO/AURA/NSF.

Although one cannot identify any given type of planetary nebula as a standard candle, one can make use of the circumstance that the *shape of the luminosity function seems to be the same in all galaxies*. As shown in Figure 10.11, there is a cut-off in the luminosity function at a specific magnitude, which can serve as a standard candle. This can be determined by detecting enough planetary nebulae in the relatively flat part of the luminosity function. The absolute magnitude for this is around -4.5, which allows one to use this method for estimating distances up to about 20 Mpc. George Jacoby and his coworkers have estimated the distance from us of the centre of the Virgo cluster to be 15.1 ± 0.9 Mpc.

(*d*) *H-II regions.* H-II regions are large domains of ionized hydrogen. These are found not only in our Galaxy but also in others. The linear diameter of the largest H-II region, or, better still, the mean linear size of the three largest H-II regions, exhibits a strong variation with the luminosity and the luminosity class (see §1.3) of the parent galaxy. For dwarf galaxies this mean size is as low as 75 pc, whereas for supergiant galaxies the size goes up to 460 pc. By comparing the angular sizes of such H-II regions in remote and nearby galaxies of similar type, we can estimate the ratio of their distances from us. Then, if the distance for the nearby galaxy is known, the distance for the remote galaxy can be estimated. Note that this method, unlike the others mentioned so far, relies on the size rather than the luminosity of the distance indicator. With this type of distance indicator, one can extend the distance ladder to about 100 Mpc.

(*e*) *Type-II supernovae.* A new technique based on type-II supernovae has recently

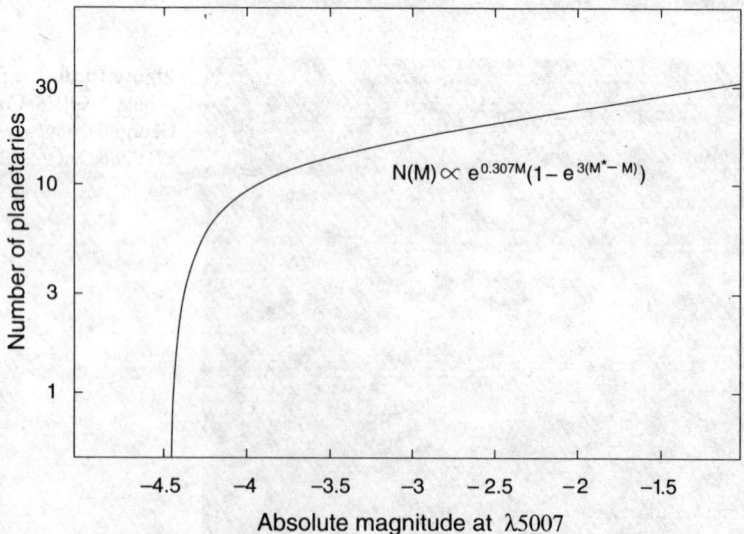

Figure 10.11 The luminosity function of planetary nebulae. See the cut-off at the bright end which serves as the luminosity of the standard candle.

shown promise of reliability and does not require many arbitrary assumptions. Basically, this method involves determining the actual flux of light at various frequencies leaving the photosphere of the exploding star and it does not depend on other step-by-step methods of distance determination. The method consists of measurements of the rate of expansion of the photosphere of the supernova and makes use of a variant of the method used by W. Baade in 1926 for variable stars.

If we approximate the photosphere (see Figure 10.12) by a blackbody of temperature T and radius R and suppose that its distance from us is D, then its angular size θ and the flux density $f(v)$ at frequency v are given by

$$\theta = R/D, \tag{10.11}$$

$$f(v) = \frac{R^2}{D^2} \frac{2\pi h v^3}{c^2 (e^{hv/(kT)} - 1)} \tag{10.12}$$

(here the redshift effects have been ignored).

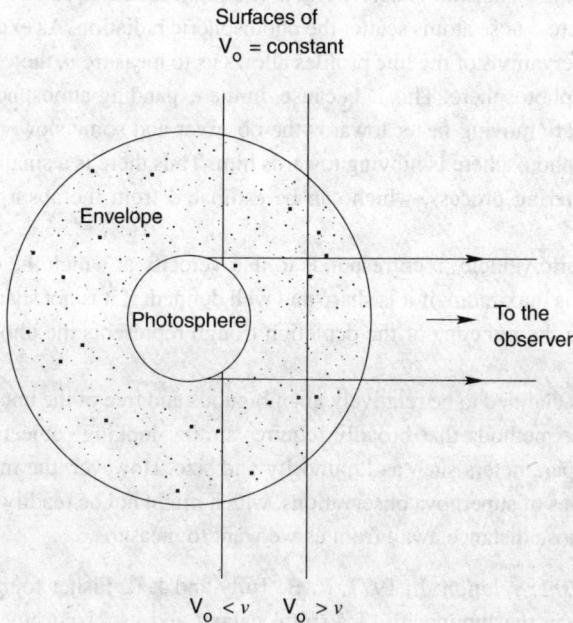

Figure 10.12 The shaded region is the expanding envelope of scattering particles around the photosphere. The surfaces of constant velocity V_0 of the particles in the envelope relative to the observer are shown (in idealized condition) as planes. In the plane on the left-hand side V_0 is less than the photospheric velocity v, whereas in the plane on the right-hand side $V_0 > v$. The switchover from $V_0 < v$ to $V_0 > v$ can be related to the extent of scattering produced by the particles in the envelope and is seen in the line profiles of the supernova in the form of varying levels of depletion. A study of the line profiles allows the astronomer to fix the value of v.

Thus we get from (10.11) and (10.12)

$$\theta = \left(\frac{f(v)c^2(e^{hv/(kT)} - 1)}{2\pi h v^3} \right)^{1/2}.$$

(10.13)

Hence, if we measure $f(v)$ and T, we can get θ. Furthermore, if we measure R, we get D from (10.11). Spectral scans of the continuum spectrum of the supernova give a good estimate of T. To measure R, R. P. Kirshner and J. Kwan suggested the following method.

In this method we approximate the rate of expansion of the photosphere by a constant value v, so that, if the expansion started at $t = t_0$ when the radius was $R = R_0$, the radius at subsequent times is given by

$$R = v(t - t_0) + R_0$$

(10.14)

(constancy of v is justified by the fact that the pressure in the interstellar medium is negligible and the expansion is nearly free). The photosphere is surrounded by a tenuous atmosphere whose atoms scatter the photospheric radiation. As explained in Figure 10.12, observations of the line profiles allows us to measure v, the velocity of expansion of the photosphere. This is because, in the expanding atmosphere, some scattering atoms are moving faster towards the observer and some slower than the rate at which the photosphere is moving towards him. Thus there is a small Doppler effect in the scattering process, which can be estimated from the absorption line profiles.

The photospheric velocity v corresponds to that velocity at which the depletion of the continuum is maximum, if it is sharp and well defined. If it is not sharp but has a flat trough, then the red edge of the depletion trough represents the photospheric velocity.

This process is claimed to be relatively unambiguous and free of the uncertainties surrounding other methods that broadly require 'similar-looking' objects to have 'equal' physical parameters such as luminosity and size. However, the method requires a good series of supernova observations, which might not be readily available for the galaxy whose distance away from us we want to measure.

(f) The Tully–Fisher relation: In 1977, R. B. Tully and J. R. Fisher found a good correlation between the luminosity of a spiral galaxy and its 21-cm line width, a correlation that does not depend on the type of galaxy. Thus, in principle, if we determine the 21-cm line profile of a remote spiral galaxy, we can estimate its luminosity. The connection between these two physical parameters may appear *ad hoc* to start with, but can be rationalized in the following way.

The 21-cm line profile can acquire a width if the galaxy is rotating. If the angle of inclination of the plane of the galaxy to our line of sight is i, and the maximum velocity of rotation of the galaxy is V_m, then the Doppler width of the line will be

$$w = \frac{2V_{\mathrm{m}} \sin i}{c}. \tag{10.15}$$

Using the gravitational force of galactic mass M at a distance equal to the galactic radius R to generate the velocity of rotation V_{m} of a typical H-I region, we get

$$M = V_{\mathrm{m}}^2 R/G. \tag{10.16}$$

Here we are identifying V_{m} with the flat-rotation curve for the galaxy (see Chapter 1). Assuming that the mass-to-light ratio of the galaxy is k, which is the same for all galaxies of similar types, we get the luminosity of the galaxy as

$$L = \frac{V_{\mathrm{m}}^2 R}{kG}. \tag{10.17}$$

We next assume that all galaxies of this class have the same surface brightness $\Sigma = L/(\pi R^2)$, so that the above relations combine to give

$$L = \frac{1}{\pi k^2 \Sigma G^2} V_{\mathrm{m}}^4 \propto V_{\mathrm{m}}^4. \tag{10.18}$$

Next, using the relation (10.15), we get the following idealized Tully–Fisher relation:

$$M_{\mathrm{G}} = -10 \log V_{\mathrm{m}} + b = -10 \log\left(\frac{w}{2 \sin i}\right) + b, \tag{10.19}$$

where M_{G} is the absolute magnitude of the galaxy and b is a constant.

An empirically derived Tully–Fisher relation is written in the form

$$M_{\mathrm{G}} = -a \log\left(\frac{w}{2 \sin i}\right) + b, \tag{10.20}$$

where a and b are constants determined from observations.

There is a practical problem associated with this method. To avoid internal absorption in the galaxy it should ideally be seen face-on. However, the line width is best determined for spirals viewed edge-on, for which the internal absorption in the galaxy is large. Thus the observer is forced to use those spirals that are viewed at an inclined angle and yet give a reasonably reliable line width. The distances of the M81 and M101 groups have been estimated this way after using data on the nearer galaxies M31 and M33 for calibration. This method also takes us as far as ~100 Mpc.

(*g*) *Faber–Jackson method and the fundamental plane.* The Tully–Fisher method works for spirals that exhibit significant rotation. What about ellipticals? They do not rotate and also they contain very little gas, so clouds of neutral hydrogen would not be available there. In 1976, Sandra Faber and Robert Jackson found a method that works for ellipticals and is roughly analogous to the Tully–Fisher method for spirals. This consists of relating the luminosity L of the galaxy to the velocity dispersion σ

found mainly in the central region of an elliptical galaxy. From the virial theorem, we know that, in dynamical equilibrium, there would be an approximate equipartition between the kinetic energy (T) of random motion and the gravitational potential energy (Φ), expressed by the relation $2T + \Phi = 0$. For a uniform distribution of total mass M and radius R, this will become

$$\sigma^2 = \frac{GM}{5R}. \tag{10.21}$$

Using the same assumptions of a constant mass-to-light ratio as given by (10.17) and a constant average surface brightness as given by (10.18), we get

$$L = \text{constant} \times \sigma^4. \tag{10.22}$$

This is the Faber–Jackson relation. By measuring σ, one can deduce L and hence the distance away from us of the elliptical galaxy.

However, application of this relation to clusters of galaxies revealed considerable scatter around the basic relation, suggesting that it may be missing some extra parameter. In 1987 George Djorgovski and Marc Davis showed that, instead of luminosity L, one should use *two* parameters: R_e, the effective radius of the elliptical galaxy, and I_e, the average surface brightness within that radius. In that case there is a tight relationship among σ, R_e and I_e, which can be written as a linear relation of their logarithms:

$$\log R_e = 1.36 \log \sigma - 0.85 \log I_e. \tag{10.23}$$

This linear relation describes a 'plane' on the logarithmic scale in the three-dimensional space of the above three quantities, a plane called the *fundamental plane* for the elliptical galaxies. By using this technique, one is therefore able to estimate R_e for an elliptical galaxy and then, from its angular size, deduce its distance away from us.

It is also possible to obtain a tighter version of the Faber–Jackson relation, by confining oneself to the central region of the elliptical galaxy. The group of astronomers titled the 'Seven Samurai', whom we will encounter in the next section, chose the region by limiting themselves to the part wherein the total average surface brightness expressed on the magnitude scale is 20.75^m per square arcsecond. Denoting the diameter of this (circular) region D_n, they found that there is a tight correlation between D_n and σ.

(*h*) *Brightest cluster galaxy.* If we consider the thousand-odd galaxies in the Virgo cluster, one galaxy, M87, stands out as being significantly brighter, more massive, larger than the rest. It is an elliptical galaxy. Allan Sandage noted that other, more distant, clusters of galaxies also contain similarly dominating elliptical galaxies. Under the assumption (supported by observations of nearby clusters) that such ellipticals have comparable luminosities, we can estimate M and hence the distance modulus of clusters as remote as 1000 Mpc away.

(*i*) *Type-Ia supernovae*. Unlike the type-II supernovae, the type-Ia supernovae arise from white dwarfs. When a white dwarf in a binary having mass close to the Chandrasekhar limit accretes matter from its companion, it develops instability and explodes. A type-II supernova explodes to leave a remnant core; but type-Ia supernova completely destroys the star. Also, in the former, the bulk of the explosion energy is carried by neutrinos, whereas in the latter the optical component is much larger. That is why type-Ia supernovae are much brighter at their peak than are other types. Figure 10.13 shows the typical light curve for such a supernova. Notice that the peak magnitude is close to $M_{max} = -19.5$.

Allan Sandage and his coworkers have carried out systematic studies of the peak magnitudes of these supernovae and made comparisons with the distance-measuring method using Cepheids. The peak magnitude has a small dispersion. For example, it may be of the order $\Delta M_{max} \approx 0.15$. These values are to be taken as indicative only, since continuing improvement of estimates is currently going on.

The type-Ia supernovae thus provide standard candles that take the distance measurements well past the 1000-Mpc mark. Their disadvantage is that these events are not all that common and so the database increases slowly. A further disadvantage is that the highly luminous phase does not last very long, as can be seen from Figure 10.13. We shall return to these distance measurements when we discuss the Hubble relation at high redshifts in the following chapter.

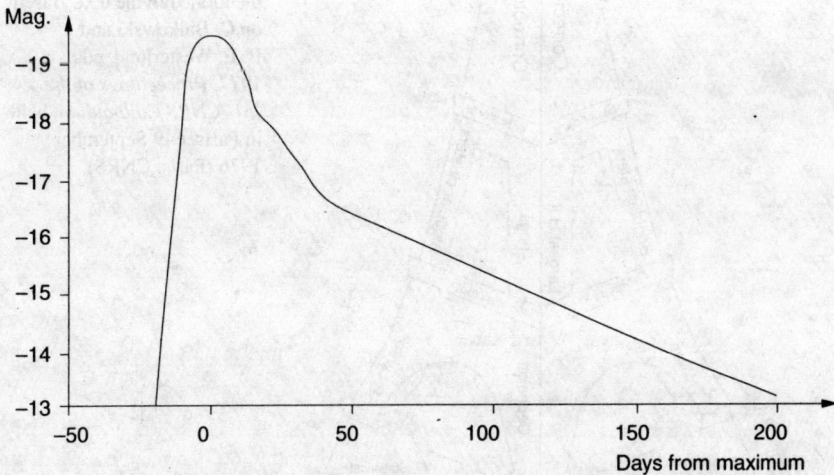

Figure 10.13 The typical light curve of a type-Ia supernova. The peak luminosity can serve as a standard candle.

10.2.4 The Hubble constant

The above account is admittedly sketchy and incomplete. For example, we have not described the surface-brightness-fluctuation method, the globular-cluster-luminosity-function method, the use of gravitational lensing, megamasers and the Sunyaev–Zel'dovich effect. It is because each method has certain in-built assumptions that may be error-prone that one cannot rely 100% on any particular method. Which is why astronomers these days like to cross check their estimates with respect to more than one method.

Such methods were summarized in an 'Eiffel Tower' diagram by the late Gérard de Vaucouleurs, which is shown in Figure 10.14. Notice that, in many cases, the distances are determined in progression from one stage to the next. At each stage

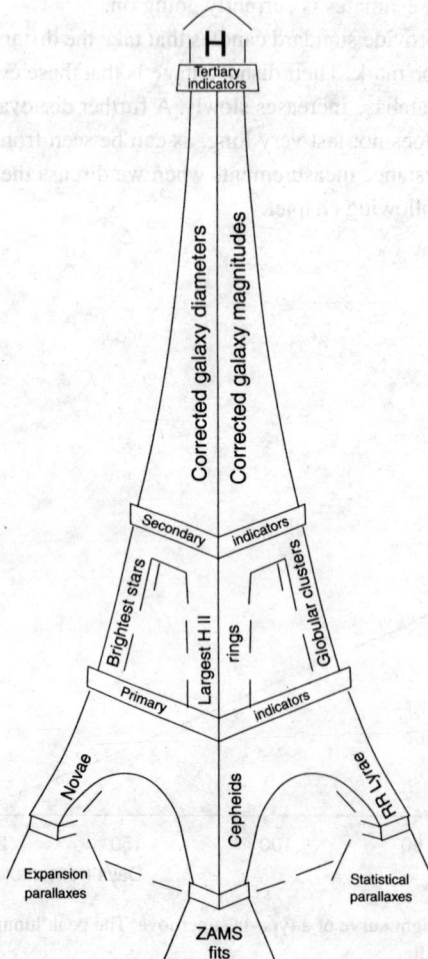

Figure 10.14 The Eiffel Tower shown here describes how cosmological distances are measured in stages. The various levels used for calibration are shown starting from the nearest at the first level and leading to the furthest at the top. ZAMS stands for 'zero-age main sequence', which refers to the method of measuring distances using the Hyades main sequence mentioned in the text. Based on C. Balkowski and B. E. Westerlund, eds., 1977, *Proceedings of the IAU/CNRS Colloquium*, held in Paris, 6–9 September 1976 (Paris: CNRS).

of the tower there is scope for errors of calibration. For example, even a revision of the stellar distance scale, such as that of the Hyades main sequence in our Galaxy, will lead to revision of all subsequent scales. This happened, for example, when stellar distances measured by the Hipparcos satellite revised the earlier estimates downwards. Such systematic errors were present in Hubble's original method and, when they were pooled together, they gave a value of $h_0 \geq 5$. For example, the Cepheid period–luminosity relation available to Hubble was incorrect. He also used too faint an absolute magnitude for the brightest stars in other galaxies. From (10.6) we see that a high value of M will lead to a high value of h_0.

In all this discussion we have been tacitly assuming that, in the velocity–distance relation, it is the distance that causes problems, namely that the velocity–redshift relationship is fairly secure. This is not necessarily so! A major source of uncertainty, at least for nearby measurements, comes in distinguishing the 'true' Hubble flow from the peculiar motions caused by other relatively local inhomogeneities. In the following section we highlight this problem.

For these reasons, it would be wise on the part of the cosmologist to be cautious about the exact value of Hubble's constant. Sandage and Tamman prefer a value of $h_0 \sim 0.5$, whereas de Vaucouleurs argued that $h_0 \sim 0.1$ and later the HST Key Project Group would go for the value $h_0 \sim 0.75$. In view of the prevailing uncertainties of various distance indicators, it is customary nowadays to say that h_0 lies between these two limits. We will take $h_0 \sim 0.65 \pm 0.10$, as is consistent with the bulk of present measurements. The holy grail of a well-determined Hubble's constant is still some distance away.

10.3 The anisotropy of local large-scale velocity fields

We now look at possible pitfalls in the determination of v in the Hubble relation $v = H_0 D$, for the velocity implied in this relation is that ascribed to the expansion of the universe and should not contain any other components. Indeed, this would be the case if the Weyl postulate were strictly valid, which, in actuality, it is not. The galaxies do have other motions superposed on it, which are also included in the spectral shift that the observer measures. For example, the first measurement of a galactic spectral shift was performed in 1912 by Slipher and it was a blueshift. He concluded that the nebula M31 was approaching our Galaxy at a speed of around 200 km s^{-1}. A few nearby galaxies do exhibit blueshifts and these are explained by the so-called random or 'peculiar' motions of the galaxies superimposed on the 'Hubble flow'. These random motions are in general of the order of ~ 300 km s^{-1} and will therefore significantly modify the total velocity of a nearby galaxy, for which, with small D, the Hubble component will be small. Only when we go to distances well beyond, say, 50 Mpc should we expect the Hubble law to dominate the velocity field.

The simple picture of a homogeneous and isotropic universe based on the Robertson–Walker line element is now beginning to look oversimplified, especially with the discovery of large-scale velocities that appear superimposed on the Hubble flow. Since, as mentioned above, the Hubble flow is small in our 'local' neighbourhood, it tends to be swamped by these other velocities. It is an immensely complicated problem to untangle the two and pinpoint the causes of the non-Hubble velocities. Here we will give a bare outline of the situation on the theoretical and observational fronts, beginning with the latter.

It was in the mid-seventies that work by V. C. Rubin, W. K. Ford and others gave a glimpse of the problem. The so-called Rubin–Ford effect showed that the Hubble constant is *not* isotropic when it is measured for the radical velocities of 184 Sc I and Sc II galaxies (I and II are the van den Bergh luminosity classes for spiral Sc galaxies with Sc I being the brightest class of galaxies.) The anisotropy was of the dipole type and could be accounted for by the assumption that our Galaxy is moving with a substantial velocity against the background of the galaxies. The velocity was

$$(454 \pm 125) \text{ km s}^{-1} \text{ towards } l = 163° \pm 15°, \qquad b = -11° \pm 14°$$

relative to the distant part of the sample and

$$(474 \pm 164) \text{ km s}^{-1} \text{ towards } l = 167° \pm 20°, \qquad b = 5° \pm 20°$$

relative to the nearer part. This was the first indication that the Galaxy is not in the cosmological rest frame.

10.3.1 The local distribution

The *Nearby Galaxies Atlas* published by Tully and Fischer contains detailed maps of the distributions and speeds of galaxies in the relatively local region. These maps are helpful in constructing the topography of the nearby region. Figure 10.15 gives a schematic plot of the distribution over a cubic region around our Galaxy with each side of the cube measuring approximately a speed differential of $10\,000 \text{ km s}^{-1}$. (That is, if H_0 is the Hubble constant, the linear size is approximately H_0^{-1} times this value. We may find it convenient to use speeds as distances in this way.)

The shaded region of the cube is the galactic zone of avoidance which is perpendicular to the supergalactic plane (see Chapter 1). One may consider the motions of these objects as made up of several components:

1. the flow towards the Great Attractor (GA) located at a distance of $\sim 4200 \text{ km s}^{-1}$ from the Local Group (the GA is located approximately at $l = 309°$ and $b = +18°$ (galactic coordinates); the two Centaurus clusters are, for example, falling into the GA with speeds of $\sim 1000 \text{ km s}^{-1}$ (away from us));

2. the infall of matter towards the Virgo cluster;

3. the 'Local Anomaly' which appears to require a bulk velocity correction of 360 km s^{-1} for a region extending from the Local Group out to distances of 700 km s^{-1}; and

4. the Hubble flow.

This multi-component model has several parameters that can be determined by the least-squares technique, by using the Tully–Fisher relation for spirals to measure their distances (see §10.2) and the redshifts for radial velocities. The model determines velocities that are compared with the observed values and the differences are minimized by the least-squares method. This technique was first used in 1988 by the 'Seven Samurai' D. Lynden-Bell, S. M. Faber, D. Burnstein, R. L. Davies, A. Dressler, R. Terlevich and G. Wegner. The broad conclusions are as follows.

The GA is a large mass attracting matter towards it, causing a large-scale streaming motion in its direction. On a smaller and nearer scale, the Virgo cluster has neighbouring galaxies falling towards its centre, including the Local Group. However, the

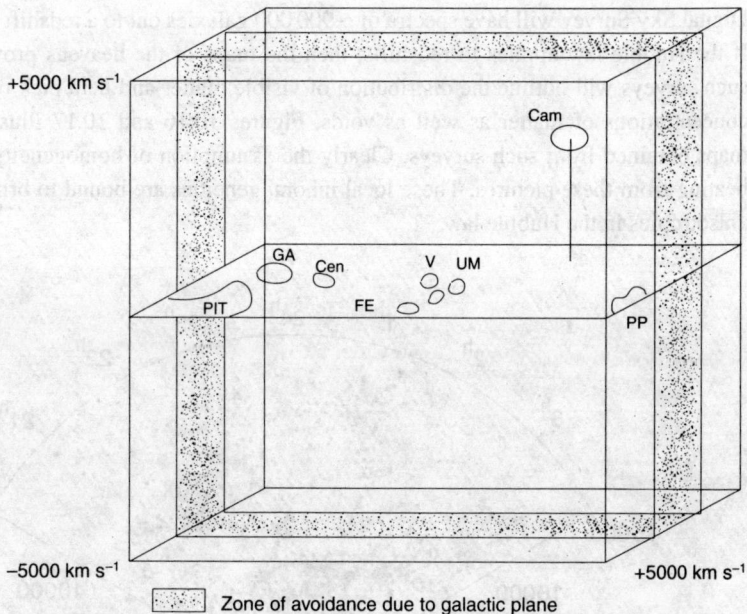

Zone of avoidance due to galactic plane

Figure 10.15 A cubic volume containing some significant large-scale structures in our neighbourhood. Shown in the cube are GA, Great Attractor; V, Virgo cluster; UM, Ursa Major cluster; Cen, Centaurus; FE, Fornax–Eridanus; Cam, Camelopardalis; PP, Perseus–Pisces; PIT, Pavo-Induiis-Telescopium. Adapted from S. M. Faber and David Burstein: 'Motions of galaxies in the neighbourhood of the Local Group' in the proceedings of the Vatican Study Week on 'Large-Scale Motions in the Universe', eds. V. C. Rubin and G. V. Coyne, Princeton University Press, 1988, p. 118.

Local Group has a further anomalous motion relative to the Virgocentric flow. It is perhaps too early to take all the numerical estimates of speeds and directions as very accurate. More observations in the future will certainly help in making these estimates more reliable.

There has been, however, considerable discussion as to whether the GA exists at all. For example, D. A. Mathewson, V. L. Ford and M. Buchhorn measured the peculiar velocities of 1355 spiral galaxies in the southern sky and used the Tully–Fischer relation to estimate their distances from us. They found no back-side infall into the GA region; rather they found a bulk flow of about 400 km s^{-1} on scales of $100h_0^{-1}$ Mpc. Since there is no visible concentration of mass in the GA region, considerable doubt was expressed about the existence of an attracting mass there. It may well be that, instead of a single large mass, the pull may have been exercised by several large clusters distributed in space in that direction.

These issues will be resolved as we get a clearer picture of the universe through the large-redshift surveys that are currently under way or have just been completed. For example the Two Degree Field Survey in Australia and the Las Campañas Redshift Survey in Chile contain a few hundred thousand redshifts. The Sloan Digital Sky Survey will have spectra of $\sim$900 000 galaxies out to a redshift of $\sim$0.2. If the Hubble law applies to redshifts, then the maps of the heavens provided by such surveys will outline the distribution of visible matter and help pick out dense concentrations of matter as well as voids. Figures 10.16 and 10.17 illustrate the maps obtained from such surveys. Clearly the assumption of homogeneity takes a beating from these pictures. These local inhomogeneities are bound to bring about anisotropies in the Hubble law.

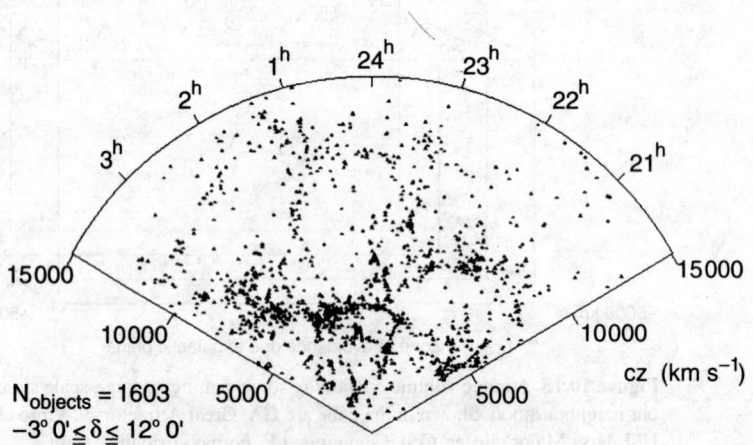

Figure 10.16 This is one of the sectors of the sky surveyed in depth by the Center for Astrophysics. The velocities can be converted to distance away from us by dividing them by the Hubble constant. Courtesy of M. G. Geller and J. P. Huchra of the Smithsonian Astrophysical Observatory.

An independent piece of information that we shall consider in §10.8 is the motion of our Galaxy with respect to the rest frame of the cosmic microwave background.

10.3.2 The Hubble constant revisited

We return to the question of why the controversy over the value of H_0 (i.e., whether $h_0 \approx 1$ or $h_0 \approx 0.5$) persists. R. B. Tully has argued that the local velocity anomaly is the culprit confusing the issue. The argument may be illustrated by a simplified example.

Imagine a local mass concentration M superimposed on a Hubble flow. At a distance R from the mass, the radially outward velocity V may be given by

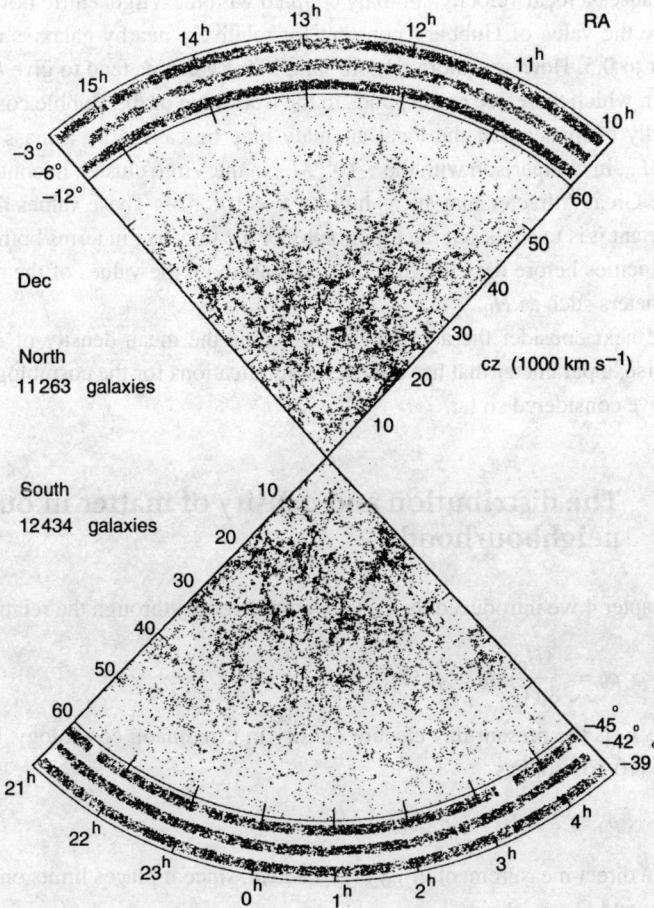

Figure 10.17 This map has been generated by the Las Campañas redshift survey. Reproduced by courtesy of Huan Lin, Steward Observatory, University of Arizona.

$$V = -\sqrt{\frac{2GM}{R}} + HR \equiv H_{\text{eff}}R. \tag{10.24}$$

The first term is an inward velocity corresponding to a zero value at infinity while the second term is the Hubble flow with the Hubble constant H. We may look upon (10.24) as a Hubble flow only, with an 'effective' Hubble constant

$$H_{\text{eff}} = H - \sqrt{\frac{2GM}{R^3}}. \tag{10.25}$$

Thus the effective Hubble constant is *smaller* than the 'true' Hubble constant, *closer* to the mass concentration. As we go away from M, the effective Hubble constant approaches the true value.

Hence the possible presence of a mass concentration in the Coma–Sculptor Cloud that causes a local velocity anomaly coupled with the Virgocentric flow manages to reduce the value of Hubble's constant for relatively nearby galaxies and make h_0 closer to 0.5. However, more remote samples of galaxies tend to give h_0 somewhat higher, which therefore corresponds to the true value of the Hubble constant.

Tully estimates that the local anomaly may be caused by a mass of the order $10^{14} M_\odot$, in comparison with the $\sim 10^{15} M_\odot$ in the Virgo cluster. In contrast the mass of the Great Attractor may be as high as $5 \times 10^{16} M_\odot$. These values illustrate how important it is to chalk out the topography of the universe in terms both of mass and of velocities before drawing firm conclusions about the values of the cosmological parameters such as H_0.

We next consider the attempts to determine the mean density of matter in the universe, a parameter that has significant implications for the cosmological theories we have considered so far.

10.4   The distribution and density of matter in our neighbourhood

In Chapter 4 we introduced the density parameter Ω_0 through the relation

$$\rho_0 = \frac{3H_0^2}{8\pi G}\Omega_0 \equiv \rho_c \Omega_0, \tag{10.26}$$

where ρ_c is the present-day closure density in Friedmann cosmology. In numerical terms (10.26) implies

$$\rho_0 = 2 \times 10^{-29} (h_0^2 \Omega_0) \text{ g cm}^{-3}. \tag{10.27}$$

Thus a direct measurement of ρ_0 is of interest, since it places limits on the parameters h_0 and Ω_0.

The present approach to the problem involves setting limits on the density of matter in the form of galaxies, clusters of galaxies and so on; that is, of matter in the standard luminous form. This is done as follows. Suppose that we know the average mass-to-light ratio for galaxies, which is conventionally expressed in solar units:

$$\left\langle \frac{M_G}{L_G} \right\rangle = \eta \frac{M_\odot}{L_\odot}. \tag{10.28}$$

Next we determine the mean luminosity density l_G of galaxies. The best value of l_G comes from the Revised Shapley Ames Catalogue and is given by

$$l_{GS} \cong 4.4 \times 10^7 L_\odot h_0 \text{ Mpc}^{-3} \qquad \text{for spiral galaxies}, \tag{10.29}$$

$$l_{GE} \cong 17.4 \times 10^7 L_\odot h_0 \text{ Mpc}^{-3} \qquad \text{for E and S0 galaxies}. \tag{10.30}$$

The total luminosity density is therefore of the order

$$l_G \cong 2.2 \times 10^8 L_\odot h_0 \text{ Mpc}^{-3}. \tag{10.31}$$

From (10.28) and (10.31) we then get the mean cosmological density in the form of galaxies in our neighbourhood as

$$\rho_G \cong 2.2 \times 10^8 \eta M_\odot h_0 \text{ Mpc}^{-3}$$
$$\cong 1.5 \times 10^{-32} \eta h_0 \text{ g cm}^{-3}. \tag{10.32}$$

What is the estimate for η? The main difficulty in estimating η lies in the measurement of galactic masses. In comparison, the measurement of luminosities is easy, the only uncertainty in the process arising from the lack of precision in H_0. We will briefly review the methods employed in the measurement of η for various types of objects before summarizing the results in Table 10.1 towards the end of this subsection.

Example

Take the luminosity function of galaxies as estimated by Schechter, viz.

$$\Phi(L) = \frac{\Phi^*}{L^*} \left(\frac{L}{L^*} \right)^{-\alpha} \exp\left(-\frac{L}{L^*} \right),$$

with $\Phi^* = 10^{-2} h_0^3 \text{ Mpc}^{-3}$, $L^* = 10^{10} h_0^{-2} L_\odot$ and $\alpha \sim 1.25$. The number of galaxies with luminosities in the range $(L, L + dL)$, is then given by $\Phi(L) \, dL$. Integrating between 0 and ∞ then gives

$$l_G = \int_0^\infty \Phi(L) L \, dL \sim 3.3 \times 10^8 h_0 L_\odot \text{ Mpc}^{-3}.$$

10.4.1 The mass-to-light ratios

Methods of measuring η for individual galaxies and for clusters of galaxies are summarized below.

Spiral galaxies

The best handle on the mass contained in a typical spiral is given by its rotation curve. Figure 10.18 illustrates the principle by means of a flat, disc-shaped object representing a circular distribution of stars moving round a common centre C. The rotation velocity v of a star S at a distance r from C is related (in an equilibrium distribution) to the gravitational force F_r acting on S towards the centre:

$$m\frac{v^2}{r} = F_r. \tag{10.33}$$

Therefore, if we have v as a function of r, we get F_r as a function of r. Then, by Newton's law of gravitation (which is applicable here because the gravitational fields are weak), we can determine the mass distribution. For example, if most of the mass were concentrated in the nuclear region around C, we would have $F_r \propto r^{-2}$ and $v \propto r^{-1/2}$. The distribution of light across a spiral galaxy does suggest the above to be a good approximation. However, in actual fact the rotation curve – the function $v(r)$ – is flat for most galaxies. That is, after rising sharply outside the nuclear region, v remains constant $= v_0$ (say). Moreover, this relation extends well beyond the visible disc. Figure 10.19 shows some examples.

The implications of this result are either that there is more mass in the outer parts of the galaxy than is indicated by its luminosity distribution, or that Newton's laws of motion and the inverse-square law of gravitation might not be valid over the galactic distance range ($\sim$ a few kiloparsecs). Taking the former (and less radical) view, astronomers have estimated the masses of spirals. S. M. Faber and J. S. Gallagher have listed the rotation velocities and masses contained within the Holmberg radius (wherein the surface brightness drops to $\sim 26.5 m_{pg}$ arcsecond^{-2}) for 39 spirals.

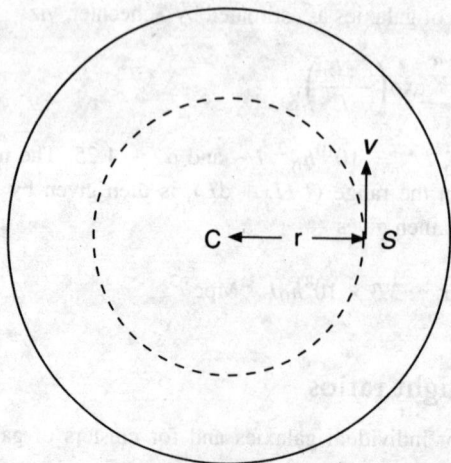

Figure 10.18 The Galactic disc approximated as a system of stars S moving in circular orbits round a common centre C. The velocity v of S is governed by Newton's laws of gravitation and motion.

Since the luminosities are also known, we can estimate the mean value of η for this sample. The result is

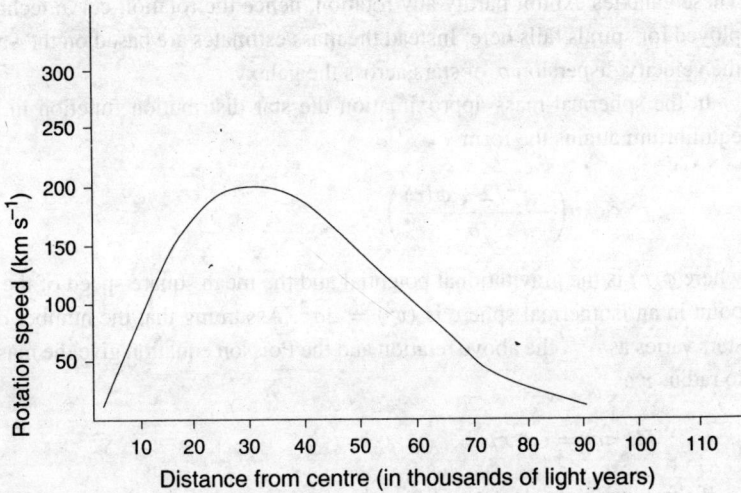

(i)

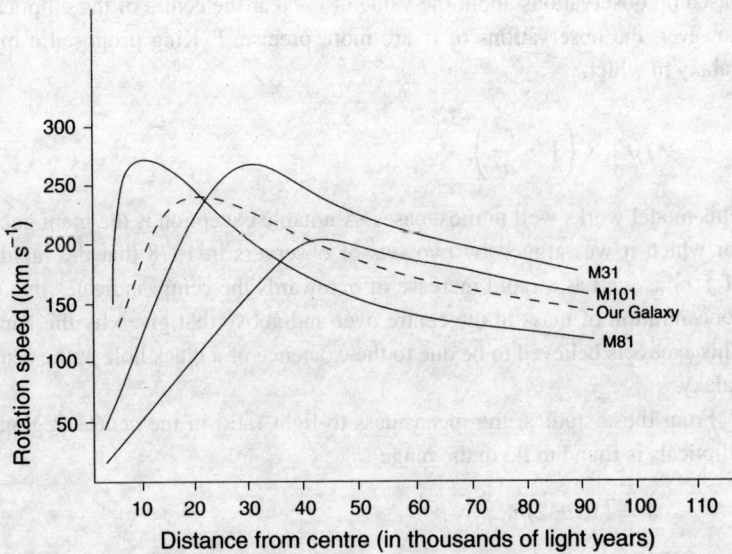

(ii)

Figure 10.19 The rotation curves of some spiral galaxies: notice that, at large distances from the centre, the typical curve is expected to drop to near-zero velocity (as shown by the curve in (i)), instead of remaining more or less constant as seen in reality in (ii).

$$\eta \cong (9 \pm 1)h_0. \tag{10.34}$$

Elliptical galaxies

These galaxies exhibit hardly any rotation, hence the rotation-curve technique employed for spirals fails here. Instead the mass estimates are based on the variation of the velocity dispersion σ of stars across the galaxy.

In the spherical-mass approximation the star distribution function in statistical equilibrium attains the form

$$f \propto \exp\left(-\frac{v^2/2 + \phi(r)}{\sigma^2}\right), \tag{10.35}$$

where $\phi(r)$ is the gravitational potential and the mean square speed of the stars at a point in an isothermal sphere is $\langle v^2 \rangle = 3\sigma^2$. Assuming that the number density of stars varies as $r^{-\epsilon}$, the above relation and the Poisson equation give the mass interior to radius r as

$$M(<r) = \epsilon\sigma^2 r/G. \tag{10.36}$$

If the luminosity density $j(r)$ varies as the number density, the galactic mass-to-luminosity ratio varies as $r^{\epsilon-2}$. It is not yet possible to make precise statements based on observations about the value of ϵ. Near the centre of the elliptical galaxy, however, the observations of σ are more precise. I. King proposed a model of a galaxy in which

$$j(r) \propto \left(1 + \frac{r^2}{a^2}\right)^{-3/2} \tag{10.37}$$

This model works well in most cases. A notable exception is the giant galaxy M87, for which it was argued by two sets of observers in 1978 that the rapid increase of $j(r)$ as well as a rapid increase of σ towards the centre indicates that there is a concentration of mass in the centre over and above that given by the King model. This excess is believed to be due to the existence of a black hole at the centre of the galaxy.

From these studies, the mean mass-to-light ratio in the central region of large ellipticals is found to lie in the range

$$\eta = (10 \pm 2)h_0. \tag{10.38}$$

Statistics of groups of galaxies

A typical catalogue of galaxies lists them by their coordinates on the celestial sphere, two galaxies with nearly the same coordinates being seen near each other. However, can we be certain that groups of apparently nearby galaxies are indeed close to one another and part of one physical system? The answer is generally sought along two different lines, both statistical in nature and both leading to estimates of η.

S. J. Aarseth, J. R. Gott and E. L. Turner adopted the approach of N-body simulations in which galaxies move under each other's gravitational pulls and tend to cluster together in small or large groups. A comparison of such distributions with real galaxy catalogues helps in identifying groups of galaxies and hence in estimating η.

The other approach, pioneered by E. L. Scott and J. Neyman and used extensively by P. J. E. Peebles and others, involves galaxy–galaxy correlation functions. We referred to it earlier in Chapter 7. In this approach the probability of finding a galaxy in a small volume δV at a distance r from a typical galaxy is defined as

$$\delta P = n\,\delta V[1 + \xi(r)], \tag{10.39}$$

where n is the number density of galaxies on the average. In a uniform distribution $\xi(r) = 0$. A positive $\xi(r)$ indicates an enhancement of the density of galaxies near the typical galaxy, hence $\xi(r)$ is called the two-point correlation function.

In actual measurements the position vector $\mathbf{r}$ from the typical galaxy has two components with respect to the observer. The radial component π can be measured from the observed difference in the redshifts of the two galaxies using Hubble's law. The transverse component σ is measured by noting the angular separation of the two galaxies and multiplying it by their mean Hubble distance. Apart from the universal velocity, however, the galaxies also have peculiar (random) velocities relative to their local cosmological rest frames. Such velocities tend to distort the radial component, with the result that, if the two components of $\mathbf{r}$ are plotted on a Cartesian coordinate system, the distribution of points tends to cluster around the axis corresponding to the radial component.

Using such plots for NGC and IC galaxies, Peebles concluded that a reasonably good estimate of $\xi(r)$ is given by

$$\xi(r) = \left(\frac{r_0}{r}\right)^{\gamma}, \qquad \gamma = 1.77, \qquad r_0 = 4.2h_0^{-1}\ \text{Mpc}. \tag{10.40}$$

Aarseth and his colleagues arrived at similar results from their computer simulations. The peculiar velocities of galaxies can be estimated from the above-mentioned concentration effect and the velocity dispersion comes out as

$$\langle v^2 \rangle^{1/2} \cong (600 \pm 250)\ \text{km s}^{-1}. \tag{10.41}$$

From this result we can estimate η as follows. The mean number of neighbours within the characteristic distance $r_0 \sim R = 5h_0^{-1}$ Mpc is given by

$$N = n \int_0^R [1 + \xi(r)]\,\mathrm{d}^3 r$$

$$= 42 \tag{10.42}$$

for $n = 0.03h_0^3$ Mpc^{-3} (estimated for bright galaxies). The peculiar velocity v_i of the ith galaxy having N_i neighbours of mass M at distance R is expected to be of the order

$$v_i^2 \sim GMN_i/R.$$

The result follows from the so-called virial theorem, which essentially states that, in an equilibrium N-body distribution, there exists an equipartition between the kinetic and potential energies. From these results and from estimates of $\langle N_i^2 \rangle$ we get

$$M \sim \frac{R\langle v_i^2 \rangle \langle N_i \rangle}{G \langle N_i^2 \rangle} \sim 5 \times 10^{12} h_0^{-1} M_\odot. \tag{10.43}$$

A detailed calculation using the data on luminosities then gives η in the following range:

$$\eta \sim (500 \pm 200)h_0. \tag{10.44}$$

Clusters of galaxies

A similar correlation-function analysis has been applied to Abell clusters up to redshifts $z \leq 0.2$. The value of η comes out close to that for the nearby groups of galaxies:

$$\eta \sim (500 \pm 100)h_0. \tag{10.45}$$

As early as 1933 F. Zwicky had pointed out what has now become well known as the *missing-mass problem* in clusters. The problem can be briefly stated as follows. If we estimate the mass of galaxies moving in one another's gravitational field in a cluster, then the virial theorem gives the mass of the cluster in terms of the velocity dispersion and the effective mean radius:

$$M = \langle v^2 \rangle \frac{R}{G}. \tag{10.46}$$

From observations of the velocity dispersion $\langle v^2 \rangle^{1/2}$ we can therefore estimate the total mass M in the cluster. This value comes out considerably higher than that estimated on the basis of mass-to-light ratios η_G of individual galaxies. That is, if we see n galaxies in the cluster and if the total luminosity of the cluster is L, then the mass in the cluster is $L\eta_G$. Zwicky was the first to point out that

$$L\eta_G \ll M. \tag{10.47}$$

For the Coma cluster, for example, $M/(L\eta_G) \sim 300$ (see Exercise 20).

Typically one arrives at a cluster mass in the neighbourhood of $10^{15}h_0^{-1}M_{\odot}$. Observations suggest that there are around 4000 large clusters within a 'local' sphere of radius $600h_0^{-1}$ Mpc. This leads to a mean density of matter in clusters of

$$\rho_{0cl} \approx 4 \times 10^{-31}h_0^2 \text{ g cm}^{-3}. \tag{10.48}$$

The density estimated for galaxies is of the same order, although not all galaxies reside in clusters. The clusters have a proportionately higher mass than that of the galaxies contained in them because the M/L ratio for them is as high as $\sim300h_0M_{\odot}/L_{\odot}$, about ten times higher than that for galaxies. This is because the clusters appear to require a greater mass of dark matter for their virial equilibrium.

Observations of X-rays from clusters (see Figure 10.20) have indicated that the emission is through Bremsstrahlung from hot gas and the amount of baryonic matter in the Coma cluster is not sufficient to account for the missing mass estimated by using the virial theorem. If the ratio of baryonic to total gravitating matter in the Coma cluster is representative of the universal value, then the total density parameter Ω_0 is constrained by the inequality

$$\Omega_0 \leq \frac{0.15h_0^{-1/2}}{1+0.55h_0^{3/2}}. \tag{10.49}$$

With this type of inequality, it is clear (i) that, if the deuterium in the universe were made primordially, then we cannot have the density parameter attain the upper limit with baryons alone; and (ii) the universe is open ($k = -1$) unless there is a large quantity of dark matter residing *outside the clusters*. Already at this stage the *known* baryonic content of cluster mass (M_B) as a fraction of the total cluster mass

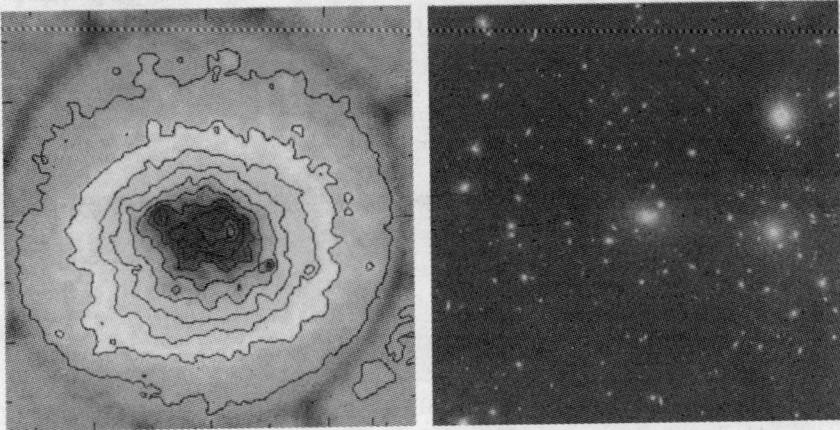

Figure 10.20 The X-ray distribution in a cluster. On the left-hand side is the ROSAT image of the coma cluster showing X-ray intensity contours. The optical image is on the right-hand side. By courtesy of Dr T. J. Ponman, University of Birmingham, UK.

(M_{tot}) threatens a contradiction with observations of the abundance of deuterium. For example, for the Coma cluster, we have

$$\frac{M_B}{M_{tot}} \approx 0.01 + 0.05 h_0^{-3/2}. \tag{10.50}$$

If this ratio were universal, it would lead to a conflict with the deuterium-abundance constraint for $\Omega_0 = 1$. In fact, if $h_0 = 0.65$, say, then setting this ratio equal to $0.01 h_0^{-2}$ for consistency with the abundance of deuterium gives $\Omega_0 \approx 0.23$. Thus, if the universal value of Ω_0 were claimed to be unity (as originally required by inflation), then the conclusion has to be that baryons are selectively located in clusters, while the non-baryonic matter fills the intercluster space. This epicyclic statement could be avoided, if one admits the possibility of a low-density universe.

The local supercluster

It was pointed out by G. de Vaucouleurs that we are situated in a region that seems to be on the outskirts of a concentration of galaxies centred on the Virgo cluster of galaxies located at a distance from us of

$$D = 11 h_0^{-1} \text{ Mpc}.$$

Estimates of the average mass-to-light ratio for the supercluster are still tentative but are believed to be in the range

$$\eta \sim (80 \pm 30) h_0. \tag{10.51}$$

Table 10.1 summarizes the above results as well as some others not discussed here.

Returning to (10.31) we now see that the density parameter can be determined, at least within broad limits, from the values given in Table 10.1. Since the estimate is based on galaxy data we will denote the estimate of Ω_0 by Ω_G.

Table 10.1 The average mass-to-light ratio per galaxy

Object	ηh_0^{-1}
Our Galaxy (inner part)	6 ± 2
Our Galaxy (outer part)	40 ± 30
Spiral galaxies	9 ± 1
Elliptical galaxies	10 ± 2
Pairs of galaxies	80 ± 20
Local Group	160 ± 80
Statistics of clustering	500 ± 200
Abell clusters	500 ± 200
Local supercluster	80 ± 30

According to Table 10.1 ηh_0^{-1} ranges between values of 4 and 700. This gives the value of Ω_G in the extreme range

$$0.003 \leq \Omega_G \leq 0.53. \tag{10.52}$$

Note that this range does not depend on h_0. The range is extreme, in the sense that it uses the maximum range of uncertainty in the universal value of the parameter η. A more realistic upper limit may be $\Omega_G \leq 0.30$.

10.4.2 Constraints from structure formation

The theories of structure formation place additional constraints on the density of matter. We will briefly review the current ideas, with the cautionary remark that they may have to be modified as these theories evolve.

In Chapter 7 we have defined the root-mean-square fluctuation of mass $\langle \delta M_R \rangle$ in a spherical region of radius R (*vide* equation (7.75)). We define the parameter

$$\sigma_8 = \left\langle \frac{\delta M_R}{M_R} \right\rangle, \qquad R = 8h_0^{-1} \text{ Mpc}. \tag{10.53}$$

The size R chosen above is characteristic of volumes containing large clusters. Normally we expect that massive clusters can form by gravitational contraction of large volumes in a low-density universe, or of smaller volumes in a high-density universe. It is found from details of the process of condensation that the relationship between σ_8 and Ω_G for reproducing the observed cluster masses is approximately given by

$$\sigma_8 \sqrt{\Omega_G} \approx 0.5. \tag{10.54}$$

Thus, for example, if we had $\Omega_G = 1$, then we need $\sigma_8 = 0.5$, whereas for a mass density a quarter of the closure value, $\sigma_8 = 1$.

Now the structure-formation theories tell us that, in high-density models, like those with $\Omega_G = 1$, the fluctuations started growing relatively recently and so their growth had to be fast. Thus, if we looked at the number density of massive clusters like Coma, we should find very few clusters at high redshifts. For example, the chance of finding a Coma-like cluster with redshift >0.5 over an area of 1000 square degrees is just about 1% (given the present-day abundance of such clusters). In contrast, the abundance of such clusters in a low-density universe would have evolved very slowly in the redshift range [0, 1]. Studies of cluster samples at various redshifts reveal an abundance distribution more consistent with the latter case, with $\Omega_G = 0.3$, in the range [0.2, 0.4]. Figure 10.21 shows how the evolution of the abundance of clusters changes with the density of the universe. Here the data in fact suggest tighter constraints, favouring a density parameter ~ 0.2.

We may compare this estimate with the mass-to-light ratios obtained earlier or from the baryonic component of clusters. All three suggest that it is hard to make the total density of matter, baryonic as well as non-baryonic, equal to the closure density without introducing highly contrived scenarios.

10.4.3   Dark matter

In the inflationary cosmology we require $\Omega_0 = 1$. Clearly Ω_G falls short of this value. Can we therefore conclude that the universe is open? Is inflation ruled out? The answers are not so simple, however. The matter budget so far looks like this.

It is already noticeable that a considerable part of the matter in the universe might be non-luminous. We have seen that, if we stick to the Newtonian inverse-square law of gravitation, the flat-rotation curves of spiral galaxies imply that there is more mass in the outer regions of these galaxies than can be observed in the form of stars. For clusters of galaxies, the virial theorem (which again is based on the

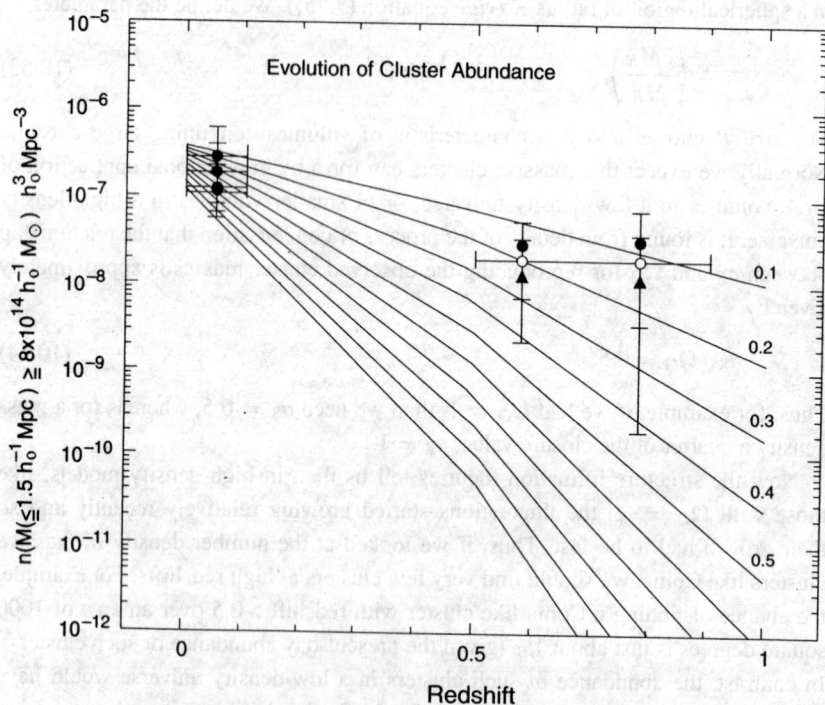

Figure 10.21 These curves show how the abundance of clusters evolves with the redshift for various values of Ω_0. The clusters with mass greater than $8 \times 10^{14} h_0^{-1} M_\odot$ within a comoving radius $1.5 h_0^{-1}$ Mpc are considered in this count per unit volume. The observed points with error bars are shown together with evolutionary curves for various mass density parameters Ω_G (=0.1, 0.2, 0.3, . . .). The data appear to favour a low-density universe with $\Omega_G \approx 0.2$. Adapted from N. A. Bahcall and X. Fan, 1998, *Ap. J*, *445*, 1.

Newtonian law of gravitation) demands even higher masses than can be directly observed.

In the latter context it is worth recalling the views of the Armenian astronomer Victor Ambartsumian.

> 'It is natural to try to uncover the secrets of nature by observing the key points where they are hidden. We can hardly achieve this aim only by theorizing. Observations produce almost innumerable items of evidence in favour of ejections and explosions [of galaxies], and are rather scant regarding the processes of condensation and collapse. The facts are pronouncing an indictment against the ideas connected with the condensation process; in the observable universe, the opposite phenomena, i.e., expansion and diffusion, are responsible for the majority of changes now taking place.'

[From: *A Life in Astrophysics – Selected Papers of V. A. Ambartsumian*, Allerton, New York (1998)]

Future observations may decide whether the clusters by and large are indeed virialized or are expanding, as Ambartsumian argued. The argument for dark matter is weakened in the latter case. For the time being we will proceed with the first alternative.

However, even taking the virial theorem to apply to clusters, we are not able to make up for the total mass density required by inflation. We would have to assume that, if we do have such matter, it is mostly distributed in the voids and is non-luminous. Also, it has to be non-baryonic.

In Chapter 5 we found that there are stringent limits on the baryonic density of the universe, limits imposed by the observations of primordial deuterium. We will review the deuterium evidence in §10.6, but will now take note of its implications for Ω_0. The large M/L ratios in Table 10.1 and the limits set by structure formation theories imply that, even within the acceptable limits, there has to be a lot of non-luminous matter. To the extent that their density does not cross the bounds set by the abundance of deuterium, baryons can account for the non-luminous matter in the following forms.

1. *Low luminosity stars and stellar remnants.* One possibility is of 'brown dwarfs', that is, stars with masses too low ($\leq 0.08 M_\odot$) for them to be able to shine through nuclear fusion of hydrogen. Such stars may form during star-formation processes but are very difficult to detect unless they are part of binaries. At the other end of stellar evolution, high-mass stars may have reached their final states of white dwarfs/neutron stars/black holes. However, as calculated by B. Carr and others, the density in such remnants cannot account for more than $\Omega_0 \approx 0.03$; otherwise their integrated light intensity would be unacceptably high.

2. *Small solid bodies like comets, asteriods, dust grains etc.* There is a limit on how much these can contribute to Ω_0 since they are mostly made of heavy

elements whose abundances together do not exceed ~ 0.01 of the abundance of hydrogen.

3. *Neutral and ionized gas in the form of hydrogen.* This, however, is too small in amount to account for dark matter. For example, the X-ray halo of M87 has been found to have a gas content of only about 3% of the total mass of the galaxy.

There are also stringent limits on the intergalactic amount of neutral hydrogen. In the spectrum of a high redshift ($z > 2$) quasar, the blue side of the Lyman α ($\lambda = 1215$ Å) line should exhibit a significant dip in the continuum as a result of absorption by neutral hydrogen *en route*. In 1965 J. E. Gunn and B. A. Peterson looked for this effect in the quasar 3C-9 and placed an upper limit (on the basis of there being no detectable effect within the limits of sensitivity of observations) of $\Omega_{H\text{-}I} \leq 4 \times 10^{-7} h_0^{-1}$.

Molecular hydrogen can also be ruled out as a possible contender for dark matter by similar tests. The Lyman-α absorption-line systems found in the quasars may be due to discrete clouds of neutral hydrogen. However, most of the hydrogen in these intergalactic clouds may have been photo-ionized by the quasar radiation and hence these data can be used to make an estimate of $\Omega_{H\text{-}II} \approx 10^{-3}$. Furthermore, the condition that the clouds have not been overheated by conduction sets a limit on the density of ionized intergalactic medium of $\Omega_{H\text{-}II} \leq 0.02 h_0^{-2}$.

4. *Massive black holes.* These with masses exceeding a few hundred solar masses might also be candidates for dark matter. Such black holes form from the collapse of massive stars that do not explode as supernovae and so do not eject heavy elements into the surrounding medium. (There cannot be too many black holes of smaller mass since they are formed by supernova explosions and hence pollute the interstellar medium with the heavy elements.) On the basis of the maximum of such effects seen, B. Carr and others have argued that the contribution to Ω_0 from such black holes is no more than $\sim 10^{-4}$. For the massive ones, however, another restriction applies. C. Canizares has argued that too many such massive black holes would exaggerate the effect of gravitational lensing on light from quasars. The absence of any significant lensing distortion makes the number of such massive black holes also negligible.

5. *Gravitational microlensing.* A clever way to detect the presence of brown dwarfs and other dark sub-stellar-mass objects was begun in the early 1990s by several groups using the concept of gravitational microlensing. In this phenomenon, light from a star in the bulge of the Galaxy or in the very nearby galaxies such as the Magellanic clouds could be gravitationally lensed as shown in Figure 10.22, by an intervening dark-matter object of the above

kind. Since, compared with the gravitational lensing of light from quasars by galaxies or clusters of galaxies, this was a phenomenon on a much smaller scale, it was called *gravitational microlensing*. It causes the star to brighten for a short duration, of the order of a few days to a month, when the lensing object crosses its line of sight. The duration of amplification tells us the mass of the lensing object. Groups with imaginative acronyms like MACHO (Massive Astrophysical Compact Halo Object), EROS (Experience de Recherche d'Objets Sombres), OGLE (Optical Gravitational Lensing Experiment), etc., have dedicated telescopes looking for such systems. Figure 10.23 shows one clear-cut case of such a rise in intensity. Of course, not every case in which a star suddenly brightens is a case of microlensing, since a star's luminosity may vary for other reasons too. Thus care is needed when one analyses the cases thrown up by automated searches.

The current microlensing programmes are sensitive to masses of 0.01–5 times the solar mass. This is because of the time scale: the time scale for a 0.1-solar-mass object is about 7 days, whereas the time scale for a five-solar-mass object is about 200 days. The monitoring programmes will not detect events if the events last much less than 7 days, or much longer than 200 days. So, if there are objects beyond this mass range, the microlensing

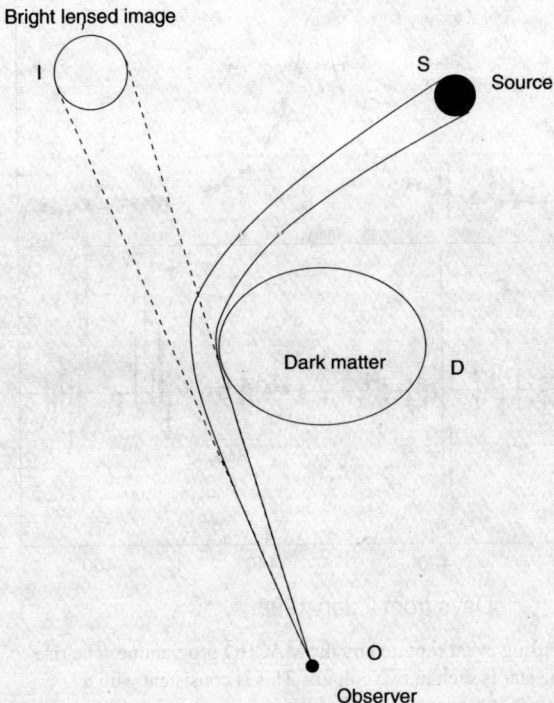

Figure 10.22 A schematic diagram of a microlensing system. The image of the star brightens when it is lensed by an intervening dark mass.

methods cannot spot them.

Microlensing can, in principle, still be used to detect objects smaller than 0.1 and larger than five solar masses. However, the programme needs to be tuned so that it is sensitive to events on longer and shorter time scales.

At this stage it is not clear to what extent MACHO-type objects contribute to the total baryonic density of the universe, largely because it is not yet possible to state unequivocally where their population resides *vis-à-vis* the Galaxy. There is the distinct possibility that some of the microlenses are not in the

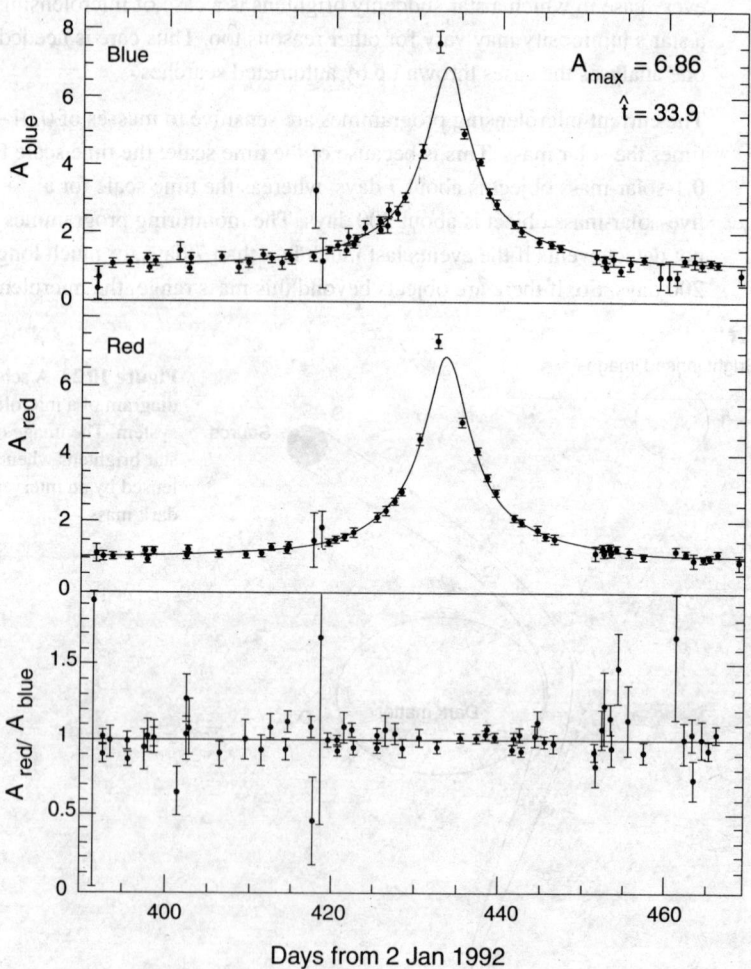

Figure 10.23 A microlensing event reported by the MACHO programme. The rise and fall in intensity of the star is seen in two colours. This is consistent with a gravitational lens effect since it acts at all wavelengths.

disc or halos of the Galaxy, but are located in the Large or Small Magellanic Cloud. Thus some estimates put the MACHO fraction of Ω_B comparable to unity, whereas in others it is negligible. A lot also depends on the mass of a typical MACHO, which is as yet uncertain. Nevertheless, this method is potentially a significant way of detecting a dark component of the universe.

6. *Non-baryonic matter.* Hence we are led to non-baryonic alternatives, which are certainly required if the inflationary cosmology with $\Omega_0 = 1$ is to be believed. We have considered various forms of non-baryonic dark matter in Chapter 7. Since they do not seem to interact with light or any known baryonic or leptonic matter, they have been dubbed 'weakly interacting massive particles' (WIMPs). As discussed earlier, dark matter could come in two possible forms: cold and hot. Their inferred interaction, through gravitation, with visible matter is supposed to play a key role in explaining the large-scale structure observed at present. Although, at the time of writing, CDM seems the favoured option, it does not seem entirely satisfactory for any scenario of structure formation. No WIMP has been confirmed to 'exist' in laboratory experiments. The nearest to experimental credibility are massive neutrinos (which belong to the HDM category), for which there are conflicting claims of a non-zero rest mass.

Can 'inflation' survive as an idea if astronomers see no direct evidence for $\Omega_0 = 1$? It can! This can be achieved by resurrecting the λ-term (see Exercise 22). This, of course, leaves the problem of fine tuning of the λ-term unsolved (see Chapter 6).

10.5   The age of the universe

The formulae (4.47), (4.61) and (4.74) give the age of the universe according to the various Friedmann models. Since these formulae depend on two parameters, H_0 and q_0 (or Ω_0), both of which have been discussed above, we are now in a position to take a look at the problem of whether the Friedmann age estimates are consistent with the various astrophysical estimates of the age of the universe. Figure 4.13 gives the range of values of the ages of the Friedmann models for purposes of comparison.

At present there are two different ways of estimating the ages of galaxies, both of which have been applied to our Galaxy. A primary requirement of consistency is, of course, that the age of a Friedmann model must exceed the age of any object in it.

10.5.1   Stellar evolution

This method, which is applied to globular clusters in our Galaxy, is based on the principle that stars become redder and brighter when they leave the main sequence to become red giants. Since the red-giant phase in the star's life lasts a comparatively short time, say up to about 10% of the time the star spends on the main sequence, the turning point from the main sequence to the giant branch provides the age of a cluster to within 10% uncertainty.

Let the *age of the cluster*, the time when the stars turn off from the main sequence, be denoted by $t_c \times 10^9$ years and let Y and Z be the abundances of helium and metals in the star at this stage. The calculations of stellar evolution then show that

$$\log t_c = 1.035 + 2.085(0.3 - Y) - 0.03(\log Z + 3). \qquad (10.55)$$

Thus the age depends critically on the abundance of helium Y. Y can be estimated from a comparison of the time a star spends on the horizontal branch with the time it spends on the red-giant branch. If this ratio is R, then calculations show that

$$Y = 0.3 - 0.39 \log(f/R), \qquad (10.56)$$

where $f = 2$ if the stellar model takes account of semi-convection and certain other effects, whereas $f = 1$ if these effects are not taken into account. R can be estimated from the observed ratio of horizontal-branch stars and red-giant stars in the cluster. Ages deduced by this method fall in the range from $\sim 13 \times 10^9$ to $\sim 18 \times 10^9$ years.

10.5.2 Nuclear cosmochronology

In 1960 F. Hoyle and W. A. Fowler demonstrated how the relative abundances of radioactive nuclei of long lifetimes can lead to estimates of the age of our Galaxy. The method was already being used for estimating the age of the Solar System. For example, current observations of the $^{87}Sr/^{86}Sr$ ratios plotted against the $^{87}Rb/^{86}Sr$ ratio for various Solar-System materials (such as meteorites) give the age accurately as $t_S \simeq 4.54 \times 10^9$ years (see Exercise 24).

As illustrated in Figure 10.24, the method of nuclear cosmochronology attempts to estimate the time elapsed before the Solar System was formed. According to this method, we start our nuclear clock at $t = 0$ with the birth of the Galaxy. The stars evolve and the more massive ones become supernovae, which manufacture long-lived radioactive nuclei in the so-called *r-process* (the rapid absorption of neutrons by heavy nuclei). The rate at which this process goes on is denoted by a function $p(t)$, which declines to a negligible value at $t = T$. Between this epoch and the formation of the Solar System there occurs a short time gap Δ, known as the *isolation time*, during which we may ignore nucleosynthesis, in particular the r-process. Thus the total nuclear age of the Galaxy is

$$t_G = T + \Delta + t_S. \qquad (10.57)$$

Briefly, T and Δ can be estimated as follows. The formalism, a variation on the earlier work of Hoyle and Fowler, is due to D. N. Schramm and G. J. Wasserburg. We consider a series of nuclei i ($i = 1, 2, \dots$) with decay constants λ_i and rates of production $P_i\,p(t)$. We also assume that the abundance N_i of nucleus i is reduced exponentially at the rate ω due to dilution of stellar matter with external gas and the cycling of matter back into stars. Thus N_i satisfies the following differential equation:

$$\frac{dN_i}{dt} = -\lambda_i N_i - \omega N_i + P_i\,p(t). \tag{10.58}$$

It is assumed that the relative rate of production P_i/P_j of two nuclei i and j is constant.

Equation (10.58) can be integrated from 0 to T to give

$$N_i(T) = P_i e^{-(\lambda_i+\omega)T} \int_0^T p(t) e^{(\lambda_i+\omega)t}\,dt. \tag{10.59}$$

Between T and $T + \Delta$ we may ignore the ω and $P_i\,p$ terms of (10.58) and deduce

$$N_i(T + \Delta) = N_i(T) e^{-\lambda_i \Delta}. \tag{10.60}$$

For long-lived nuclei $\lambda_i T \gg 1$ and certain approximations can be made. Define

$$R_{ij} = \frac{P_i N_j(T + \Delta)}{P_j N_i(T + \Delta)}, \tag{10.61}$$

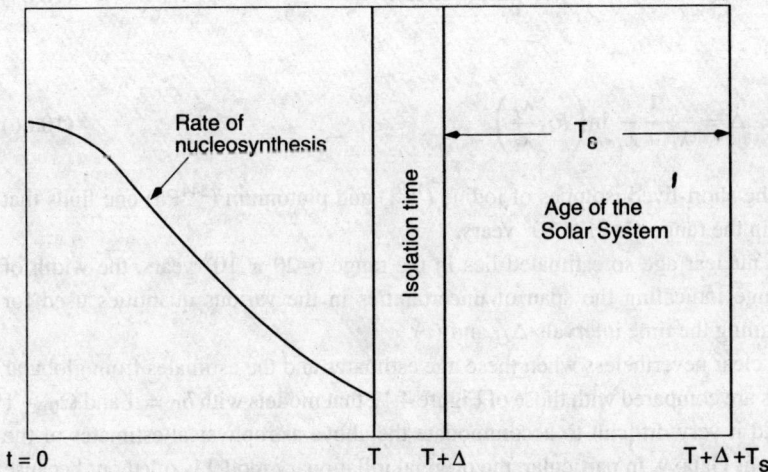

Figure 10.24 A time chart showing how the age of the Galaxy is estimated. The details are explained in the text. Based on J. Audouze, 1979, 'Ages of the universe', in R. Balian, J. Audouze and D. N. Schramm, eds., *Physical Cosmology*, Les Houches Lectures Session XXXII, p. 195 (Amsterdam: North Holland).

$$\langle \tau \rangle = \frac{\int_0^T t p(t)\, dt}{\int_0^T p(t)\, dt}. \tag{10.62}$$

It is easy to see that, for $p(t) = $ constant, $\langle \tau \rangle = T/2$, whereas for $p(t) \propto \delta(t)$, $\langle \tau \rangle = 0$. The value of $\langle \tau \rangle$ will in general lie between these two extreme limits. Simple algebra and calculus then give from (10.59) and (10.62)

$$T = \langle \tau \rangle + \Delta_{ij} - \Delta, \tag{10.63}$$

where

$$\Delta_{ij} = \frac{\ln R_{ij}}{\lambda_i - \lambda_j}. \tag{10.64}$$

Radioactive isotopes of thorium (^{232}Th) and uranium (^{238}U) and, more recently, the osmium (^{187}Os)–rhenium (^{187}Re) pair have been used to estimate Δ_{ij} and hence T and t_G. The decay constants λ_i and λ_j and the quantity R_{ij} are required. The ratio $N_i(T + \Delta)/N_j(T + \Delta)$ in R_{ij} is that prevailing at the time of formation of the Solar System, which can be estimated from the present ratio in meteorites and from knowledge of t_S. The ratio P_i/P_j is taken from theories of nucleosynthesis.

Short-lived isotopes ($\lambda_i T \ll 1$) are used to estimate Δ. We have from (10.59) and (10.60)

$$N_i(T + \Delta) = \frac{P_i}{\lambda_i} p(T) \exp(-\lambda_i \Delta). \tag{10.65}$$

Hence

$$\Delta = \frac{1}{\lambda_i - \lambda_j} \ln\left(R_{ij} \frac{\lambda_j}{\lambda_i} \right). \tag{10.66}$$

From the short-lived isotopes of iodine (^{129}I) and plutonium (^{244}Pu) one finds that Δ lies in the range $(1\text{--}2) \times 10^8$ years.

The nuclear age so estimated lies in the range $6\text{--}20 \times 10^9$ years, the width of this range indicating the span of uncertainties in the various quantities used for determining the time intervals Δ_{ij} and $\langle \tau \rangle$.

It is clear nevertheless when these age estimates and the estimates from globular clusters are compared with those of Figure 4.12 that models with $h_0 = 1$ and $\Omega_0 \geq 1$ will find it very difficult to accommodate the above astrophysical estimates of the age of our Galaxy. In particular, the original inflationary model is ruled out because it predicts $\Omega_0 = 1$ unequivocally. One needs the cosmological constant.

To make the problem easier for the conventional point of view, attempts to see whether the stellar and radioactive ages can be brought down significantly are being made. For example, if significant loss of mass occurs during the main-sequence stage of stellar evolution, then the time spent by the star on the main sequence is reduced. (For it started with higher mass and evolved faster.) Arguing in this way, L. A. Willson, G. H. Bowen and C. Struck-Marcell claim that it may be possible to reduce the ages of globular clusters to values as low as $(7-10) \times 10^9$ years. Likewise W. A. Fowler and C. C. Meisl have recalculated the nuclear age of the Galaxy using a time-dependent model for nucleosynthesis in which an early 'spike' is followed by a uniform synthesis. They claim that the age then comes down to $(11 \pm 1.6 \,(1\sigma)) \times 10^9$ years. Even these exercises, however, do not help the inflationary model if $h_0 \approx 1$.

Observationally also, M. Feast *et al.*, working with the Hipparcos data on stellar parallaxes, came up with a likely way of reducing stellar ages; they argued that there have been systematic increases in the revised stellar distances, so that the stellar luminosities are increased and the evolutionary time scales reduced. This could certainly help in reducing the gap between stellar and cosmological ages, but it is doubtful that the discrepancy can be completely eliminated in this way.

For these reasons, the resurrection of the cosmological constant has helped the big-bang cosmology. For, as we saw in Chapter 4, the λ-term can be suitably chosen to make the age of the universe as long as we please. Figure 10.25, for example, shows how the age of the flat Friedmann model changes as the magnitude of the cosmological constant (Ω_Λ as defined in Chapter 4) is increased. However, the introduction of this constant increases the cosmological distances and thereby increases the probability of a distant light source being gravitationally lensed. From

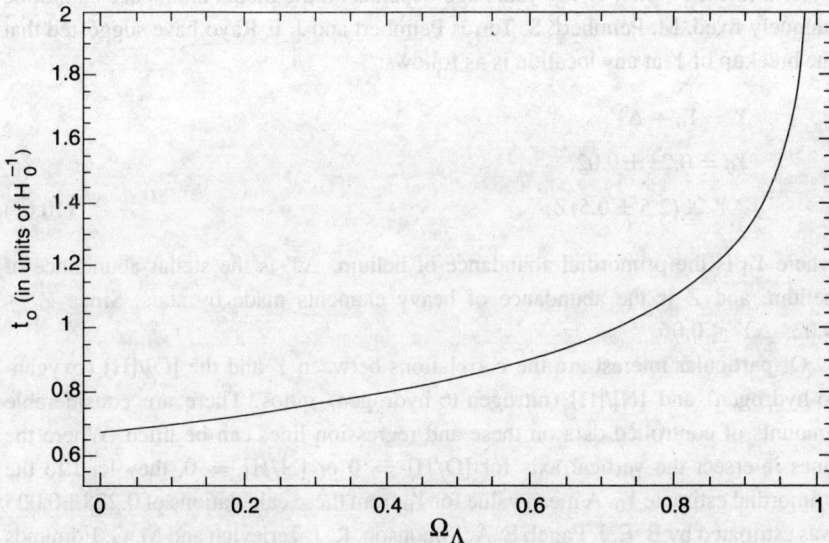

Figure 10.25 The ages of various Friedmann models in units of H_0^{-1} plotted against Ω_Λ. The age can be increased arbitrarily by a fine tuning of the cosmological constant.

the frequencies of lensed objects, upper limits have been placed on the dimension-less parameter Ω_Λ: it is generally agreed that Ω_Λ cannot exceed 0.7.

10.6   The abundances of light nuclei

It is generally recognized that nuclei with atomic masses $A \geq 12$ are synthesized in stars through various processes discussed in theories of stellar evolution. The nuclei ^{6}Li, ^{9}Be, ^{10}B and possibly ^{11}B could be produced in galactic cosmic rays by the break-up of heavy nuclei as they travel through the interstellar gas. It is the lighter nuclei, in particular ^{2}H, ^{3}He, ^{4}He and ^{7}Li, that appear to pose difficulties regarding their production in stars in the amounts observed. Furthermore, their abundances are such that they could have been produced in the big-bang nucleosynthesis. We will therefore discuss what constraints observations of their abundances place on standard cosmology.

10.6.1 ^{4}He

The observed abundances of helium (always denoted by the mass fraction Y) in the universe are quoted as lying in the broad range $0.13 \leq Y \leq 0.34$. The scatter is wide because of the uncertainties of various observational estimates. Furthermore, the estimate of the amount of primordial helium in the Sun at the time the Solar System formed $\sim 4.54 \times 10^9$ years ago depends on the model and hence cannot be uniquely fixed. M. Peimbert, S. Torres Peimbert and J. F. Rayo have suggested that the breakup of Y at any location is as follows:

$$Y = Y_0 + \Delta Y,$$

$$Y_0 = 0.23 \pm 0.02,$$

$$\Delta Y \cong (2.5 \pm 0.5)Z, \tag{10.67}$$

where Y_0 is the primordial abundance of helium, ΔY is the stellar abundance of helium, and Z is the abundance of heavy elements made by stars. Since $Z \leq 0.02$, $\Delta Y \leq 0.06$.

Of particular interest are the correlations between Y and the [O]/[H] (oxygen-to-hydrogen) and [N]/[H] (nitrogen-to-hydrogen) ratios. There are considerable amounts of controlled data on these and regression lines can be fitted. Where the lines intersect the vertical axis for [O/H] $= 0$ or [N/H] $= 0$, they lead to the primordial estimate Y_0. A mean value for Y_0 from these calculations of 0.228 ± 0.005 was estimated by B. E. J. Pagel, E. A. Simonson, R. J. Terlevich and M. G. Edmunds in 1992.

E. Terlevich, R. Terlevich, E. Skillman, J. Stepanian and V. Lipovetskii have looked for the helium contents in extremely metal-poor galaxies since it would be closer to the primordial value. In the first sample of such galaxies they found that the galaxy SB5 0335–052 has $Y = 0.215 \pm 0.01$.

Clearly the survival of the theory of big-bang nucleosynthesis depends on such low values of Y becoming exceptions rather than the rule. Since ^{4}He once it has been produced and ejected into the interstellar medium is difficult to get rid of, low-Y objects have to be explained as arising from inhomogeneities in the primordial set-up. What is the tolerable range for the standard big-bang nucleosynthesis?

It is helpful to go back to Chapter 5 and to recall Figure 5.5, reproduced here

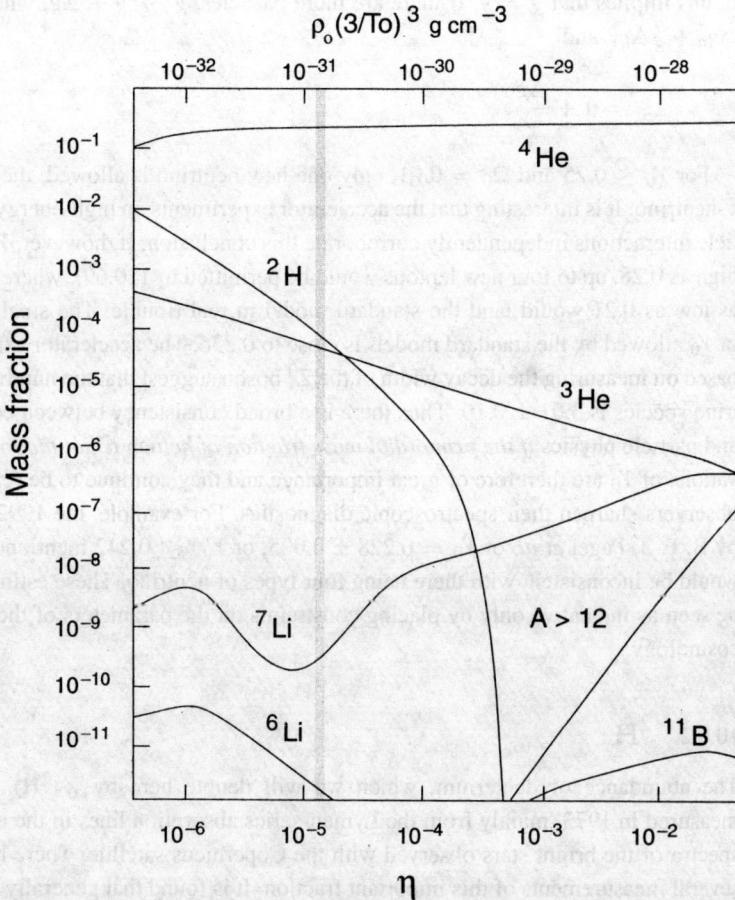

Figure 10.26 Primordial abundances of light nuclei as functions of the present-day density of matter in the universe. The relation between ρ_0 and η is given by (5.56). The shaded band shows the range of baryon densities tolerated by the present data. Adapted from R. V. Wagoner, 1979, 'The early universe', in R. Balian, J. Audouze and D. N. Schramm, eds., *Physical Cosmology*, Les Houches Lectures Session XXXII, p. 395 (Amsterdam: North Holland).

as Figure 10.26. We note that, in the primordial picture, Y_0 is relatively insensitive to h_0 and Ω_0. However, the introduction of new light leptons would push up the neutron-to-proton ratio and hence the value of Y_0. The following formula due to R. V. Wagoner summarizes this result for the fraction η defined in (5.56) exceeding $\sim 10^{-5}$:

$$Y_0 = 0.333 + 0.0195 \log \eta + 0.380 \log \xi. \tag{10.68}$$

Here the fraction $\xi = 1$ if no new particles except those considered in Chapter 5 are assumed to be present in the early universe. In terms of our notation of Chapter 6, this implies that $g = 9$. If there are more particles, $g \rightarrow g + \Delta g$, where $\Delta g = \Delta g_b + \frac{7}{8} \Delta g_f$, and

$$\xi^2 = 1 + \frac{\Delta g}{g}. \tag{10.69}$$

For $Y_0 \leq 0.25$ and $\Omega_0 = 0.01$, only one new neutrino is allowed, the so-called τ-neutrino. It is interesting that the accelerator experiments on high-energy interparticle interactions independently corroborate this conclusion. If, however, Y_0 were as high as 0.28, up to four new leptons would be permitted by (10.69), whereas a value as low as 0.21 would land the standard model in real trouble. The smallest value of Y_0 allowed by the standard models is close to 0.236. The accelerator experiments based on measuring the decay width of the Z^0 boson suggest that the number of neutrino species is 3.01 ± 0.10. Thus there is a broad consistency between cosmology and particle physics *if the primordial mass fraction of helium is not too low.* Observations of Y_0 are therefore of great importance and they continue to be reported, as observers sharpen their spectroscopic diagnostics. For example, the 1992 estimate by B. E. J. Pagel *et al.* of $Y_0 = 0.228 \pm 0.005$, or $Y_0 \leq 0.242$ mentioned earlier, would be inconsistent with there being four types of neutrino. These estimates may be seen as indicative only by placing constraints on the parameters of the standard cosmology.

10.6.2　^{2}H

The abundance of deuterium, which we will denote here by $X(^2H)$, was first measured in 1973, mainly from the Lyman-series absorption lines in the ultraviolet spectra of the bright stars observed with the Copernicus satellite. There have been several measurements of this important fraction. It is found that generally

$$9 \times 10^{-6} \leq X(^2H) \leq 3.5 \times 10^{-5}.$$

Although a mean interstellar value of $X(^2H) \simeq 2 \times 10^{-5}$ is often quoted, there is considerable variation in its value from cloud to cloud. It is not clear whether these variations are due to partial destruction of primordial deutrium through various processes. It has to be destruction, since so far no satisfactory stellar scenario for production of deuterium is known. Thus the primordial value would correspond to the upper end of the range of observations. At least we expect it to exceed $\sim 2 \times 10^{-5}$. (Contrast this situation with that for ^{4}He, for which there is no destruction mechanism but processes of production go on in stars.)

Referring back to Figure 10.26, we see that a primordial abundance of $X(^2\text{H}) \geq 2 \times 10^{-5}$ implies that the baryonic density at present cannot exceed 4×10^{-31} g cm^{-3}, which in turn sets an upper limit on the present-day baryon-density parameter $(\Omega_B)_0$:

$$h_0^2(\Omega_B)_0 \leq 0.02. \tag{10.70}$$

It is interesting to note that, in 1996, from measurements of the abundance of deuterium in clouds around a high-redshift quasar, Tytler, Fan and Burles placed limits on the baryon-density parameter:

$$h_0^2(\Omega_B)_0 \approx 0.024 \pm 0.006.$$

Thus, if matter in the universe is predominantly in baryons, the universe must be open. (Notice that, since black holes are expected to be made of baryons, the hypothesis that most of the deficit between Ω_B and unity is made of unseen matter in the form of black holes is not tenable.) The missing mass or the unseen mass could be non-baryonic, as discussed earlier.

The limit on the abundance of deuterium cited above can be cross checked with the abundance of helium, which is not so sensitive with respect to the density of baryons. We then find that Y_0 is as high as 0.248. Whether this is in conflict with the observed values described above continues to be a matter of debate.

There are, however, fine tunings involved here. For the restriction on density of baryons implies a relatively tight relation between density and temperature, i.e., the constant of proportionality in the relation $\rho_B \propto T^3$ in the relation (5.57) holding during big-bang nucleosynthesis has to be correctly chosen for the model to give the right answer. One may therefore ask whether this can be claimed as a success of the big-bang cosmology.

10.6.3 ^{7}Li and ^{3}He

The ^{7}Li-abundance curve has a plateau with a dip, touching a minimum value of [Li]/[H] of 10^{-10} for $\eta = 3.2 \times 10^{-10}$. The observed data rule out a plateau value of $\sim 10^{-9}$. Even the minimum value is only marginally consistent with the observations. The upper limit on [Li]/[H] by number was placed at 0.8×10^{-10} by K. C. Sahu, M. Sahu and S. R. Pottasch by observing interstellar absorption in the direction of the Large Magellanic Cloud.

The predictions of standard hot-big-bang nucleosynthesis have the merit of being well defined. If there are discrepancies, what does one do? One way tried is to consider an earlier epoch when the nucleons had not formed: that is, when the matter existed in the form of a quark–gluon plasma. The introduction of inhomogeneity at this stage can lead to some parts of the universe being 'proton-rich' while some parts become 'neutron-rich' relative to the standard neutron-to-proton ratio. It is then possible to have an additional parameter to provide better agreement between the observed and predicted abundances of light nuclei. This method works partially successfully but cannot explain away the above problem with ^{7}Li.

The ^{3}He nucleus does not provide a powerful check on cosmological models because it could be produced in the observed amounts in stars. Thus, by and large, only ^{2}H and ^{4}He (and possibly ^{7}Li) give us the most stringent limits on the parameters of the early universe.

10.7   The evidence for antimatter

In Chapter 6 we briefly discussed baryogenesis, that is, recent attempts to account for the predominance of matter over antimatter in the universe, attempts that make use of the grand unified theories (GUTs). How firm is the evidence that the universe is indeed made up only of matter? During the late 1950s and 1960s, H. Alfven and O. Klein produced cosmological models that start off with perfect symmetry between matter and antimatter. In their model, which we will not discuss here in detail, the symmetric components of the plasma that make up the universe are subsequently separated into matter-dominated and antimatter-dominated regions by a hydromagnetic process. Baryon symmetric big-bang models were also discussed by R. Omnes, F. W. Stecker and others in the late 1960s and 1970s.

In Chapter 6 we found that, unless specific symmetry-breaking techniques such as those proposed by the GUTs are employed, the standard big-bang universe would end up with a nett baryon number of zero. GUTs attempt to explain not only why there is a nett number of baryons in the whole universe but also why the photon-to-baryon number ratio is of the magnitude implied by (5.69). In contrast, in the baryon-symmetric cosmology there is separation between regions of matter and antimatter, while the overall number of baryons in the universe is zero.

Theoretical speculations apart, what is the direct evidence for antimatter in the universe? Space probes in the Solar System appear to rule out the existence of antimatter there. Interaction with the solar wind would have produced strong γ-rays had any of the planets been made of antimatter. Since observations beyond the Solar System are largely based on electromagnetic radiation, which treats matter and antimatter alike, it is hard to obtain a firm answer to the above question for a star or a galaxy. Cosmic rays do bring nuclei from the distant parts of the Galaxy (and even from beyond the Galaxy). However, intensive searches have failed to detect significant amounts of antimatter nuclei in cosmic rays. A few antiprotons (one part in $\sim 10^4$) are found, but these could be produced by the interaction of cosmic rays with interstellar matter. Nevertheless, heavy antinuclei cannot be produced in this way and hence their detection in cosmic rays would confirm the existence of antimatter in the universe. The present evidence is somewhat tentative, though it cannot rule out the possibility that there is a substantial antimatter component in extragalactic cosmic rays.

Faraday rotation is one form of indirect evidence. This is the rotation of the plane of polarization of light passing through a medium containing charged particles and a magnetic field. Because they are light, electrons (rather than protons) contribute most to Faraday rotation. If positrons were also present they would also produce Faraday rotation, but in the opposite sense. Since a nett Faraday rotation is observed in radiation from sources in and outside the Galaxy, G. Steigman has interpreted this result as showing that there is an imbalance between the abundances of electrons and positrons. However, this conclusion is based on the magnetic field retaining the same sign throughout. If the field changes sign as radiation enters an antimatter region, the Faraday rotation produced by positrons will be of the same sign as that produced by electrons.

Other indirect evidence could come from observations of the γ-ray background. Such a background can arise from various astrophysical causes – such as primordial black holes, blackbody radiation and the inverse Compton process – in addition to the annihilation of nucleons and antinucleons. Each process, however, has its own signature and imposes its own limits on the magnitudes of the physical quantities involved. From an analysis of the γ-ray spectrum over the energy range of ~ 1–10^2 MeV, F. W. Stecker had concluded that the interpretation involving matter–antimatter annihilation is the one that fits the data best. Such regions of matter and antimatter would have to be separated from each other. However, Steigman criticized this claim on the grounds that the fit is based on a number of parameters that could be adjusted to fit any spectrum of γ-rays.

The symmetry between matter and antimatter was also considered by G. R. Burbidge and F. Hoyle in the 1950s in the context of the steady-state universe. If newly created particles were also accompanied by newly created antiparticles, the symmetry in the universe would be preserved. However, it turned out that the γ-ray background resulting from the annihilation of particles and antiparticles would be very strong – far above that observed today.

In the quasi-steady-state cosmology, it is assumed that the scalar field linked with the creation mechanism has forever broken the matter–antimatter symmetry in favour of the former, the argument being that, in a long-range Machian interaction, a predominance of matter over antimatter in the previous QSSC cycles will perpetuate itself. This assumption, however, needs to be rigorously established with the help of high-energy particle physics.

10.8   The microwave background

We now come to an observation that in its importance to standard cosmology ranks second only to Hubble's discovery of nebular redshifts. This important discovery was first made in an unexpected fashion in 1965 by A. A. Penzias and R. W. Wilson, scientists at the Bell Telephone Laboratory. While looking for radio-wave intensities in the plane of the Milky Way with the help of an antenna having a 20-foot horn reflector of low noise, Penzias and Wilson decided to use the wavelength 7.35 cm because at this wavelength the noise from the Galaxy was negligible. After making measurements in various directions and allowing for numerous unknown causes of radiation, they discovered that an unexplained isotropic noise remained. Was this radiation background genuine? If so, what was its cause? Not knowing the answers to these questions, they hesitated before announcing their discovery.

Penzias and Wilson would not have waited to publish this result had they been aware of the prediction George Gamow's colleagues Alpher and Herman had made some 17 years earlier. This was the prediction that, if the universe had a hot phase soon after the big bang, it should now possess a cooled-down relic radiation background. In 1948, Alpher and Herman had estimated the present-day background temperature to be around 5 K, whereas Gamow, a few years later, had made a guess of ∼7 K. Penzias and Wilson had assigned a temperature of 3.3 ± 0.33 K to the background radiation they observed on the assumption that it represented blackbody radiation.

While Penzias and Wilson were puzzling over their discovery, the news reached Princeton, where P. J. E. Peebles, himself a leading worker in the early-universe calculations, grasped its significance. Indeed, the Princeton group, including Peebles and R. H. Dicke, P. G. Roll and D. T. Wilkinson, had already set up an experiment to measure this relic radiation. Although their own measurement at the wavelength of 3.2 cm came in late 1965, it was thus anticipated by the announcement of the discovery of Penzias and Wilson on 13 May 1965.

10.8.1   The spectrum

The background temperature has since been measured at several wavelengths by ground-based radiometers at frequencies upwards from 0.015 cm^{-1} and by balloon-, rocket- or satellite-borne instruments at higher frequencies. The results are summarized in Table 10.2, which is not claimed to be exhaustive. It is often convenient to express the frequencies in units per centimetre by dividing the frequency expressed in hertz by c. Thus 3 cm$^{-1} \equiv 9 \times 10^{10}$ Hz. The observed flux is expressed in the form of a temperature of the blackbody radiation with the corresponding flux in the given frequency range.

The entries against the CN-molecule experiment in Table 10.2 were obtained as follows. The ground state of the CN molecule has rotational levels $J = 0, 1, 2,$

3, ... The transition from $J = 0$ to $J = 1$ is effected by incident radiation of frequency 3.79 cm^{-1}, whereas that from $J = 1$ to $J = 2$ is caused by incident radiation at a frequency of 7.58 cm^{-1}. Observations of CN molecules in interstellar space show that the upper levels are partially populated, thus indicating the presence of a radiation field. The ambient radiation temperature can be determined from the degree of excitation of these levels (see Exercise 33). Such observations were first made by McKellar, as long ago as 1941, but their significance was not appreciated at the time. They in fact tell us that the microwave background extends beyond our local neighbourhood and could be considered the first discovery of the microwave background.

To check the true blackbody character of the radiation it is necessary to have detectors above the Earth's atmosphere since ground-based measurements do not reach the peak wavelengths of the expected blackbody curve due to atmospheric absorption. There were several early attempts using balloons and rockets. However, some of these reported departures from the Planckian spectrum later turned out to be false alarms. The most accurate and exhaustive study came in 1990 with the COBE satellite, and is reported at the end of Table 10.2.

The Cosmic Background Explorer Satellite (COBE) was launched in 1989 and obtained a beautiful spectrum, shown in Figure 10.27. The COBE measurements

Table 10.2 Measurements of the microwave background

Type of experiment	Frequency (cm^{-1})	Temperature (K)	Observers	Reference
Ground-based radiometers	0.0136–0.0207	3.7 ± 1.2	T. F. Howell and J. R. Shakeshaft	*Nature*, **216**, 753 (1967)
	0.079	2.70 ± 0.07	N. Mandolesi *et al.*	*Astrophys. J.*, **310**, 561 (1986)
	0.136	3.3 ± 0.33	A. A. Penzias and R. W. Wilson	*Astrophys. J.*, **142**, 419 (1965)
	0.31	$2.69 \pm ^{0.16}_{0.21}$	R. A. Stokes, R. B. Partridge and D. T. Wilkinson	*Phys. Rev. Lett.*, **19**, 1199 (1967)
	0.313	3.0 ± 0.5	P. G. Roll and D. T. Wilkinson	*Phys. Rev. Lett.*, **16**, 405 (1966)
	0.413	2.783 ± 0.025	D. G. Johnson and D. T. Wilkinson	*Astrophys. J. Lett.*, **313**, L1 (1986)
	0.633	$2.78 \pm ^{0.12}_{0.17}$	R. A. Stokes, R. B. Partridge and D. T. Wilkinson	*Phys. Rev. Lett.*, **19**, 1199 (1967)
	0.667	2.0 ± 0.4	W. J. Welch, S. Keachie, D. D. Thornton and G. Wrixon	*Phys. Rev. Lett.*, **18**, 1068 (1967)

Table 10.3 Continued.

Type of experiment	Frequency (cm^{-1})	Temperature (K)	Observers	Reference
	1.08	3.16 ± 0.26	M. S. Ewing, B. F. Burke and D. H. Staelin	*Phys. Rev. Lett.*, **19**, 1251 (1967)
	1.17	$2.56 \pm {}^{0.17}_{0.22}$	D. T. Wilkinson	*Phys. Rev. Lett.*, **19**, 1195 (1967)
	1.22	2.9 ± 0.7	V. J. Puzanov, A. E. Salomonovich and K. S. Sankevich	*Sov. Phys. – Astronomy*, **11**, 905 (1968)
	1.89	2.70 ± 0.04	D. M. Meyer and M. Jura	*Astrophys. J.*, **297**, 119 (1985)
	2.79	2.4 ± 0.7	A. G. Kislyakov, V. I. Chernyshev, Yu. V. Lebskii, V. A. Maltsev and N. V. Serov	*Sov. Astronomy – A. J.*, **15**, 29 (1971)
	3.0	2.61 ± 0.25	M. F. Miller, M. McColl, R. J. Pederson and F. L. Vernon Jr	*Phys. Rev. Lett.*, **26**, 919 (1971)
	3.0	$2.46 \pm {}^{0.40}_{0.44}$	P. E. Boynton, R. A. Stokes and D. T. Wilkinson	*Phys. Rev. Lett.*, **19**, 462 (1968)
	3.0	$2.46 \pm {}^{0.40}_{0.54}$	P. E. Boynton and R. A. Stokes	*Nature*, **247**, 528 (1974)
CN molecule	3.79	2.93 ± 0.06	P. Thaddeus	*Ann. Rev. Astron. Astrophys.*, **10**, 305 (1972)
	7.58	$2.9 \pm {}^{0.4}_{0.5}$	D. J. Hegyi, W. A. Traub and N. P. Carlton	*Astrophys. J.*, **190**, 543 (1974)
Rocket	1.67–33.3	$3.8 \pm {}^{0.8}_{1.9}$	K. D. Williamson, A. G. Blair, L. L. Catlin, R. D. Hiebert, E. G. Lloyd and H. V. Romero	*Nature*, **241**, 79 (1973)
	7.69–25	$.3.4 \pm {}^{0.7}_{3.4}$	J. R. Houck, B. T. Soifer, M. Harwit and J. L. Pipher	*Astrophys. J.*, **178**, L29 (1972)

Table 10.4 Continued.

Type of experiment	Frequency (cm^{-1})	Temperature (K)	Observers	Reference
	4.31	2.795 ± 0.0018	T. Matsumoto and S. Hayakawa	*Astrophys. J.*, **329**, 567 (1988)
	7.05	2.963 ± 0.017	H. Murakami and K. Sato	
	10.4	3.150 ± 0.026	A. E. Lange and P. L. Richards	
	3–16	2.736 ± 0.017	H. Gush, M. Halpern and E. Wishnow	*Phys. Rev. Lett.*, **65**, 537 1990
Balloon	1–11.5	$2.5 \pm ^{0.25}_{0.45}$	D. Muehlner and R. Weiss	*Phys. Rev. Lett.*, **30**, 757 (1973)
	2.38–13.53	$2.96 \pm ^{0.13}_{0.08}$	D. P. Woody and P. L. Richards	*Phys. Rev. Lett.*, **42**, 925 (1979) and *Astrophys. J.*, **248**, 18 (1981)
Satellite (COBE)	1–20	2.735 ± 0.06	J. C. Mather *et al.*	*Astrophys. J. Letts.*, **354**, L37 (1990)

gave a very precise Planckian spectrum with a blackbody temperature of

$$T_0 = 2.735 \pm 0.06 \text{ K}. \tag{10.71}$$

The overall sensitivity and accuracy of the experiment made it clear that some of the earlier claims of significant departures from the Planckian spectrum at high frequencies (e.g. by Woody and Richards and by Matsumoto *et al.* in Table 10.2) were erroneous. Indeed, even laboratory experiments are not known to produce a Planckian spectrum of this level of accuracy.

10.8.2 Anisotropy

If the microwave background radiation (MBR hereafter) is indeed of primordial origin, its anisotropies can tell us a lot about the present and the past history of the universe. The early developments after the epoch of decoupling imprinted their signature on the radiation background, imprints that are expected to survive to this day. Observations of anisotropies are discussed below; but first we write down the formalism for quantifying them.

The angular power spectrum

When we look at the distribution of a physical quantity across the celestial sphere, its anisotropies can be best described with the help of spherical harmonics. The quantity describing the MBR is its temperature $T(\theta, \phi)$, written as a function of two spherical coordinates (such as declination and right ascension). We may accordingly write

$$\frac{\Delta T(\theta, \phi)}{T} = \sum_{l=1}^{\infty} \sum_{m=-l}^{m=l} a_{lm} Y_{lm}(\theta, \phi). \qquad (10.72)$$

The sum over l begins with 1 instead of zero, for the zeroth perturbation is isotropic over the whole sky and can be absorbed into T. The $l = 1$ term is the so-called *dipole-anisotropy* term, which, as we shall see, arises from the motion of the Earth relative to the rest frame of the MBR. Henceforth we will not include this term also in the above series. The next, $l = 2$, mode is the *quadrupole* mode.

The *angular power spectrum* is defined by quantities C_l defined by

$$C_l \equiv \langle |a_{lm}|^2 \rangle, \qquad (10.73)$$

where the averaging is with respect to all realizations of the sky and summed over all m. Thus each C_l tells us the relative strength of the lth harmonic in the overall distribution.

In general we will be interested in looking at $\Delta T/T$ over a certain angular scale ϑ. Thus, if we take two directions denoted by unit vectors $\mathbf{e}_1$ and $\mathbf{e}_2$ enclosing this angle between them, we get

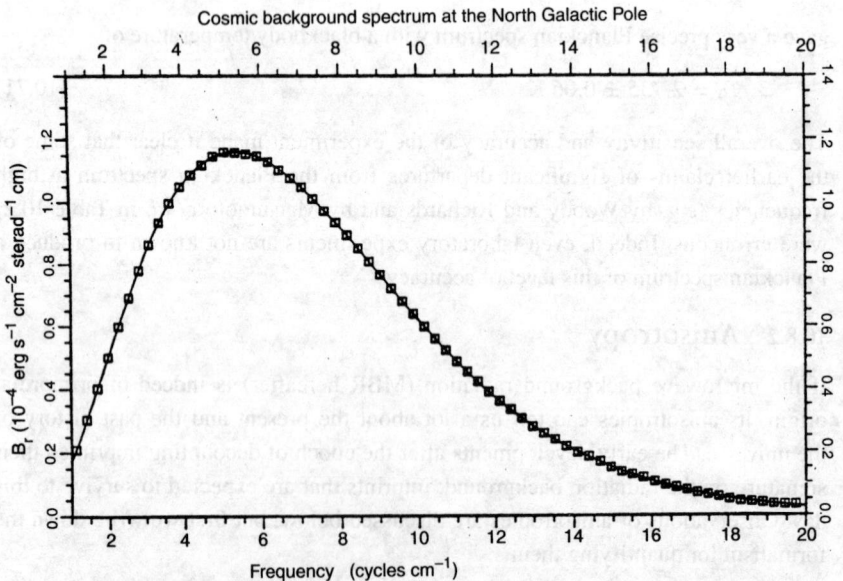

Figure 10.27 The COBE measurements of the microwave background radiation at various frequencies. The continuous curve represents the best-fit Planckian curve to the data points. (See Table 10.2 for the source.) Courtesy of the NASA Goddard Space Flight Center and the COBE Science Working Group.

$$\mathbf{e}_1 \cdot \mathbf{e}_2 = \cos \vartheta. \tag{10.74}$$

Now define the *autocovariance function* which tells us how the temperature fluctuations compare over directions separated by the angle ϑ:

$$C(\vartheta) = \left\langle \frac{\Delta T(\mathbf{e}_1)}{T}, \frac{\Delta T(\mathbf{e}_2)}{T} \right\rangle, \tag{10.75}$$

which, for stationary fluctuations can be expressed in the form

$$C(\vartheta) = \frac{1}{4\pi} \sum_{l=2}^{\infty} (2l+1) C_l P_l(\cos \vartheta). \tag{10.76}$$

Suppose that, from observations of a single sky, we have obtained the estimate of the autocovariance function as $C(\vartheta)$:

$$\hat{C}(\vartheta) = \frac{1}{4\pi} \sum_{l=2}^{\infty} |\hat{a}_{lm}|^2 P_l(\cos \vartheta), \tag{10.77}$$

where the a_{lm} are determined from a single observation of the sky. In this case one needs to estimate the *cosmic variance* of the quantity $C(\vartheta)$. This can be shown to be

$$\langle |\hat{C}(\vartheta) - C(\vartheta)|^2 \rangle = \left(\frac{1}{4\pi}\right)^2 \sum_{l=2}^{\infty} (2l+1) C_l^2 P_l^2(\cos \vartheta). \tag{10.78}$$

In practice the details are considerably intricate when one attempts to extract a signal from the actual sky data. We will not go into those details here. We point out, however, that the Legendre polynomials $P_l(\cos \vartheta)$ contain the following information: the typical angular scale of anisotropy corresponding to the index l is of the order $180°/(\pi l)$. We will now look at the evidence for anisotropies on various scales and the physical causes associated with them.

The dipole anisotropy

These anisotropies can arise from two sources. There could in principle be an intrinsic dipole component in the MBR arising from anisotropy in the universe. The other possibility is a consequence of the Earth moving relative to the rest frame of the MBR in which the radiation is isotropic. Measurements over large angles give no evidence for anisotropy of the first kind; but they do provide evidence for the second effect.

In early experiments E. S. Cheng, P. R. Saulson, D. T. Wilkinson and B. E. Corey observed the anisotropy described by a variation in temperature with direction of the following kind:

$$T = T_0 + T_1 \cos \theta, \tag{10.79}$$

with $T_1 \sim 3 \times 10^{-3}$ K. Such a variation can be explained by the assumption that the centre of the Galaxy has a velocity of ~ 540 km s^{-1} in the direction

$l = 280°$, $b = 30°$. Another set of measurements by G. F. Smoot, M. V. Gorenstein and R. A. Muller indicated a similar effect, but the corresponding velocity of the Galaxy is $\sim$630 km s^{-1} in the direction $l = 261°$, $b = 33°$. The
first group used balloons for measurement, whereas the second group used a U-2
plane. The COBE measurements of 1992 also tend to agree with this conclusion: a dipole amplitude of 3.372 ± 0.007 mK, with a maximum in the direction
$(l, b) = (264.14° \pm 0.30°, 48.26° \pm 0.30°)$, implying a velocity of the Galaxy
of 547 ± 17 km s^{-1} relative to the background. Note that this motion is different
from the large-scale streaming motions discussed in §10.3. Thus the kinematical
rest frame of galaxies in our neighbourhood does not agree with the rest frame of
the microwave background.

This discrepancy could be due to another large-scale streaming motion of the
cluster population relative to the rest frame of the MBR. Further studies are needed
in order to resolve this issue.

The anisotropy caused by the particle horizon

In Chapter 6 we discussed the limitations on homogeneity imposed by the particle
horizon. The particle horizon at decoupling subtends an angle θ_H at the observer today. It can be shown that measurements of radiation in different directions separated
by angles large relative to

$$\theta_H \cong 2\sqrt{\frac{2q_0}{z_R}} \sim 5° \sqrt{q_0}, \tag{10.80}$$

should, in principle, reveal differences reflecting the early inhomogeneities on the
scale of the particle horizon at $z = z_{dec}$. However, no such inhomogeneities are seen:
the MBR appears homogeneous over such scales. The issue is resolved by appealing
to inflation, which removes the horizon problem, as seen in Chapter 6. (See Exercise
22 in Chapter 7 for the derivation of (10.80).)

Small-angle anisotropies

If the search is for a specific angular level θ, a large antenna with beam width
$B \approx \theta$ is pointed at a fixed angle relative to the Earth and swept across the sky
by the rotation of the Earth. The temperature of the antenna T_A undergoes a small
fluctuation ΔT_A composed of the intrinsic fluctuation of the background as well as
receiver noise. Thus ΔT_A sets an upper bound on the intrinsic fluctuation. If $B \gg \theta$
then we may look upon the beam as covering $\sim(B/\theta)^2$ patches of angular size θ.
Detailed calculations then show that the intrinsic fluctuation is less than

$$\left(1 + \frac{B^2}{\theta^2}\right)^{1/2} \Delta T_A. \tag{10.81}$$

Table 10.3 gives the pre-1992 data on small-angle fluctuations. These were all upper
limits with no positive detection on any angular scale larger than a few arcminutes.

The fact that $\Delta T/T$ is less than $\sim 10^{-5}$ on the scale of a few arcminutes posed severe difficulties for theories of galaxy formation. For, according to our discussion of Chapter 7, larger fluctuations in temperature than this should have been observed in the relic background today.

It was against this background that, in 1992, the COBE experiment referred to earlier also provided for the first time a positive detection of anisotropy of the MBR. The COBE map of the sky in Figure 10.28 shows the patches of temperature fluctuations $\Delta T/T \sim 6 \times 10^{-6}$.

The typical inhomogeneity shown in the COBE map of Figure 10.28 is of angular scale $10°$ and magnitude $\Delta T \sim 30$ µK. The quadrupole anisotropy ($l = 2$) is of the order 17 µK. The COBE data were collected for four years and the experiment was switched off in 1994. The analysis of the radiation anisotropy, when it is convoluted with inhomogeneities in the distribution of matter, gives a spectral index for the fluctuation scale k that seems consistent with $n = 1$. However, detailed analyses show that the standard CDM and HDM models of structure formation do not explain the evolution of large-scale structure starting with the COBE data (at the 'initial' stage of structure formation) and ending with the observed clustering of galaxies today (the 'final' stage of structure formation).

Such a high degree of isotropy also posed difficulties for a theory that attempts to explain the microwave background as arising from superpositions of radiation from discrete sources. As discussed in Exercise 34, the sources would have to be more numerous and more closely spaced than galaxies. We have, however, discussed the possibility of producing the MBR in an altogether different fashion in the quasi-steady-state cosmology. We will review the observational evidence for that model in

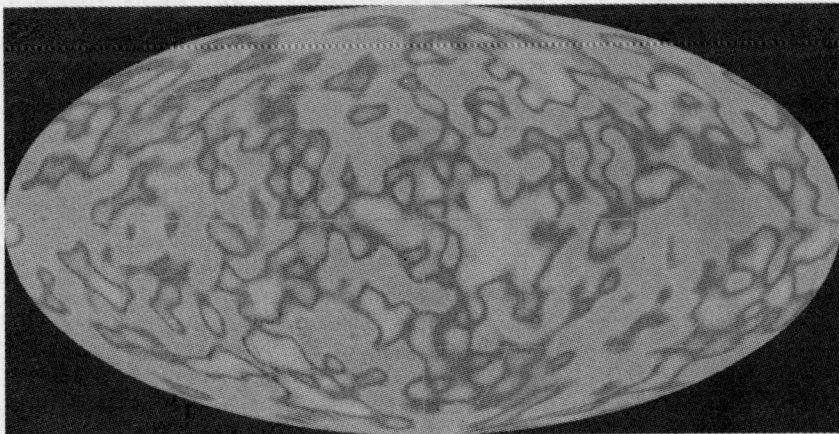

Figure 10.28 The 1992 COBE map of the first positive detection of fluctuations of the temperature of the MBR. Courtesy of the NASA Goddard Space Flight Center and the COBE Science Working Group.

the final chapter. We continue our discussions here within the framework of standard cosmology.

The COBE data and a few subsequent important detections of anisotropy are listed in Table 10.4. Table 10.4 lists the average CMB temperature anisotropies measured on various angular scales by combining the detections within the corresponding multipole bands, l_{band}. The values in the table are merely illustrative of the broad picture emerging from the band-power estimates of 145 CMB anisotropy detection data points (Sept. 2000, Ratra & Souradeep).

The value of these measurements lies in constraining the theories of structure formation and through them the cosmological parameters. It is too early to say what the final picture is going to be like. We will content ourselves with listing a few possible *causes* of anisotropies of the MBR so that their signal may be looked for in such measurements. The smallness of angles implies that we are looking at higher harmonics in the range ~ 10 to $\sim 10^4$.

Table 10.5 The small-angle anisotropy of the microwave background (early, pre-1992 measurements)

Frequency (cm^{-1})	Angular scale (arcminutes)	$\Delta T/T$	Observers	Reference
0.090	10–20	$<1.5 \times 10^{-4}$	K. C. Stankevich	*Sov. Astron.*, **18**, 126 (1974)
0.278	>2	$<7 \times 10^{-4}$	R. L. Carpenter, S. Gulkis and T. Sato	*Astrophys. J.*, **182**, L61 (1973)
0.0357	10	$<1.8 \times 10^{-3}$	E. K. Conklin and R. N. Bracewell	*Nature*, **216**, 777 (1967)
0.325	4.5	$<4.5 \times 10^{-5}$	J. M. Uson and D. T. Wilkinson	*Astrophys. J. Lett.*, **277**, L1 (1984)
0.34	480	$<4 \times 10^{-5}$	R. D. Davies *et al.*	*Nature*, **326**, 462 (1987)
0.357	>5	$<8.0 \times 10^{-5}$	Y. N. Parijskij	*IAU Symp. No. 79*, 315 (1978)
0.500	>1.25	$<7.0 \times 10^{-4}$	J. C. Pigg	*IAU Symp. No. 79*, 317 (1978)
0.66	2	$<1.7 \times 10^{-5}$	A. C. S. Redhead *et al.*	*Astrophys. J.*, **346**, 566 (1989)
1.03	7	$<8.0 \times 10^{-5}$	R. B. Partridge	*Astrophys. J.*, **235**, 681 (1980)
3.0	20	$<3.5 \times 10^{-5}$	P. R. Meinhold and P. M. Lubin	*Astrophys. J.*, **370**, L11 (1991)
7.692	30	$<1.2 \times 10^{-4}$	N. Caderni, V. De Cosmo, R. Fabbri, B. Melchiorri, F. Melchiorri and V. Natale	*Phys. Rev.* D **16**, 2424 (1977)

The Sachs–Wolfe effect

This measures the metric fluctuations near the surface of last scattering. For example, if there is inhomogeneity of matter (clumping/voids) in a given region, this would lead to fluctuation of g_{ik} from the homogeneous Robertson–Walker form. In Newtonian terms we may argue that the photons making up the radiation background come from wells of different potential (φ), which would produce a change of energy and hence of T, given by

$$\left.\frac{\Delta T}{T}\right|_{\text{energy}} = \frac{\delta\varphi}{c^2}. \tag{10.82}$$

In addition to this there is time dilatation, so that the photons emerging from a potential well are delayed in relation to surface photons and therefore encounter the scale factor S during a later epoch. For the Einstein–de Sitter universe $S \propto t^{2/3}$ and the fluctuation in T is given by

$$\left.\frac{\Delta T}{T}\right|_{\text{time delay}} = -\frac{\delta S}{S} = -\frac{2}{3}\frac{\delta t}{t} = -\frac{2}{3}\frac{\delta\varphi}{c^2}, \tag{10.83}$$

because the gravitational redshift produces the above time delay. On adding the two effects we get

$$\frac{\Delta T}{T} = \left.\frac{\Delta T}{T}\right|_{\text{energy}} + \left.\frac{\Delta T}{T}\right|_{\text{time delay}} = \frac{1}{3}\frac{\delta\varphi}{c^2}. \tag{10.84}$$

In addition to this there can be *tensor* fluctuations, which will produce small contributions to $\Delta T/T$. Since these fluctuations are associated with time-dependent

Table 10.6 Measurements of anisotropies of the microwave background

l_{band}	ΔT_l (μK)	N_{expts}
2.5 ± 1	13 ± 3	2
4.5 ± 1	30 ± 5	2
8 ± 2	26 ± 2.5	5
13 ± 4	36 ± 3	6
19 ± 7	35 ± 8	2
52 ± 12	29 ± 3	4
63 ± 9	34 ± 3.5	10
87 ± 12.5	42 ± 3	13
118 ± 17	49 ± 3	25
156 ± 22	57 ± 2	25
241 ± 69	67 ± 2	30
420 ± 123	43 ± 2	12
629 ± 219	44 ± 3	7
2544 ± 1081	20 ± 5	2

changes in the metric tensor, they are essentially caused by gravitational waves. Some inflationary models predict gravitational-wave-type fluctuations, which are potentially detectable.

Finally, another subtle effect, called the *Rees–Sciama effect*, arises from changes in a gravitational well while the photon is inside it. Thus, on crossing through such a well, the photon will experience a change of energy and hence of frequency. This can lead to non-linear effects in $\Delta T/T$.

The Sunyaev–Zel'dovich effect

This effect suggests that the photons of the MBR entering a cluster with hot gas will be 'kicked upstairs' to higher (X-ray) energy by the Thomson scattering from high-energy electrons. Thus, if we observe in the direction of the cluster, we should find a drop in the intensity of radiation. A crude approximation, modelling the cluster as an isothermal sphere of radius R_c, gives a fractional drop in the temperature of the MBR as

$$\frac{\Delta T}{T} = -\frac{4R_c n_e k T_e \sigma_T}{m_e c^2},\qquad(10.85)$$

where n_e is the density of electrons in the cluster, T_e is the electron temperature, m_e is the mass of the electron and σ_T is the Thomson scattering cross section. So far there are several claims of positive detection of this effect, in clusters ranging up to redshifts $\gtrsim 2$ (see Table 10.5). This not only shows that the MBR extends that far but also, by giving an estimate of R_c, allows a determination of Hubble's constant. The values of H_0 determined in this way are of the order ≤ 40 kms^{-1} Mpc^{-1}. Clearly, measurement of this effect, even though it is not directly linked with the formation of large-scale structure, is nevertheless useful for cosmologists.

Sakharov oscillations

Another way of measuring the anisotropy of the MBR is through velocity effects arising from acoustic oscillations of perturbations inside the horizon at the surface of last scattering. These oscillations lead to fluctuations in number of photons and in their temperature, both being related to the wavelength of the oscillations. So one may see periodic behaviour, giving a peak in the C_l coefficients of the power spectrum estimated at

$$l_{peak} \approx 200\Omega_0^{-1/2}.\qquad(10.86)$$

An announcement of the detection of such a peak (incongruously called the 'Doppler peak', since oscillations of matter rather than velocities are responsible for the effect) was made in 2000 by the 'BOOMERANG' (Balloon Observations of Millimetric Extragalactic Radiation and Geomagnetics) group of experimentalists. They found a peak amplitude $\Delta T_{200} = (69 \pm 8)$ μK at $l_{peak} = 197 \pm 6$, which is consistent with $\Omega_0 = 1$. The group in fact measured the angular power spectrum at $l = 50$–600. At the time of writing these results are being analysed further.

Future experiments

After COBE several ground-based detections of anisotropy of the MBR were made. Figure 10.29 shows a compilation that was up to date at the time of writing, but which keeps growing with time! In any case it serves as an indication of the boost COBE's positive detection provided for such efforts.

Following COBE two more ambitious satellite-borne experiments have been in preparation. These are MAP and Planck. The former was launched by NASA in July, 2001 to measure the microwave sky at five different frequencies in the range 22–90 GHz with a resolution of $\sim$20 arcminutes and a sensitivity of $\sim$35 µK per $0.3° \times 0.3°$ pixel. By combining the three highest-frequency channels, the sensitivity can be increased to $\sim$20 µK per pixel. MAP will map the MBR power spectrum up to $l \sim 800$.

The ESA's Planck Surveyor will be launched in 2007 and will measure the radiation at frequencies in the range 30–100 GHz with the Low-Frequency Instrument and in the range 100–90 GHz with the High-Frequency Instrument. The expected resolution is $\sim$10 arcminutes, with a sensitivity of $\Delta T/T \sim 2 \times 10^{-6}$.

To summarize, the MBR is being looked upon as a mine of information by the big-bang cosmologists. Since it is regarded as a relic of the early universe, dating back at least to the surface of last scattering, its spectrum and anisotropies should contain valuable information about the past developments of the universe, much like an archaeological site contains information about its past history.

Table 10.7 Clusters with reliable Sunyaev–Zel'dovich effects

Cluster	Redshift	ΔT^a (mK)
Abell 478	0.0900	-0.38 ± 0.03
Abell 665	0.1810	-0.37 ± 0.07
Abell 697	0.282	-0.13 ± 0.02
Abell 773	0.1970	-0.31 ± 0.04
Abell 990	0.144	-0.13 ± 0.03
Abell 1413	0.1427	-0.15 ± 0.02
Abell 1656	0.0232	-0.27 ± 0.03
Abell 1689	0.1810	-1.87 ± 0.32
Abell 2142	0.0899	-0.44 ± 0.03
Abell 2163	0.201	-1.62 ± 0.22
Abell 2218	0.1710	-0.40 ± 0.05
Abell 2256	0.0601	-0.24 ± 0.03
CL 0016 + 16	0.5455	-0.43 ± 0.03

[a] Recent measurements. For detailed references see M. Birkinshaw, *Physics Reports*, 1999, **310**, 97–195, Tables 4 and 7.

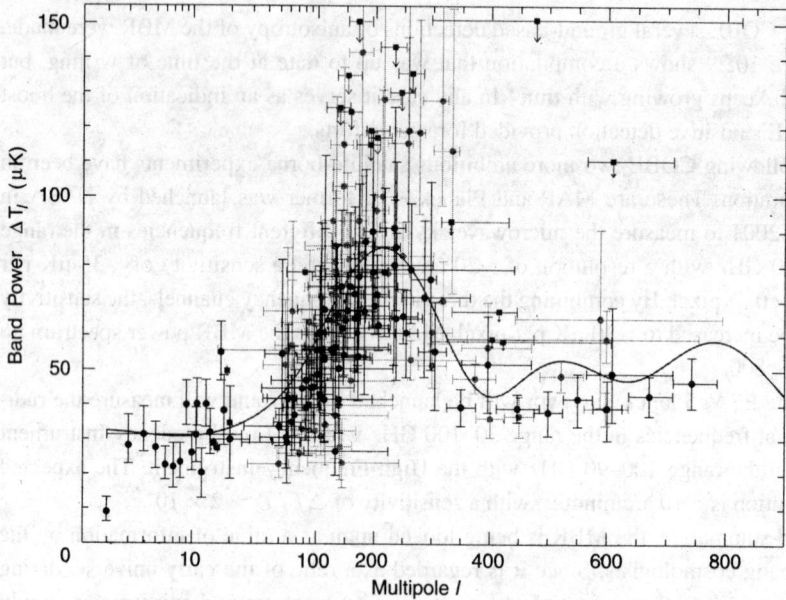

Figure 10.29 The band-power estimates of the CMB anisotropy measurements as of September 2000. The detections are plotted as solid circles. The squares denote 2σ upper limits. The location in l corresponds to the peak of the average zero-lag window function. The horizontal error bar denotes region of l space probed by the window (within $\sqrt{e}$ of the peak sensitivity). For detections the 1σ error bar for the band power includes known systematic uncertainties. Where they are known, uncertainties in beam width and calibration have been accounted for and foreground contamination removed. Detections at $l > 900$ and upper limits above $\Delta T_l = 150\ \mu K$ have been omitted. The curve is the predicted spectrum for the cosmological-constant-dominated, high-baryon-density, flat model preferred by recent data. Courtesy of Tarun Souradeep and Bharat Ratra.

This concludes our discussion of local tests of cosmological importance. We will postpone a survey of the overall observational situation until we have looked at the surveys and tests relating to the large-scale structure of the universe in the following chapter.

Exercises

1 A galaxy has an apparent magnitude of 18 and an absolute magnitude of −17. Show that its distance from us is 100 Mpc. Under what circumstances can this conclusion be wrong?

2 Define the distance modulus suitable for cosmological distances. Show that an uncertainty of 1.5 magnitudes in the distance modulus can lead to an uncertainty by a factor of two in the estimate of the Hubble constant.

3 Comment on the way galactic extinction affects the measurement of extragalactic distances. If this effect is ignored, will the estimate of Hubble's constant be higher or lower than the true value?

4 Discuss the galactic-extinction models currently in force. Show that de Vaucouleurs model always leads to a higher value for the extinction parameter A than does the Sandage–Tamman model. Estimate this difference for galactic latitudes $b = 30°$ and $b = 60°$.

5 The distance from us of a nearby galaxy at $b = 30°$ is being estimated by observing Cepheids in it and using the period–luminosity relation. Show that its distance estimated by using de Vaucouleurs extinction model will be smaller than that estimated by using the Sandage–Tamman model, after correcting for galactic extinction. What is the corresponding ratio of the Hubble constants measured with the two models?

6 The period P (days) and the absolute visual magnitude M of galactic Cepheids are related by

$$M = -1.18 - 2.90 \log P \qquad (3 < P < 50).$$

A Cepheid in a nearby galaxy has a period of 10 days and an apparent magnitude (corrected for galactic extinction) of 20. Estimate the distance from us of the galaxy from these data.

7 In the supernova-expansion method of determining distance, the estimates of v, the photospheric velocity, are v_1 and v_2 at times t_1 and t_2. If the angular radii at t_1 and t_2 are θ_1 and θ_2, show that an estimate of the distance D of the supernova from us is given by

$$\frac{v_2(t_2 - t_1) + R_0(1 - v_2/v_1)}{\theta_2 - \theta_1(v_2/v_1)},$$

where the radius follows the law $R = v(t - t_0) + R_0$.

8 A supernova in NGC 1058 had a photospheric velocity of 8.6×10^8 cm s^{-1} on Julian date (JD) 2 440 568, whereas on JD 2 440 589 its photospheric velocity was 6×10^8 cm s^{-1}. The angular radii of the supernova on these dates were 0.039×10^{15} and 0.115×10^{15} cm Mpc^{-1}, respectively. Show, with the help of Exercise 7 and ignoring R_0, that the distance from us of the supernova is about 12 Mpc and that its outward expansion started on JD 2 440 558. (The Julian date is counted from January 1, 4713 BC.)

9 Outline the observational difficulties that stand in the way of a precise determination of Hubble's constant.

10 In the Newtonian framework applicable to our local neighbourhood, the isotropic Hubble law may be expressed as the velocity–distance relation

$$\mathbf{V}(\mathbf{r}) = H_0 \mathbf{r},$$

r being the position vector of a galaxy relative to the origin. If the observer at the origin has a peculiar velocity $\mathbf{w}$, he observes an anisotropic velocity–distance relation given by

$$\mathbf{V}'(\mathbf{r}) = \mathbf{V}(\mathbf{r}) - \mathbf{w} = H_0\mathbf{r} - \mathbf{w}.$$

Show that the effective Hubble constant $H(\theta)$ in a direction making an angle θ with the direction of the observer's peculiar velocity is given by

$$H(\theta) = H_0 - \frac{w\cos\theta}{r}.$$

Thus $H(\theta)$ is maximum at the antapex ($\theta = \pi$) and minimum at the apex ($\theta = 0$).

11 Imagine that the Great Attractor exists at a distance r_0. This will pull galaxies in its local neighbourhood towards itself. Show that the velocity–distance relation observed from our Galaxy would have an 'S' shape as a result of this perturbation.

12 Comment on the fact that, although the redshift of a nearby extragalactic source is measurable very accurately, its interpretation as the velocity to be used in Hubble's velocity–distance relation is likely to contain errors.

13 Using the information of §10.2 on extragalactic distance scales, deduce that the luminosity density of galaxies scales as h_0. Show also that Ω_G determined from the mass-to-light ratio of luminous objects is independent of h_0.

14 Let $\sigma(r)$ denote the surface mass density at a point P located at a distance r from the centre of a thin, disc-shaped galaxy. Show that the gravitational force F_r at P is directed towards the centre of the galaxy and is given by

$$F_r = G\int_0^\infty \sigma(rx)x\,\mathrm{d}x \int_0^{2\pi} \frac{(1 - x\cos\theta)\,\mathrm{d}\theta}{(1 - 2x\cos\theta + x^2)^{3/2}}.$$

15 Show that the integral in Exercise 14 can be evaluated for $\sigma(r) \propto r^{-1}$ and that it gives flat-rotation curves

$$v^2 = 2\pi G r\sigma(r) = \text{constant}.$$

16 Discuss the implications of the flat-rotation curves of elliptical galaxies. If there is no unseen mass involved, but Newton's laws are modified, how is the gravitational force expected to behave with distance?

17 In a spherical mass distribution in an SO galaxy, the star distribution function is given by (10.35). Assuming that all stars have equal mass and that their number density varies as $r^{-\epsilon}$ ($\epsilon < 0$), show that the mass contained in a sphere of radius r concentric with the galaxy is given by (10.36).

18 Discuss qualitatively how peculiar velocities of galaxies in a cluster distort the distribution of points on a two-dimensional plot for galaxies in a group, a plot that gives the radial separation of galaxies from a typical member against their transverse separation.

19 Let σ and π denote the components of the separation vector of a typical galaxy G from a fixed galaxy G_0, as seen by a remote observer perpendicular and parallel to his line of sight. The difference in redshift between G and G_0, $\pi H_0/c$, is made up of the cosmological component and the Doppler component due to a peculiar velocity w. If w has a distribution function $f(w)$, then show that the two-point correlation function $\xi(\sigma, \pi)$ is related to the spatial correlation function $\xi(r)$ by the relation

$$\xi(\sigma, \pi) = \int_{-\infty}^{\infty} f(w)\xi\left[\sqrt{\sigma^2 + \left(\pi - \frac{w}{H_0}\right)^2}\right] dw.$$

20 In the Coma cluster of galaxies the observed velocity dispersion is ~ 861 km s^{-1}, while the radius of the cluster is $\sim 4.6h_0^{-1}$ Mpc. Show that the mass of the cluster given by the virial theorem is $\sim 2.3 \times 10^{15} h_0^{-1} M_\odot$. The total luminosity of the cluster is estimated to be $\sim 75 \times 10^{12} h_0^{-2} L_\odot$. Show that the mass-to-light ratio parameter η for the cluster is $\sim 300h_0$.

21 Discuss the missing-mass problem for clusters and galaxies.

22 Show that, if $\Omega_0 < 1$, a closed universe requires a λ-term exceeding the value

$$\frac{3H_0^2}{c^2}(1 - \Omega_0).$$

23 In a globular cluster the metal content is $Z \sim 10^{-3}$ and the ratio of horizontal branch stars to red giants is 0.9. Show that, in the $f = 1$ model, the age of the globular cluster is around 11.9×10^9 years, whereas in the $f = 2$ model it is increased to around 2.0×10^{10} years.

24 The nucleus ^{87}Rb decays to ^{87}Sr with a half-life of $\tau = 4.7 \times 10^{10}$ years. Let $X(t)$ and $Y(t)$ denote the numbers of these nuclei in a meteorite at any time t, so that the quantity $X(t) + Y(t)$ is conserved. Let t_0 denote the epoch when the Solar System was formed. Show that a plot of relative abundances $X(t)/Z$ against $Y(t)/Z$, where Z is the number of ^{86}Sr nuclei (which remains unchanged) leads to a straight line whose slope is given by

$$\exp(\lambda t_0) - 1,$$

where $\lambda = \tau^{-1} \ln 2$.

25 Deduce (10.63) from (10.59)–(10.62).

26 Comment on the statement that very low values for the abundance of ^{4}He (for example, $Y \leq 0.15$) are embarrassing for the standard picture of big-bang nucleosynthesis. Contrast this situation with that of the abundance of deuterium.

27 Discuss how many leptons are permitted in the primordial era by the observed abundance of ^{4}He. Why is this deduction considered a success for big-bang nucleosynthesis?

28 What limits does the present-day abundance of deuterium place on the density of baryons of the universe? Suppose that we are given that the universe is closed. What modifications to the standard big-bang picture can you suggest in order to reconcile the two results?

29 Determine from (10.70) the minimum value of λ necessary in order to reconcile the present abundance of deuterium with a closed universe.

30 Discuss with the help of Figure 10.26 whether the limits on the density of baryons needed for producing deuterium and lithium in the observed amounts are consistent with each other.

31 Using the theory describing the passage of a plane-polarized electromagnetic wave of frequency ν travelling through a medium containing n charged particles of mass m and charge e and a magnetic field, show that, if $H_{\parallel}$ is the component of the magnetic field along the direction of propagation of the wave, the plane of polarization turns through an angle

$$\Delta\theta = \frac{ne^3}{2\pi m^2 c^2} H_{\parallel} \, \Delta l \, \nu^{-2}$$

as the wave traverses a distance Δl. Comment on how this result (known as Faraday rotation) has been used to argue about the possible existence of antimatter in the universe.

32 Discuss the evidence for or against the presence of antimatter in the universe.

33 The ratio of occupied levels for $J = 1$ and $J = 0$ states for the CN molecule in the star ζ-Ophiuchi is 0.55 ± 0.05 and that in the star ζ-Persei is 0.48 ± 0.15. The difference in energy between the two levels is equal to kT, $T = 5.47$ K and the ratio of occupation weights is $g_1/g_0 = 3$. Deduce that the temperatures of the incident radiation lie in the respective ranges 3.22 ± 0.15 and 3.00 ± 0.6 K.

34 Let n be the number density of sources generating a cosmic radiation background. Construct a cone of angle 2θ at the observer with the requirement that, if such a cone is extended to cosmological distances, it contains typically one source. Then θ denotes the typical angle over which the generated background would exhibit patchiness. Show that $\theta \sim (H_0/c)^{3/2} n^{-1/2}$. Apply this result to the small-angle anisotropy of the microwave background to set a limit on n.

35 Using the formulae of the Lorentz transformation of flux density from one inertial frame to another, show that, if an observer is moving with a uniform velocity V relative to the cosmological rest frame, then, in the approximation $|V| \ll c$, he will measure a deviation in temperature in the direction of the unit vector $\mathbf{k}$ of magnitude

$$\frac{\Delta T}{T} \cong -\frac{\mathbf{V} \cdot \mathbf{k}}{c}.$$

36 It is given that a high-energy proton loses 13% of its energy in a single collision with a microwave photon that produces a pion. In the microwave background there are ~550 photons cm^{-3} and the mean cross section of the above interaction is ~2×10^{-28} cm^2. Show that the proton loses most of its energy over a distance of ~4×10^{25} cm. Discuss the implications of this result for high-energy cosmic rays of extragalactic origin, showing that it sets a limit on the distance such rays can travel.

Chapter 11

Observations of distant parts of the universe

11.1 The past light cone

Technically speaking, all our observations of the universe are confined to the past light cone. Nevertheless, it is possible to make a distinction between the observational tests of cosmology described in Chapter 10 and those to be described here. This distinction is illustrated with the help of Figure 11.1. This diagram describes schematically the past light cone of a present observer in terms of the cosmological redshift. The observations described in Chapter 10 fall in the shaded region I with $z \leq 0.1$. In this region it is usually possible to go over to the locally inertial frame and use Newtonian physics and special relativity (see §2.4). Although most cosmological models sink their geometrical differences close to the observer, *it is still possible to test their physical differences in this region*. For example, we can attempt to measure q_0 and Ω_0 from observations of galaxies (see §10.4).

Region II, $0.1 \leq z \leq 1.0$, has been a traditional hotbed of cosmological controversies. In this region observations of galaxies and radio sources have been used to determine the geometrical nature of the universe. We will examine this region more closely in §§ 11.2–11.4.

Region III, $1 \leq z \leq 5.0$, consists mostly of quasars and some high-redshift galaxies. Whatever geometrical differences among the various cosmological models exist in region II are magnified in region III. For this reason quasars were expected to be useful probes of cosmology. However, a few astronomers still hold reservations regarding the cosmological origin of quasar redshifts. We will discuss these reservations in the final chapter, while taking the more conventional view in this one.

436

Region IV, which extends from $z = 5.0$ to $z \cong 1000$, has so far proved inaccessible to cosmological observations, while region V, which takes us right back to the big bang ($z = \infty$), is in principle unobservable, since it is supposed to be optically thick (see Chapters 5–7).

Of course, the classification shown in Figure 11.1 has been guided by the standard hot-big-bang picture. In some alternative models like the quasi-steady-state cosmology, for example, regions IV and V do not exist. Distant objects beyond $z \sim 5$ do exist in this cosmology, but they do not have high redshifts. We will discuss the QSSC in the final chapter.

In Chapter 10 we had limited ourselves manifestly to region I, although the relic interpretations of the MBR, light nuclei and large-scale structure indirectly took us to the epochs lying in regions II–V. Here, however, we will confine ourselves to direct observations of region II. With this background we will first describe four classical tests for region II:

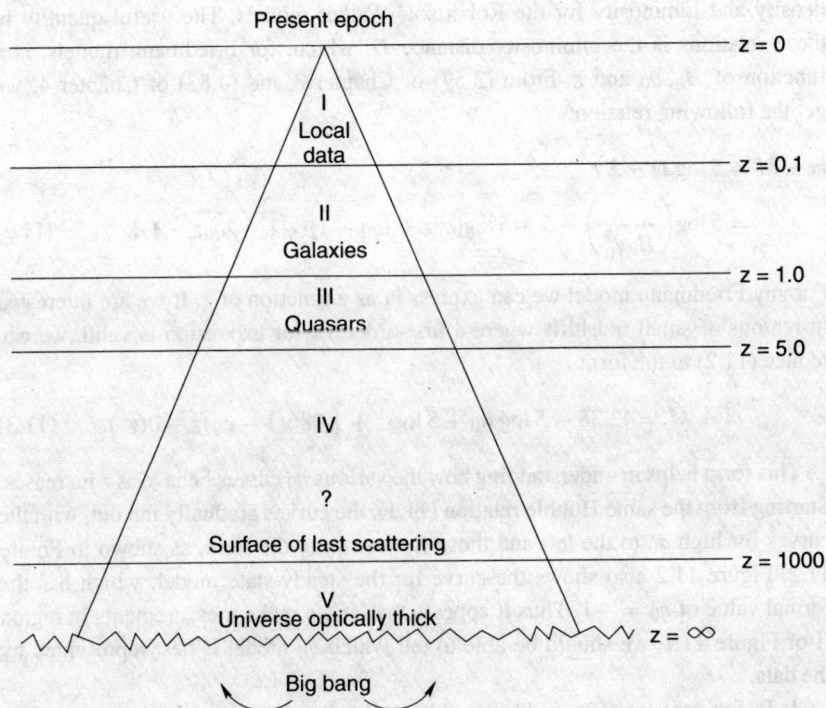

Figure 11.1 A schematic description of the status of cosmological observations along the past light cone. Regions I–V, described in the text, are marked out by the epochs of redshifts $z = 0, 0.1, 1.0, 5.0, 1000$ and ∞.

1. the redshift–magnitude relation,
2. the number counts of extragalactic objects,
3. the variation of angular size with distance and
4. the variation of surface brightness with redshift.

In addition we will discuss a few related issues, including the use of quasars as probes of the distant parts of the universe.

11.2　The redshift–magnitude relation

Basically this is an extension of Hubble's relation to region II. In Chapter 10 we saw that, for the nearby region, the relation is described by (10.6), which is reproduced below:

$$m - M = 42.38 - 5 \log h_0 + 5 \log z. \tag{11.1}$$

What is the form of this relation in general, when the redshifts are not small?

The answer to this question is provided by the relation (3.56) between the flux density and luminosity for the Robertson–Walker models. The useful quantity in these relations is the luminosity distance D, which, for Friedmann models, is a function of H_0, q_0 and z. From (3.57) of Chapter 3 and (4.83) of Chapter 4, we get the following relation:

$$m - M = 5 \log D - 5$$
$$= 5 \log \left(\frac{c}{H_0 q_0^2} \right) - 5 + 5 \log[q_0 z + (q_0 - 1)(\sqrt{1 + 2q_0 z} - 1)]. \tag{11.2}$$

For any Friedmann model we can express m as a function of z. If we are interested in regions of small redshifts where a first-order Taylor expansion is valid, we can reduce (11.2) to the form

$$m - M = 42.38 - 5 \log h_0 + 5 \log z + 1.086(1 - q_0)z + 0(z^2). \tag{11.3}$$

This form helps in understanding how the various q_0 curves behave as z increases. Starting from the same Hubble relation (11.1), the curves gradually fan out, with the curves for high q_0 to the left and those for low q_0 to the right, as shown in Figure 11.2. Figure 11.2 also shows the curve for the steady-state model, which has the formal value of $q_0 = -1$. Thus it appears that, if we make measurements in region II of Figure 11.1, we should be able to tell which q_0 model is best represented by the data.

A. R. Sandage (see Figure 11.3) and his colleagues have spent a number of years on this cosmological test in the hope that the correct geometry of the universe would be revealed. Although in the 1960s Sandage often quoted a value of $q_0 \approx 1$, it gradually became clear that several uncertainties combine to make this test rather inconclusive. The various issues that arise in practical applications of this test are discussed below. Some of the issues have been understood and partially resolved; others continue to be difficult to settle.

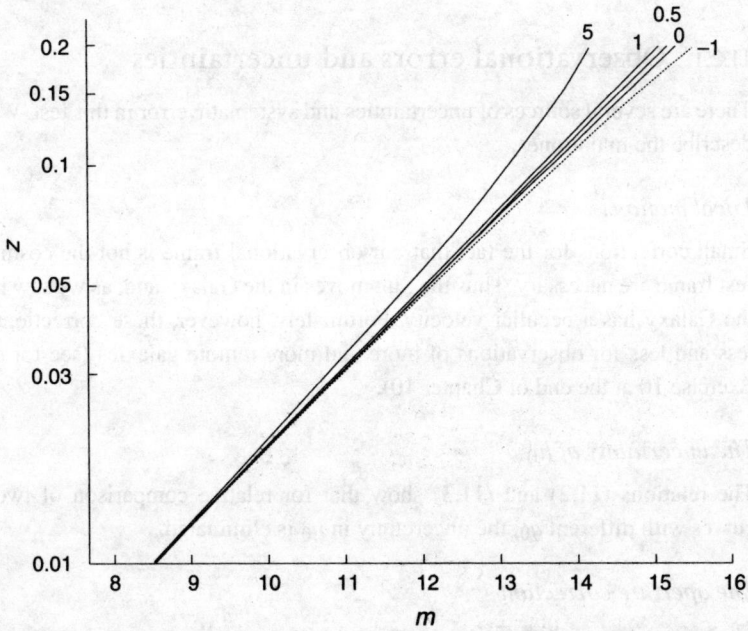

Figure 11.2 A set of theoretical (m, z) curves for $M = -23.88$ and $h_0 = 1$, drawn according to the approximate relation (11.3) for the cases $q_0 = 0, 0.5, 1$ and 5. The dotted curve represents the steady-state model with $q_0 = -1$.

Figure 11.3 Allan Sandage (1926–).

However, in the late 1990s a fresh attempt to revive this test was made when it was realized that type-Ia supernovae can be used to estimate m relatively unambiguously at redshifts as high as ~ 1. We will come to these recent attempts after first describing the classical efforts led by Sandage.

11.2.1   Observational errors and uncertainties

There are several sources of uncertainties and systematic error in this test. We briefly describe the major ones.

Local motions

Small corrections for the fact that our observational frame is not the cosmological rest frame are necessary. Thus the Sun moves in the Galaxy and, as we saw in §10.3, the Galaxy has a peculiar velocity. Fortunately, however, these corrections matter less and less for observations of more and more remote galaxies (see for example Exercise 10 at the end of Chapter 10).

The uncertainty of h_0

The relations (11.2) and (11.3) show that for relative comparison of two model curves with different q_0, the uncertainty in h_0 is eliminated.

The aperture correction

The importance of this effect was realized only gradually. It arises from the fact that galaxies are not objects with sharp boundaries; they tend to fade gradually into the background light of the sky. Therefore the amount of light received from the galaxy in relation to the background depends on the aperture of the telescope.

J. E. Gunn and J. B. Oke have suggested the following recipe for aperture correction. Suppose that we are measuring magnitudes in a wavelength band centred on λ_0. Then (3.55) gives the flux per unit waveband at λ_0 as

$$\mathcal{F}(\lambda_0) = \frac{LI[\lambda_0/(1+z)]}{4\pi(1+z)D^2(q_0, z)}, \tag{11.4}$$

where $D(q_0, z)$ is the luminosity distance in the appropriate model. Suppose that the luminosity L is distributed across the galaxy according to a power law with respect to the projected radius ρ from the centre of the galaxy:

$$L\,(\leq \rho) = L_0 \left(\frac{\rho}{\rho_*}\right)^\alpha, \qquad \rho_* = \text{constant}, \tag{11.5}$$

where α is a number of the order of unity; $\alpha = 0.7$ is a good approximation.

Now, if we lived in the Einstein–de Sitter universe, the projected radius ρ_0 would subtend an angle at our location that is given by (see (4.89))

$$\gamma = \frac{H_0 \rho_0 (1+z)^2}{cD(\frac{1}{2}, z)}. \tag{11.6}$$

In a general Friedmann model, the same angle will be subtended at the observer by a radius ρ given by

$$\gamma = \frac{H_0 \rho (1+z)^2}{cD(q_0, z)}. \tag{11.7}$$

A comparison of (11.6) and (11.7) shows that, if the astronomer decides to measure the apparent magnitude of the galaxy within a given angular radius γ, he collects light from within different radii (ρ_0 and ρ) of the galaxy, depending on which Friedmann model is being used. If we standardize with respect to the Einstein–de Sitter model, we must correct for the luminosity according to (11.5). Thus, instead of a fixed L, we must have

$$L = L_0 \left(\frac{D(q_0, z)}{D(1/2, z)} \right)^\alpha, \tag{11.8}$$

so that (11.4) is changed to

$$\mathcal{F}(\lambda_0) = \frac{L_0 I [\lambda_0/(1+z)]}{4\pi (1+z) D(q_0, z)^{2-\alpha} D(\frac{1}{2}, z)^\alpha}. \tag{11.9}$$

To fix their ideas Gunn and Oke used $\rho_0 = 16$ kpc for $h_0 = 0.6$.

The formula (11.9) corrects for the fact that by fixing γ, we make allowance for light coming from a smaller ($q_0 > \frac{1}{2}$) or larger ($q_0 < \frac{1}{2}$) region of the galaxy than that with $\rho = \rho_0$ for $q_0 = \frac{1}{2}$.

The K-correction

This effect was briefly hinted at towards the end of §3.7 through the relation (3.55). It arises from the fact that, when an astronomer measures the magnitude of a galaxy of large redshift z at a wavelength λ_0, he is receiving light from the galaxy at the emission wavelength of $\lambda_0 (1+z)^{-1}$. Hence, for a comparison of $m(\lambda_0)$ of two galaxies of different redshifts, we must allow for the fact that their absolute magnitudes are being observed at different wavelengths.

Taking the logarithm of (11.4) and converting to magnitudes, we get

$$m(\lambda_0) - M_{\text{bol}} = -2.5 \log\{I[\lambda_0/(1+z)]\} + 2.5 \log(1+z) + 5 \log D - 5. \tag{11.10}$$

If we apply the standard bolometric correction appropriate for wavelength λ_0 we would get

$$m_{\text{bol}} = m(\lambda_0) + \Delta m(\lambda_0).$$

However, from (11.10) we see that, if we wish to use (11.2), we must make a further correction and write instead

$$m_{\text{bol}} = m(\lambda_0) + \Delta m(\lambda_0) - K(\lambda_0), \qquad (11.11)$$

where

$$K(\lambda_0) = 2.5 \log(1 + z) - 2.5 \log\{I[\lambda_0/(1 + z)]\}. \qquad (11.12)$$

Thus it is necessary to know the intensity distribution function $I(\lambda)$ for the source galaxy. Oke and Sandage estimated this effect in 1968 in a quantitative way. More recently, ultraviolet (UV) astronomy is giving more information about $I(\lambda)$ for galaxies at UV wavelengths. For large z, these wavelengths are redshifted to the observed optical wavelengths (see Exercise 6).

Oke and Sandage also pointed out that the correct estimate of the K-term eliminates the so-called *Stebbins–Whitford effect* observed in the 1950s. This effect was based on the observation that galaxies appear to be redder and redder as their redshift increases. If it were genuine, such an effect would imply that the more distant galaxies – that is, those of an earlier epoch – were systematically redder than the galaxies of the present epoch and hence that the universe must be evolving. This was therefore used as an argument against the steady-state universe, an argument that has now been shown to be invalid.

The Malmquist bias

If we use some average luminosity of galaxies in a distant cluster as a standard candle, this bias creeps in. For, as we examine more and more remote clusters of galaxies, we would tend to miss out larger and larger fractions of the intrinsically faint ones. So, in a magnitude-limited sample, the luminosity distribution becomes truncated at the lower end, the effect becoming more and more severe for more and more remote clusters.

The Scott effect

This effect, which was first pointed out by E. Scott, is of the type against which caution was expressed above. Since the brightness distribution of galaxies in clusters has no sharp upper limit, we would tend to pick out more and more intrinsically bright galaxies as we look further and further away. This effect leads to an overestimate of q_0.

Intergalactic absorption

In 1976 S. M. Chitre and the author estimated the effect of absorption by intergalactic dust on the measurements of q_0. Since absorption by dust leads to overestimation of magnitudes, this effect leads to underestimation of q_0. The effect may be considerable for even a minuscule proportion of intergalactic dust (see Exercise 8). We will discuss this again in the context of the m–z relation in the quasi-steady-state cosmology in Chapter 12.

The evolution of luminosity

One of the most serious difficulties in the redshift–magnitude test arises from the uncertainty in corrections to the observed luminosities of galaxies to take account of the evolution of luminosity. If galaxies all formed during an epoch $t_G (< t_0)$ and, as they grew older, their luminosity $L(t)$ changed as a function of $t - t_G$, then the present-day luminosity $L(t_0)$ might not give a reliable estimate of $L(t)$. As we know, for any redshift z the epoch of emission is given by

$$1 + z = \frac{S(t_0)}{S(t)}. \tag{11.13}$$

The interval $t_0 - t$ is called the *look-back time* of the galaxy. For very small z ($\ll 1$), $t \approx t_0$ and $L(t) \approx L(t_0)$. However, for z in region II, $L(t)$ may differ substantially from $L(t_0)$ and hence our extrapolations based on nearby clusters (which give $L(t_0)$) need not be good for remote clusters of galaxies. Beatrix Tinsley was the first to appreciate the importance of this effect and to work it out quantitatively. Basically, the evolution of $L(t)$ comes from the evolution of stars in the galaxy. We will not go into the details of Tinsley's arguments here but simply state the empirical rule that seems to emerge from them.

The stellar population in giant ellipticals of the type used by Sandage is predominantly very old and metal-rich like the population-II stars in the disc of our Galaxy. To estimate their integrated luminosity it is necessary to know the 'initial mass function' (IMF). The IMF essentially specifies the relative number distribution of stars (in a cluster) within various ranges of mass at the time of formation. Since the rate of evolution of a star depends on its mass, the IMF is important in determining the future composition of the cluster. In the visual region the integrated luminosity of such a population satisfies the law

$$L(t) \propto t^{-1.3+0.3x} \tag{11.14}$$

where x is the slope of the IMF for stars in the mass range $0.8 M_\odot \leq M \leq 1.2 M_\odot$. (Since $t_G \ll t$ for region II, (11.14) shows L as a function of t rather than of $t - t_G$.) The Salpeter IMF in the solar neighbourhood has $x \approx 1.35$. Thus the t dependence of $L(t)$ is close to t^{-1} in (11.14).

11.2.2 The Hubble diagram

All these effects taken together pose a formidable array of problems from which it is very difficult to extract the 'true' value of q_0. For example, taking the aperture effect and the effect of the evolution of luminosity together, the first order m–z relation becomes, instead of (11.3), the following:

$$m - M = 42.38 - 5 \log h_0 + 5 \log z$$
$$+ 1.086z \left(\frac{2 - \alpha}{2} \right) \left[q_0 - \frac{2}{2 - \alpha} \frac{1}{H_0} \frac{1}{L_0} \left(\frac{dL}{dt} \right)_{t_0} \right]$$
$$+ 1.086 \frac{\alpha}{4} z + \mathcal{O}(z^2). \tag{11.15}$$

Here we have performed a Taylor expansion to first order to estimate $L(t)$. From (11.15) it is easy to see that, for an evolution of luminosity given by (11.14), a first-order correction to q_0 is

$$\Delta q_0 = (2 - 0.5x)(H_0 t_0)^{-1}. \tag{11.16}$$

The product $H_0 t_0$ depends on q_0 (and λ if λ cosmology is used). It is clear, however, that the 'true' q_0 is less than the observed q_0 by an amount Δq_0.

Allan Sandage has implemented a long programme of studying the Hubble relation out to larger and larger redshifts, as the technology of observing improved. Sandage found that, insofar as remote clusters were concerned, a good standard candle was provided by the brightest and most massive elliptical galaxy in the cluster. The luminosity variation of such elliptical galaxies from cluster to cluster is found to be remarkably small for nearby clusters. This is important, since we might notice spurious effects simply by observing a systematic variation of the standard candle.

Figure 11.4 illustrates what a Hubble diagram looks like for galaxies in region II after making corrections for various effects (with the exception of evolution of luminosity). For comparison, a number of theoretical curves of Figure 11.2 are superposed on the galaxy data. Although the Hubble diagram has a reasonably small scatter, it is not tight enough to rule out (with a great deal of confidence) any of the theoretical curves. Furthermore, since negative q_0 is also permitted, the accelerating universes of λ cosmologies cannot be ruled out.

We conclude this section by briefly mentioning two alternative cosmologies. In the 1960s Sandage's claims of $q_0 \approx 1$ went against the steady-state prediction of $q_0 = -1$. However, the uncertainties of the m–z relation are such that the value of q_0 estimated by such techniques has large error bars. Indeed, as we shall see, today's observations generally favour a *negative* q_0. As we shall see in the final chapter, these recent data are well explained by the quasi-steady-state cosmology by invoking intergalactic dust in the form of metallic whiskers.

11.2.3 The Hubble diagram obtained using type-Ia supernovae

During the 1980s, it was realized that type-Ia supernovae can serve as standard candles in the following way. The light curve of such a supernova has an approximately symmetric characteristic rise and fall over ~30 days, followed by a much slower decline. The maximum luminosity of a type-Ia supernova exhibits an almost uniform value for this population, the dispersion being no more than 0.15 magnitude. We have already remarked in the previous chapter how this property has helped in the measurement of Hubble's constant. Going beyond that, however, we now see that, because of their high peak luminosity, such supernovae can be spotted in distant galaxies. Thus they are suitable for determining the m–z relation out to redshifts of ~1 or even more.

In 1988 the Supernova Cosmology Project (SCP) was launched and systematic searches for and observations of such supernovae were carried out by several ob-

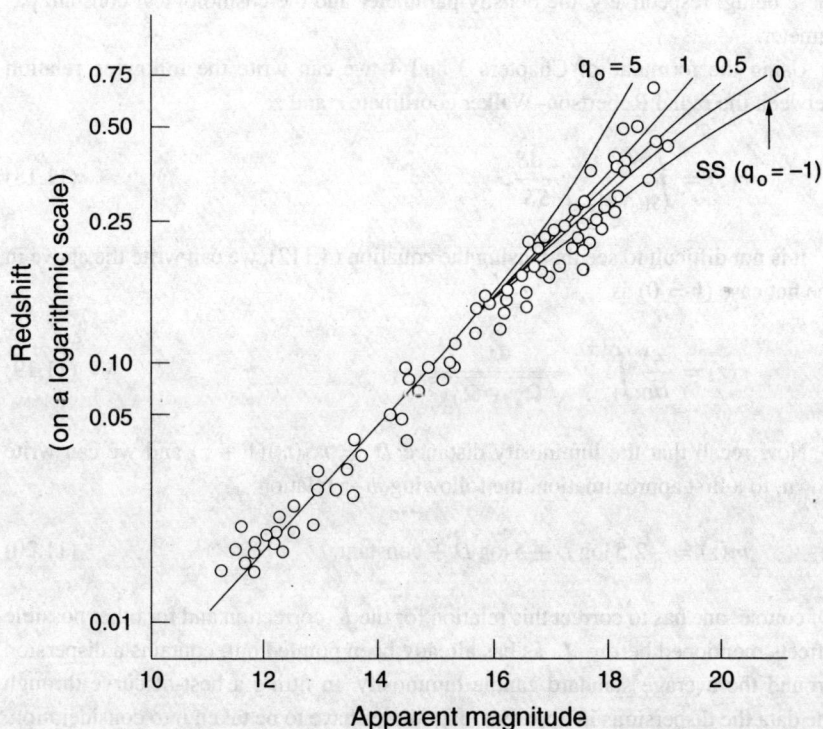

Figure 11.4 The redshift–magnitude relation for the brightest members of a cluster. Several theoretical curves ($q_0 = 5$, 1, 0.5, 0 and −1) are superposed on the data. SS stands for the steady-state model. Based on J. Kristian, A. Sandage and J. A. Westphal, 1978, 'The extension of the Hubble diagram – III', *Ap. J. 221*, 383.

servatories around the world, using telescopes in the 4-m class. The Keck I and II telescopes were used for measurements of redshifts and spectral identifications together with the ESO's 3.6-m telescope. The database continues to grow.

In 1999 Perlmutter *et al.* used 60 supernovae to draw up the Hubble plot. Of these 18 came from the work on nearby supernovae by Hamuy *et al.* These were used essentially to set the zero point of the plot, with the remaining 42 coming from the SCP with redshifts starting from 0.18 and going as far as 0.83.

Theoretical Friedmann models can be fitted to such data using the formulae for $D(z, q_0)$ from Chapter 4. However, the fits were not very satisfactory and the parameter space had to be expanded to include the cosmological constant. We briefly discuss the theoretical aspect of these models, showing how the m–z relation can be derived numerically. The dimensionless parameters in question are

$$\Omega_0 = \frac{8\pi G\rho_0}{3H_0^2}, \qquad \Omega_\Lambda = \frac{\lambda c^2}{3H_0^2}, \tag{11.17}$$

these being, respectively, the density parameter and the cosmological constant parameter.

Using the formulae of Chapters 3 and 4, we can write the following relation between the radial Robertson–Walker coordinate r and z:

$$r(z) = \int_{S(t_0)/(1+z)}^{S(t_0)} \frac{c\,dS}{S\dot{S}}. \tag{11.18}$$

It is not difficult to see that, using the equation (4.112), we can write the above in the flat case ($k = 0$) as

$$r(z) = \frac{c}{H_0} \int_1^{1+z} \frac{dx}{\sqrt{\Omega_\lambda + \Omega_0 x^3}}. \tag{11.19}$$

Now recall that the luminosity distance $D = rS(t_0)(1 + z)$ and we can write down, to a first approximation, the following m–z relation:

$$m(z) = -2.5 \log L + 5 \log D + \text{constant.} \tag{11.20}$$

Of course, one has to correct this relation for the K-correction and for other possible effects mentioned before. L, as has already been pointed out, contains a dispersion around the average standard candle luminosity. In fitting a best-fit curve through the data the dispersions in apparent magnitudes have to be taken into consideration. Figures 11.5 and 11.6 show the situation of how well the various theoretical models match the observations.

Perlmutter *et al.* found that the simplest Friedmann model, namely the *flat* Einstein–de Sitter model, does not give a statistically satisfactory fit. The flat model, however, does fit well if a non-zero cosmological constant is allowed. That is, consistently with (4.115), i.e.,

$$\Omega_\lambda + \Omega_0 = 1,$$

the model with $\Omega_0 = 0.28$, gives the best fit. *This implies that a non-zero cosmological constant as high as ≈ 0.7 is needed.* In other words, the universe has a negative $q_0 \approx -0.6$, if we use the relation (4.113).

Clearly, in view of the profound significance of such a finding, careful follow-up is being done. Several questions arise. How sure are we that there is no evolution in supernovae that spoils their standard-candle interpretation? Some four or five supernovae in the data have to be left out of the curve-fitting exercise since they lie far from the best-fit curve: why? Could some other explanation rather than the cosmological constant account for the extra dimming found for the supernovae? The

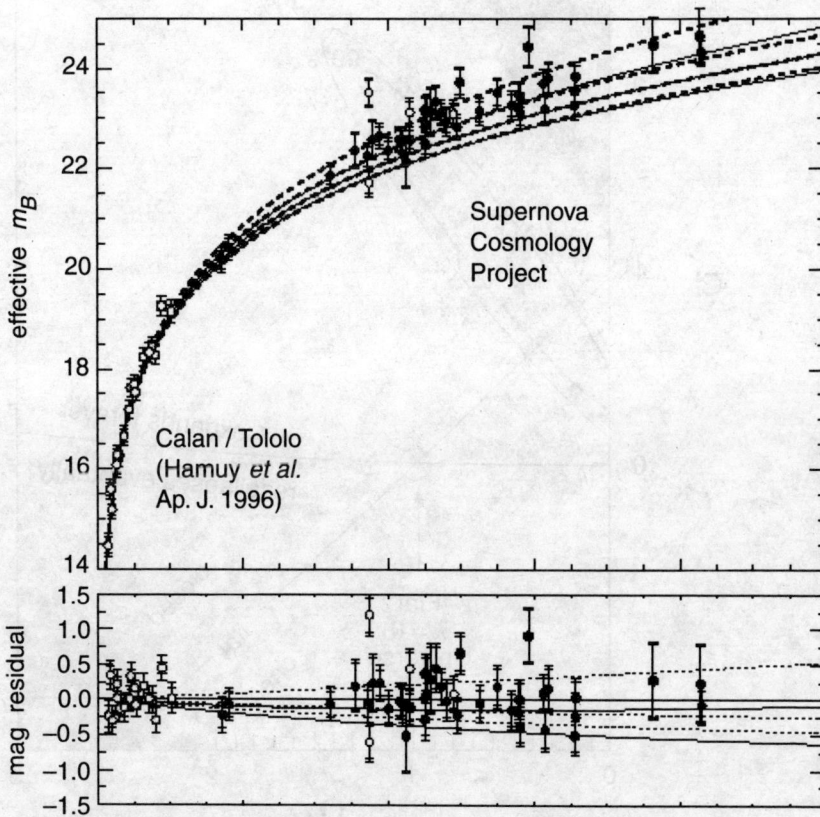

Figure 11.5 The redshift–magnitude relation obtained by Perlmutter *et al.*, 1999, *Ap. J. 517*, 565, using the type-Ia supernova data. The dashed curves with non-zero cosmological constant fit the data better.

possible effects of dust in standard cosmology have also been invoked by some. We mentioned this in §11.2.1. The progressive effect on the spectrum, e.g., reddening, has been suggested as a distinguishing feature for further studies. We will discuss the role of intergalactic dust in the context of the quasi-steady-state cosmology when we return to this test in the next chapter. We will also discuss in the next chapter how this test serves a complementary role in the parameter space in relation to the cosmic microwave background.

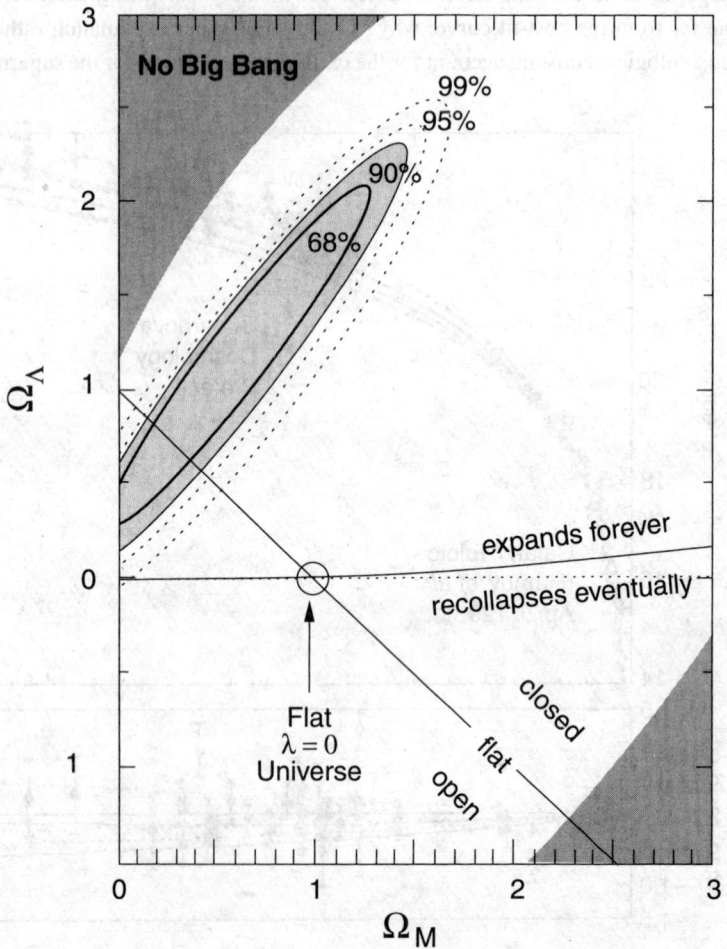

Figure 11.6 The various confidence regions in the $[\Omega_0, \Omega_\Lambda]$ plane based on the work of Perlmutter *et al.*, 1999, *Ap. J. 517*, 565. Notice that the zero-λ models, even open ones, lie away from the allowed regions.

11.3 Number counts of extragalactic objects

So far considerable work has been done on number counts of three types of ex-tragalactic objects: (1) galaxies, (2) radio sources and (3) quasars. Of these we will defer the discussion of quasars to §11.6 and consider only galaxies and radio sources, although the latter contain quasars in some cases.

The basic idea behind these tests is to find out whether the number counts reveal the non-Euclidean nature of the spacetime geometry of the universe assumed by most models. Suppose that we have a class of objects that (1) are uniformly distributed in space and (2) have the same luminosity L. If we further assume that (3) the universe is of Minkowski type, that is, with Euclidean spatial geometry, the number of sources up to a given distance R will go as

$$N \propto R^3, \tag{11.21}$$

while the flux density from the faintest of the sources up to distance R goes as

$$S \propto R^{-2}. \tag{11.22}$$

Eliminating R between these relations, we get

$$N^2 S^3 = \text{constant}, \qquad \text{that is} \quad \frac{\mathrm{d}\log N}{\mathrm{d}\log S} = -1.5. \tag{11.23}$$

Thus (11.23) tells us how N and S are related under our three assumptions (1)–(3). Under these assumptions N measures the volume and $S^{-1/2}$ the radius of a spherical region centred on the observer; and (11.23) is simply the volume radius relation in Euclidean geometry.

In §§3.11 and 4.8 we saw how the volume–radius relation differs from the cube law of Euclidean geometry when we consider Robertson–Walker models. We also saw in these sections how to work out the corresponding relations in non-Euclidean geometries. It is therefore possible, in principle, to test whether the observed relation agrees with one of the various cosmological models. Unfortunately, as with the m–z test, various uncertainties prevent us from drawing a clear-cut conclusion, as we shall see with the counts of galaxies and radio sources below.

11.3.1 Galaxies

In 1936 Hubble attempted number counts of galaxies in order to distinguish among model universes. However, he had to abandon the test because the number of galaxies to be counted is very large and, unless one goes fairly deep into space, one cannot detect any significant departures from Euclidean geometry. Since the optical astronomer measures fluxes in magnitudes, the relation (11.23) may be re-expressed as a number–magnitude relation:

$$\frac{\mathrm{d}\log N}{\mathrm{d}m} = 0.6. \tag{11.24}$$

Hubble's programme has been revived in recent years by a number of workers, who now have at their disposal many electronic and solid-state devices to facilitate galaxy counts out to very faint magnitudes ($m \sim 24$). For example, in 1979 J. A. Tyson and J. F. Jarvis first used techniques of automated detection and classification of galaxies on plates. The main problem at faint magnitudes was that of how to distinguish stars from galaxies.

In the two decades since then, galaxy imaging and spectroscopy at faint levels have progressed considerably and it is possible to get more accurate galaxy counts by allowing for the effects discussed earlier in this chapter, like aperture correction and the K-correction. It is also necessary to decide to what extent the luminosity function of galaxies evolves with time. For example, the Schechter luminosity function given by

$$\Phi(L) \propto \left(\frac{L^*}{L}\right)^{5/4} \exp\left(-\frac{L}{L^*}\right) \tag{11.25}$$

serves well during the present epoch. That is, galaxies do not have a fixed L, but a distribution of L. The constant L^* corresponds to an absolute magnitude of -20.6. The question to be answered is this: if galaxies formed at a more or less unique epoch (or during a relatively narrow time span), how does the evolution of stars in them affect their luminosities and hence their luminosity function?

To take account of the K-correction we need to know the intensity distribution $I(\lambda)$ for galaxies in the ultraviolet. The K-corrections are still unknown, so it is therefore possible to accommodate a wide range of Friedmann models as well as the steady-state model within the uncertainty band. However, if we take $I(\lambda) \propto \lambda^\alpha$, then we get

$$K(z, \lambda_0) = 2.5(\alpha + 1)\log(1 + z). \tag{11.26}$$

In a blackbody-type spectrum, which may crudely apply to galaxies, at the long-wavelength end we have $\alpha = -4$, whereas towards short wavelengths it tends towards -2. However, the actual values would depend on the spectrum of real galaxies.

Figure 11.7 shows the results of such counts. Notice that they tend to flatten at faint magnitudes, as is expected under convergence of the brightness of the sky in an expanding universe (*vide* the Olbers paradox mentioned in Chapter 4). However, the 'no evolution' lines drawn through the points for $\Omega = 1$ show that there is a lack of agreement at the faint end. The discrepancy is most marked for the blue-band counts, indicating that there is an excess of blue galaxies.

There has been considerable discussion in the literature on these faint blue galaxies. Ideas on star formation suggest that these are galaxies in which *starburst*, that is, active star formation, is going on. It is thought that these represent an excess over the normal population expected on the basis of the Schechter luminosity function and various ideas have been suggested. They could be dwarf galaxies at low redshifts ($z \lesssim 0.4$), which fade away during the present epoch. Or we may argue that there were more galaxies in the past, many of which merged, reducing the present-day numbers. The last word has not yet been said in this respect.

We briefly touch upon related issues that require further probes into the range of epochs $1 < z < 4$ for galaxies. How do the colours of galaxies evolve, as the stars inside them age? Redshift surveys in the near infrared help in coming to grips with this problem. For example, the mass (and light) of galaxies is dominated by giants, which are the stars more massive and evolved earlier. These generate the near-infrared luminosity in a galaxy. Although the composition and hence colours of galaxies generally depend on their morphological (Hubble class) make-up, the near-infrared luminosity, being dominated by old stars, remains more or less the same for all types of galaxy. So, for near-infrared redshift surveys, the morphological mix does not change with the redshift.

High-redshift galaxies in the above redshift range also give information on the

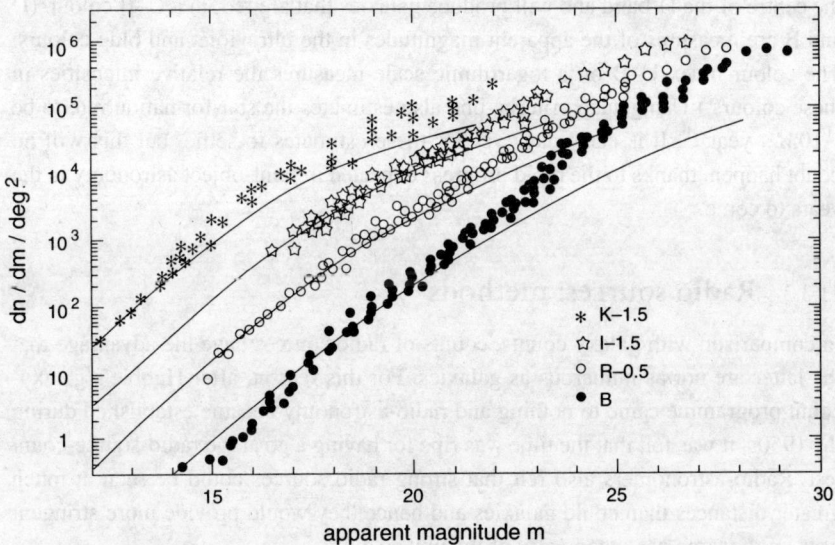

Figure 11.7 Differential counts of galaxies in four bands, B, R, I and K are shown here for the magnitude range ~12–28. Notice that the counts flatten at high magnitudes, as is necessary if the brightness of the sky is to remain finite. The lines show counts as per no-evolution models with $\Omega = 1$. The most significant departure from the no-evolution line is for the line of filled circles, indicating an excess population of faint blue galaxies.

early stages of star formation and stellar evolution. Estimating the age of a galaxy at, say, $z = 3$ tells us when the earliest star formation took place. In the Einstein–de Sitter cosmology, such a galaxy is being seen when the universe was an eighth of its present age. Such studies may therefore put constraints not only on the cosmological models but also on the theories of structure formation. The colour–magnitude relations of ellipticals also help in placing constraints of this kind.

Another useful relationship is that between the abundance of metals Z and the optical/infrared background. Knowing that metals are produced only in stars, from abundance of metals one can estimate the background radiation by using the efficiency (~ 0.01) of the production of energy in stellar nucleosynthesis. If one has further details of complete samples of galaxies in specified redshift shells, one can infer how much radiation was produced by massive stars (which burn fast and produce metals in a relatively short time) and then relate the result to the rate of star formation during those epochs. The first such attempts in 1996 by Lilley *et al.* gave star-formation rates at redshift $z \sim 1$ approximately ten times the present star-formation rate. At higher redshifts of ~ 3, the Lyman-limit galaxies (i.e. those for which the Lyman-α wavelength of 912 Å has moved into the visible) are useful. For, because of absorption by intervening clouds of neutral hydrogen, the spectra of these galaxies exhibit dips or truncation. An active star-forming galaxy will have a flat spectrum down to the Lyman limit. The Lyman break for $z = 3$ will move into the centre of the U-band and will produce galaxies that are red, in a U–B colour. (U and B are measures of the apparent magnitudes in the ultraviolet and blue colours. The colour index U–B on a logarithmic scale measures the relative intensities in these colours.) Using such criteria, one also estimates the star-formation rate to be $\sim 10 M_\odot$ year^{-1}. It is early days yet for these estimates to settle, but this will no doubt happen, thanks to the rapid progress expected in faint-object astronomy in the years to come.

11.3.2 Radio sources: methods

In comparison with galaxy counts, counts of radio sources have the advantage that the latter are not as numerous as galaxies. For this reason, after Hubble's galaxy-count programme came to nothing and radio-astronomy became established during the 1950s, it was felt that the time was ripe for having a go at the radio-source-count test. Radio-astronomers also felt that strong radio sources could be seen at much greater distances than could galaxies and hence they would provide more stringent tests on the large-scale geometry of the universe.

M. Ryle at Cambridge, B. Mills at Sydney and J. Bolton at Caltech did pioneering work on the source-count programme. Since the radio-astronomer measures S over a specified band width, he tends to plot $\log N$ against $\log S$, where S is the *flux density*, the flux S received over a frequency band divided by the band width. The usual unit for S is the Jansky (Jy) (named after Karl G. Jansky, who did pioneering work on radio-astronomy in the 1930s), which equals 10^{-26} W m^{-2} Hz^{-1}. Similarly, the *power* of the radio source is defined as the luminosity over a unit frequency band per unit solid angle and is expressed in units of watts per hertz per steradian (W Hz^{-1} sr^{-1}).

There are several ways of plotting the radio-source-count data. Since the astronomical literature contains references to all of them, they are only briefly outlined below.

The $\log N$–$\log S$ relation

This is the form discussed earlier. The Euclidean geometry makes the prediction that

$$\frac{d \log N}{d \log S} = -1.5. \tag{11.27}$$

The Friedmann models and the steady-state model give a flattening tendency, as shown in Figure 11.8(a).

The $\log N/N_0$–$\log S$ relation

Instead of plotting N against S, it is often convenient to plot N as a fraction of the number N_0 expected in Euclidean geometry against S. Figure 11.8(b) shows how such plots are expected to look for standard cosmology and the steady-state theory. In Figure 11.8(b) we have plotted not the ratio N/N_0 but the ratio $\Delta N/\Delta N_0$ of differential counts. That is, we denote by ΔN_0 the number of sources expected in Euclidean geometry in a given range of flux density $(S, S + \Delta S)$, while ΔN denotes the actual number of sources found (or expected to be found with a given model). Thus $\Delta N_0 \propto S^{-5/2} \Delta S$ and we expect $\Delta N/\Delta N_0$ to decrease steadily from unity as S decreases.

The luminosity–volume test

Instead of plotting N against S, it is often more interesting to approach the source-count problem in the following way. Suppose that S_m denotes the minimum flux density that can be picked up in a given survey. Let a source with $S > S_m$ be found and suppose that we know its distance from us d (from its redshift, say). We then know its luminosity. We can then ask how much further away the source could be moved for it to be barely detected in our survey.

Thus we have two volumes, V and V_m. The volume V is the volume of the spherical region centred on our location and with radius d at whose boundary the source actually exists. The volume V_m is that of the limiting sphere on whose boundary the same source would be barely detectable in the survey. If the source were randomly distributed within the limiting sphere, the average value of the ratio V/V_m would be $\frac{1}{2}$. Thus, in any cosmology with the assumption (1) of uniform distribution, the average value of the ratio V/V_m is expected to be $\frac{1}{2}$.

In practice, if we know the redshifts of the radio sources, we can perform a luminosity–volume test and compute the average value

$$f = \left\langle \frac{V}{V_m} \right\rangle. \tag{11.28}$$

This computation will of course depend on the cosmological model chosen. If $f = \frac{1}{2}$ our assumption of a uniform distribution is confirmed. If $f > \frac{1}{2}$ the survey implies

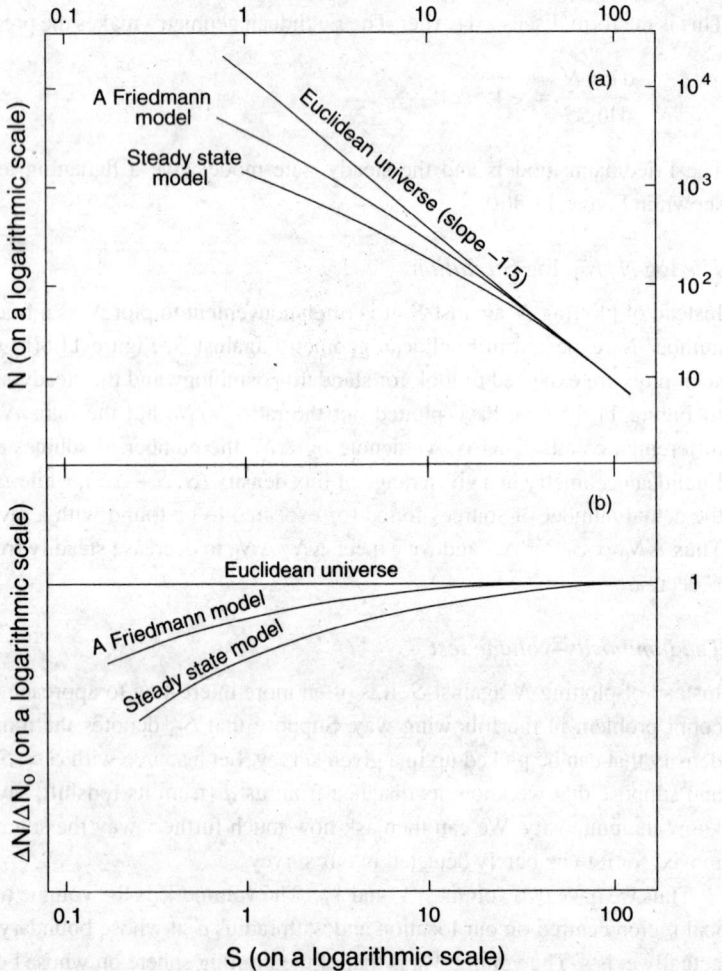

Figure 11.8 Schematic plots of (a) $\log N$ against $\log S$ and (b) $\log(\Delta N / \Delta N_0)$ against $\log S$. The curves shown for a Friedmann model are representative of a number of curves that lie between the steady-state and the Euclidean curves for various values of q_0.

that we are finding relatively more distant than nearby sources. Since distant sources are seen in earlier epochs, this result implies an evolution of density with more sources in the past than there are now. Similarly $f < \frac{1}{2}$ indicates an evolution of density of the opposite kind. Maarten Schmidt was the first to suggest this test in 1968 and to apply it to radio sources and quasars.

The maximum-likelihood method

This method of analysing N–S data was suggested by D. F. Crawford, D. L. Jauncey and H. S. Murdoch in 1970 and it can be described in brief as follows.

Suppose that the N–S relation is of the form

$$N(S) = kS^{-\alpha}, \qquad S \geq S_{\mathrm{m}}. \tag{11.29}$$

Write $\sigma = S/S_{\mathrm{m}}$ so that $\sigma \geq 1$ and let σ_0 denote the maximum value of σ. Let $\mathrm{d}p(\sigma)$ denote the probability that a source is found in the range $(\sigma, \sigma + \mathrm{d}\sigma)$. Then, from (11.29), we have

$$\mathrm{d}p = \frac{\alpha\sigma^{-(\alpha+1)}\mathrm{d}\sigma}{1 - \sigma_0^{-\alpha}}. \tag{11.30}$$

If we make ranges of σ small enough that each range contains at most one source, we may have M such ranges, say. Denoting them by the label i $(i = 1, 2, \ldots, M)$ and the corresponding probabilities by p_i, we write the likelihood function $\mathcal{L}$ as

$$\mathcal{L} = \sum_i \ln p_i. \tag{11.31}$$

The method consists of maximizing $\mathcal{L}$ with respect to α. Using (11.30) for p_i, a simple calculation (see Exercise 15) gives

$$\alpha = \frac{M}{\sum_i \ln \sigma_i}. \tag{11.32}$$

In the usual $\log N$–$\log S$ plot, the number N is built out of additions of numbers from successive flux intervals, during which process errors tend to add up. Thus the various values of N in such a plot are not independent and the estimate of α based on them is not quite reliable. The maximum-likelihood method treats each observation independently, so the estimate of α given by (11.32) is therefore free of cumulative systematic errors.

11.3.3 Radio sources: the data

Before we come to the actual data the following points need to be made.

1. The number-count test is a test for the volume–distance relationship. The measure of the distance of a galaxy is its redshift. The radio-astronomer is,

however, unable to measure redshifts directly. If the radio source is optically identified with a galaxy of known redshift, only then do we have a measure of its distance. The flux density S cannot be a reliable indicator of distance unless we are sure that all sources have approximately the same luminosity. In practice the powers of radio sources vary over the range from 10^{23} to 10^{28} W Hz^{-1} sr^{-1} from weak to strong sources. Thus it is possible to mistake a nearby weak source for a strong but distant source.

2. Even if all redshifts in a survey were known, we would not have a complete sample to test the volume–redshift relation. This is because a sample that is complete with respect to a minimum flux density S_m is not necessarily complete with respect to a given maximum redshift z_m and *vice versa*. In the former sample, very weak sources of moderate redshift are missed, whereas in the latter sample very strong and very distant sources are missed. For practical reasons the radio-astronomer has complete $S \geq S_m$-type samples rather than complete $z \leq z_m$-type samples. The former samples suffer from the difficulty mentioned in point 1.

3. It has become clear, over the years, that simply counting objects as 'black boxes' can be misleading. We should know some basic features of what we are counting.

With this background we turn to observations of the radio-source counts. The source counts have been obtained at various frequencies and, as an illustrative example, Figure 11.9 gives the differential source counts (relative to Euclidean values) as in Figure 11.8(b), for four surveys at frequencies of 408, 1420, 2700 and 5000 MHz. Two important things are immediately noticeable from this diagram.

First, the surveys give different results at different frequencies. The (negative) curvature of the source-count curve declines in magnitude as the frequency rises. The main reason for this discrepancy among curves is as follows. For a typical radio source the intensity–frequency function (see §3.7) has the form

$$J(\nu) \propto \nu^{-\alpha}, \qquad 0 \leq \alpha \leq 1, \tag{11.33}$$

where α is called the *spectral index*. Thus, for a large α, the flux density falls rapidly with frequency, with the result that a source is more easily picked up in a low-frequency survey than it is in a high-frequency one. For low α this effect is less noticeable. For this reason the ratio of steep-spectrum sources (high α) to flat-spectrum sources (low α) in a survey declines as the frequency of the survey increases. The predominance of flat-spectrum sources in high-frequency surveys thus explains the expected behaviour of Figure 11.9.

The second point relates to the preliminary rise of $\Delta N / \Delta N_0$ as S decreases from the highest flux value. This effect is most noticeable in the low-frequency survey at 480 MHz and was the cause of a controversy between Ryle and Hoyle in the early 1960s. We describe it briefly for its historical interest.

If we compare the topmost point P with the highest-flux-density point Q in Figure 11.9 we can draw one of two possible conclusions:

1. There is a significant rise in the number ΔN relative to ΔN_0 as we go from Q to lower flux densities. If these sources are very distant and powerful ones, we are seeing an evolutionary effect, implying that the number density of radio sources was much greater in the past epochs than it is during the present epoch.

2. Compared to P, the point Q shows a deficit of high-flux-density sources in relation to the Euclidean value. If the sources are by and large not very strong, this deficit is a local one and simply indicates that we are in a 'hole' with fewer radio sources than average. Such effects could arise if there are inhomogeneities on the scale ~ 50 Mpc in the universe, as were expected, for example, in the 'hot-universe' version of the steady-state model (see Chapter 8).

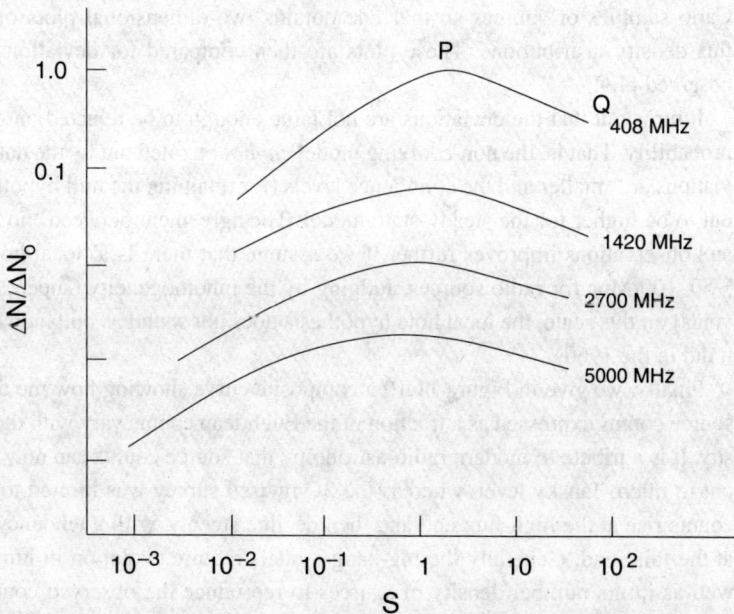

Figure 11.9 Differential source counts at four different frequencies plotted against S on a logarithmic scale. The positioning of the four curves is arbitrary with respect to the $\Delta N/\Delta N_0$-axis. However, each marked interval shows an increase in $\Delta N/\Delta N_0$ by a factor of ten as we move upwards along this axis. Adapted from J. V. Wall and D. J. Cooke, 1975, 'Source counts at high spatial densities from pencil beam observations of background fluctuations', *Mon. Not. Roy. Astron. Soc. 171*, 9.

Hoyle subscribed to the second viewpoint, whereas Ryle and his collaborators at Cambridge thought the first possibility to be correct. The Cambridge view, as it is now called, implied a strong evolution in number density and hence a disproof of the steady-state theory. It is clear that the rise in number density cannot continue indefinitely, otherwise the radio background would be too high. So various functions describing the variation of the number density of sources with redshift z have been considered in order to fit the observed data at all flux levels. However, such a parameter-fitting exercise makes the test ineffective as a means of distinguishing among various q_0 cosmologies, since the geometrical differences are masked by the proposed evolutionary functions.

To what extent is evolution really proved? The 3CR sample of radio sources has now been almost entirely optically identified and the redshifts of its members have been determined. Assuming that the redshifts are indicators of distance, it is now possible to re-examine the number counts of that sample. In 1989 P. Das Gupta, G. Burbidge and the author carried out this exercise. They started with the hypothesis of 'no evolution' in the luminosity function and number density. On the basis of the redshift data, this null hypothesis allows one to construct this radio-luminosity function (RLF). Using this RLF, it is then possible to examine theoretical Monte Carlo samples of sources so that one obtains two-dimensional plots of number–flux-density distributions. These plots are then compared for deviations from the observed plot.

It turns out that the deviations are not large enough to be rejected on grounds of probability. That is, the non-evolving model *cannot* be ruled out by the data. The deviations are smaller and the confidence levels (for retaining the null hypothesis) turn out to be higher for the steady-state model. The agreement between 'no evolution' and observations improves further if we assume that there is a 'local hole' of size ~50–100 Mpc for radio sources. Judging by the inhomogeneity (superclusters and voids) on this scale, the local hole hypothesis does not sound as outlandish today, as it did in the 1960s.

Finally, we give in Figure 11.10 a composite curve showing how the differential source counts expressed as a fraction of the Euclidean counts vary with the flux density. It is a tribute to modern radio-astronomy that source counts can now be carried out to micro-Jansky level, whereas the 3C revised survey was limited to 9 Jy. The counts rise at the high-flux end and then decline steeply, with a tendency to flatten at the faint end. Generally the big-bang models require evolution in luminosity as well as in the number density of sources to reproduce the observed counts. In the next chapter, however, we will show how the counts can indeed be reproduced in the quasi-steady-state cosmology *without* assuming evolution of any kind.

11.4   The variation of angular size with distance

This test was briefly discussed in Chapters 3 and 4 (see §§3.9 and 4.7), where we saw that the angular size of an object of fixed projected linear size does not steadily decrease with its spatial distance from us. Figure 4.10 showed how the angular size changes with the redshift of the object. In 1958 F. Hoyle first suggested that this property of non-Euclidean geometries could in principle be tested by astronomical observations. (The classic book by Tolman written in the mid-1930s, however, contains a derivation of this result.)

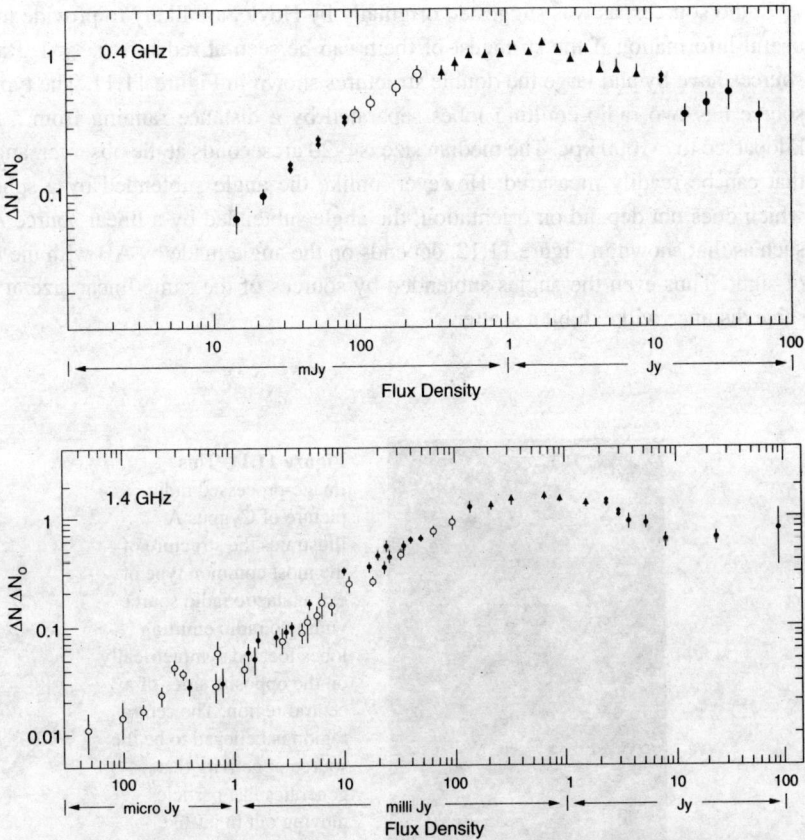

Figure 11.10 Differential source counts relative to the Euclidean values at a range of flux densities show that after a rise for a small stretch at the bright end, the counts drop steadily and this generally requires evolution to be postulated in order to fit any Friedmann model. The top frame shows the counts at 408 MHz, whereas the bottom one shows the same at 1.4 GHz. There is a tendency for the count curve to flatten at the faint end.

Such a test could be performed for galaxies in the optical region. However, the redshifts of galaxies do not go far enough (that is, to $z \geq 1$) to make the predicted effects observable. For instance, in 1975 R. J. Dodd, D. H. Morgan, K. Nandy, V. C. Reddish and H. Seddon examined the images of 3000 faint galaxies down to B magnitude 23 on the 48-inch UK Schmidt telescope in Australia. Instead of redshifts, which are not expected to exceed 0.5, they plotted the number N of galaxies larger than a specified angular size θ. The observed curve was, however, consistent with a broad range of Friedmann models ($0 \leq q_0 \leq 13$) as well as with $q_0 = -1$, corresponding to the steady-state model. This result shows that galaxies are not likely to be useful as sensitive probes for determining q_0 within a narrow range.

Radio sources, as was suggested originally by Hoyle, are likely to provide more useful information if the strongest of them can be seen at redshifts $z \geq 1$. Radio sources have by and large the double structures shown in Figure 11.11. The typical source has two radio-emitting lobes separated by a distance ranging from a few kiloparsec to ~1000 kpc. The median size is ~20 arcseconds at the observer, angles that can be readily measured. However, unlike the angle subtended by a sphere, which does not depend on orientation, the angle subtended by a linear source AB, such as that shown in Figure 11.12, depends on the angle made by AB with the line of sight. Thus even the angles subtended by sources of the same linear size at the same distance will exhibit a scatter.

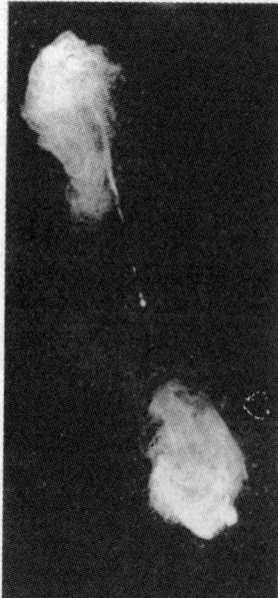

Figure 11.11 This image-processed radio picture of Cygnus A illustrates the structure of the most common type of extragalactic radio source with two radio emitting lobes located symmetrically on the opposite sides of a central region. The central region is believed to be the source of activity that generates fast particles moving out in jet-like structures that create the two radio emitting lobes after impinging on the intergalactic medium. By courtesy of R. Perley, J. Dreher and J. Cowan.

The data published in 1974 by J. F. C. Wardle and G. K. Miley for quasars exhibited enormous scatter, partly due to the above-mentioned projection effects and partly because the linear size d itself is not fixed. In short there is no such thing as a 'standard rod'. Even so, the upper envelope of the plot revealed a dependence of the form $\theta \propto z^{-1}$ for z up to 2.5, which is difficult to explain.

Figure 11.13 shows a plot of median angular size against redshift for radio sources, with the $q_0 = 0$ Friedmann model curves superposed on it. The observed points are from the 1979 work of J. Katgert-Merkelijn, C. Lari and L. Padrielli. The two theoretical curves are for median linear sizes of $125h_0^{-1}$ and $165h_0^{-1}$ kpc. The agreement is not bad, although we cannot expect the data to single out a particular q_0 with any degree of confidence.

Because redshifts of radio sources are not directly measurable but have to be obtained by the process of optical identification, some radio-astronomers have preferred to plot the angular size θ against the flux density S. Since radio sources have varying luminosities, this procedure adds a further source of scatter to the observations. However, R. D. Ekers at Groningen as well as V. K. Kapahi and G. Swarup at Ootacamund did extensive work on this project between 1974 and 1975. In their work the median value of θ did not exhibit the expected upturn at low flux densities, but instead tended to level off. This could imply one of the following interpretations:

1. Low S means large z. Since, at large z, θ should begin to increase according to the Friedmann models with $q_0 > 0$, a flattening of θ implies evolution. In particular the linear size d must decrease as z increases, implying that radio sources were smaller in the past than they are now.

2. Low S need not imply large z. We may simply be seeing sources of low luminosity. If there is a correlation between the size and luminosity of a source, the θ–S observations could be reproduced.

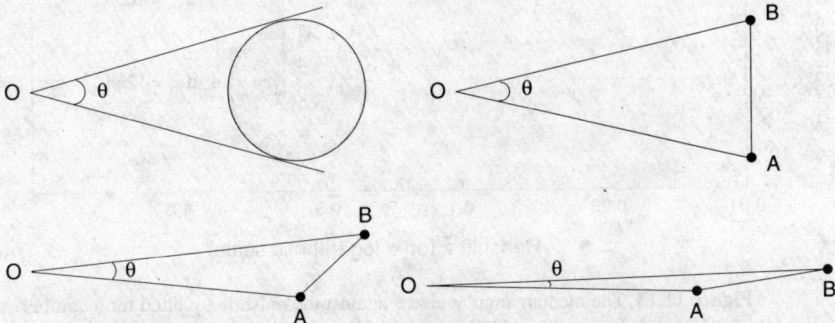

Figure 11.12 The linear source AB subtends an angle at the observer O. This angle is large if the source is transverse to the line of sight OC and small if the source is almost aligned with respect to the line of sight. In contrast, a spherical source makes no such difference to θ since it presents the same aspect from all directions.

Kapahi preferred interpretation 1 and argued that a size evolution of the form

$$d \propto (1+z)^{-\alpha}, \qquad \alpha \approx 1 \tag{11.34}$$

can explain the θ–S observations. The argument given was that, at a large redshift, the intergalactic medium (IGM) was denser than it is today and it could have restricted the size of a typical source formed at the time more efficiently than would be the case if it were formed today. (Recall that the source may have been created by jets emerging as in Figure 11.11 from a central source object and forming the radio emitting lobes after impinging on the IGM.)

S. M. Chitre and the author took the second view, arguing that the correlation between intrinsic size and luminosity may emerge when the physics of radio sources is better understood. Which view is correct will come from the study of the structural properties of radio sources and from the measurements of their redshifts.

Subsequently Kapahi considered four complete samples of radio galaxies with the following ranges of flux density and redshifts:

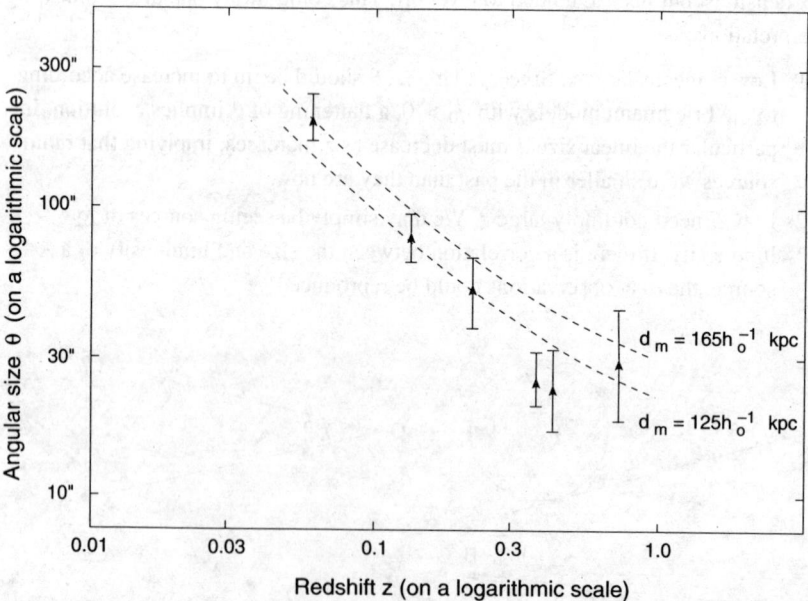

Figure 11.13 The median angular size θ against the redshift z plotted for a number of radio sources, together with the theoretical curves for $q_0 = 0$ and median linear sizes of $165h_0^{-1}$ and $125h_0^{-1}$ kpc. The error bars seem to permit a wide range of values of q_0, although even a Euclidean result $\theta \propto z^{-1}$ cannot be ruled out. Based on J. Katgert-Merkelijn, C. Lari and L. Padrielli, 1979, 'Statistical properties of radio sources of intermediate strength', *Astron. Astrophys. Suppl. 40*, 91.

1. the BDFL sample of A. H. Bridle, M. M. Davis, E. B. Fomalont and
   J. Lequeux with flux density at 1.4 GHz > 2 Jy and redshifts
   $0.075 < z < 0.2$;

2. the GB/GB2 sample of J. Machalski and J. J. Condon with flux density at
   1.4 GHz > 0.55 Jy and redshifts $0.15 < z < 0.4$;

3. the same sample as that above but with flux density at 1.4 GHz > 0.2 Jy and
   redshifts $0.25 < z < 0.6$; and

4. Leiden–Berkeley Deep Survey (LBDS) sample of R. A. Windhorst,
   G. M. van Heerde and P. Katgert with flux density at 1.4 GHz > 0.01 Jy and
   redshifts $z \geq 0.8$.

Spectroscopic redshifts are known for most of the sources only in the brightest
BDFL sample. For the GB/GB2 samples they have been *estimated* from optical
magnitudes by using the Hubble relation. The LBDS sample has sources that are
either identified with galaxies of F magnitude > 22 or have no optical counterparts,
implying that they have optical magnitudes fainter than the plate limit. Thus there is
no direct information on redshifts but the expectation that redshifts exceed 0.8 and
are probably less than 2.

The plot of median angular size versus redshift for these four samples again
revealed a steady decrease of θ_m as z^{-1} (see Figure 11.14). To explain this Kapahi
invoked evolution of linear size with redshift, with sources at large redshifts being
systematically smaller.

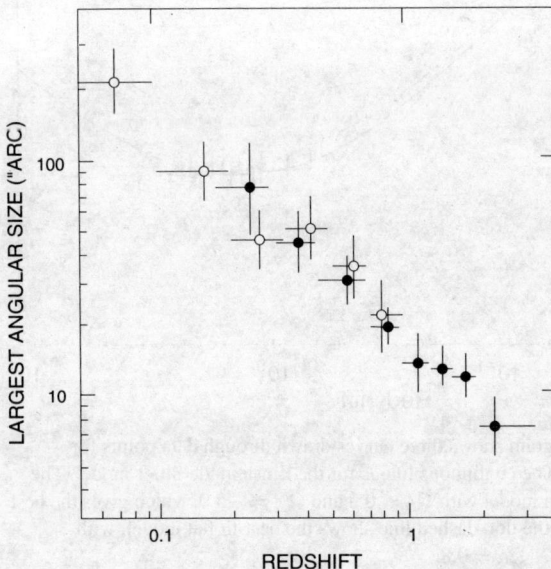

Figure 11.14 Kapahi's analysis of the angular-size–redshift relation produces the above diagram, with the line drawn close to the median points of the four component surveys. The line approximates to $\theta \propto z^{-1}$ on the log–log plot.

To eliminate or at least to minimize the evolutionary effect of the IGM, Keller-mann in 1993 suggested that the test be applied to the very tiny inner components of quasars which are seen through very-long-baseline interferometry (VLBI). His preliminary studies gave a result in broad agreement with the Einstein–de Sitter model. However, a more thorough analysis of a sample of 256 ultracompact sources with redshifts in the range 0.5–3.8 by J. C. Jackson and Marina Dodgson showed that this model is in fact ruled out and, for better fits, one needs to invoke the cosmological constant. Figure 11.15 shows the θ–z curves for three types of models fitted by them to the data. The points shown represent median values of data divided into 16 bins.

In the next chapter we will apply the same data to the quasi-steady-state cosmol-ogy. Evidently, if it can be established clearly that the evolutionary effect on linear size is eliminated in this way, then such a sample is capable of giving information on cosmological parameters.

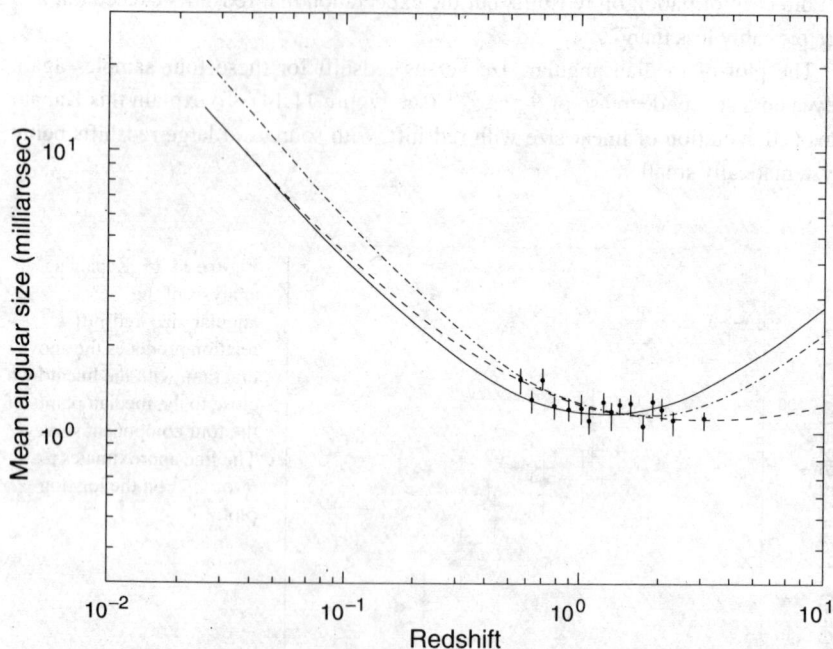

Figure 11.15 This diagram shows three curves drawn through data points for ultracompact sources. The continuous line is for the Einstein–de Sitter model. The dashed line represents a model with $\Omega_0 = 0.2$ and $\Omega_\Lambda = -3.0$, which gives the best fit to the data, whereas the dot–dashed line shows the best-fit flat model, with $\Omega_0 = 0.2$ and $\Omega_\Lambda = 1 - \Omega_0 = 0.8$.

11.5 The surface-brightness test

A test that combines the magnitude and angular sizes involves the measurement of surface brightness as a function of redshift. In the Robertson–Walker world models, the formulae (3.57) and (3.70) of Chapter 3 give the apparent brightness of a source and its angular area, respectively, as

$$\mathcal{F}_{bol} = \frac{L_{bol}}{4\pi r_1^2 S^2(t_0)(1+z)^2},\tag{11.35}$$

$$A \equiv \frac{\pi}{4}(\Delta\theta_1)^2 = \frac{\pi d^2(1+z)^2}{4r_1^2 S^2(t_0)}.\tag{11.36}$$

Dividing (11.35) by (11.36) gives the surface brightness of the source as

$$\sigma = \frac{L_{bol}}{\pi^2 d^2(1+z)^4}.\tag{11.37}$$

Notice that σ does *not* depend on r_1; neither does it depend on q_0 – the parameter that differentiates among cosmological models. It depends only on $1+z$ as its negative fourth power. Sandage has emphasized that this fourth-power law is a signature of Hubble expansion and as such could be used to distinguish the expanding world models from other types of theories in which the redshift does not come from expansion.

Although we have not discussed such theories in Chapter 8, we should mention that there are such cosmologies. For example, there are the 'tired light' theory of J.-C. Pecker and J. P. Vigier, the chronometric cosmology of J. E. Segal and so on.

Sandage has found that the $(1+z)^{-4}$ law seems to be obeyed by first-ranked cluster numbers with a fairly narrow scatter, thus confirming the expanding-universe picture for galaxy redshifts.

11.6 Quasars as probes of the history of the universe

The tests described above assume that the redshifts of the objects used for the tests are of cosmological origin. This assumption is fairly sound in the case of galaxies, for which, at least in the case of the first-ranked cluster numbers, the Hubble relationship is fairly tight.

In contrast, quasars have considerably larger redshifts. At the time of writing this text around 15 000 quasars have been listed in various catalogues. These objects would therefore belong to region III of §10.1 and should be considerably valuable as cosmological probes. We will assume here that the quasar redshifts are due to the expansion of the universe and so obey Hubble's law. To begin with, we discuss the evidence for this assumption, which we shall refer to as the cosmological hypothesis (CH).

The Hubble diagram

A plot of $\log z$ against m should, according to Hubble's law, give a slope $d \log z / dm = 5$ after correction for cosmological effects at large z. As early as 1966 G. R. Burbidge and F. Hoyle pointed out that the Hubble diagram for quasars is a scatter diagram with no apparent correlation between $\log z$ and m. This conclusion survived as more and more quasars were found. Figure 11.16 shows the Hubble diagram for $\sim$5000 quasars in the Hewitt–Burbidge list. Certainly Hubble, or for that matter any astronomer encountering these data in isolation, would not have concluded that there is any relationship between redshifts and magnitudes of quasars.

However, historically quasars were discovered at a time when Hubble's law for galaxies was well established and none of the other modes of redshifts – the Doppler and gravitational – were known to produce redshifts as high as the $z = 0.1$ common for galaxies. Thus it was natural to assume that quasar redshifts are also cosmological.

The conventional view when confronted with a scatter Hubble diagram has therefore been that the scatter is due to the vast spread in quasar luminosities. J. N. Bahcall and R. E. Hills argued in 1973 that a tight Hubble relationship for quasars is revealed when (1) corrections for various selection effects are made, (2) the quasar sample is divided into small redshift intervals (bins) and (3) the brightest quasar in each redshift bin is chosen. This conclusion has, however, been challenged by Burbidge and S. O'Dell, who find that their analysis along similar lines leads to much flatter slopes for $d \log z / dm$: slopes in the range 2–3 instead of the expected slope of 5.

Whatever the outcome of such calculations, it is clear that the Hubble diagram cannot be taken as a proof of the correctness of the CH; at best, arguments of the Bahcall–Hills type might make it compatible with the CH. Certainly there seems no hope at present of using quasars to measure q_0 with the help of their Hubble diagram.

We next discuss other tests involving quasars.

Number counts

Using the luminosity–volume test, M. Schmidt concluded that the average $\langle V / V_m \rangle$ for radio quasars in the 3CR catalogue was as high as 0.64 (compared with the Euclidean value of 0.50). Similarly high values emerge in other surveys. On this basis it is usually argued that the number density of quasars has been strongly evolving, namely that it was considerably higher in the past than it is now. Models of the evolution of luminosity as well as density with enhancement factors like $(1 + z)^n$ $(n > 1)$ or exponential functions in the look-back time are used to fit the quasar-number-count data. It is also argued that steep-spectrum radio quasars have stronger evolution than do flat-spectrum radio quasars.

There have also been number counts of optical quasars, which give a super-Euclidean slope of the $\log N$–m relation (a slope of ~ 0.8 as opposed to 0.6 for the Euclidean universe) for the bright quasars. The $\log N$–m curve flattens beyond the B magnitude ~ 20. It is argued that the evolution of luminosity is responsible for this steepness. The numbers, however, begin to fall off significantly beyond $z \approx 3$.

Curiously enough, $\langle V/V_m \rangle$ for radio quasars turns out to be close to 0.5 if we assume that quasars are local (in region I) and uniformly distributed. This was found by R. Lynds and D. Wills in their examination of several complete samples of radio quasars.

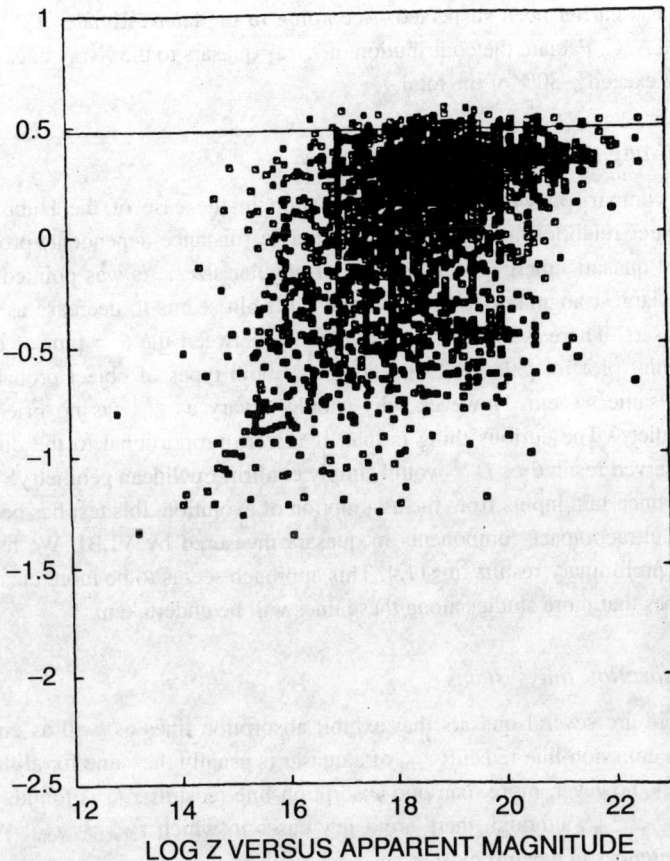

Figure 11.16 The Hubble diagram for quasars in the Hewitt–Burbidge catalogue. It looks like a scatter diagram with no perceptible relationship between m and $\log z$. Reproduced with permission from G. Burbidge and A. Hewitt.

In 1979 a new dimension was added to the source-count problem with the discovery (due largely to the Einstein Observatory) that X-ray emission is a characteristic feature of many quasars. Thus, in principle, it is possible to perform a $\log N$–$\log S$ test for X-ray quasars. The early data suggested that the X-ray and optical luminosities of quasars are correlated. Hence, if optical number counts of quasars were to be taken as the basis of number counts of X-ray quasars also, then, by using formulae like (4.102), it would be possible to estimate the overall contribution to the X-ray background from quasars alone.

The optical number counts of quasars in 1979 suggested that there is a steep rise in the number density of faint quasars, which led to the so-called X-ray-background catastrophe. The quasars alone seemed to contribute over 100% to the X-ray background. However, it is now known that the number density of quasars does not rise as had earlier been suspected. According to the later estimate of A. K. Kembhavi and A. C. Fabian, the contribution of X-ray quasars to the X-ray background should not exceed ~30% of the total.

The angular-size–redshift relation

In comparison with the chaotic situation in the case of the Hubble diagram, a clearer relationship between an observable (distance-dependent) property and the z of quasars emerges from a study of angular sizes. As was pointed out in §11.4, the largest angular size in a given redshift bin seems to decrease as z^{-1} for radio quasars. There is also a rough continuity between the θ–z plot of quasars and a similar plot for galaxies, suggesting that both types of object probably belong to the same system. However, why should θ vary as z^{-1}, as no Friedmann model predicts? The curious thing is that, if z were proportional to the distance D, the observed result $\theta \propto D^{-1}$ would simply confirm Euclidean geometry! To get around the uncertain inputs from the assumption of evolution, this test has been performed for ultracompact components in quasars measured by VLBI. We have discussed the preliminary results in §11.4. This approach seems to be more clear-cut and one hopes that more studies along these lines will be undertaken.

Absorption-line systems

There are several quasars that exhibit absorption lines as well as emission lines. The emission-line redshift z_{em} of a quasar is usually the same for all lines. In some cases, however, more than one absorption-line redshift z_{abs} is found. Also, mostly $z_{abs} < z_{em}$, although there are a few cases in which $z_{abs} > z_{em}$. Why do these differences in redshift exist?

In principle, the difference between the emission-line redshift and the absorption-line redshift could be accounted for by (1) a relative motion between the emitting and the absorbing region, (2) a small contribution of gravitational redshift/blueshift between the two regions, or (3) the difference between the cosmological redshifts of the emitting and absorbing regions. Both (1) and (2) arise in the source, whereas (3) generally requires absorption to occur *en route* from the source to the observer. Note also that, whereas (1) could be adjusted to have both $z_{abs} \geq z_{em}$, (2) and (3) require $z_{abs} < z_{em}$.

It is not clear from the work so far whether entirely satisfactory mechanisms exist in (1) to account for the various absorption-line systems within the object. In the case of the quasar 3C 286, $z_{em} = 0.85$ while $z_{abs} = 0.69$; 21-cm observations of the source reveal a very small difference in velocity (~ 3 km s^{-1}) across a distance of ~ 300 pc in the source. This result was quoted as a stumbling block to the theory, which seeks to explain the difference $z_{em} - z_{abs}$ as arising from high-speed gas driven outwards from within the quasar.

The more popular explanation of absorption-line systems comes from (3), with the absorbers being intergalactic clouds or halos of galaxies situated *en route* from the QSO to the observer. Typically there are three types of absorption lines:

1. The broad absorption lines (BAL) or trough systems of C IV, Si IV, N V, O VI etc., in addition to Ly α. The troughs are located on the blue side of the corresponding emission lines and have widths corresponding to a velocity $\leq 0.10c$.

2. The heavy-element systems containing sharp lines due to H and to heavier elements that may arise in a tenuous gas of near-solar composition. Here the difference between the emission and absorption redshifts corresponds to a velocity towards the quasar of $\leq 0.8c$.

3. The Ly α systems appearing with greater density on the blue side of the emission line.

Considerable work has been done to argue that the majority of the absorption lines arises from randomly distributed intervening cosmological objects. In some cases in which an absorption line at a specific redshift has been found, a cluster of galaxies with the same redshift has been reported in the vicinity. Such studies will provide useful probes of the properties of the intergalactic medium. As discussed in Chapter 10, however, the Gunn–Peterson test looking for substantial intergalactic neutral hydrogen *en route* to quasars did not yield a positive signal. This was interpreted more as evidence against the intergalactic hydrogen than against the quasars being at their redshift distances. On the basis of this finding, cosmologists now have to argue that, sometime after the epoch of last scattering but before the epoch of redshift ~ 5 say, the IGM was mostly re-ionized, so that the Ly-α absorption was confined to clouds of neutral hydrogen.

Gravitational lensing

In 1979 two quasars, 0957+561 A and B, with the same redshift of 1.4 and identical spectra were discovered (see Figure 11.17). Their similarity led to the suspicion that they are two images of a single object produced by a gravitational lens. A cluster of galaxies with a redshift of 0.36 has since been identified as a probable candidate for such a lens.

Since the quasars are separated by only ~6 arcseconds, such an interpretation seems plausible and therefore provides support for the CH, in the sense that the 'lensed' quasar is shown to be further away than the 'lensing' galaxy. However, there have been very few such clear-cut cases of gravitational lensing.

Assuming, however, that the quasars are distant as per their redshifts, their lensing can provide useful probes for cosmological models. For example, the twin quasar 0957+561 A and B exhibits a fluctuation of luminosity with time in both its images. Now, if the two images correspond to the same source, then we are seeing the source fluctuations *via two different paths*. If the path from B is longer than that from A, then at any given time we are seeing B at an earlier epoch than A. So whatever fluctuations were seen in A will be repeated in B *later*, after an interval corresponding to $\tau = \Delta L/c$, ΔL being the path difference. There are indications

Figure 11.17 The two quasar images shown above are labelled 0.957+56 A and B. They are believed to be the gravitationally lensed images of the *same* quasar. See the text for details. Image by HST created with support from the Space Telescope Science Institute operated by the Association of Universities for Research in Astronomy. Reproduced with permission from AURA/STScI.

that, in the case of the above quasar, the 'time delay' is ~415 days. So, if we have a model for the lens and a model for the universe, then we have a theoretical value for this time delay, in the form

$$\tau = \frac{1}{H_0} F, \qquad (11.38)$$

where F is calculated for the lensing system. Thus, by comparing this result with the observed value, one can determine the Hubble constant. Notice, however, that the computation of τ is model-dependent, i.e., it depends on the geometry of the lensing system, which might not be unique.

Another useful constraint from gravitational lensing is that on the value of Ω_Λ, from the statistics of gravitationally lensed quasars with large angular separations. As the value of this parameter is increased, the distance scale for the same redshift increases, which increases the probability of such lensing. The present analysis suggests that there is a strong *upper limit* of ~0.65 on this parameter. This will be an important consideration in limiting the overall parameter space for the standard models.

It should be pointed out in passing that the first suggestion that distant nebulae can act as gravitational lenses came from Fritz Zwicky, as early as in 1937. Over four decades elapsed before the first case of 0957+567 A and B was observed.

Variability

Quasars exhibit rapid variability in radio and optical wavelengths as well as for X-rays, as is indicated by recent data. A rule of thumb is that, if τ is the characteristic time scale of variation, the physical radius R of the object should not exceed $c\tau$. This leads to an energy-generation problem that was first pointed out by Hoyle, Burbidge and W. L. W. Sargent in 1966. The difficulty is briefly described as follows.

Since we measure the flux $\mathcal{F}$ from a quasar, its total luminosity is deduced from its distance. Under the CH, the distances are large and hence the luminosity is large. The quasar must therefore generate large quantities of energy in a small volume limited by the linear size $c\tau$. In the usual energy-production scenario, the so-called synchrotron process, relativistic electrons radiate in magnetic fields. However, as Hoyle, Burbidge and Sargent pointed out, this process results in the production of a very large density of photons. These photons collide with electrons, causing a very large Compton scattering, which degrades the energy of the fast electrons. Thus it is not possible to sustain energy production over distances even comparable to $R \simeq c\tau$; the electrons lose energy long before they have travelled a distance of this order.

The kinematical difficulty of whether quasars can manage to be confined to the radius $R \leq c\tau$ is partially alleviated by the following idea proposed by M. J. Rees. If an object expanding relativistically with a large Lorentz factor γ is viewed from a distance, it appears to increase its radial size at the rate γc. Thus the observed time scale of variation γ may be too small and the real inequality on R is $R \leq \gamma c\tau$ (see Exercise 20). For $\gamma \gg 1$, the problem of confinement is made easier.

A compact size implied by the short time scale of variability is sometimes invoked in support of the idea that a quasar's energy is derived from a supermassive black hole. For example, the X-ray quasar OX 169 exhibited a significant drop in its X-ray luminosity within 100 min. The size limit implied by this time scale can accommodate a black hole of mass $\leq 10^8 M_\odot$.

Superluminal separation

VLBI observations have revealed a curious phenomenon for a number of quasars. In the central region of such a quasar, two radio components are observed to separate from each other very quickly. Since angular separations of the order of a few milli-arcseconds are measurable by VLBI techniques, observations over a few months or a few years are sufficient to give a detectable effect. Thus it is found that the separation angle θ is changing (increasing) with time in such a way that the projected linear distance must change at speeds considerably faster than the speed of light – provided that the quasars are at distances specified by the CH. Clearly, if the distances are much smaller than these, the speeds of separation become subluminal and the discrepancy with relativity disappears.

To retain the CH in spite of such data requires the conclusion that the observed separation is illusory. Various ways out have been suggested and three of these are illustrated in Figure 11.18. Figure 11.18(a) shows the Christmas-tree effect, which creates an illusion of motion by sequential lighting of stationary light bulbs. Figure 11.18(b) illustrates the so-called relativistic beaming, a variant of the idea proposed by Rees and described above under variability. The model illustrated in Figure 11.18(c) invokes a gravitational screen in the form of an intervening galaxy or cluster of galaxies that bends the light rays (or radio waves) from the two components differentially so that their virtual images appear to separate at superluminal speed.

Morphology

It is argued that quasars and the nuclei of Seyfert galaxies are basically similar objects and that in general we may think of a quasar as a galactic nucleus that is so bright that the rest of the galaxy either is not visible or is too faint to be seen. According to this argument, if the CH is correct, a large redshift implies a large distance and at that distance only the bright nucleus would be visible as a quasar. In some cases, such as the quasar Ton 256, it is argued that a fuzz surrounding the quasar has the luminosity distribution of an elliptical galaxy. To establish this line of evidence, which goes in favour of the CH, it would be necessary to show that a galaxy of stars indeed exists around the quasar. Absorption lines characteristic of stars in elliptical galaxies would be conclusive evidence for this purpose.

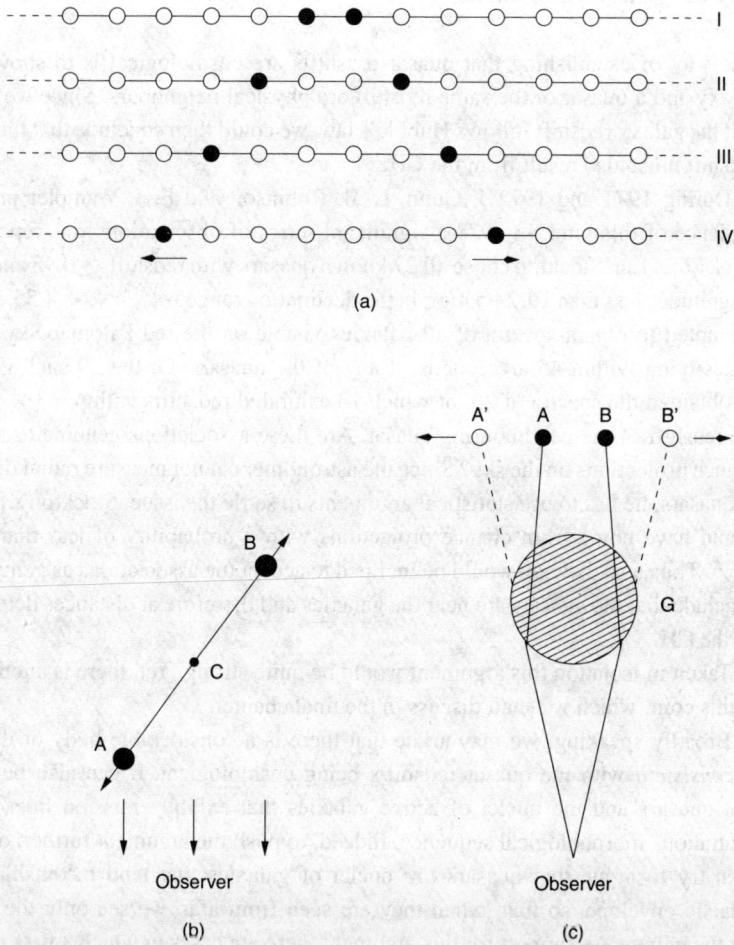

Figure 11.18 Three ways of creating illusions of separation at speeds faster than the speed of light. (a) A row of lights is denoted by circles. The filled circles are lights that are lit. In stages I–IV the lighting is so contrived that a remote observer may think that two sources of light are moving outwards from the centre. (b) The observer sees the two actually separating components at different times; for light from the nearer component A leaves later than light from the farther component B. C is the central nucleus at rest. An illusion of superluminal separation between A and B is created provided that the line AB is almost aligned with the line of sight. This is very rare. Moreover, A is blueshifted and hence should be much brighter than B, which is redshifted with respect to the observer. It is usually assumed that B is at rest and A is beamed at the observer. (c) Here an intervening galaxy bends the rays from A and B so that the observer sees their images A′ and B′, which could separate at superluminal speed even if A and B were moving apart with subluminal relative velocity. For this to happen the galaxy G must occupy a rather special position between the source and the observer.

Quasar–galaxy associations

One way of establishing that quasar redshifts are cosmological is to show that a galaxy and a quasar of the same redshift are physical neighbours. Since we believe that the galaxy redshift follows Hubble's law, we could then conclude that the quasar redshift must also result from the CH.

During 1971 and 1972 J. Gunn, L. B. Robinson and E. J. Wampler produced evidence of this kind. In 1978 a significant series of observations was reported by A. N. Stockton. Stockton chose all 27 known quasars with redshifts ≤ 0.45 and visual magnitudes less than $19.2 + 5 \log z$ in the declination range $-15 < \delta < +55$. He then attempted to obtain spectra of all galaxies visible on the red Palomar Sky Survey plates lying within 45 arcseconds of any of the quasars. Of the 29 such galaxies, he obtained the spectra of 25, of which 13 exhibited redshifts within $\sim 3 \times 10^{-3}$ of the redshift of the neighbouring quasar. Are these associations genuine or are they chance projections on the sky? Since the astronomer cannot measure radial distances of quasars, he has to use statistical arguments to settle the issue. Stockton's pairings would have come from chance projections with a probability of less than 1.5×10^{-6}. Thus a statistician would be inclined to accept the associations as genuine and conclude that the quasars are near the galaxies and therefore at distances determined by the CH.

Taken in isolation this argument would be quite strong. Yet, there is another side to this coin, which we shall discuss in the final chapter.

Broadly speaking, we may argue that there is a considerable body of data *that is consistent* with the quasar redshifts being cosmological. It can also be argued that quasars and the nuclei of active galaxies that exhibit emission lines form a continuous morphological sequence. Indeed, to push the argument further, one may even try to argue that quasars *are* nuclei of galaxies that tend to outshine their galactic envelope, so that, when they are seen from afar, we see only the quasar, not the galaxy. As support for this argument there are cases in which a fuzz is found surrounding some low-redshift quasars.

However, we have as yet no direct proof that all the very-high-redshift quasars are in fact very far away, as the cosmological distance formula would have us believe. The nearest to direct evidence is the association between quasars and galaxies of the same redshift found by Stockton. Indirect evidence comes from gravitational lensing and time-delay studies, which are consistent with the CH. With regard to the latter, it is worth noting that the lensing configurations are scalable with the distance from us of the source (and the lens). If the source is nearer, a lens of smaller mass at a lesser distance will produce the same effect and so the time-delay test is crucial in deciding whether the quasar is at its Hubble redshift distance or considerably nearer.

Until a reliable distance indicator independent of Hubble's law becomes available for quasars and it shows that the Hubble law is applicable to them, for the CH support must rest largely on consistency arguments or just simple faith.

11.7 Observational constraints on cosmological parameters

In 1975, in an article in *Nature*, Jim Gunn and Beatrix Tinsley reviewed the then-available data in cosmology to conclude

'New data on the Hubble diagram, combined with constraints on the density of the universe and the ages of galaxies, suggest that the most plausible cosmological models have a positive cosmological constant, are closed, too dense to make deuterium in the big bang, and will expand for ever ... '.

Thanks to new technology, as we have seen in this and previous chapters, observations and fresh inputs from particle physics as well as cosmology have since advanced both on the observational front and on the theoretical front. The standard hot-big-bang model has, if at all, become more deeply rooted in cosmology today than it was in 1975. It is therefore opportune that, in the closing parts of this volume, we take fresh stock of the cosmological situation today and examine the observational and theoretical constraints as they are now. Not surprisingly, some of the issues discussed by Gunn and Tinsley continue to be relevant today, whereas fresh ones have replaced the rest. Our purpose here is to carry out a similar exercise in the modern cosmological framework. The bottom line is going to be that, *despite the availability of the cosmological constant as an extra parameter for flat Friedmann models, the allowed parameter space for such models has shrunk drastically.* The observations that we will consider here include the ages of globular clusters, measurement of the Hubble constant, the abundance of rich clusters of galaxies, the fraction of mass contributed by baryons in rich clusters and the abundance of high-redshift objects. In addition we will take into consideration the ideas on structure formation.

The standard big-bang model has no clear mechanism for generating small inhomogeneities in the early universe. It is, however, possible to come up with such a mechanism if one invokes the hypothesis that the universe went through an inflationary phase during a very early epoch. The models involving inflation generically lead to two predictions: (i) that the total density parameter $\Omega_{\text{total}} = \Omega_0 + \Omega_\Lambda = 1$ and (ii) that the initial power spectrum of inhomogeneities has the form $P_{\text{in}}(k) \propto k^n$ with $n \simeq 1$. Over the years, the idea of inflation has undergone several modifications to meet observational challenges and it is now possible to find a model that will provide almost any value for Ω_{total} and any form for $P_{\text{in}}(k)$. For the sake of definiteness we will work only with $n = 1$ models. Observations of microwave background radiation are consistent with the index n being equal to unity. As the fluctuations grow, the power spectrum becomes modified at small scales by various physical processes and this change is described by a transfer function. We shall work with the transfer function suggested by Efstathiou, Bond and White, parametrized by $\Gamma \equiv \Omega_0 h_0$ (see Chapter 7). The power spectrum is normalized with the COBE DMR observations that find $Q_{\text{rms-ps}} = 20 \pm 3$ μK. Here $Q_{\text{rms-ps}}$ is the amplitude of fluctuations in the quadrupole inferred from fluctuations in the higher moments.

Here we will look at constraints on two models, namely those with (i) $\Omega_0 + \Omega_\Lambda = 1$ (flat models) and (ii) $\Omega_0 < 1$ and $\Omega_\Lambda = 0$ (models with negative spatial curvature). The first one is consistent with the inflationary models, though it requires an extreme fine-tuning of the cosmological constant, which is contrary to the spirit of the inflationary scenario. (We shall comment more on this later.) The second model may be thought of as an 'observer's model' in the sense that it tries to use what is known observationally. The amplitude of fluctuations for open models is obtained by rescaling the $\Omega_0 = 1$ model. Effects of curvature are not important since we are interested only in scales much smaller than the curvature scale.

We first list the constraints arising from theory as well as observations, giving possible sources of error for each constraint. Then we merge constraints to study the allowed regions in the parameter space defined by the density parameter for matter (Ω_0) and the Hubble constant (h_0). It is instructive to look at Figures 11.19–11.21 for these discussions.

Figure 11.19 shows the constraints on the density parameter contributed by all types of matter, Ω_0, and the Hubble constant h_0 arising from (i) ages of globular clusters, (ii) measurements of the Hubble constant and (iii) the abundance of rich clusters. The last was discussed briefly in Chapter 10. The mass per unit volume contained in rich clusters can be estimated from the observed number density of such clusters and their average mass estimated by various methods like those using the virial theorem, gravitational lensing, X-ray studies and so on. This number may be represented as the parameter $\Omega_{\text{clusters}}^{\text{obs}}$ and successful models should satisfy Ω ($> M_{\text{clusters}}$) $= \Omega_{\text{clusters}}^{\text{obs}}$. A comparison of observations with theory can be carried out by converting the number density of clusters into the amplitude of density fluctuations, which is then scaled to $8h_0^{-1}$ Mpc assuming a power law for the RMS fluctuations σ_8 of density perturbations, the power-law index being chosen as per the model considered. This result is then expressed as a constraint on σ_8. It is more likely that the masses may have been overestimated and corrections of any errors will lower the values of the parameter $h_0\Omega_0$.

The top frame shows the constraints for a model with $\Omega_0 + \Omega_\Lambda = 1$, $\Omega_\Lambda \neq 0$. The lower frame is for the model with $k = -1$, $\Omega_\Lambda = 0$ and $\Omega_0 \leq 1$. Lines of constant age are shown as dashed lines for the range of values of Ω_0 and h_0. Dotted lines mark the band-enclosing value of the local Hubble constant ($0.63 < h_0 < 0.97$) obtained from HST measurements. We have also shown the assumed lower limit for its global value ($h_0 = 0.5$). Inner unbroken lines enclose the region which is permitted by the observed abundance of clusters. Outer unbroken lines show the extent to which this region can shift due to uncertainty in the COBE normalization of the power spectrum. Note that these three constraints rule out large regions in the parameter space. In particular, it is clear that the $\Omega_0 = 1$ model is ruled out.

Next, in Figure 11.20, are shown the constraints on the density parameter contributed by matter, Ω_0, and the Hubble constant h_0 arising from (i) ages of globular clusters, (ii) measurements of the Hubble constant, (iii) the abundance of high-redshift objects, (iv) the fraction of mass contributed by baryons in clusters and primordial nucleosynthesis and (v) measurement of the deceleration parameter. The top frame shows the constraints for a model with $\Omega_\Lambda \neq 0$ and $\Omega_0 + \Omega_\Lambda = 1$. The

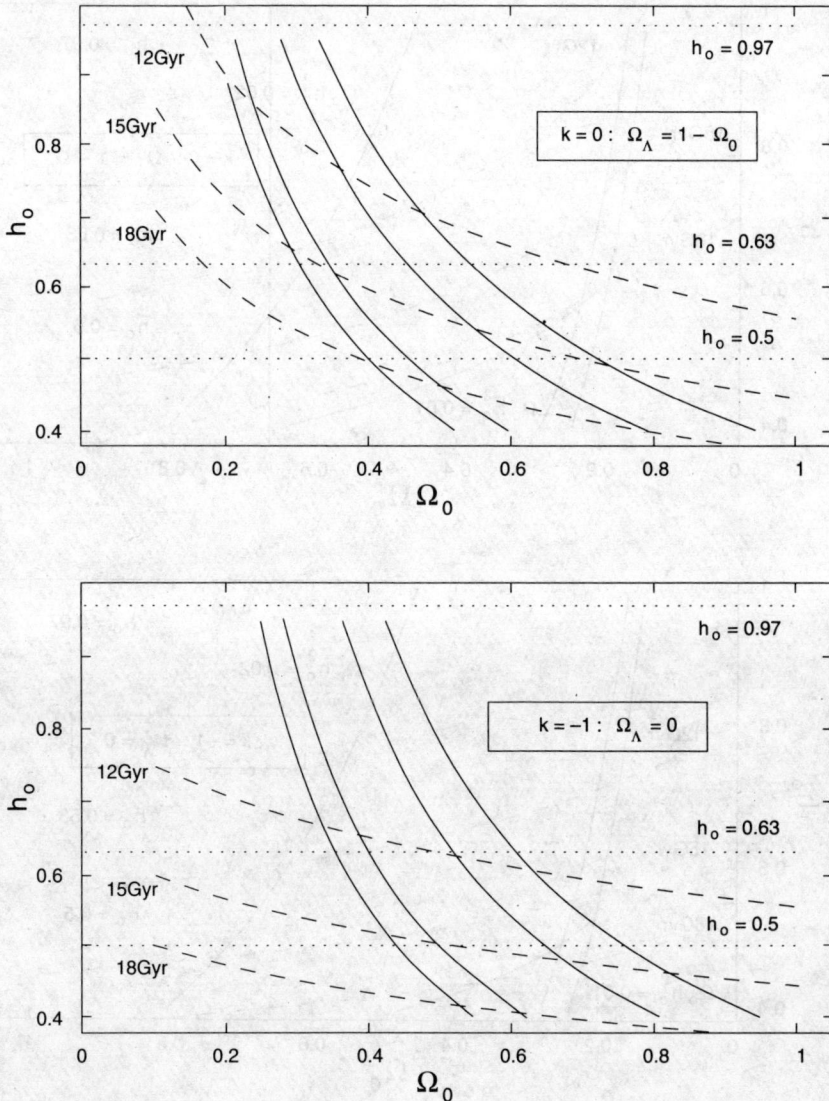

Figure 11.19 See the text for details.

lower frame is for the model with $\Omega_\Lambda = 0$ and $\Omega_0 \leq 1$. Lines of constant age are shown as dashed lines for the range of values of Ω_0 and h_0. Dotted lines mark the band enclosing value of the local Hubble constant ($0.63 < h_0 < 0.97$) obtained from HST measurements. We have also shown the assumed lower limit for its global value ($h_0 = 0.5$). The thick left-to-right line is a lower bound on permitted values of h_0 from the abundance of high-redshift objects. This line depicts the requirement that

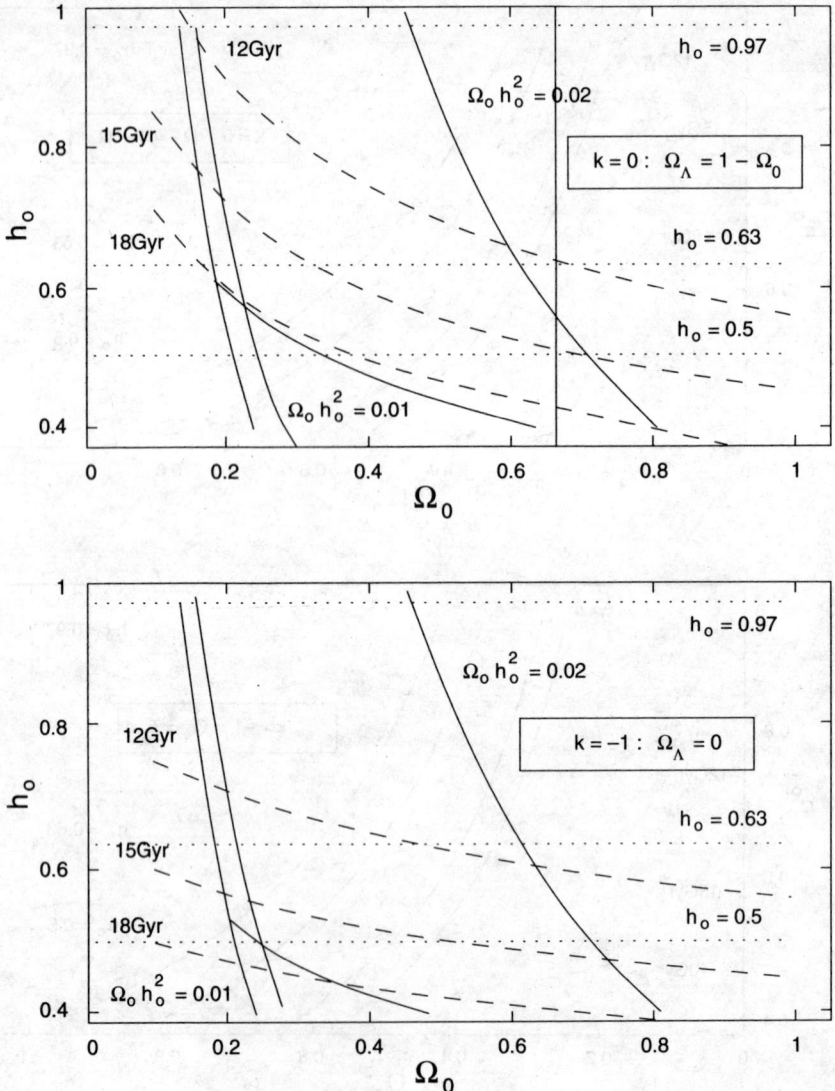

Figure 11.20 See the text for details.

the amplitude of density perturbations at mass length $\sim 10^{11} M_\odot$ was of order unity at $z \simeq 2$. Note that this constraint implies that a flat universe cannot be much older than 18 Gyears. The nearly vertical lines mark the extreme upper limits allowed by primordial nucleosynthesis and the fraction of mass contributed by baryons in clusters. For a given Ω_0, allowed values of h_0 lie below this curve; conversely, for a given h_0, allowed values of Ω_0 lie to the left of this curve. The second such line shows the upper bound implied by observation of the abundance of deuterium at high redshifts.

Finally, Figure 11.21 summarizes all the constraints plotted in the previous two figures. The shaded region is permitted for $t_0 > 12$ Gyr, $h_0 > 0.5$ and other constraints being satisfied. The cross-hatched area shows the region with $t_0 > 15$ Gyears and the abundance of clusters in the allowed region without taking uncertainty in normalization of the COBE data into account. If the uncertainties in the observations are pushed to the extreme limits, then the allowed parameter space corresponds to the shaded region. A somewhat less conservative interpretation of observations will lead to a much smaller allowed region than that shown here as the cross-hatched area. Here we have not used the bounds arising from values of the deceleration parameter and observation of the abundance of deuterium at high redshift.

This brief review highlights the new developments in cosmology since the review of Gunn and Tinsley. Although the constraints of 'age' have been with the big-bang cosmology for several decades, only now are they coming into focus, thanks to the greater precision in the measurements of the Hubble constant and an improvement in our understanding of stellar evolution. Even allowing for errors on both fronts, the inescapable conclusion today is that the standard big-bang models *without* the cosmological constant are effectively ruled out. Even with the cosmological constant, one has to remember that gravitational lensing limits the value of Ω_Λ to less than 0.65.

In this connection it is also interesting to incorporate two other observations into a diagram plotting Ω_Λ against Ω_0 (see Figure 11.22). These are the constraints imposed by the m–z relation for type-Ia supernovae and the observed Doppler peak at $l \approx 200$ from the Boomerang data. The two intersecting regions represent regions of 95% confidence. The overlap includes the line segment along $\Omega_\Lambda + \Omega_0 = 1$, for a flat universe. Again, the cosmological constant is sorely needed.

The constraints from structure formation, abundances of clusters, primordial nucleosynthesis and high-redshift objects are all relatively recent; but they additionally constrain the models *even with the cosmological constant*. Indeed, with the present understanding of extragalactic astronomy, very little parameter space is now left for the standard model with or without the cosmological constant. While a big-bang supporter may claim that the narrowing of the permissible window represents a convergence towards THE model of the universe, this raises another issue that theoreticians need to worry about.

This is the issue of 'fine-tuning' raised earlier in this book. If we take the requirement of no fine-tuning to imply the dictum 'all dimensionless parameters should be of the order of unity', then we would consider $\Omega_{total} = 1$ models natural. (Any other model would require fine-tuning of this parameter in the early universe, a difficulty usually called the 'flatness problem'.) By the same token we would have insisted that $\Omega_\Lambda = 0$. Such a model is clearly ruled out by the observations. It is indeed hard to understand why the leftover cosmological constant is such as to conform exactly

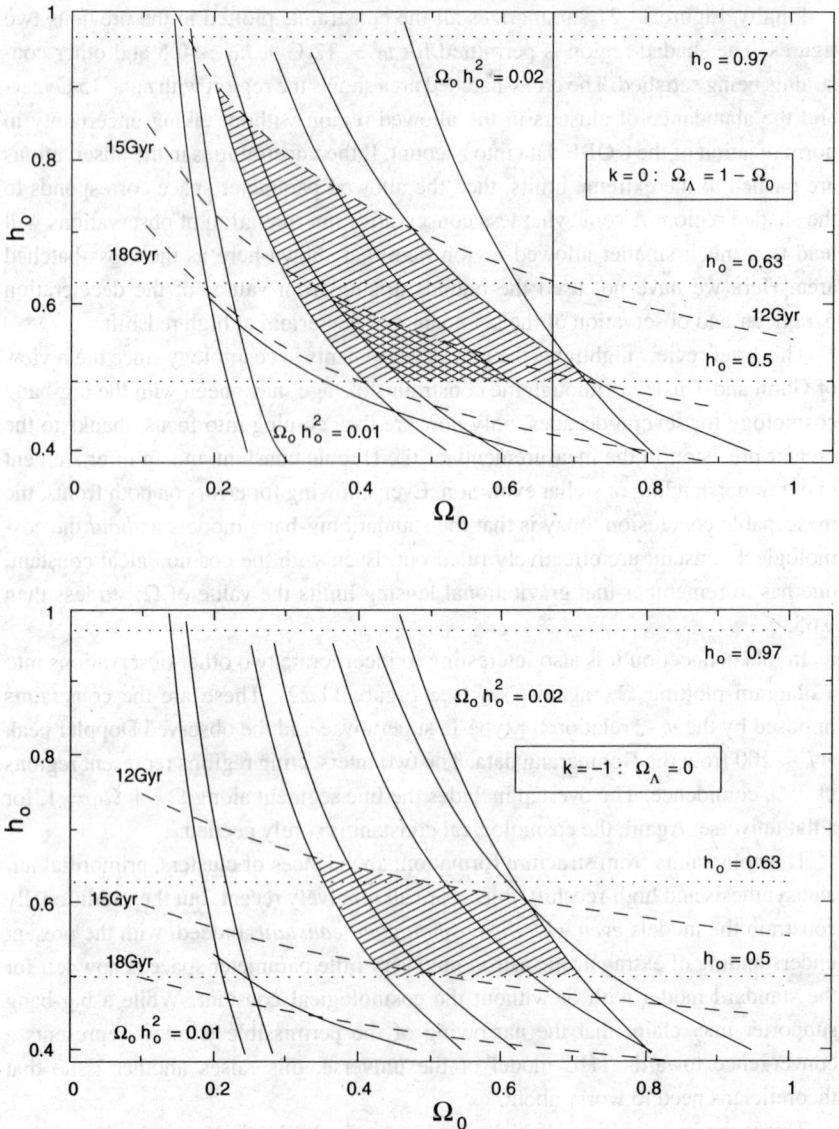

Figure 11.21 See the text for details.

to the flatness condition. As pointed out by Steven Weinberg in 1989, this requires fine-tuning to one part in 10^{108}. There have been attempts (e.g. quintessence) to invoke a dynamically evolving cosmological constant in order to circumvent this difficulty; however, none of these models has any compelling feature. At present, we must conclude that there is indeed a crisis in cosmology.

11.8 The variation of fundamental constants

Standard cosmology is based on the conservative assumption that physics as we know it here and now can be extrapolated to apply to the large-scale structure of the universe. Such an assumption is justified on the basis of the economy of postulates, or Occam's razor. Among the non-standard cosmologies, only the perfect cosmo-

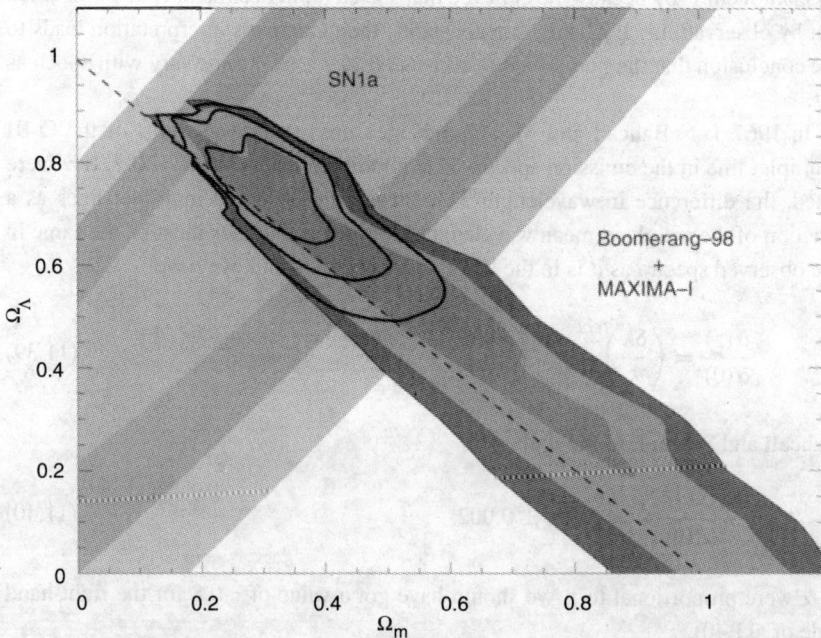

Figure 11.22 The plot obtained from Jaffe *et al.* 2000 (astro-ph/0007333) shows the current constraints from anisotropy of the cosmic microwave background (CMB) and SN1a in the $(\Omega_m, \Omega_\Lambda)$ parameter plane. Shaded contours nearly parallel to the dotted line given by $\Omega_m + \Omega_\Lambda = 1$ are the 1σ, 2σ and 3σ limits (defined as the equivalent likelihood ratio for a two-dimensional Gaussian distribution) of the joint likelihood of the measurements of the anisotropy of the CMB by COBE, Boomerang and the MAXIMA-I experiments. The contours labelled 'SN1a' are similar likelihood contours from observations of high-redshift supernovae. The heavy contours are the combined constraints from the supernovae and CMB-anisotropy data given by the product of the two probability distributions.

logical principle guarantees the validity of this assumption. On empirical grounds there is no reason to believe that the assumption must otherwise hold. Thus it is possible to have fundamental constants like c, $\hbar$, e and G and masses of particles varying with space and time. In Chapter 8 we encountered cosmologies in which it is assumed that the last two items on the above list may vary with the epoch. We will consider the evidence relevant to these issues in this section.

11.8.1 The variation of $\alpha = e^2/(\hbar c)$

We have noticed in the context of the large-numbers hypothesis (LNH) that $e^2/(Gm_p m_e)$ should vary with the cosmic time t. Dirac assumed this to imply $G \propto t^{-1}$ with e, m_p and m_e constants. There is also the alternative conclusion to be drawn from the LNH that $e^2 \propto t$, with G, m_p and m_e constants. This was suggested by Gamow in 1967 because he believed that such a rapid decline in G as t^{-1} is ruled out by observations. If $\hbar$ and c are constants, then Gamow's interpretation leads to the conclusion that the fine-structure constant $\alpha \equiv e^2/(\hbar c)$ must vary with epoch as $\sim t$.

In 1967 J. N Bahcall and M. Schmidt measured the wavelengths of the O-III multiplet line in the emission spectra of five radio galaxies with $z \sim 0.2$. If α were fixed, the difference in wavelength $\delta\lambda$ between the observed multiplet lines as a fraction of the weighted mean wavelength λ of one of the lines must be the same in the observed spectra as it is in the laboratory spectra. If not, we have

$$\frac{\alpha(z)}{\alpha(0)} = \left(\frac{\delta\lambda}{\lambda}\right)^{1/2}_{\text{observed}} \left(\frac{\delta\lambda}{\lambda}\right)^{-1/2}_{\text{laboratory}} \tag{11.39}$$

Bahcall and Schmidt found that

$$\frac{\alpha(z = 0.2)}{\alpha(0)} = 1.001 \pm 0.002. \tag{11.40}$$

If α were proportional to t, we should have got a value of ~ 0.8 for the right-hand side of (11.40).

In 1977 M. S. Roberts compared the redshifts measured at optical wavelengths and at 21 cm for extragalactic sources to find that

$$\left|\frac{\dot{\alpha}}{\alpha}\right| \le 4 \times 10^{-12} \text{ year}^{-1}. \tag{11.41}$$

Again this is an order of magnitude lower than the predicted rate of variation $\alpha \propto t$. These early observations already showed the robustness of the assumption that α is constant.

11.8.2 The variation of G

This is an important observation, since the constancy of G is the basis of general relativity, on which standard cosmology is based. On the other hand, several non-standard cosmologies predict $|\dot{G}/G| \sim H_0$ in the present epoch. We summarize below the direct and indirect evidence for the variation (or lack of variation) of G.

Radar observations

In 1976 I. I. Shapiro and his colleagues reported the result of an analysis of several thousand observations of radar signals bounced off the inner planets between 1966 and 1975. Taking the other data from the Moon and the outer planets, the radar results gave

$$\left|\frac{\dot{G}}{G}\right| < 10^{-10} \text{ year}^{-1}.$$

The radar technique has improved over time and the limit has become more stringent. For example, lunar laser ranging now puts the value of $\dot{G}/G$ at $(0 \pm 8) \times 10^{-12}$ year^{-1}. These values are consistent with zero, given the experimental errors.

From pulsars

The limits on $\dot{G}/G$ also come from pulsar observations. However, caution in interpreting data is needed. For normally one approximates linearly by writing for the gravitational constant $G = G_0 + G_1 t$, where G_0 and G_1 are constants, and then working out the dynamics as per Newtonian methods. However, in certain theories of gravity, such as the Brans–Dicke theory, the inertia and other dynamical quantities also are modified and these effects are not taken into account in the values quoted. Thus the binary-pulsar limits on $\dot{G}/G$ of $(11 \pm 11) \times 10^{-12}$ or the upper limit of 55×10^{-12} from the studies of the pulsar PSR 0655+64 are to be taken as indicative only. However, values of this order rule out the Dirac cosmologies and imply a high value of ω for the Brans–Dicke cosmology.

Lunar mean motion

T. C. Van Flandern had examined Earth–Moon–Sun observations over several years using two time scales, namely atomic time, as measured by atomic clocks, and ephemeris time, derived from the Sun's motion around the Earth. The basis of these observations is as follows.

Suppose that a body goes around another much more massive body in a circular orbit of radius r and mean angular velocity n. If M is the mass of the central body, Newtonian mechanics gives the following two relations:

$$GM = r^3 n^2, \qquad r^2 n = \text{constant} = h. \tag{11.42}$$

If we now introduce a slow variation of G with time, it is easy to deduce from the above two relations that

$$\frac{\dot{G}}{G} = \frac{\dot{n}}{2n}.$$ (11.43)

Thus the Earth's mean angular velocity around the Sun measured in terms of ephemeris time will slow down at the fractional rate of $2\dot{G}/G$ if G decreases with time. A similar equation should hold for the Moon, except that the Moon's motion is also affected by the tidal friction of the Earth–Moon system. Thus we have for the Moon

$$\frac{\dot{n}_M}{n_M} = \left(\frac{\dot{n}_M}{n_M}\right)_{tidal} + \frac{2\dot{G}}{G}.$$ (11.44)

If $\dot{n}_M/n_M$ is measured in terms of atomic time, we should get (11.44). Measuring the same quantity in terms of ephemeris time will, however, take out the $2\dot{G}/G$ term arising from (11.43) and will measure only the tidal part. Thus the difference between the two observations should give us $2\dot{G}/G$.

The main uncertainty in this method has always been in obtaining a reliable estimate of the tidal effect. If the errors quoted in the various determinations of $\dot{n}$ are reliable, then there is a genuine contribution of the $\dot{G}/G$ term towards the Moon's motion. Early assessment of the data by Van Flandern suggested a value of the order $\dot{G}/G \sim (-6.9 \pm 2.4) \times 10^{-11}$ year^{-1}, when it is considered within the framework of the Dirac cosmology. This rate is consistent with the Hoyle–Narlikar cosmology but is too high for the Brans–Dicke theory with $\omega \geq 30$.

The Viking lander

A very accurate measurement of the rate of change of G was reported in 1983 from an analysis of the range data from the Viking landers on Mars. The experiment conducted by R. W. Hellings, P. J. Adams, J. D. Anderson, M. S. Keesey, E. L. Lau, E. M. Standish, V. M. Canuto and I. Goldman used the range measurements from the Viking landers and the Mariner 9 spacecraft in orbit around Mars, the radar-bounce range measurements from the surfaces of Mercury and Venus, the lunar laser range measurements and optical position measurements of the Sun and planets. A least-squares fit of the parameters of the Solar-System model to the data shows that

$$\dot{G}/G = (0.2 \pm 0.4) \times 10^{-11} \text{ year}^{-1}.$$

Thus the result is certainly consistent with *zero variation* of G.

Stellar evolution

If G were greater in the past than it is now, stellar evolution would have proceeded at a faster rate, which would lead to modifications of the m–z relation. In 1980 Canuto and the author showed that insofar as the G-varying HN cosmology is concerned, the then data on m–z relation were consistent with the theoretical prediction. However, the uncertainties of the m–z relation are such that it cannot tell us definitely whether variation of G is taking place. In any case, such an exercise may need to be repeated now for the database of type-Ia supernovae.

Biological evolution of the Earth

If G had been higher in the past, the Sun would have been brighter and the Earth closer to it than it is now. The Solar constant (the flux of radiation from the Sun outside the Earth's atmosphere, at present $\approx 1.388 \times 10^6$ erg cm^{-2} s^{-1}) therefore must have been considerably higher at the time of formation of the Earth than it is today. As estimated by Hoyle on the basis of the G-varying cosmology, at the time when life began, say around 3×10^9 years ago, this constant may have been about three times its present-day value. Would life have been possible under such circumstances? Again, it can be shown that the variation of G in the HN cosmology is not inconsistent with the biological evolution of the Earth, although such evidence also cannot be used *to prove* that G does vary with epoch.

Exercises

1 Calculate the past light cone for Friedmann models by expressing $D(q_0, z)$ as a function of z. Plot these cones for $q_0 = 0$, $\frac{1}{2}$ and 1 as well as for the steady-state model.

2 Discuss how the m–z curves for the various cosmological models branch out for different values of q_0. Why does the uncertainty in the value of H_0 not hamper the test of the value of q_0?

3 What are the various issues that need to be considered before the m–z plot can lead to something of cosmological significance?

4 Discuss the aperture correction. In what way does it depend on q_0?

5 Show that

$$\frac{D(1, z)}{D(\frac{1}{2}, z)} = \frac{z}{2[(1+z) - \sqrt{1+z}]}$$

and deduce that, for $\alpha = 0.7$ the aperture correction introduces a magnitude difference of $\sim 0.09^\mathrm{m}$ at $z = 0.7$.

6 Suppose that $I(\lambda) \propto \lambda^2$ in the range 2500 Å $< \lambda <$ 5000 Å. A galaxy of redshift 0.5 is being observed in a wavelength band centred on 5000 Å. Another galaxy of redshift 0.7 is also observed at 5000 Å. Show that the K-terms for the two galaxies will differ by $\sim 0.41^\mathrm{m}$.

7 Discuss the Stebbins–Whitford effect. Is it eliminated by taking due account of the K-correction? Why?

8 Let $K(\lambda)$ denote the cross section of absorption per unit mass of intergalactic dust at wavelength λ. Show that, in a Friedmann model of given q_0, H_0, the apparent

magnitude of a galaxy of redshift z_0, is increased due to intergalactic absorption by an amount (at the measured wavelength λ_0)

$$\Delta m = 2.5 \log_{10} e \, \frac{c\rho_0}{H_0} \int_0^{z_0} K\left(\frac{\lambda}{1+z}\right) \frac{(1+z)\,dz}{\sqrt{1+2q_0 z}},$$

where ρ_0 is the present density of intergalactic matter.

Taking $K(\lambda) = (6400/\lambda_{\text{ångström}}) \times 10^4$ cm g^{-1}, $h_0 = 0.5$, $q_0 = \frac{1}{2}$ and $\rho_0 \equiv 2.45 \times 10^{-33}$ g cm^{-3}, show that at $\lambda_0 = 6400$ Å, Δm is of the order of 1^{m} for a galaxy of redshift unity.

9 Show how the evolution of luminosity introduces uncertain corrections to the value of q_0. Using (11.16) with $x = 1.35$, compute the 'true' values of q_0 for the measured values $q_0 = 1$ and $q_0 = \frac{1}{2}$.

10 For the luminosity function of galaxies given by (11.25), show that the N–m relation in the $q_0 = 0$ Friedmann model is given by

$$N\,(<m) \propto \int_0^\infty \frac{x^{1/4} e^{-x}\,dx}{[x^{1/2} + \text{dex}(4.658 - m/5)]^3}$$

where dex $y = 10^y$. Show that, for small m, the above result becomes the same as for Euclidean geometry.

11 Show why the K-correction is necessary for the number counts of faint galaxies.

12 A radio galaxy of redshift $z = 0.1$ has a spectral index $\alpha = 1$ and a luminosity of 10^{44} erg s^{-1} over the frequency range 150 Mhz $\leq \nu \leq$ 1500 MHz. For $h_0 = 1$ show that the flux density of the galaxy is $\sim$350 Jy at 1000 MHz and $\sim$1750 Jy at 200 MHz. (Neglect any cosmological effects.)

13 Express the radio power of the source in Exercise 12 in units of watts per megahertz per steradian at the frequencies of 200 and 1000 MHz, respectively.

14 Suppose that the probability that the ratio V/V_m lies in the range $(x, x + dx)$ for $0 \leq x \leq 1$ is proportional to $x^n\,dx$. Estimate n from the observed value of $\langle V/V_m \rangle = \frac{2}{3}$.

15 Suppose that, for small enough intervals $d\sigma_i$, we have at most one source per interval. Writing (11.30) in the form

$$p_i = \frac{\alpha \sigma_i^{-(\alpha+1)}\,d\sigma_i}{1 - \sigma_0^{-\alpha}}$$

and maximizing with respect to α the expression

$$\mathcal{L} = \sum \ln p_i,$$

show that, for $\sigma_0 \gg 1$, we get

$$\alpha = M \Big/ \sum \ln \sigma_i.$$

16 Let $f(L)\,dL$ denote the number of radio sources per unit volume in the luminosity range $(L, L + dL)$. Suppose that, for small redshifts, the plot of $\log z$ against $\log L$ follows a straight line of slope $\frac{1}{2}$. Also assume that the number of points in equal intervals of $\log L$ is found to be constant. Using Euclidean geometry with distance $\propto z$, deduce from these observations that $f(L) \propto L^{-2.5}$.

17 Discuss why a sample of radio sources complete with respect to a minimum flux density is not necessarily complete with respect to a maximum redshift and *vice versa*.

18 A radio-source survey gives $N = 10$ at $S = 12.5$ Jy and $N = 93$ and $S = 5$ Jy. Show that, in a Euclidean universe, the above counts imply either a deficit of 13 sources at the high-flux end or an excess of 53 sources at the low-flux end. Use this example to comment on the Ryle–Hoyle controversy of the 1960s.

19 Show that, in the Einstein–de Sitter model, the number of galaxies intervening between a quasar of cosmological redshift z and the observer is given approximately by

$$N = 0.006\left(\frac{R}{3 \text{ kpc}}\right)^2\left(\frac{N_g}{0.1 \text{ Mpc}^{-3}}\right)[(1 + z)^{3/2} - 1],$$

where $h_0 = 1$, R is the radius of the typical galaxy (assumed to be spherical) and N_g is the number density of galaxies. Use this number to estimate the probability of a quasar of redshift z_0 having n absorption lines. What can you say about the intervening-galaxy interpretation of quasar absorption lines on the basis of this result?

20 A spherical explosion leads to the expansion of an object with radial velocity V in the rest frame of a remote observer O. By considering the shape of the surface of simultaneity seen by O, deduce that the object appears to expand laterally with speed $V(1 - V^2/c^2)^{-1/2}$.

21 Show that a supermassive black hole of mass $10^8 M_\odot$ has a characteristic time scale of $\sim$15 min. (In an accretion-disc scenario, the disc may extend out to $\sim 10^3$ times the radius of the black hole, thus increasing the above time scale by a factor of a thousand.)

22 For the quasar 3C 345 an angular separation of central components was observed to increase from $\sim$0.6 milli-arcseconds in 1970 to $\sim$1.6 milliarcseconds in 1975. The redshift of 3C 345 is 0.595. Show that, if the redshift is cosmological, the speed of separation must be at least $\sim$6.6 c for $h_0 = 1$.

23 Show that, if the fine-structure constant varies as t, then at a redshift of 0.2 the fine-structure constant should be 77% of its present-day value in the Einstein–de Sitter model.

24 Deduce, that for a slow variation of G in Newtonian mechanics, the angular speed n of a particle going around a massive body in a circular orbit changes as follows:

$$\frac{\dot{n}}{n} = \frac{2\dot{G}}{G}.$$

25 Discuss the uncertainty introduced into the measurement of $\dot{G}/G$ by the tidal force between the Earth and the Moon.

Chapter 12

A critical overview of cosmology

12.1 Cosmology as a science

The preceding chapters describe the attempts of present-day cosmologists to study their subject within the discipline imposed by science. From the days when it was a subject of philosophical speculations and religious dogma, cosmology has now developed into a subject to which the scientific method of investigation can be applied. This change has resulted from improvements in techniques for observing the large-scale structure of the universe and from the wide applicability of the laws of physics. In this final chapter we will take stock of all the available material to see how theories have performed *vis-à-vis* observations. However, a cautionary remark may be in order.

By claiming to describe the universe as a whole, cosmology transcends the realms of all other branches of science. Any conclusions about the universe are bound to be profound and hence must be drawn with caution. This caution is often found to be missing from statements about cosmology. All too often the investigator (whether a theoretician or an observer) is tempted to mistake the model of the universe for the real thing. Categorical remarks about the state of the universe are often found upon closer examination to be model-dependent. For example, if an inhomogeneity of harmonic l is discovered in the microwave background, it may have one interpretation in one theory and a different one in another. If both interpretations are consistent with observations, it is unfair to claim that one theory alone is proved.

In the last two chapters we have looked at the big-bang cosmology from various observational points of view. In §11.7 we have summarized the observational constraints on the big-bang parameters and shown that very little, if any parameter space remains viable. Before proceeding further, we will briefly review the quasi-steady-state cosmology in the light of the same observations, noting the above caveat, however, that the same observation may be interpreted differently by the QSSC from the way it is interpreted by the standard (big-bang) cosmology, which we shall refer to in brief as SC hereafter.

12.2 Observational constraints on the QSSC

In Chapter 9 we have discussed the theoretical and observational aspects of the QSSC. We will supplement those ideas with the inputs of observations of the kind discussed in Chapters 10 and 11. Notice, however, that, in terms of Figure 11.1, the QSSC has regions I, II and III, but does not have the equivalents of earlier regions of the SC. This is because, in terms of redshifts, this cosmology does not have a monotonic behaviour with respect to time. As the universe oscillates, when we look back beyond $z_{max} \sim 5$, the redshifts start decreasing and give way to modest blueshifts, which again give way to redshifts. Beyond a few cycles backwards in time, however, there are no blueshifts to be found, although the redshifts continue to oscillate between finite ranges in each cycle.

First we will consider the tests of the distant universe, which go up to redshifts of the order of ~ 5, corresponding to regions II and III. We may mention right away that, since the QSSC uses the expanding-universe hypothesis, it is consistent with the findings of the surface-brightness test. The other three tests work out as follows.

12.2.1 The redshift–magnitude relation

How does the QSSC explain the m–z test using type Ia supernovae? The important difference between the standard m–z relation and that in the QSSC comes from the presence of whisker-like dust grains (see §9.13.2). This produces an additional dimming, which progressively increases with redshift according to the formula

$$\Delta m(z) = 2.5 \log_{10} e \times \int_0^{x(z)} \kappa \rho_g \, dl, \tag{12.1}$$

where κ is the mass absorption coefficient produced by the grains, ρ_g is the density of grains, and $x(z)$ is the proper distance traversed through the intergalactic dust up to redshift z. One has to *add* this extra magnitude to the theoretically computed value for a dustless universe.

Figure 12.1 shows the theoretical 'best-fit' curve for the data. To obtain the best fit, there were two parameters to vary: (i) the absolute magnitude of the supernova at the peak and (ii) the present-day value of ρ_g. The χ^2 value for the best fit compares favourably with the best-fit value for standard cosmology optimized with respect to Ω_Λ. Note that the fit shown in Figure 12.1 *does not* optimize with respect to the other cosmological parameters, which are taken as $z_{max} = 5, \eta = 0.811, h_0 = 0.65$ and $\Omega_\lambda = -0.36$. It is interesting that the best-fit value of ρ_g turns out to be 3.3×10^{-34} g cm^{-3}, which is in the right range for the dust to be able to thermalize the relic starlight into the microwave background. Had the model been along wrong or unphysical lines, we could have got a wildly different value of this parameter.

The QSSC can also suggest an answer to the question of why some supernovae look exceptionally dim. This is because the dust (which is formed from the metallic vapours expelled by a supernova) lingers around the supernova to create more extinction of light.

12.2.2 The radio-source counts

In the QSSC these are generated by a combination of various populations of radio sources. Consider for example the two classes as follows:

$$\text{Class I}: L = 10^{26} \text{ W Hz}^{-1}, \qquad n_0 = A = \text{constant}, \qquad z \geq 0,$$

$$\text{Class II}: L = 5 \times 10^{26} \text{ W Hz}^{-1}, \qquad n_0 = A/4, \qquad z \geq 0.14.$$

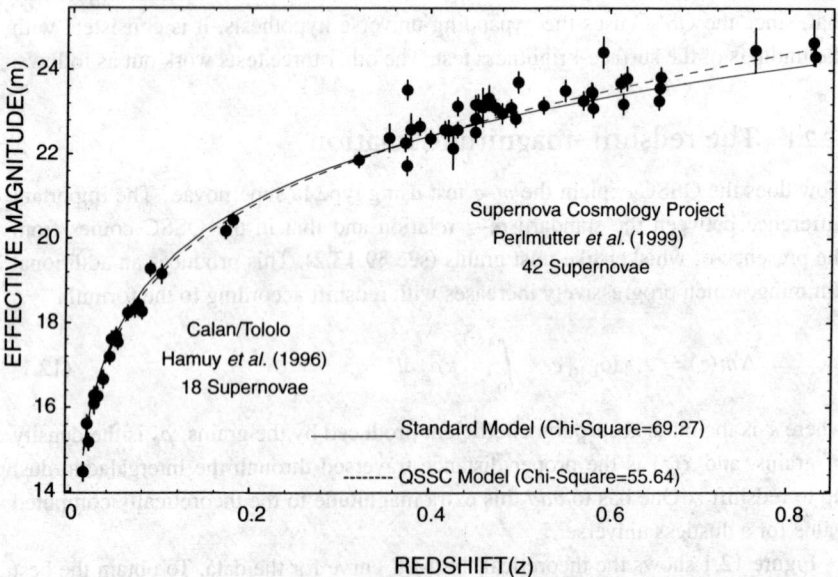

Figure 12.1 The redshift–magnitude data for the type-Ia supernovae are fitted with the QSSC, with a density-optimized population of whisker grains in the intergalactic space causing extinction. The SC model with $\Omega_0 = 1$ and $\Omega_\Lambda = 0$ is shown for comparison.

Here L is the luminosity and n_0 the present-day number density of a population. So the sources in the second class are more powerful but less populous and they die out at redshifts less than 0.14. However, with this simple combination, we arrive at a source count curve in Figure 12.2 that agrees very well with the data shown in Figure 11.10.

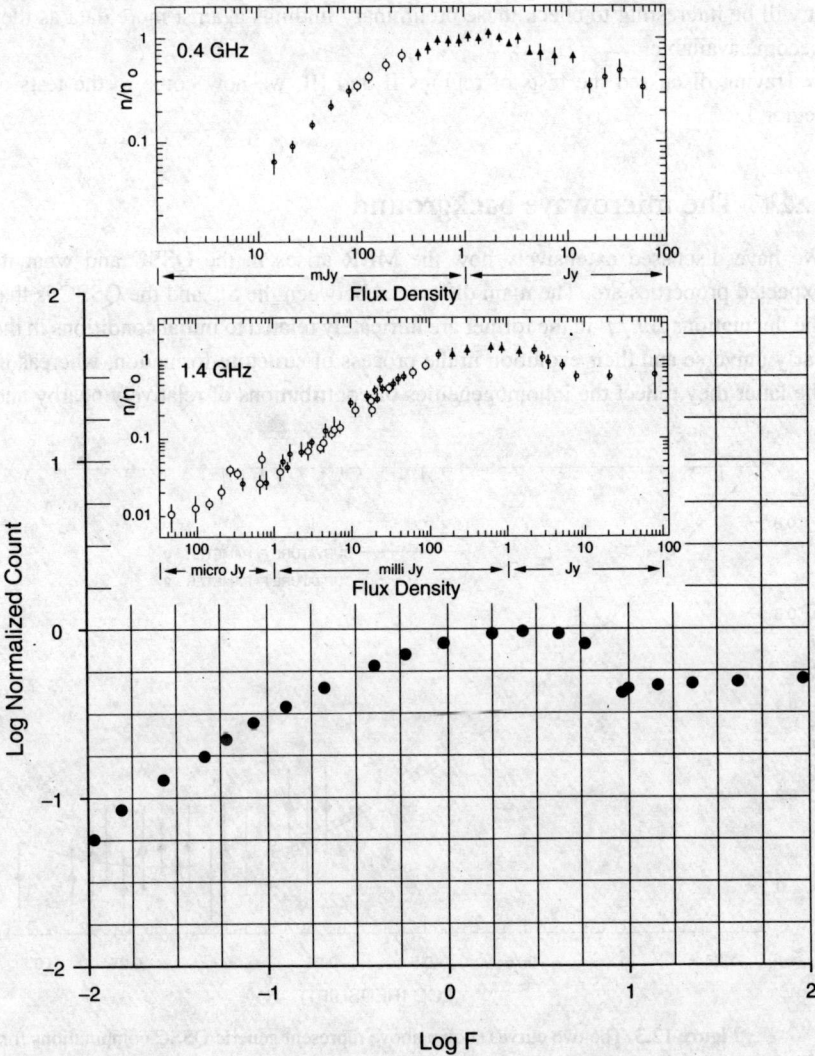

Figure 12.2 A theoretical QSSC computation of the differential radio source counts expressed as a ratio of the Euclidean counts, based on a mix of two populations of radio sources. This may be compared with the observed curves of Figure 11.10, which are reproduced here in the upper half of the figure.

12.2.3 The angular-size–redshift relation

The data on ultracompact radio sources discussed in the previous chapter have been fitted with the QSSC models. The resulting fits for two typical cases are shown in Figure 12.3. The two cases represent flat ($k = 0$) and negative-curvature ($k = -1$) models. The curvature parameter referred to in Figure 12.3 is $K_0 = kc^2/(H_0^2 S_0^2)$. Although both types of model give statistically good fits, the latter performs better. It will be interesting to check these preliminary findings against more data as they become available.

Having discussed the tests of regions II and III, we now come to the tests of region I.

12.2.4 The microwave background

We have discussed extensively how the MBR arises in the QSSC and what its expected properties are. The main difference between the SC and the QSSC is that the fluctuations $\Delta T/T$ in the former are intricately related to initial conditions in the early universe and their evolution in the process of structure formation, whereas in the latter they reflect the inhomogeneities of contributions of relatively nearby and

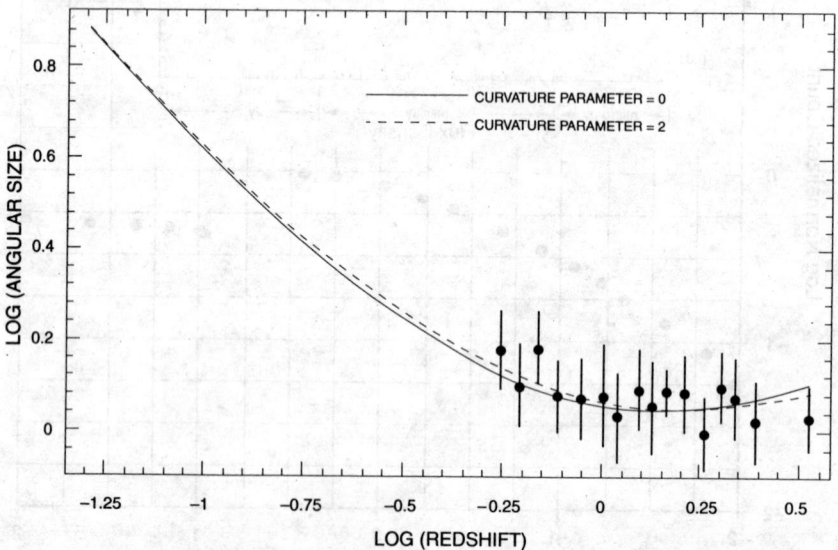

Figure 12.3 The two curves shown above represent generic QSSC computations for zero and negative spatial curvatures. Although both models give a satisfactory fit to the angular-size and redshift data, the latter performs somewhat better. The curvature parameter K_0 used here is a scaled version of k, as defined in the text. Based on the paper by S. K. Banerjee and J. V. Narlikar, 1999, *Mon. Not. Roy. Astron. Soc.*, *307*, 73.

recent origin. The latter have still to be worked out, and may require further details of the cluster/void distribution in the universe.

For example, if the inhomogenesity of the distribution of dust in the present cycle is on the scale of a rich cluster of galaxies, it will cause inhomogeneity in the contribution of thermalized relic starlight of the present cycle, on the scale of angles subtended by such objects, at redshifts $z_{max} \sim 5$. These correspond to l in the range 200–250. Thus the bump seen by 'Boomerang' receives a different interpretation here.

12.2.5 Light nuclei

These have been discussed in the context of the QSSC in Chapter 9. There we argued that the light nuclei form as a result of the decay of the Planck particle in an energetic fireball. However, in 1998, Geoffery Burbidge and Fred Hoyle argued that almost all nuclei can be made in stars. Thus, although the stellar contribution to helium in the present cycle is only $\sim 10\%$ of that observed, the stars in the previous cycles also contributed. Integrating over all previous cycles, we get a total contribution of stars of all previous cycles towards the present-day abundance of ^{4}He that is of the right order.

The fact that there can be two possible ways of getting light nuclei may be something of an embarrassment to the theory, for then one has to understand why one rather than the other process dominated. The stellar option is aesthetically better since it involves only one process for the creation of *all* nuclei, light or otherwise. Also it relates the resulting starlight quite naturally to the microwave background. The scenario involving minicreation events, on the other hand, provides the right quantity of deuterium, for which no explicit stellar option is yet known. It may well be that the typical minicreation event produces only baryons and possibly deuterium, of which the former are then used in a star-formation process.

12.2.6 Dark matter

Unlike the big-bang cosmology, the QSSC does not seem to place any restriction on the baryonic density of the universe, either from considerations concerning light nuclei or from the size of imprints of structure formation on the microwave background. Thus dark matter, to the extent that it is required, can be all baryonic. Since the typical minicreation event in the QSSC generates large-scale structure, it is likely that the clusters seen today retain some expansion velocities dating from their origin and therefore need not be virialized. In that case the amount of dark matter in them may have been overestimated.

Amongst baryonic dark matter, the possibility of very old (say, ~ 30–40 Gyears) relic white dwarfs may provide an option, together with brown dwarfs and Jupiter-like planets. In this sense, the findings of 0.5 Solar-mass objects from MACHO-type surveys may be significant.

Having described the present-day observational constraints on the QSSC, we will now enter into a brief debate between the adherents of the standard cosmology and its opponents.

12.3　The case for standard cosmology

An ardent supporter of standard cosmology will mention the following points in favour of the hot-big-bang models.

1. The models are based on Einstein's general theory of relativity. To the extent that it has been possible to test this theory, its predictions have always been borne out by observations. Thus we have confidence that the framework of our models is based on a sound theory.

2. The standard models are the simplest solutions of Einstein's equations. It is remarkable therefore that they are able to reproduce such a profound observation as Hubble's law. Moreover, these models *predicted* this law rather than coming as afterthoughts. This is clearly an indication that we are working along the right lines.

3. So far there is no satisfactory alternative to the theory of primordial nucleosynthesis for explaining the abundances of light nuclei, especially ^{4}He and ^{2}H. The agreement between the observed abundances and the theoretical predictions is good enough to generate confidence in the hot big bang.

4. The observation of the microwave background radiation and its Planckian spectrum is a striking confirmation of the early hot phase in the history of the universe. Again, as in point 2, it is to the credit of the picture that the observation had been *predicted* by the theory.

5. The recent successes of grand unified theories (GUTs) applied to the very early universe suggest that such a scenario must have some germ of truth in it. For example, the expectation based on primordial nucleosynthesis that there cannot be more than three species of neutrino appears to be borne out by experiments with particle accelerators. In any case, the physical conditions under which the three basic forces of nature unite could have existed only in the very early universe. Since it is believed that redundant laws do not exist in nature, the situation leading to a GUT must have operated sometime; hence the very early universe is the logical choice for the stage at which the unified laws enacted their roles.

6. A logical consequence of the GUT phase transition is 'inflation', which has turned out to be a fruitful input to the standard hot-big-bang cosmology and

promises to resolve some of its outstanding problems, including the provision of suitable initial conditions from which the large-scale structure evolved.

7. The number counts and angular sizes of discrete source populations such as galaxies, radio sources and quasars exhibit evolution with epoch on the characteristic time scale of the expansion of the universe. The evolutionary models demand a greater density of quasars in the past, which is consistent with the predictions of the standard models that the universe was denser in the past than it is at present.

8. The success in detecting very tiny inhomogeneities in the microwave background has led to a very flourishing effort at understanding the evolution of large-scale structure in terms of its interaction with radiation. Such an exercise helps in understanding the physics of the universe as it went through various past epochs even prior to those when discrete sources could be seen.

12.4 The case against standard cosmology

The agnostic may use the following counter-arguments in the cosmological debate.

1. General relativity has been tested only in the weak-field approximation. We have no empirical evidence regarding how the theory fares for the strong-field scenarios of cosmology. The standard models therefore are to be looked upon as nothing more than extrapolations into speculative regions.

2. Relativistic cosmology in general and standard models in particular have the curious and unsatisfactory feature of a spacetime singularity. The appearance of infinities is considered disastrous in any physical theory. In general relativity it is worse, since the singularity refers to the structure of spacetime and the physical content of the universe itself. Moreover, some adherents sometimes seek to elevate this defect out of the reach of physics. Thus one is not supposed to worry that the big bang violates all basic conservation laws of physics, such as the laws of conservation of matter and energy. Rather one is asked to look upon the event as beyond the scope of science.

3. There is a discrepancy between the astrophysical age estimates and the Hubble age of the standard models. The discrepancy is made worse if h_0 is close to unity rather than $\frac{1}{2}$, $q_0 \geq \frac{1}{2}$ rather than $q_0 \sim 0$. The recent revival of the cosmological constant is a symptom of this problem.

4. The photon-to-baryon ratio of $\sim 10^8$ is not explained by the standard models and the present temperature of ~ 3 K of the radiation background remains to be derived from a purely theoretical argument. Although GUTs provide one way of explaining N_γ/N_B, the present explanation still has the character of a postdicting and parameter-fitting exercise.

5. Despite numerous attempts by so many experts in the field, the formation of large-scale structure in the universe is ill-understood, especially in the context of the extraordinary smoothness of the microwave background, on the one hand, and the observed large-scale streaming motions relative to the Hubble flow observed today on the other.

6. On a somewhat epistemological issue one feels uncomfortable at the way the research on the very early universe is being carried out. Because so much brainpower is currently being devoted to this field and has been for the last decade or more, this issue needs to be stressed somewhat more forcefully. A comparison with other branches of physics would be helpful for understanding the problem.

In general, physics (or, for that matter, science in general) progresses with a close interplay between theoretical ideas and observed facts. Sometimes theory is speculative and is checked by firm observations. On other occasions theory is well founded but observations need to be further sharpened. In the work on the very early universe neither scenario holds: here one is dealing with theoretical speculations side by side with no direct observational evidence.

When the electro-weak theory was formulated it was an exercise in theoretical speculations in gauge theories. It might or might not have worked. That it does work was eventually demonstrated by accelerator experiments. This is an example of how the scientific method works in particle physics. In Gamow's work on the early universe well-established physics was used in a cosmological scenario that was speculative (there are no astronomical observations of the universe when it was ~1 s old). However, the ultimate predictions of the work can be compared with hard facts: the abundances of light elements and the radiation background.

Neither of these conditions holds vis-à-vis the very early universe. No one can deny that the theoretical work on GUTs is still highly speculative. Neither can the theories be dynamically tested with particle accelerators. To capture the full flavour of a grand unified theory one needs to attain particle energies of $\sim 10^{15}$ GeV, which are far beyond the capabilities of present-day technology. On the cosmological side, the physics of the standard model with or without inflation at $t \sim 10^{-36}$ s is also entirely speculative as those epochs cannot be observed.

Thus one is matching one speculation with another. There is no harm in doing that, provided that one keeps reminding oneself that at best the exercise can claim consistency of this matching with what is observed today. Instead, very definitive claims are often made about what the universe was like at these epochs.

7. Furthermore, the requirement of 'repeatability of an experiment' is not satisfied in this picture. The GUT phase transition, inflation etc. happened once only and conditions conducive to them would not occur again. We may contrast this situation with nucleosynthesis in stars. This is an ongoing process with each star as an independent experiment.

8. The role of non-baryonic dark matter is highly reminiscent of the 'Emperor's New Clothes' – the story by Hans Christian Anderson. Except for neutrinos (whose massiveness is still open to question) no other form of such matter has been established experimentally. Yet the existence of various esoteric particles is taken for granted uncritically. Perhaps the cosmologists think that the physicists have established their existence on a firm footing while the physicists think that such particles must exist because cosmologists tell them so. The hard fact is that there are no hard facts on either side!

So the sceptic may be permitted to remark that 'The work on the very early universe, inflation, dark matter etc. is certainly very interesting, but is it physics?'.

12.5   The outlook for the future

In the light of what has been presented so far, we may ask a specific question both of the SC and of the QSSC: 'What test can be performed that could in principle disprove this cosmology?'. This question is in the spirit of Karl Popper's view of a scientific theory, namely that it should be disprovable. Thus, if such a test is performed and its results disagree with the prediction of the theory, then the theory is considered disproved. If the theory seeks survival by adding an extra parametric dimension, then that is against the spirit of this question.

12.5.1   Tests that can disprove the SC

With regard to the SC we suggest the following tests, which we feel to be decisive in disproving the current versions of the SC.

1. *Blue shifts.* If we find that a faint population of galaxies exhibits blueshifts, then the SC cannot be sustained. The QSSC, on the other hand, does predict such a population, namely the galaxies which are observed for the epochs close to the last maximum of the scale factor. The expected shifts are small, however, not exceeding ~ 0.1. It should be noted that, while finding such a population would disprove the SC, *it does not prove* the QSSC: it is merely consistent with the theory. Likewise, not finding a blueshifted spectrum does not *prove* the SC, but is consistent with it.

There are, however, problems with such a test. The obvious selection effect that an astronomer looks for line identifications on the short wavelength side of the observed line works against finding a blueshift. As the continuum itself gets brightened by blueshifting, the relatively weak blueshifted lines may be hard to detect against it. If one goes by the QSSC predictions, one needs to perform spectroscopy with galaxies fainter than $\sim 27^m$ to find blueshifts, which is by no means easy.

2. *Very old stars.* The QSSC expects very old stars, born in the previous cycle, to be found in the Galaxy. They could be low-mass ($\sim 0.5 M_\odot$) stars just off the main sequence, which could be in the giant stage, as shown in Figure 12.4, or seen as horizontal branch stars *without the helium flash*, or they could be very old white dwarfs. With ages as great as ~ 40–50 Gyears, these stars cannot be accommodated within the SC framework even with the cosmological constant.

3. *Baryonic dark matter.* If, through studies of clusters of galaxies containing hot gas, intergalactic space containing dust and MACHO-type microlensing

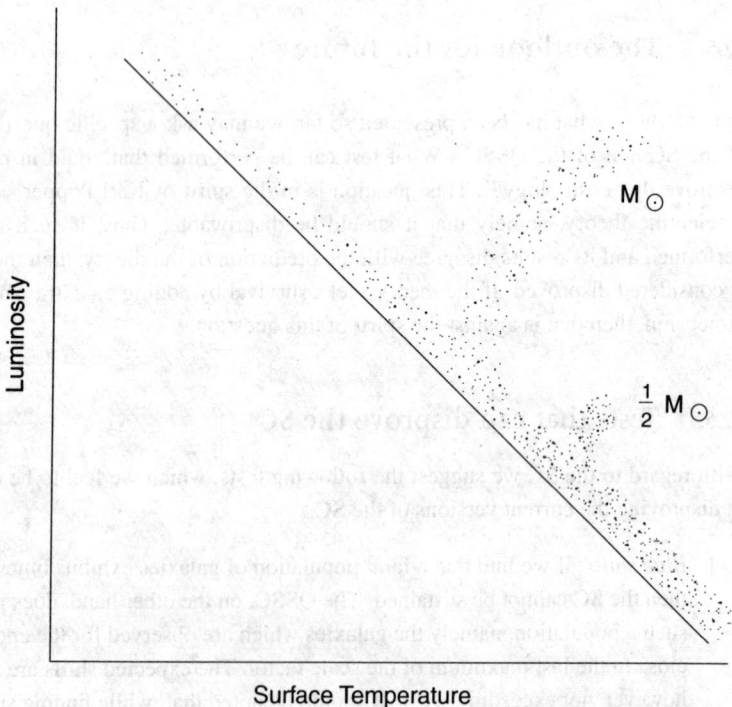

Figure 12.4 In the HR diagram two stars with masses $M_\odot$ and $M_\odot/2$ are shown branching off the main sequence onto the giant branch. A star of half the Solar mass branching off as a red giant can be 40–50 Gyears old.

observations, it is shown that the baryonic density parameter Ω_{bar} exceeds, say $0.02h_0^{-2}$, then the SC stands disproved, for it will have lost its one major asset, namely the ability to account for the observed abundance of deuterium. A large amount of baryonic matter will also pose difficulties for the structure-formation scenarios, which aim at explaining the observed inhomogeneity of galactic matter and its large-scale motions while keeping the anisotropies in the microwave background at the microkelvin level.

4. *The Ω_0–h_0 diagram.* We have already seen how this diagram helps constrain the SC. Further observations may help tighten the error bars on various parameters of the SC and thereby ultimately eliminate any permissible window in such a diagram. The MAP and PLANCK studies are expected to determine the parameters of the SC, together with other tests such as the *m*–*z* test. These will help to constrain the parameter space.

5. *Ages of stars and galaxies.* Regardless of item 2 above, more precise determination of the ages of stars in the globular clusters in the Galaxy can in principle rule out many versions of the SC, if some ages turn out close to, say, 18 Gyears. Likewise, nuclear cosmochronology can also in principle pose problems for many SC models by producing nuclear ages of the same order.

In addition, if one improves the age–colour relationship for high-redshift galaxies, one can in principle disprove many SC models (i) via the total age criterion, viz. the look-back time plus the age of the galaxy exceeding the age of the universe, or (ii) by detecting fully formed mature galaxies too early in the universe.

12.5.2 Tests that can disprove the QSSC

We next outline a few tests that have the potential to disprove the QSSC.

1. *The discovery of epochs of ultra-high redshifts.* As we have seen, the QSSC model has a maximum redshift in the present cycle. In the typical case described here, z_{max} was taken as 5. There is sufficient flexibility in the model to make z_{max} somewhat higher, say up to 10–15. However, any direct evidence that the universe had passed through an epoch of much higher redshift, say ≥ 30, would bring the credibility of the QSSC into question. (Abundances of light nuclei and the microwave background known today *do not* constitute such evidence because these are so interpreted within the SC framework: they have a different interpretation in the QSSC.)

2. *Non-detection of old matter.* Just as detection of old matter goes against the SC, so will the non-detection of such matter go against the QSSC. Since the QSSC claims that the observable universe contains very old stars, dedicated searches for such objects are important in order to test the theory. In this

connection, it is worth noting that current findings from gravitational microlensing would rule out the existence of white dwarfs of age 10–12 Gyears since they would be luminous: however, they are consistent with the existence of white dwarfs as old as 40–50 Gyears.

3. *The absence of metallic whiskers.* The dust grains which act as thermalizers of the relic stellar radiation so as to produce the microwave background, viz., the metallic whiskers in interstellar and intergalactic space, hold the lifeline to the QSSC. Narlikar *et al.* (1997) have discussed how these are produced and distributed in space, pointing out preliminary evidence consistent with their existence. Such evidence needs to be critically examined to see whether such dust indeed exists. In this connection determination of the m–z relation using type-Ia supernovae out to redshifts exceeding unity can play a crucial role. So can high-redshift quasars with substantial luminosity in the millimetre wavelengths. The finding of such quasars means either that, because of absorption by dust, they must be abnormally luminous in millimetre wavelengths, or that their redshifts are substantially non-cosmological, a possibility we will briefly touch upon in the final section. Failing these two alternatives, the QSSC loses one of its main arguments.

4. *Finding accretion of black holes.* The QSSC claims that the pockets of high-energy emission in the universe such as the active galactic nuclei are explosive events pouring new matter into the universe. It questions the black-hole paradigm which invokes infalling matter circulating in an accretion disc. As observational methods improve, the nuclear region can be examined more critically to see which of the two alternatives is correct. Since the SC is not related to the black-hole/accretion-disc paradigm, finding *versus* not finding such accretion will not affect it seriously. However, the QSSC is more critically linked with the explosive-creation paradigm, so finding a clear instance of accretion by a black hole at the centre of an energy source will be an embarrassment to the theory. Here again it is worth stressing that finding evidence *consistent with* accretion of black holes does not constitute a proof of the phenomenon.

12.6   Concluding remarks

Cosmology is currently going through a critical stage. This statement is more than the cliché that it sounds like. We will elaborate on it with a few examples of outstanding questions that need to be settled clearly one way or the other.

1. *How homogeneous is matter in the universe?*

Technology has reached a stage at which automated observations on a large scale are a reality. Thus redshift surveys are expected to churn out spectra and redshifts of galaxies by the million. It therefore makes sense to await a better comprehension of large-scale structure in our immediate vicinity before theorizing on the basis of a homogeneous and isotropic model. In particular, we need to know whether above the scale of, say, 200 Mpc, the universe can safely be regarded as homogeneous.

It has been claimed, for example, that there is no ultimate scale of inhomogeneity and that the universe is fractal in nature. If so, the present Robertson–Walker models fail and one may have to do the modelling afresh. Less drastically, one may find inhomogeneity on a larger scale of, say, 400 Mpc. To what extent then are our present models reliable?

2. *Are the clusters dynamically relaxed?*

A major part of the present argument for dark matter comes from the virial theorem for clusters. If the peculiar velocity distribution for clusters can be determined or estimated with greater accuracy than it can at present, we may be able to ascertain whether the clusters are virialized or rather are expanding as Ambartsumian so strongly claimed in the 1960s.

The QSSC suggests that the clusters may have been created in explosive creation processes and that they are expanding. This will mean that the virial theorem is not valid for clusters and this takes away the strength of the present argument for dark matter. Hence it is necessary to find independent checks on whether the velocity distributions of galaxies in clusters are in statistical equilibrium.

3. *Does the universe have a cellular structure?*

Analyses on larger scale have led to claims in 1998 by Einasto *et al.* of a cellular/periodic structure of the universe. There are also claims for a fractal universe for which no 'mean density' can be defined. Neither of these investigations can be easily accommodated in the standard cosmology or in the QSSC, although Hoyle and Burbidge have attempted to explain periodic redshifts in the QSSC by a modification of the Machian relationship relating the inertia of a particle to the rest of the particles in the universe. So it is important to check with more detailed studies whether the claimed cellular structure in the large-scale distribution of galaxies really exists.

4. *Does Hubble's law hold for all extragalactic objects?*

Throughout this book we have taken it for granted that the redshift of an extragalactic object is cosmological in origin, i.e., that it is due to the expansion of the universe. In Chapter 11 we described this assumption as the cosmological hypothesis (CH). There we commented on the fact that, whereas the Hubble diagram on which the CH is based gives a fairly tight m–z relationship for first-ranked galaxies in clusters, a corresponding plot for quasars has enormous scatter. Although we discussed the cosmological tests on the basis of CH for quasars as well as galaxies, we found that, in some cases, special efforts are needed to make the CH consistent with data on quasars. These included, apart from the Hubble diagram, the superluminal motion of quasars, rapid variability, the absence of a Ly-α absorption trough, etc.

To what extent is the CH valid for quasars? Let us begin with the type of data Stockton had collected, in which quasars and galaxies were found in pairs or groups of close neighbours in the sky. The argument was that, if a quasar and a galaxy are found to be within a small angular separation of one another, then it is very likely that they are physical neighbours and, according to the CH, their redshifts must be nearly equal.

This argument is based on the fact that the quasar population is not a dense one and, if we consider an arbitrary galaxy, the probability of finding a quasar projected *by chance* within a small angular separation from it is very small. If the probability is <0.01, say, then the null hypothesis of projection by chance has to be rejected. In that case the quasar must be physically close to the galaxy. This was the argument Stockton used.

While Stockton found evidence that in such cases the redshifts of the galaxy and the quasar, z_G and z_Q, say, were nearly the same, there have been data of the other kind also. In two books listed in the bibliography H. C. Arp has described numerous examples in which the chance-projection hypothesis is rejected but $z_G \ll z_Q$. Over the years four types of such discrepant cases have emerged.

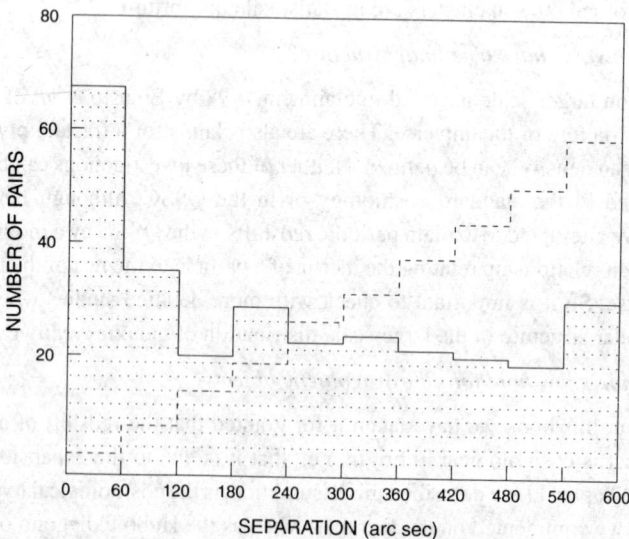

Figure 12.5 A histogram of the distribution of separations of 300 quasar–galaxy pairs. If the quasars were randomly distributed with respect to bright galaxies then their numbers should have increased in proportion to the square of the angular separation. Instead, there is a peak within 60 arcseconds. The quasars are all of considerably higher redshifts than galaxies. Adapted from G. Burbidge, A. Hewitt, J. V. Narlikar and P. Das Gupta, 1990, 'Association between quasi-stellar objects and galaxies', *Ap. J. Suppl.* 74, 679.

1. There is growing evidence that large-redshift quasars are preferentially distributed closer to low-redshift bright galaxies (see Figure 12.5).

2. There are alignments and similarities of redshift among quasars distributed across bright galaxies (see Figure 12.6).

3. Close pairs or groups of quasars of discrepant redshifts are found more frequently than would occur due to chance projection (see Figures 12.6 and 12.7).

4. There are filaments connecting pairs of galaxies with discrepant redshifts (see Figures 12.8(a) and (b)).

It is worth recording that there are continuing additions to the list of anomalous cases. They are not limited to optical and radio sources only, but are also found among X-ray sources, as seen for example in Figure 12.9. The reader may find it interesting to go through the controversies surrounding these examples. The supporters of the CH like to dismiss all such cases as either observational artefacts or selection effects. Or, they like to argue that the excess number density of quasars near bright galaxies could be due to gravitational lensing. While this criticism or resolution of discrepant data may be valid in some cases, it is hard to see why this should hold in *all* cases.

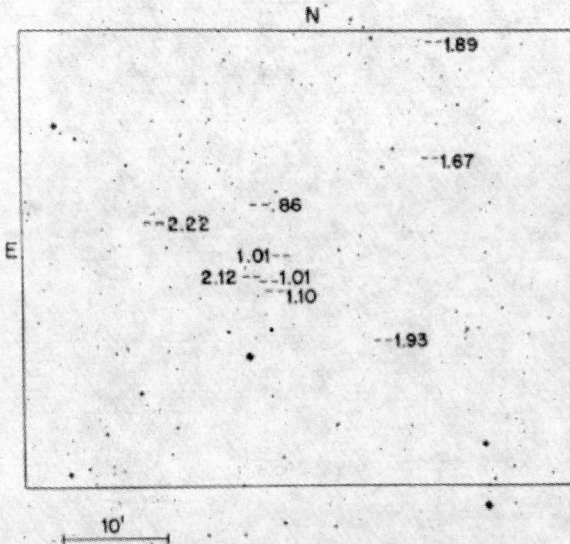

Figure 12.6 The concentration of quasars with various redshifts (marked in the figure) in the vicinity of the region of right ascension $11^h 46^m 14^s$ and declination $11°11'42''$ found by H. Arp and C. Hazard (1980, *Ap. J. 240*, 726) Reproduced by courtesy of H. C. Arp.

Another curious effect that was first noticed by G. Burbidge in the late 1960s concerns the apparent periodicity in the distribution of redshifts of quasars. The periodicity of $\Delta z \cong 0.06$ first found by Burbidge for about 70 QSOs is still present with the population multiplied 30-fold (see Figure 12.10). What is the cause of this structure in the z distribution? Various statistical analyses have confirmed that the effect is significant. Another claim, first made by Karlsson in 1977, is that $\log(1+z)$ is periodic with a period of 0.206. This also is very puzzling and does not fit into the simple picture of the expanding universe that we have been working with in this book.

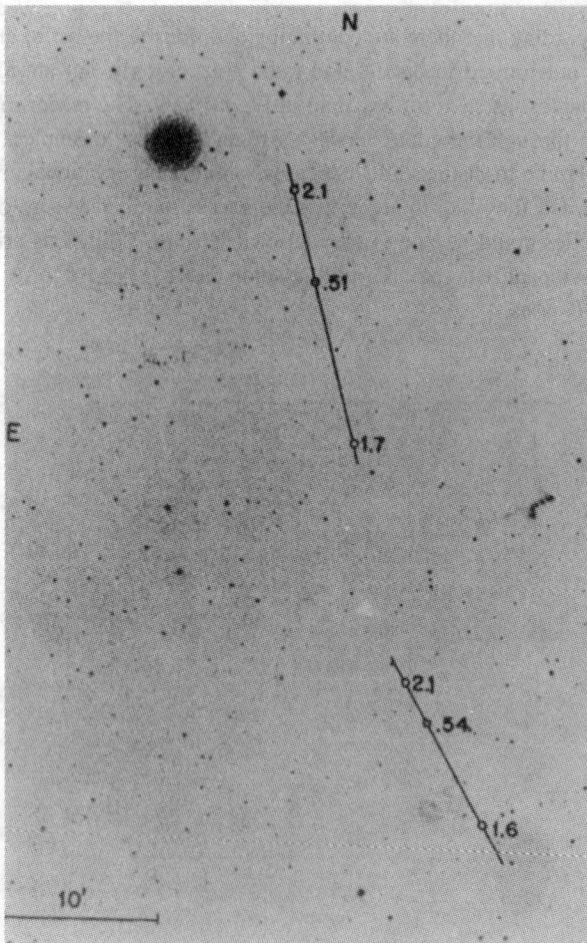

Figure 12.7 These two triplets of quasars found on the same photographic plate show examples of exact alignments of quasars of various redshifts (shown in the figure). The chance of this alignment by random projection is less than 10^{-4}. Reproduced by courtesy of H. C. Arp.

On a much finer scale W. Tifft has been discovering a redshift periodicity $c\,\Delta z = 72$ km s^{-1} for differential redshifts for double galaxies and for galaxies in groups. The data have been refined over the years with accurate 21-cm redshift measurements. If the effect were spurious, it would have disappeared. Instead it has grown stronger and has withstood fairly rigorous statistical analyses (see Figure 12.11).

(a)

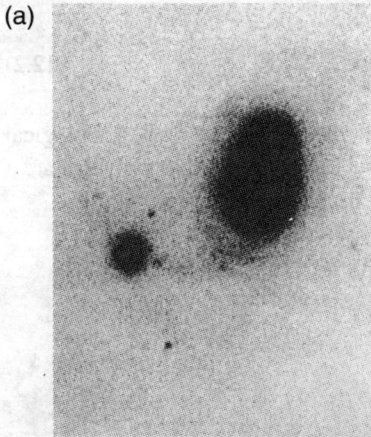

(b)

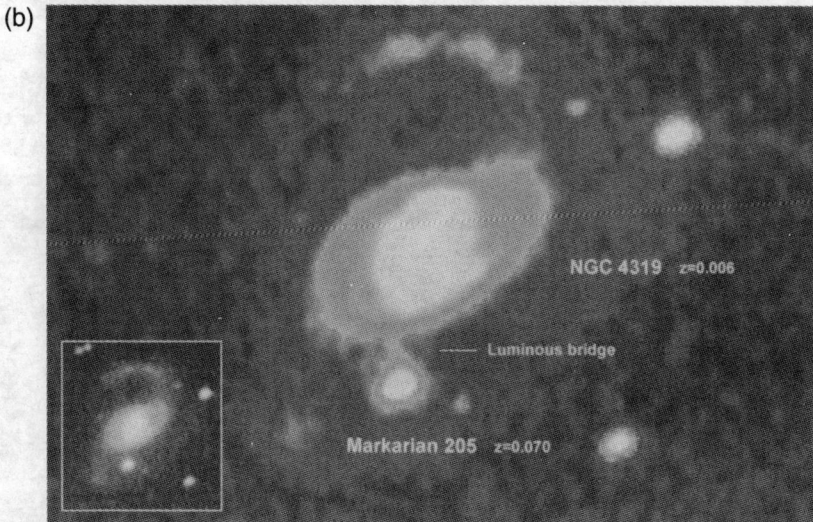

Figure 12.8 (a) The large galaxy NGC 7603 ($cz = 8700$ km s^{-1}) appears connected to a compact companion ($cz = 16\,900$ km s^{-1}). (b) The luminous connection first found by Arp between NGC 4319 and Markarian 205 with redshifts $z = 0.0056$ and 0.07, respectively, has been confirmed by J. Sulentic with CCD observations. Images reproduced by courtesy of H. C. Arp.

For a universe regulated by Hubble's law, it is hard to fit in these results. The tendency on the part of the conventional cosmologist is to discount them in the hope that, with more complete data, they may disappear. At the time of writing this account the data show no such tendency!

It is probable that the effects are genuine and our reluctance to ignore them also stems from the lack of any reasonable explanation. The explanation may bring in other non-cosmological components in the observed redshift z. Thus we should write

$$1 + z = (1 + z_C)(1 + z_{NC}). \tag{12.2}$$

The cosmological component z_C obeys Hubble's law while the non-cosmological part z_{NC} exhibits the anomalous behaviour. What could z_{NC} be due to? There are a few possibilities, none of which has thoroughly been tested.

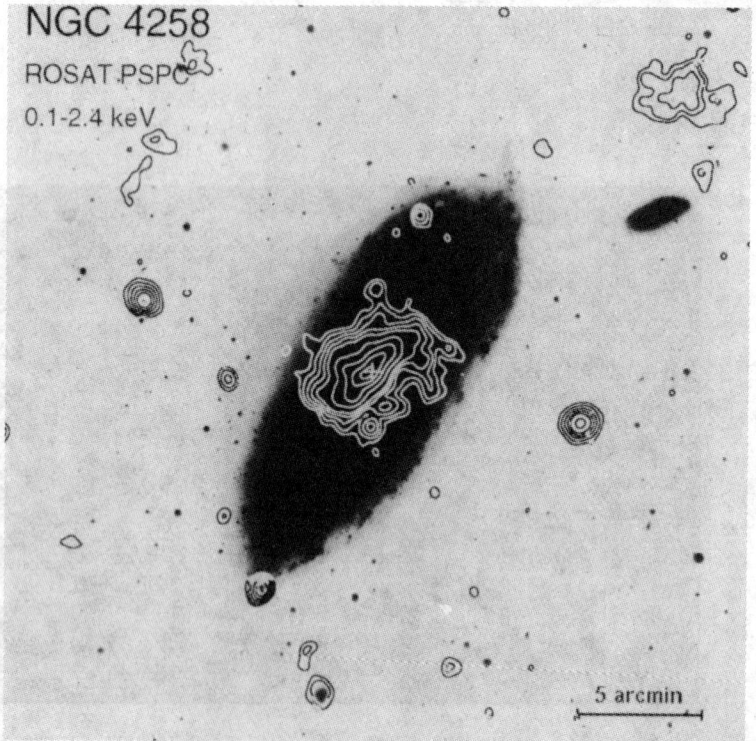

Figure 12.9 Two X-ray sources aligned across NGC 4258 were conjectured to be ejecta from the galaxy. They turned out to be quasars with redshifts 0.4 and 0.65, which are considerably higher than the redshift of the galaxy, which is 0.002. Picture reproduced by courtesy of H. C. Arp.

1. The Doppler effect arises from peculiar motions relative to the cosmological rest frame. It is a well-known phenomenon in physics.

2. The gravitational redshift arises from compact massive objects, as discussed in Chapter 2.

3. The spectral coherence discussed by E. Wolf causes a frequency shift in propagation when fluctuations of light at the source are correlated.

4. In the tired-light theory a photon of non-zero rest mass loses energy while propagating through space.

5. In the variable-mass hypothesis arising from the Machian theory of F. Hoyle and the author, particles may be created in small and large explosions and those created more recently will have smaller masses and hence larger redshifts.

To what extent can these alternatives provide explanations for the discrepant data? Would the discrepancies dwindle away as observations improve or would they grow

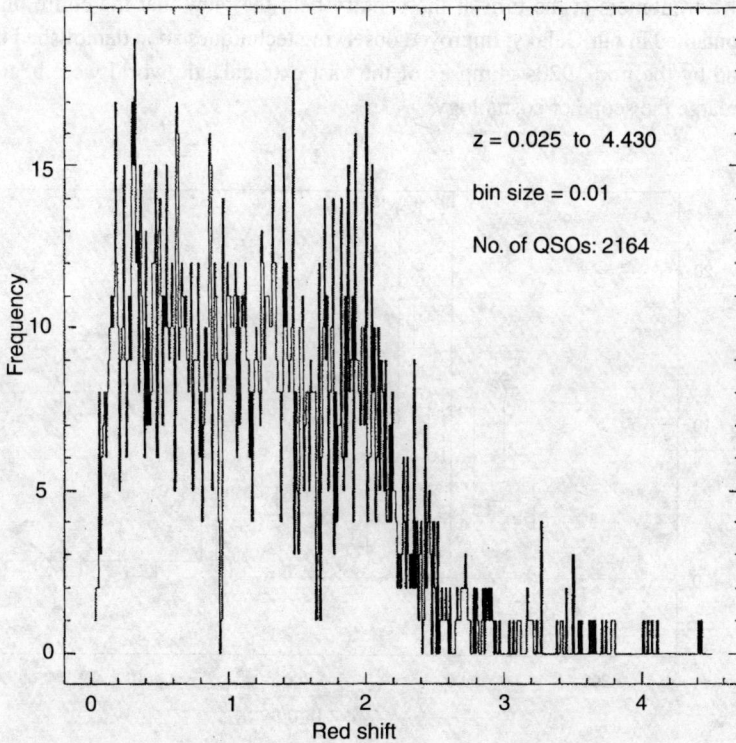

Figure 12.10 A histogram of quasar redshifts showing peaks at approximate multiples of 0.06. The peaks are confirmed by power-spectrum and other analyses carried out by D. Duari, P. Das Gupta and J. V. Narlikar.

in significance? Clearly these issues have enormous implications for Hubble's law in particular and for cosmology in general.

What theory will replace general relativity on the Planck scale?

Observations apart, there are new concepts coming through theory also. The general theory of relativity which has served cosmology well throughout the previous century is now being scrutinized to see how it fits within a wider framework of unification of all interactions. Quantum gravity, the loops approach of Ashtekar *et al.*, or the string-theory approach are all being tried and it is too early to predict what the generally accepted outcome will be. Will these new approaches throw some fresh light on Mach's principle or the large-numbers hypothesis? Whatever the eventual perception may be, it is bound to throw fresh light on the spacetime singularity and the notion of a big bang. It may also substantially modify the early-universe scenarios being talked about today.

The above discussion should be sufficient to convince the reader that the subject of cosmology is still very much open. As discussed in Chapter 1, the majority of astronomers at the turn of the century held the view that the entire universe is contained in our Galaxy. Improved observing techniques soon demolished that view and by the mid-1920s glimpses of the vast extragalactic world were beginning to enlarge the scope of cosmology.

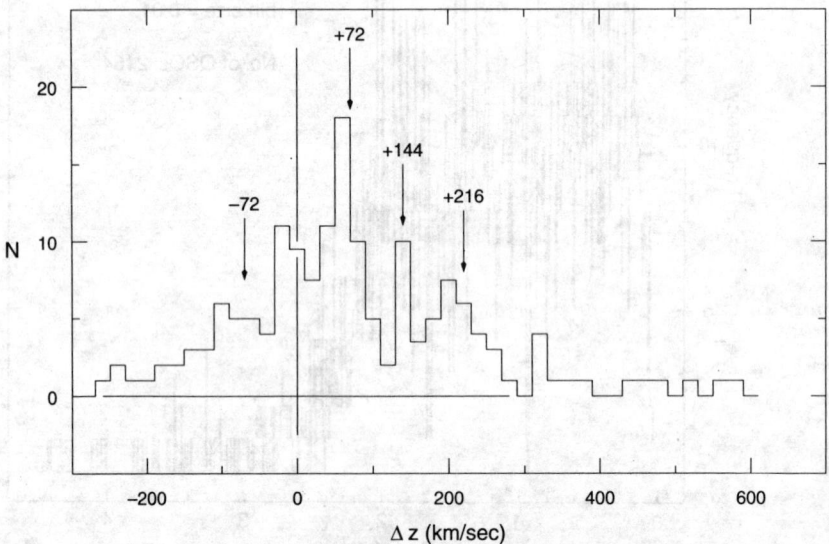

Figure 12.11 This histogram of differences between redshifts of dominant and companion galaxies shows that peaks occur at multiples of 72 km s^{-1}. From H. Arp, 1987, *Quasars, Redshifts and Controversies* (Berkeley: Interstellar Media).

Today we are similarly on the brink of another observational breakthrough with many new-technology telescopes in the offing, both on the ground and in space. It may well be that observations from these will confirm the standard big-bang picture. It is more likely that the bigger and better telescopes of the future will reveal unexpected new phenomena in the universe, phenomena that will provide the greatest intellectual challenges to whatever ideas we hold sacrosanct today.

We conclude with this prognostication.

Exercises

1 For $\kappa \cong 6.11 \times 10^4$ cm^2 g^{-1}, make a rough estimate of ρ_g so as to get $\Delta m(z)$, the additional dimming produced by intergalactic dust out to redshift 0.5 as 0.2^m.

2 What is the maximum blueshift that one expects to find in the QSSC, given its parameters P, Q, η and t_0? Obtain a numerical estimate for the model given by (9.117).

3 Name an observation that can disprove the standard big-bang model, clarifying the details that, if they were to be found, could constitute an unequivocal disproof? Do a similar exercise for the QSSC.

4 Suppose that quasars are located at a fixed distance D from their companion galaxies. Show that this assumption leads to a $\log \theta$–$\log z_G$ plot with a mean slope of -1. Why is this assumption inconsistent with the CH? (θ is the angular separation between the quasars and the galaxy.)

5 Let $n(m)$ denote the number of quasars per unit square degree brighter than apparent magnitude m. If two quasars are found with angular separation θ arcminutes, but with very different redshifts, what statistical test will you apply to decide whether the pair constitutes an example of an anomalous redshift?

Table of constants

This table is taken from the following sources:

I. M. Ryshik and I. S. Gradstein, 1957, *Tables of Series, Products and Integrals* (Berlin: Veb Deutscher Verlag der Wissenschaften).

Particle Data Group, 1980, 'Review of particle properties', *Rev. Mod. Phys. 52*, No. 2

A. H. Wapstra and N. B. Gove, 1971, 'The 1971 Atomic Mass Evaluation', *Nuclear Data Tables 9A*, 265.

C. W. Allen, 1973, *Astrophysical Quantities* (London: Athlone Press).

Figures given in parentheses represent 1σ uncertainty in the last digits of the main numbers.

Mathematical constants

$\pi = 3.14159,$ $e = 2.71828,$ $\zeta(3) = 1.20206$

$\ln 2 = 0.69315,$ $\ln 10 = 2.30259,$ $\log e = 0.43429$

1 arcsecond $= 4.8481 \times 10^{-6}$ radians

1 steradian $= 3.2828 \times 10^3$ square degrees

The number of square degrees on a sphere is 41 252.961 24

Physical constants

Speed of light	$c = 2.99792458(1.2) \times 10^{10}$ cm s^{-1}
Planck's constant	$\hbar = 1.0545887(57) \times 10^{-27}$ erg s
	$= 6.582173(17) \times 10^{-16}$ eV s
	$h \equiv 2\pi\hbar = 6.62620 \times 10^{-27}$ erg s
Electron volt	$1 \text{ eV} = 1.6021892(46) \times 10^{-12}$ erg
Gravitational constant	$G = 6.6720(11) \times 10^{-8}$ dyn cm^2 g^{-2}
Charge of the electron	$e = 4.803242(14) \times 10^{-10}$ esu
Fine structure constant	$\alpha \equiv e^2/(\hbar c) = [137.03604(11)]^{-1}$
Planck length	$\sqrt{\dfrac{G\hbar}{c^3}} = 1.6 \times 10^{-33}$ cm
Planck time	$\sqrt{\dfrac{G\hbar}{c^5}} = 5.4 \times 10^{-44}$ s
Planck mass	$\sqrt{\dfrac{c\hbar}{G}} = 2.2 \times 10^{-5}$ g
Electron mass	$m_e = 9.109534(47) \times 10^{-28}$ g
Electron-mass energy	$m_e c^2 = 0.5110034(14)$ MeV
Proton-mass energy	$m_p c^2 = 938.2796(27)$ MeV
Neutron-mass energy	$m_n c^2 = 939.5731(27)$ MeV
Planck energy	$\sqrt{\dfrac{c^5\hbar}{G}} = 1.2 \times 10^{19}$ GeV
Thomson cross section	$\dfrac{8\pi e^4}{3m_e^2 c^4} = 0.06652448(33) \times 10^{-24}$ cm^2
Boltzmann constant	$k = 1.380662(44) \times 10^{-16}$ erg K^{-1}
	$k^{-1} = 11604.50(36)$ K eV^{-1}
Radiation constant	$a = \dfrac{8\pi^5 k^4}{15c^3 h^3} = 7.5641 \times 10^{-15}$ erg cm^{-3} K^{-4}
Number density of photons in blackbody radiation of temperature T	$\dfrac{2\zeta(3)}{\pi^2}\left(\dfrac{kT}{c\hbar}\right)^3 \cong 20.3\, T^3$ cm^{-3}
Weak-interaction constant	$\mathcal{G} = 1.02 \times 10^{-5}\left(\dfrac{\hbar^3}{m_p^2 c}\right)$
Binding energy of deuterium	$= 2.22464(4)$ MeV
Binding energy of helium	$= 28.2969(4)$ MeV

Astronomical constants

Light year	1 light year $= 9.4605 \times 10^{17}$ cm
Parsec	1 pc $= 3.0856(1) \times 10^{18}$ cm $\cong 3.26$ light years
Radius of the Sun	$R_\odot = 6.959 \times 10^{10}$ cm
Mass of the Sun	$M_\odot = 1.989(1) \times 10^{33}$ g
Luminosity of the Sun	$L_\odot = 3.826(8) \times 10^{33}$ erg s^{-1}
Mass/light ratio for the Sun	$M_\odot/L_\odot \cong 0.51$ g erg^{-1} s
Luminosity of a star of zero absolute magnitude ($M_{bol} = 0$)	$L_0 = 2.97 \times 10^{35}$ erg s^{-1}
Flux from a star of zero apparent magnitude	$l_0 = 2.48 \times 10^{-5}$ erg cm^{-2} s^{-1}
Radio flux density	1 Jy $= 10^{-26}$ W m^{-2} Hz^{-1}
Hubble constant	$H_0 = 100 h_0$ km s^{-1} Mpc^{-1}, $0.5 \leq h_0 \leq 1$
Hubble age	$\tau_0 = H_0^{-1} \cong 9.8 h_0^{-1} \times 10^9$ years

Glossary of symbols and abbreviations

3C, 4C, ...	Cambridge catalogue of radio sources
CDM	Cold dark matter
CH	Cosmological hypothesis (for quasar redshifts)
GA	Great Attractor
GeV	Giga-electron-volt (unit for energy)
H	Hubble's constant
h_0	H measured in units of 100 km s^{-1} Mpc^{-1}
HDM	Hot dark matter
HM	Hubble modulus
HN	Hoyle–Narlikar
IMF	Initial mass function
L	Luminosity
LG	Local Group
LNH	Large-numbers hypothesis
M	Absolute magnitude
m	Apparent magnitude
MeV	Mega-electron volt (unit for energy)
NGC	New General Catalogue
PCP	Perfect cosmological principle
PSR	Pulsar catalogue label
q	Deceleration parameter
QSSC	Quasi-steady-state cosmology
$SU(n)$	Special unitary group of n dimensions
$U(n)$	Unitary group of n dimensions
VLBI	Very-long-baseline interferometry

Bibliography

The reference material listed below is divided into three categories, (i) textbooks (ii), review articles and proceedings of scientific meetings and (iii) lists of papers of a pioneering nature. In a rapidly growing subject like cosmology it is not possible to give an exhaustive list of references. It is hoped that the above distribution of sources of information will meet the varying needs of the readers.

1. Textbooks

This is a representative (but by no means complete) list of books that cover part of the subject matter of the present book, with the numbers at the end of each entry in boldface referring to the chapters of the present book where there is an overlap.

I. J. R. Aitchison and A. J. G. Hey: *Gauge Theories in Particle Physics* (Bristol, Adam Hilger Ltd, 1982). (**6**)

*H. C. Arp: *Quasars, Redshifts and Controversies* (Berkeley, Interstellar Media, 1987). (**12**)

*H. C. Arp: *Seeing Red* (Montréal, Apeiron, 1998). (**12**)

J. Barbour and H. Pfister: *Mach's Principle: From Newton's Bucket to Quantum Gravity* (Boston, Birkhauser, 1995). (**8, 9, 12**)

J. Binney and S. Tremaine: *Galactic Dynamics* (Princeton, Princeton University Press, 1987). (**6, 7, 10**)

J. Binney and M. Merryfield: *Galactic Astronomy* (Princeton, Princeton University Press, 1998). (**6, 7, 10**)

*H. Bondi: *Cosmology* (Cambridge, Cambridge University Press, 1960). (**3, 4, 10, 11**)

G. R. Burbidge and E. M. Burbidge: *Quasi-Stellar Objects* (New York, W. H. Freeman and Company, 1967). (**11, 12**)

P. Coles and F. Lucchin: *Cosmology* (Chichester, Wiley, 1996). (**7**)

E. M. Corson: *Introduction to Tensors, Spinors and Relativistic Wave Equations* (London, Blakie and Sons, 1953). (**2, 6**)

L. P. Eisenhart: *Riemannian Geometry* (Princeton, Princeton University Press, 1926). (**2, 3**)

*E. R. Harrison: *Cosmology: The Science of the Universe* (Chichester, Wiley 1996). (**1, 3, 4, 10, 11**)

S. W. Hawking and G. F. R. Ellis: *The Large Scale Structure of Space-time* (Cambridge, Cambridge University Press, 1973). (**2, 4**)

F. Hoyle, G. Burbidge and J. V. Narlikar: *A Different Approach to Cosmology* (Cambridge, Cambridge University Press, 2000). (**3–6, 9**)

F. Hoyle and J. V. Narlikar: *Action at a Distance in Physics and Cosmology* (New York, W. H. Freeman and Company, 1974). (**8, 9, 12**)

E. W. Kolb and M. S. Turner: *The Early Universe* (New York, Addison–Wesley Publishing Company, 1990). (**5–7**)

L. D. Landau and E. M. Lifshitz: *Statistical Physics* (Oxford, Pergamon Press, 1970). (**5**)

L. D. Landau and E. M. Lifshitz: *The Classical Theory of Fields* (Oxford, Pergamon Press, 1975). (**2**)

*J. V. Narlikar: *The Primeval Universe* (Oxford, Oxford University Press, 1988). (**3–6**)

T. Padmanabhan: *Large Scale Structure of the Universe* (Cambridge, Cambridge University Press, 1993). (**7, 10**)

J. V. Peacock: *Cosmological Physics* (Cambridge, Cambridge University Press, 1999). (**4–7, 10, 11**)

P. J. E. Peebles: *Physical Cosmology* (Princeton, Princeton University Press, 1971). (**4, 5, 7**)

P. J. E. Peebles: *The Large Scale Structure of the Universe* (Princeton, Princeton University Press, 1980). (**7, 9**)

A. K. Raychaudhuri: *Theoretical Cosmology* (Cambridge, Cambridge University Press, 1979). (**3–6**)

*S. Webb: *Measuring the Universe* (Cambridge, Cambridge University Press, 1999). (**10**)

S. Weinberg: *Gravitation and Cosmology* (New York, John Wiley and Sons, 1972). (**2–5**)

Ya. B. Zeldovich and I. D. Novikov: *The Structure and Evolution of the Universe* (Chicago, University of Chicago Press, 1983). (**2–5, 9, 10**)

* These texts are less technical and mathematical than the present book.

2. Reviews and proceedings

The reader wishing to delve deeper into specific topics may wish to begin with the relevant entries from the following list. In most cases the titles are self-explanatory; however, the chapter of the present book to which the work quoted is relevant is shown in boldface at the end.

E. S. Abers and B. W. Lee: *Gauge Theories. Physics Reports 9* 1, 1973. (**6**)

N. A. Bahcall: *Large-Scale Structure in the Universe Indicated by Galaxy Clusters. Ann. Rev. Astron. Astrophys. 26*, 631, 1988. (**7, 10**)

R. Balian, J. Audouze and D. N. Schramm (Eds.): *Physical Cosmology.* Les Houches Lectures Session XXXII (Amsterdam, North Holland, 1979). (**5, 10**)

F. Bertola, J. W. Sulentic and B. F. Madore (Eds.): *New Ideas in Astronomy* (Cambridge, Cambridge University Press, 1988). (**12**)

P. Crane (Ed.): *Light Element Abundances*, ESO-EPIC Workshop Proceedings (Berlin, Springer Verlag, 1995). (**5, 9, 10**)

A. Dresler: *The Great Attractor: Do Galaxies Trace the Large-Scale Mass Distribution? Nature 350*, 391, 1991. (**7, 10**)

S. M. Faber and J. S. Gallagher: *Masses and Mass to Light Ratios of Galaxies. Ann. Rev. Astron. Astrophys. 17*, 135, 1979. (**10**)

A. Hewitt, G. Burbidge and L. Z. Fang (Eds.): *Observational Cosmology.* IAU Symposium No. 124 (Dordrecht, D. Reidel, 1987). (**10–12**)

J. V. Narlikar: *Noncosmological Redshifts. Space Science Reviews 50*, 523, 1989. (**9, 12**)

J. V. Narlikar and A. K. Kembhavi: *Non-standard Cosmologies.* In *Handbook of Astronomy, Astrophysics and Geophysics, Vol. II*, V. M. Canuto and B. G. Elmgreen (Eds.) (New York, Gordon and Breech, 1988). (**8, 9, 12**)

J. V. Narlikar and T. Padmanabhan: *Inflation for Astronomers. Ann. Rev. Astron. Astrophys. 29*, 325, 1991. (**6, 7**)

B. E. J. Pagel, E. A. Simonson, R. J. Terlevich and M. G. Edmunds: *The Primordial Helium Abundance from Observations of Extragalactic H II Regions. Mon. Not. R. Astr. Soc. 255*, 325, 1992. (**5, 10**)

A. Sandage: *Observational Tests of World Models. Ann. Rev. Astron. Astrophys. 26*, 561, 1988. (**10, 11**)

K. Sato: *Cosmological Parameters and the Evolution of the Universe.* IAU Symposium No. 183 (Dordrecht, Kluwer, 1999). (**10, 11**)

D. Schramm and R. V. Wagoner: *Element Production in the Early Universe. Ann. Rev. Nucl. Sci. 27*, 37, 1977. (**5, 10**)

S. F. Shandarin and Ya. B. Zeldovich: *The Large Scale Structure of the Universe: Turbulence, Intermittency, Structures in Self Gravitating Medium. Rev. Mod. Phys. 61*, 185, 1989. (**7**)

V. Trimble: *The Existence and Nature of Dark Matter in the Universe. Ann. Rev. Astron. Astrophys. 25*, 425, 1987. (**7, 10**)

S. Weinberg: *Beyond the First Three Minutes. Physica Scripta 21*, 773, 1980. (**6, 7**)

S. Weinberg: *The Cosmological Constant Problem. Rev. Mod. Phys. 61*, 1, 1989. (**6, 12**)

C. M. Will: *The Confrontation between General Relativity and Experiment. Lecture Notes, SLAC Summer School on Particle Physics*, gr-qc/9811036, 1998. (**2, 8, 11**)

3. References to early work

Although the papers listed chapterwise below are from a historical perspective, many of them are classics that continue to be readable today.

Chapter 1

Classification of galaxy types

E. Hubble. 1926. 'Extragalactic nebulae' *Ap. J. 64*, 321.

W. W. Morgan. 1958. 'A preliminary classification of the forms of galaxies according to their stellar spectra' *Publ. Astron. Soc. Pac. 70*, 364.

S. van den Bergh. 1960. 'The luminosity classification of galaxies and stellar evolution' *Mém. Soc. Roy. Liège (Belgium) Série Cinquième, Vol. III.*

Discovery and identification of Cygnus A

J. S. Hey, S. J. Parsons and J. W. Phillips. 1946. 'Fluctuations in cosmic radiation at radio frequencies' *Nature 158*, 234.

W. Baade and R. Minkowski. 1954. 'Identification of radio sources in Cassiopeia, Cygnus A and Puppis A' *Ap. J. 119*, 206.

The first two quasars

C. Hazard, M. B. Mackey and A. J. Shimmins. 1963. 'Investigation of the radio source 3C273 by the method of lunar occultations' *Nature 197*, 1037.

J. L. Greenstein and M. Schmidt. 1964. 'The quasi stellar radio sources 3C48 and 3C273' *Ap. J. 140*, 1.

Clustering and superclustering of galaxies

C. D. Shane and C. A. Wirtanen. 1954. 'A distribution of extragalactic nebulae' *A. J. 59*, 285.

F. Zwicky. 1957. *Morphological Astronomy* (Berlin: Springer-Verlag).

G. O. Abell. 1958. 'The distribution of rich clusters of galaxies' *Ap. J. Suppl. 3*, 211.

G. de Vaucouleurs. 1961. 'Recent studies of clusters and superclusters' *A. J. 66*, 629.

D. Lynden-Bell, S. M. Faber, D. Burstein, R. L. Davies, A. Dressler, R. J. Terlevich and G. Wegner. 1988. 'Photometry and spectroscopy of elliptical galaxies' *Ap. J. 326*, 19.

Nebular redshifts

V. M. Slipher. 1914. 'Spectrographic observations of nebulae' Paper presented at the 17th meeting of the AAS, August 1914.

E. Hubble. 1929. 'A relation between distance and radial velocity among extragalacitic nebulae' *Proc. Nat. Acad. Sci. (USA) 15*, 168.

E. Hubble and M. Humason. 1931. 'The velocity–distance relation among extragalactic nebulae' *Ap. J. 74*, 35.

Chapter 2

Formulation of the field equations of general relativity

A. Einstein. 1915. 'Zur allgemeinen Relativitätstheorie' *Preuß. Akad. Wiss. Berlin, Sitzber.*, 778.

A. Einstein. 1915. 'Zur allgemeinen Relativitätstheorie (Nachtrag)' *Preuß. Akad. Wiss. Berlin, Sitzber.*, 799.

A. Einstein. 1915. 'Die Feldgleichungen der Gravitation' *Preuß. Akad. Wiss. Berlin, Sitzber.*, 844.

General relativity from an action principle

D. Hilbert. 1915. 'Die Grundlagen der Physik' *Konigl. Gesell. Wiss. Göttingen, Nachr. Math.-Phys. Kl.*, 395.

The Schwarzschild solution

K. Schwarzschild. 1916. 'Über das Gravitationsfeld eines Maßpunktes nach der Einsteinschen Theorie' *Sitzber. Deut. Akad. Wiss. Berlin, Kl. Math.-Phys. Tech.*, 189.

Birkhoff's theorem

G. D. Birkhoff. 1923. *Relativity and Modern Physics* (Cambridge, MA: Harvard University Press).

Gravitational redshift

R. V. Pound and G. A. Rebka. 1960. 'Apparent weight of photons' (Laboratory experiment) *Phys. Rev. Lett. 4*, 337.

W. L. Wiese and D. E. Kelleher. 1971. 'On the cause of redshifts of white dwarf spectra' (40 Eridani B) *Ap. J. 166*, L59.

J. L. Greenstein, J. B. Oke and H. L. Shipman. 1971. 'Effective temperature, radius and gravitational redshift of Sirius B' *Ap. J. 169*, 563.

Perihelion advance

A. Einstein. 1915. 'Erklärung der Perihelbewegung des Merkur aus der allgemeinen Relativitätstheorie' *Preuß. Akad. Wiss. Berlin, Sitzber.*, 831.

G. M. Clemence. 1947. 'The relativity effect in planetary motions' *Rev. Mod. Phys. 19*, 361.

C. W. Will. 1975. 'Periastron shifts in the binary system 1913+16: theoretical interpretation' *Ap. J. 196*, L3.

J. H. Taylor, L. A. Fowler and R. M. McCullach. 1979. 'Measurements of general relativistic effects in the binary pulsar PSR 1913+16' *Nature 277*, 437.

Bending of visible light, radio waves and microwaves

F. W. Dyson, A. S. Eddington and C. Davidson. 1920. 'A determination of the deflection of light by the sun's gravitational field, from observations made at the total eclipse of May 29, 1919' *Phil. Trans. Roy. Soc. A 220*, 291.

C. C. Counselman III, S. M. Kent, C. A. Knight, I. I. Shapiro and T. A. Clark. 1974. 'Solar gravitational deflection of radio waves measured by very long baseline interferometry' *Phys. Rev. Lett. 33*, 1621.

E. B. Fomalont and R. A. Sramek. 1975. 'A confirmation of Einstein's general theory of relativity by measuring the bending of microwave radiation in the gravitational field of the Sun' *Ap. J. 199*, 749.

K. W. Weiler, R. D. Ekers, E. Raimond and K. S. Wellington. 1975. 'Dual frequency measurement of the solar gravitational microwave deflection' *Phys. Rev. Lett. 35*, 134.

Radar echo delay

J. D. Anderson, P. B. Esposito, W. Martin, C. L. Thornton and D. O. Muhleman. 1975. 'Experimental test of general relativity time-delay data from Mariner 6 and 7' *Ap. J. 200*, 221.

R. D. Reasenberg, I. I. Shapiro, P. E. MacNeil, R. B. Goldstein, J. C. Breidenthal, J. P. Brenkle, D. L. Cain, T. M. Kaufman, T. A. Komarck and A. I. Zygielbaum. 1979. 'Verification of signal retardation by Solar gravity' *Ap. J. 234*, L219.

The principle of equivalence

I. I. Shapiro, C. C. Counselman and R. W. King. 1976. 'Verification of the principle of equivalence for massive bodies' *Phys. Rev. Lett. 36*, 555.

J. G. Williams, R. H. Dicke, P. L. Bender, C. O. Alley, W. E. Carter, D. G. Currie, D. H. Eckhardt, J. E. Faller, W. M. Kaula, J. D. Mulholland, H. H. Plotkin, S. K. Poultney, P. J. Shelus, E. C. Silverberg, W. C. Sinclair, M. A. Slade and D. T. Wilkinson. 1976. 'New test of the equivalence principle from lunar laser ranging' *Phys. Rev. Lett. 36*, 551.

The precession of a gyroscope

L. I. Schiff. 1960. 'Motion of a gyroscope according to Einstein's theory of gravitation' *Proc. Nat. Acad. Sci. (USA) 46*, 871.

Gravitational radiation

C. W. Misner, K. S. Thorne and J. A. Wheeler. 1973. *Gravitation* (New York: W. H. Freeman).

Chapter 3

Newton's attempt to construct a model of the universe

I. Newton. 1692 and 1693. Letters to Richard Bentley dated December 10, 1692 and January 17, 1693. In D. T. Whiteside, ed., 1976, *Mathematical Papers of Isaac Newton*, Vol. 7, pp. 233, 238 (Cambridge: Cambridge University Press).

Newtonian cosmology

H. Seeliger. 1895. *Astr. Nachr.* cxxxvii, 129 and 1896. *Münch. Ber. Math. Phys. Kl.*, 373.

C. Neumann. 1896. *Allgemeine Untersuchungen über das Newtonsche Prinzip der Fernwirkungen* (Leipzig).

W. H. McCrea and E. A. Milne. 1934. 'Newtonian universes and the curvature of space' *Q. J. Math. 5*, 73.

The Einstein universe

A. Einstein. 1917. 'Kosmologische Betrachtungen zur allgemeinen Relativitätstheorie.' *Preuß. Akad. Wiss. Berlin, Sitzber.*, 142.

De Sitter's universe

W. de Sitter. 1917. 'On the relativity of inertia: remarks concerning Einstein's latest hypothesis' *Proc. Koninkl. Akad. Weteusch. Amsterdam 19*, 1217.

Weyl's postulate

H. Weyl. 1923. 'Zur allgemeinen Relativitätstheorie' *Z. Phys. 24*, 230.

The Robertson–Walker line element

H. P. Robertson. 1935. 'Kinematics and world structure' *Ap. J. 82*, 248.

A. G. Walker. 1936. 'On Milne's theory of world-structure' *Proc. Lond. Math. Soc. (2) 42*, 90.

For the remaining topics covered in Chapter 3, see references listed under Chapter 4, under the corresponding topics.

Chapter 4

The Friedmann models

A. Friedmann. 1922 and 1924. 'Über die Krummung des Resumes' *Z. Phys. 10*, 377 and *Z. Phys. 21*, 326.

The Einstein–de Sitter model

A. Einstein and W. de Sitter. 1932. 'On the relation between the expansion and the mean density of the universe' *Proc. Nat. Acad. Sci. (USA) 18*, 213.

The luminosity distance

W. Mattig. 1958. 'Über den Zusammenhang zwischen Rotverschiebung und scheinbarer Helligkeit' *A. N. 284*, 109.

Variations of angular sizes

R. C. Tolman. 1933. *Relativity, Thermodynamics and Cosmology* (Oxford: Oxford University Press).

F. Hoyle. 1959. 'The relation of radioastronomy and cosmology' In R. N. Bracewell, ed., *Paris Symposium on Radio Astronomy*, p. 529 (Palo Alto, CA: Stanford University Press).

Number counts of galaxies

E. P. Hubble and R. C. Tolman. 1935. 'Two methods of investigating the nature of the nebular red-shift' *Ap. J. 82*, 302.

The Olbers paradox

E. Halley. 1720. 'Of the infinity of the sphere of fixed stars.' *Phil. Trans. Roy. Soc. Lond. 31*, 22. (The first known discussion of the paradox)

H. W. M. Olbers. 1826. 'Über die Durchsichtigheit des Weltraumes.' *Bode Jahrbuch 110*.

The λ cosmologies

A. S. Eddington. 1930. 'On the instability of Einstein's spherical world' *Mon. Not. Roy. Astron. Soc. 90*, 668.

Abbé G. Lemaître. 1931. 'A homogeneous universe of constant mass and increasing radius accounting for the radial velocity of extragalactic nebulae' *Mon. Not. Roy. Astron. Soc. 91*, 483. (Translated from the original paper in *Annales de la Société Scientifique de Bruxelles, XLVII A*, 49 (1927))

Chapter 5

The early work on primordial nucleosynthesis

G. Gamow. 1946. 'Expanding universe and the origin of elements' *Phys. Rev. 70*, 572.

R. A. Alpher and R. C. Hermann. 1948. 'Evolution of the universe' *Nature 162*, 774.

R. A. Alpher, H. A. Bethe and G. Gamow. 1948. 'The origin of chemical elements' *Phys. Rev. 73*, 80. (This paper, with the sequence of authors Alpher/Bethe/Gamow, led to the name '$\alpha/\beta/\gamma$ theory')

Stellar nucleosynthesis

G. R. Burbidge, E. M. Burbidge, W. A. Fowler and F. Hoyle. 1957. 'Synthesis of the elements in stars' *Rev. Mod. Phys. 29*, 547.

Later work on primordial nucleosynthesis

C. Hayashi. 1950. 'Proton–neutron concentration ratio in the expanding universe at the stages preceding the formation of the elements' *Progr. Theor. Phys. (Japan) 5*, 224.

F. Hoyle and R. J. Tayler. 1964. 'The mystery of cosmic helium abundance' *Nature 203*, 1108.

P. J. E. Peebles. 1966. 'Primordial helium abundance and the primordial fireball' *Ap. J. 146*, 542.

Ya. B. Zel'dovich. 1966. 'The 'hot' model of the universe' *Usp. Fiz. Nauk 89*, 647.

R. V. Wagoner, W. A. Fowler and F. Hoyle. 1967. 'On the synthesis of elements at very high temperatures' *Ap. J. 148*, 3.

The abundance of helium and types of neutrino

J. Yang, D. Schramm, G. Steigman and R. T. Rood. 1979. 'Constraints on cosmology and neutrino physics from big bang nucleosynthesis' *Ap. J. 227*, 697.

Discovery of the microwave background

A. McKeller. 1941. *Pub. Dom. Astrophys. Observatory, Victoria, B.C.,* 7, 251

A. A. Penzias and R. W. Wilson. 1965. 'Measurement of excess antenna temperature at 4080 Mc/s' *Ap. J. 142*, 419.

Chapter 6

Steps towards a unified theory of basic interactions

J. C. Maxwell. 1864. 'A dynamical theory of the electromagnetic field' *Phil. Trans. Roy. Soc. 155*. (Paper read on 8 December 1864)

S. Weinberg. 1967. 'A model of leptons' *Phys. Rev. Lett. 19*, 1264. (The electro-weak interaction)

A. Salam. 1968. 'Weak and electromagnetic interactions' In N. Swartholm, ed., *Elementary Particle Physics*, p. 367 (Stockholm: Almquist and Wiksells).

H. Georgi and S. L. Glashow. 1974. 'Unity of all elementary-particle forces.' *Phys. Rev. Lett. 32*, 438. (The $SU(5)$ framework)

The excess of baryons in the early universe

G. Steigman. 1976. 'Observational tests of antimatter cosmologies' *Ann. Rev. Astron. Astrophys. 14*, 339.

M. Yoshimura. 1978. 'Unified gauge theories and the baryon number of the universe' *Phys. Rev. Lett. 41*, 281.

S. Weinberg. 1979. 'Baryon–lepton non-conserving processes' *Phys. Rev. Lett. 43*, 1566.

The inflationary universe

D. Kazanas. 1980. 'Dynamics of the universe and spontaneous symmetry breaking' *Ap. J. 241*, L59.

A. H. Guth. 1981. 'Inflationary universe: a possible solution to the horizon and flatness problems' *Phys. Rev. D23*, 347.

K. Sato. 1981. 'First order phase transition of a vacuum and the expansion of the universe' *Mon. Not. Roy. Astron. Soc. 195*, 467.

The new inflationary universe

A. Linde. 1982. 'A new inflationary universe scenario' *Phys. Lett. B108*, 389.

A. Linde. 1982. 'Scalar field fluctuations in the expanding universe and the new inflationary scenario' *Phys. Lett. B116*, 335.

A. Albrecht and P. J. Steinhardt. 1982. 'Cosmology for grand unified theories with radiatively induced symmetry breaking' *Phys. Rev. Lett. 48*, 1220.

Chaotic inflation

A. Linde. 1983. 'Chaotic inflation' *Phys. Lett. B129*, 177.

Primordial black holes

S. W. Hawking. 1974. 'Black hole explosions?' *Nature 248*, 30.

B. J. Carr. 1975. 'The primordial black hole mass spectrum' *Ap. J. 201*, 1.

Quantum cosmology

J. Hartle. 1988. 'Quantum cosmology' In B. R. Iyer, A. Kembhavi, J. V. Narlikar and C. V. Vishveshwara, eds., *Highlights in Gravitation and Cosmology*, p. 144 (Cambridge: Cambridge University Press).

J. V. Narlikar. 1984. 'The vanishing likelihood of spacetime singularity in quantum conformal cosmology' *Found. Phys.,14*, 443

Chapter 7

The Jeans mass

J. H. Jeans. 1902. 'The stability of a spiral nebula' *Phil. Trans. Roy. Soc. 199A*, 49.

Growth of fluctuations

E. Lifshitz. 1946. 'On the gravitational instability of the expanding universe' *J. Phys. (USSR) 10*, 116.

R. H. Dicke and P. J. E. Peebles. 1968. 'Origin of the globular clusters' *Ap. J. 154*, 891.

The scale-invariant spectrum

E. R. Harrison. 1970. 'Fluctuations at the threshold of classical cosmology' *Phys. Rev. D1*, 2726.

Ya. B. Zel'dovich. 1972. 'A hypothesis, unifying the structure and the entropy of the universe' *Mon. Not. Roy. Astron. Soc. 160*, 1P.

J. M. Bardeen, P. J. Steinhardt and M. S. Turner. 1983. 'Spontaneous creation of almost scale-free density perturbations in an inflationary universe' *Phys. Rev. D28*, 679.

Dark matter and structure formation

J. R. Bond, G. Efstathiou and J. Silk. 1980. 'Massive neutrinos and the large scale structure of the universe' *Phys. Rev. Lett. 45*, 1980.

P. J. E. Peebles. 1982. 'The peculiar velocity around a hole in the galaxy distribution' *Ap. J. 258*, 415.

Non-linear evolution of structures

Ya. B. Zel'dovich. 1970. 'Gravitational instability: an approximate theory for large density perturbations' *Astron. Astrophys. 5*, 84.

S. D. M. White, C. S. Frenk and M. Davis. 1983. 'Clustering in a neutrino dominated universe' *Ap. J. 274*, L1.

S. J. Aarseth. 1985. 'Direct *N*-body calculations' In J. Goodman and P. Hut, eds., *Dynamics of Star Clusters. IAU Symposium No. 113*, p. 251 (Dordrecht: Reidel).

M. Davis, G. Efstathiou, C. S. Frenk and S. D. M. White. 1985. 'The evolution of large-scale structure in the universe dominated by cold dark matter' *Ap. J. 292*, 371.

Massive neutrinos and cosmology

R. Cowsik and J. McClelland. 1972. 'An upper limit on the neutrino rest mass' *Phys. Rev. Lett. 29*, 669.

R. Cowsik and J. McClelland. 1973. 'Gravity of neutrinos of nonzero mass in astrophysics' *Ap. J. 180*, 7.

S. Tremain and J. E. Gunn. 1979. 'Dynamical role of light neutral leptons in cosmology' *Phys. Rev. Lett. 42*, 407.

Chapter 8

Inertial forces and the absolute space

I. Newton. 1687. *Philosophiae Naturalis Principia Mathematica*, 1st edn (London: Streater). (English translation by A. Motte (1729) revised by A. Cajori (1934) (Berkeley: University of California Press))

E. Mach. 1893. *The Science of Mechanics* (Chicago: Open Court).

Sciama's theory of inertia

D. W. Sciama. 1953. 'On the origin of inertia' *Mon. Not. Roy. Astron. Soc. 113*, 34.

The Brans–Dicke theory of gravity

C. Brans and R. H. Dicke. 1961. 'Mach's principle and a relativistic theory of gravitation' *Phys. Rev. 124*, 125.

R. H. Dicke. 1962. 'Mach's principle and invariance under transformation of units' *Phys. Rev. 125*, 2163.

Solar-System tests of the Brans–Dicke theory

R. H. Dicke and H. M. Goldenberg. 1967. 'Solar oblateness and general relativity' *Phys. Rev. Lett. 18*, 313.

H. A. Hill and R. T. Stebbins. 1975. 'The intrinsic visual oblateness of the Sun' *Ap. J. 200*, 471.

Cosmological solutions for the Brans–Dicke theory

R. H. Dicke. 1968. 'Scalar tensor gravitation and the cosmic fireball' *Ap. J. 152*, 1.

Variation of G in the Brans–Dicke theory

R. H. Dicke. 1962. 'Implications for cosmology of stellar and galactic evolution rates' *Rev. Mod. Phys. 34*, 110.

The Hoyle–Narlikar theory of gravity

F. Hoyle and J. V. Narlikar. 1964. 'A new theory of gravitation' *Proc. Roy. Soc. A282*, 191.

F. Hoyle and J. V. Narlikar. 1966. 'A conformal theory of gravitation' *Proc. Roy. Soc. A294*, 138.

The electromagnetic response of the universe

J. A. Wheeler and R. P. Feynman. 1945. 'Interaction with the absorber as the mechanism of radiation' *Rev. Mod. Phys. 17*, 157.

J. E. Hogarth. 1962. 'Cosmological considerations of the absorber theory of radiation' *Proc. Roy. Soc. A267*, 365.

F. Hoyle and J. V. Narlikar. 1963. 'Time symmetric electrodynamics and the arrow of time in cosmology' *Proc. Roy. Soc. A277*, 1.

Spacetime singularity in HN cosmology

A. K. Kembhavi. 1978. 'Zero mass surfaces and cosmological singularities' *Mon. Not. Roy. Astron. Soc. 185*, 807.

Variation of G in HN cosmology

F. Hoyle and J. V. Narlikar. 1972. 'Cosmological models in a conformally invariant gravitation theory I & II' *Mon. Not. Roy. Astron. Soc. 155*, 305 and 323.

The significance of large dimensionless numbers

P. A. M. Dirac. 1937. 'The cosmological constants' *Nature 139*, 323.

The large-numbers hypothesis

P. A. M. Dirac. 1938. 'A new basis for cosmology' *Proc. Roy. Soc. A165*, 199.

Dirac cosmology with two types of creation

P. A. M. Dirac. 1973. 'Long range forces and broken symmetries' *Proc. Roy. Soc. A333*, 403.

P. A. M. Dirac. 1974. 'Cosmological models and the large numbers hypothesis' *Proc. Roy. Soc. A338*, 439.

Chapter 9

Steady-state theory (first proposed)

H. Bondi and T. Gold. 1948. 'The steady state theory of the expanding universe' *Mon. Not. Roy. Astron. Soc. 108*, 252.

F. Hoyle. 1948. 'A new model for the expanding universe' *Mon. Not. Roy. Astron. Soc. 108*, 372.

C-field cosmology

F. Hoyle and J. V. Narlikar. 1962. 'Mach's principle and the creation of matter' *Proc. Roy. Soc. A270*, 334. (Continuous creation)

J. V. Narlikar. 1973. 'Singularity and matter creation in cosmological models' *Nature 242*, 135. (Explosive creation)

The hot universe

T. Gold and F. Hoyle. 1958. 'Cosmic rays and radio waves as manifestations of a hot universe' In R. N. Bracewell, ed., *Paris Symposium on Radio Astronomy*, p. 583 (Palo Alto, CA: Stanford University Press).

The bubble universe/galaxy formation

F. Hoyle and J. V. Narlikar. 1966. 'A radical departure from the steady state concept in cosmology' *Proc. Roy. Soc. A290*, 162.

F. Hoyle and J. V. Narlikar. 1966. 'On the formation of elliptical galaxies' *Proc. Roy. Soc. A290*, 177.

The quasi-steady-state cosmology

F. Hoyle, G. Burbidge and J. V. Narlikar. 'A quasi-steady state cosmological model with creation of matter' *Ap. J. 410*, 437.

For the various aspects of this cosmology see the book which lists all the early papers on this theory:

F. Hoyle, G. Burbidge and J. V. Narlikar. 2000. *A Different Approach to Cosmology* (Cambridge: Cambridge University Press).

Chapter 10

Measurement of H_0

E. Hubble. 1929. 'A relation between distance and radial velocity among extragalactic nebulae' *Proc. Nat. Acad. Sci. (USA) 15*, 168.

M. L. Humason, N. U. Mayall and A. R. Sandage. 1956. 'Redshifts and magnitudes of extragalactic nebulae' *Ap. J. 61*, 97.

The period–luminosity relation of Cepheids

H. Leavitt. 1912. 'Periods of twenty-five variable stars in the small Magellanic cloud'
  Harvard College Observatory Circular No. 173.

Anisotropy of Hubble flow

V. C. Rubin, N. Thonnard and W. K. Ford Jr. 1976. 'Motion of the Galaxy and the Local
  Group determined from the velocity anisotropy of distant Sc I galaxies, I & II' *A. J. 81*,
  687 and 719.

Mass distributions in galaxies

I. King. 1975. 'The structure of round stellar systems: observation and theory' In
  A. Hayli, ed., *Dynamics of Stellar Systems*, p. 99 (Dordrecht: Reidel).

Statistics of groups of galaxies

E. L. Scott. 1961. 'Distribution of galaxies on the sphere' In G. C. McVittie, ed.,
  Problems in Extragalactic Research, p. 269 (New York: Macmillan).

J. Neyman. 1961. 'Alternative stochastic models of the spatial distribution of galaxies' In
  G. C. McVittie, ed., *Problems in Extragalactic Research*, p. 294 (New York:
  Macmillan).

H. Totsuji and T. Kihara. 1969. 'The correlation function for the distribution of galaxies'
  Publ. Astron. Soc. (Japan) 21, 221.

P. J. E. Peebles. 1974. 'The gravitational instability picture and the nature of distribution
  of galaxies' *Ap. J. 189*, L51.

Missing mass in clusters of galaxies

F. Zwicky. 1933. 'Die Rotverschiebung von extragalaktischen Neblen' *Helv. Phys. Acta 6*,
  10.

The local supercluster

G. de Vaucouleurs. 1961. 'Recent studies of clusters and superclusters' *A. J. 66*, 629.

q_0 from deceleration of nearby galaxies

A. Sandage, G. Tamman and A. Yahil. 1979. 'The velocity field of bright nearby galaxies
  I–IV' *Ap. J. 232*, 352 and subsequent papers in later issues of *Ap. J.*

The age of the Galaxy

W. A. Fowler and F. Hoyle. 1960. 'Nuclear cosmochronology' *Ann. Phys. (New York) 10*,
  280.

I. Iben Jr. 1974. 'Post main-sequence evolution of single stars' *Ann. Rev. Astron.
  Astrophys. 12*, 215.

Abundances of light nuclei

R. V. Wagoner. 1973. 'Big bang nucleosynthesis revisited' *Ap. J. 179*, 343.

Microwave background measurements

See references cited in Tables 10.2 and 10.3.

Dipole anisotropy of microwave background

G. F. Smoot, M. V. Gorenstein and R. A. Muller. 1977. 'Detection of anisotropy in the cosmic black body radiation' *Phys. Rev. Lett. 39*, 898.

E. S. Cheng, P. R. Saulson, D. T. Wilkinson and B. E. Corey. 1979. 'Large scale anisotropy in the 2.7 K radiation' *Ap. J. 232*, L139.

The Sunyaev–Zel'dovich effect

R. A. Sunyaev and Ya. B. Zeldovich. 1970. 'Small-scale fluctuations of relic radiation' *Astrophys. Space . Sci. 7*, 3.

The microwave background and high-energy cosmic rays

K. Greisen. 1966. 'An end to the cosmic ray spectrum?' *Phys. Rev. Lett. 16*, 748.

Chapter 11

q_0 from the m–z relation

A. Sandage. 1972. 'The redshift distance relation, II' *Ap. J. 178*, 1.

J. Kristian, A. Sandage and J. A. Westphal. 1978. 'The extension of the Hubble diagram, II' *Ap. J. 221*, 383.

A. Reiss *et al.* 1998. 'Observational evidence from supernovae for an accelerating universe and a cosmological constant' *A. J. 116*, 1009.

S. Perlmutter *et al.* 1999. 'Measurements of Ω and Λ from 42 high redshift supernovae' *Ap. J. 517*, 565.

Corrections to the Hubble diagram

K. G. Malmquist. 1920. 'A study of stars of spectral type A' *Medd. Lunds. Obs. Ser. II* No. 22.

E. L. Scott. 1957. 'The brightest galaxy in a cluster as a distance indicator' *A. J. 62*, 248. (The Scott effect)

A. Sandage. 1961. 'The ability of the 200-inch telescope to distinguish between selected world models' *Ap. J. 133*, 355.

J. B. Oke and A. Sandage. 1968. 'Energy distributions, K corrections and the Stebbins–Whitford effect for giant elliptical galaxies' *Ap. J. 154*, 21. (K-correction)

J. E. Gunn and J. B. Oke. 1974. 'Spectrophotometry of faint cluster galaxies and the Hubble diagram: an approach to cosmology' *Ap. J. 195*, 255. (Aperture correction)

B. M. Tinsley. 1975. 'Nucleochronology and chemical evolution.' *Ap. J. 198*, 145. (The evolution of luminosity)

S. M. Chitre and J. V. Narlikar. 1976. 'The effect of intergalactic dust on the measurement of the cosmological deceleration parameter q_0' *Astrophys. Space Sci.* *44*, 101. (Intergalactic absorption)

Optical counts of galaxies
E. P. Hubble. 1936. 'Effects of redshifts on the distribution of nebulae' *Ap. J.* *84*, 517.

Radio-source surveys and counts
J. R. Shakeshaft, M. Ryle, J. E. Baldwin, B. Elsmore and J. H. Thomson. 1955. 'A survey of radio sources between declinations $-38°$ and $+83°$' *Mem. Roy. Astron. Soc.* *67*, 97.

B. Y. Mills, O. B. Slee and E. R. Hill. 1958. 'A catalogue of radio sources between declinations $+10°$ and $-20°$' *Aus. J. Phys.* *11*, 360.

D. O. Edge, J. R. Shakeshaft, W. B. McAdam, J. E. Baldwin and S. Archer. 1959. 'A survey of radio sources at 159 Mc/s' *Mem. Roy. Astron. Soc.* *68*, 37.

J. G. Bolton. 1960. 'The discrete sources of cosmic radio emission' *Comptes Rendus de l'Assemblé Générale de l'URSI, Londres, 1960, Session V.*

A. S. Bennett. 1962. 'The revised 3C-catalogue of radio sources' *Mem. Roy. Astron. Soc.* *68*, 163.

The luminosity–volume test
M. Schmidt. 1968. 'Space distribution and luminosity functions of quasi-stellar radio sources' *Ap. J.* *151*, 393.

The maximum-likelihood method
D. F. Crawford, D. L. Jauncey and H. S. Murdoch. 1970. 'Maximum likelihood estimation of the slope from number–flux density counts of radio sources' *Ap. J.* *162*, 405.

Source-count interpretations for cosmology
P. F. Scott and M. Ryle. 1961. 'The number flux density relation for radio sources away from the galactic plane' *Mon. Not. Roy. Astron. Soc.* *122*, 389.

F. Hoyle and J. V. Narlikar. 1962. 'On the counting of radio sources in steady state cosmology' *Mon. Not. Roy. Astron. Soc.* *123*, 133.

D. L. Jauncey. 1967. 'Reexamination of the source counts for the 3C revised catalogue' *Nature 216*, 877.

F. Hoyle. 1968. 'Review of recent developments in cosmology' *Proc. Roy. Soc. A308*, 1.

Variation of angular size and cosmology
R. C. Tolman. 1934. *Relativity, Thermodynamics and Cosmology*, p. 467 (Oxford: Clarendon Press).

F. Hoyle. 1959. 'The relation of radioastronomy and cosmology' In R. N. Bracewell, ed., *Paris Symposium on Radio Astronomy*, p. 529 (Palo Alto, CA: Stanford University Press).

The surface-brightness test

A. Sandage and J.-M. Perelmuter. 1990. 'The surface brightness test for the expansion of the universe I & II' *Ap. J. 350*, 481 and *361*, 1.

Quasar catalogues

A. Hewitt and G. Burbidge. 1987. 'A new optical catalogue of quasi-stellar objects' *Ap. J. Suppl. 63*.

A. Hewitt and G. Burbidge. 1989. 'The first addition to the new optical catalogue of quasi-stellar objects' *Ap. J. Suppl. 69*, 1.

M.-P. Veron-Cetty and P. Veron. 1985. 'A catalogue of quasars and active galactic nuclei' *ESO Sci. Rep.* No. 4.

The Hubble diagram of quasars

G. R. Burbidge and F. Hoyle. 1966. 'Relation between the redshifts of quasi-stellar objects and their radio and optical magnitudes' *Nature 210*, 1346.

Number counts of quasars

M. Schmidt. 1968. 'Space distribution and luminosity functions of quasi-stellar radio sources' *Ap. J. 151*, 393.

A. K. Kembhavi and A. Fabian. 1982. 'X-ray quasars and the X-ray background' *Mon. Not. Roy. Astron. Soc. 198*, 921.

Gravitational lens

F. Zwicky. 1937. 'Nebulae as gravitational lenses' *Phys. Rev. 51*, 290

D. Walsh, R. F. Carswell and R. J. Weymann. 1979. '0957+561 A, B: twin quasistellar objects or gravitational lens?' *Nature 279*, 381.

Lyman-α absorption

J. E. Gunn and B. A. Peterson. 1965. 'On the density of neutral hydrogen in the intergalactic space' *Ap. J. 142*, 1633.

The problem of energy production in quasars

F. Hoyle, G. R. Burbidge and W. L. W. Sargent. 1966. 'On the nature of quasi-stellar sources' *Nature 209*, 751.

M. J. Rees. 1967. 'Studies in radio source structure, I' *Mon. Not. Roy. Astron. Soc. 135*, 345.

Superluminal separation in quasars

K. I. Kellarmann and D. B. Shaffer. 1977. 'Superlight motion in radio sources and its implications for the distance scale problem' In C. Balkowski and B. E. Westerlund, eds., *Proceedings of the IAU/CNRS Colloquium*, held in Paris, 6–9 September 1976, p. 347 (Paris: CNRS). (For observations and some theoretical models)

J. V. Narlikar and S. M. Chitre. 1980. 'Gravitational screens and superluminal separation in quasars' *Ap. J. 235*, 335.

Variation of $e^2/(\hbar c)$

J. N. Bahcall and M. Schmidt. 1967. 'Does the fine structure constant vary with time?' *Phys. Rev. Lett. 19*, 1294.

M. S. Roberts. 1977. 'High redshift 21-cm lines' In C. Balkowski and B. E. Westerlund, eds., *Proceedings of the IAU/CNRS Colloquium*, held in Paris, 6–9 September 1976, p. 501 (Paris: CNRS).

A. D. Tubbs and A. M. Wolfe. 1980. 'Evidence for large-scale uniformity of physical laws' *Ap. J. 236*, L105.

Variation of G

F. Hoyle. 1972. 'The early history of the Earth' *Q. J. Roy. Astron. Soc. 13*, 328.

I. I. Shapiro. 1976. 'Bounds on the secular variation of the gravitational constant' *B. A. A. S. 8*, 308.

Chapter 12

For standard cosmology

P. J. E. Peebles, D. N. Schramm, E. L. Turner and R. G. Kron. 1991. 'The case for the hot relativisitic big bang cosmology' *Nature 352*, 769.

J. Silk. 1999. 'The case for the big bang' *Comptes Rendus 327*, Series II-b, 829.

Against standard cosmology

H. C. Arp, G. Burbidge, F. Hoyle, J. V. Narlikar and N. C. Wickramasinghe. 1990. 'The extragalactic universe: an alternative view' *Nature 346*, 807.

J. V. Narlikar. 1999. 'The case against the big bang' *Comptes Rendus 327*, Series II-b, 841.

Index